Physics and Properties of Narrow Gap Semiconductors

MICRODEVICES

Physics and Fabrication Technologies

Series Editors: Arden Sher, *SRI International/Stanford University*
Marcy Berding, *SRI International*

Recent volumes in this series:

COMPOUND AND JOSEPHSON HIGH-SPEED DEVICES
Takahiko Misugi and Akihiro

ELECTRON BEAM TESTING TECHNOLOGY
John T.L. Thong

FIELD EMISSION IN VACUUM MICROELECTRONICS
George Fursey

ORIENTED CRYSTALLIZATION ON AMORPHOUS SUBSTRATES
E.I. Givargizov

PHYSICS OF HIGH-SPEED TRANSISTORS
Juras Pozela

THE PHYSICS OF MICRO/NANO-FABRICATION
Ivor Brodie and Julius J. Murray

PHYSICS OF SUBMICRON DEVICES
Ivor Brodie and Julius J. Murray

THE PHYSICS OF SUBMICRON LITHOGRAPHY
Kamil A. Valiev

RAPID THERMAL PROCESSING OF SEMICONDUCTORS
Victor E. Borisenko and Peter J. Hesketh

SEMICONDUCTOR ALLOYS: PHYSICS AND MATERIAL ENGINEERING
An-Ben Chen and Arden Sher

SEMICONDUCTOR DEVICE PHYSICS AND SIMULATION
J.S. Yuan and Peter Rossi

SEMICONDUCTOR MATERIALS
An Introduction to Basic Principles
B.G. Yacobi

PHYSICS AND PROPERTIES OF NARROW GAP SEMICONDUCTORS
Junhao Chu and Arden Sher

Junhao Chu Arden Sher

Physics and Properties of Narrow Gap Semiconductors

Springer

Junhao Chu
National Laboratory for Infrared Physics
Shanghai Institute of Technical Physics
500 Yu Tian Road
Shanghai 200083
China
jhchu@mail.sitp.ac.cn

Arden Sher
SRI International (Ret.)/Stanford University
Menlo Park, CA
USA

ISBN: 978-0-387-74743-9 e-ISBN: 978-0-387-74801-6

Library of Congress Control Number: 2007937089

Printed on acid-free paper.

9 8 7 6 5 4 3 2 1

springer.com

Preface

The physics of narrow-gap semiconductors is an important branch of semiconductor science. Research into this branch focuses on a specific category of semiconductor materials which have narrow forbidden band gaps. Past studies on this specific category of semiconductor materials have revealed not only general physical principles applicable to all semiconductor technology, but also those unique characteristics originating from the narrow band gaps, and therefore have significantly contributed to science and technology. Historically, developments of narrow-gap semiconductor physics have been closely related to the development of the science and technology of infrared optical electronics as narrow-gap semiconductors have played a vital role in the field of infrared-radiation detectors and emitters, and other high speed devices. The present book is dedicated to the study of narrow-gap semiconductors and their applications. It is expected that the present first volume will be valuable to not only the fundamental science of narrow-gap semiconductors but also to the technology of infrared optical electronics.

There have been several books published in this field over the past few decades. In 1977, a British scientist, D.R. Lovett, published a book *Semimetals and Narrow-Band Gap Semiconductors* (Pion Limited, London). Later, German scientists, R. Dornhaus and G. Nimtz, published a comprehensive review article in 1978, whose second edition, entitled, *The Properties and Applications of the HgCdTe Alloy System, in Narrow Gap Semiconductors*, was reprinted by Springer in 1983 (Springer Tracts in Modern Physics, Vol. 98, p. 119). These two documents included systematic discussions of the physical properties of narrow-gap semiconductors and are still important references of the field. In 1980, the 18th volume of the series *Semiconductors and Semimetals* (edited by R.K. Willardson and Albert C. Beer) in which very useful reviews were collected, was dedicated to HgCdTe semiconductor alloys and devices. In 1991, a Chinese scientist, Prof. D.Y. Tang published an important article, "Infrared Detectors of Narrow Gap Semiconductors" in the book *Research and Progress of Semiconductor Devices* (edited by S.W. Wang, Science Publish, Beijing, pp. 1–107), in which the fundamental principles driving HgCdTe-based infrared radiation detector technology were comprehensively discussed. In addition, a handbook, *Properties of Narrow Gap Cadmium-Based Compounds* (edited by P. Capper), was published in the United Kingdom in 1994. In this handbook, a number of research articles about the physical and the chemical properties of HgCdTe narrow-gap semiconductors were collected and various data and references about Cd-based semiconductors can be found.

This book *Narrow Gap Semiconductors* is being divided into two volumes. The first volume is subtitled *(Vol. I): Materials Physics and Fundamental Properties*. The second volume subtitled, *(Vol. II): Devices and Low-Dimensional Physics*, will follow. Volume II will have the following table of contents:

- Chapter 1 Introduction
  1.1 Brief Description of Volume I
  1.2 Devices on Narrow Band Gap Systems
  References
- Chapter 2 Impurities and Defects
  2.1 Conduction and Ionization Energy of Impurities and Defects
  2.2 Shallow Impurities
  2.3 Deep Levels
  2.4 Resonant Defects States
  2.5 Photoluminescence of Impurities and Defects
  References
- Chapter 3 Recombination
  3.1 Recombination Mechanisms and Life Times
  3.2 Auger Recombination
  3.3 Shockley-Read Recombination
  3.4 Radiative Recombination
  3.5 Measurement of Minority Carrier Lifetimes
  3.6 Surface Recombination
  References
- Chapter 4 Surface Two-Dimensional Electron Gases
  4.1 MIS Heterostructures
  4.2 Theoretical Model of Subband Structures
  4.3 Experimental Methods for Subband Structure Investigations
  4.4 Dispersion Relations and Landau Levels
  4.5 Surface Accumulation Layers
  4.6 Surfaces and Interfaces
  References
- Chapter 5 Superlattices and Quantum Wells
  5.1 Semiconductor Low Dimensional Structures
  5.2 Band Structure Theory of Semiconductor Low Dimensional Structures

The present book (*Narrow Gap Semiconductors: (Vol. I): Materials Physics and Fundamental Properties*) and the forthcoming book (*Narrow Gap Semiconductors (Vol. II): Devices and Low-Dimensional Physics*) aim, in the two volumes, at characterizing a variety of narrow-gap semiconductor materials and revealing the intrinsic physical principles that govern their behavior. The discussions dedicated to narrow-gap semiconductors presented in this book evolved within the larger framework of semiconductor physics, in combination with the progresses in the specific field of narrow-gap semiconductor materials and devices. In particular, a unique property of this book is the more extensive collection of results than ever previously assembled of the research results deduced by Chinese scientists, including one author of this book. These results are integrated into the larger body of knowledge from the world literature. In organizing the book, special attention was paid to bridging the gap between basic physical principles and frontier research. This is achieved through extensive discussions of various aspects of the frontier theoretical and experimental scientific issues and connecting them to device related technology. It is expected that both the students and the researchers working in relevant fields will benefit from this book.

The book was encouraged and advised by Prof. D.Y. Tang. One of the authors (J. Chu) is most grateful to Prof. D.Y. Tang's critical reading of the manuscript and invaluable suggestions and comments. The co-author (A. Sher) is indebted to Prof. A.-B. Chen for invaluable suggestions. The authors are also grateful to numerous students and colleagues who over the years have offered valuable support during the writing of this book. They are Drs.: Y. Chang, B. Li, Y.S. Gui, X.C. Zhang, S.L. Wang, Z.M. Huang, J. Shao, X. Lu, Y. Cai, K. Liu, L. He, M.A. Berding, and S. Krishnamurthy. We are indebted to Professor M.W. Muller for his careful reading of Chaps. 1–4 of the English manuscript. The electronic files of the whole camera ready manuscripts were edited by Dr. H. Shen and Dr. X. Lu.

The research of one author's group (J. Chu) that is presented in this book was supported by the National Science Foundation of China, The Ministry of Science and Technology of the People's Republic of China, the Chinese Academy of Science, and the Science and Technology Commission of the Shanghai Municipality.

Junhao Chu

He is a member of CAS, directs the National Laboratory for Infrared Physics, Shanghai Institute of Technical Physics, and is at the East China Normal University.

Arden Sher

He is retired from SRI International and Stanford University.

2007-7-12

Contents

1 Introduction

Narrow gap semiconductors are members of the semiconductor family with narrow forbidden fundamental bands. Generally materials with forbidden bandwidth E_g smaller than 0.5 eV, or equivalently those with forbidden bandwidth corresponding to an infrared absorption cut-off wavelength over 2 μm, are taken as narrow gap semiconductors (Long and Schmit 1973; Dingyuan Tang 1976).

1.1 Narrow Gap Semiconductors

Characteristics of the energy bands of narrow gap materials include strongly non-parabolic conduction bands, and spin–orbit splittings even larger than the fundamental band gaps. The early energy band electronic structures of the narrow gap semiconductors were based on a model theory of the InSb semiconductor put forward by Kane in 1957 (Kane 1957, 1966). While this theory properly predicted tends, it was not sufficiently accurate to serve as a reliable engineering tool. Newer theories based on "density functional theory," or the simpler "hybrid psudopotential-tight-binding theory (HPTB)" (Chen and Sher 1995), now are sufficiently accurate for such use.

HgCdTe alloys and InSb are the typical narrow gap semiconductor materials. In intrinsic infrared detectors, radiation excites electrons in states near the top of the valence band of the narrow gap semiconductor to electron states near the bottom of the conduction band. This results in non-equilibrium electron–hole populations, which change the electrical transport properties of the material. In photoconductive devices, the conductivity increases, whereas in photovoltaic devices, a photo-voltage is generated in response to the incident photon flux. With their small effective electron masses, high electron mobility, and long carrier lifetimes, these narrow gap semiconductors are sensitive and fast infrared photo detector materials, suitable for both infrared scene sensing and high data rate communication applications (Kane 1981). Thus the development of narrow gap semiconductor physics is closely related to infrared detector applications.

Infrared detectors are the core components of modern infrared imaging systems and the demand for and development of infrared detectors for high performance systems has promoted the fabrication of improved narrow gap semiconductors and driven their physics research.

The development of narrow gap semiconductor physics has experienced three stages. The first stage started from the 1940s when PbS, PbSe and PbTe detectors were the main infrared detector materials. The 1950s saw the first use of InSb, InAs and Hg doped Ge (Ge:Hg) materials. Experimentally, advances were made in the preparation and research on the properties of InSb. Theoretically clear and definite results were also derived from research on the properties of Ge and Si energy band structures. Based on this group IV material work, E. O. Kane calculated, by adding $\boldsymbol{k}\cdot\boldsymbol{P}$ perturbation theory to the model, the InSb energy bands. His work pioneered the understanding of the mechanisms responsible for the energy bands of narrow gap semiconductors. The theory well describes the observed trends in the relationship between energy bands and wave vectors, the dispersion of InSb near the Γ-point of the k-space Brillouin zone. It serves as the basis for describing various processes, such as carrier transport and photo-electronic excitation, etc. Consequently it laid the theoretical foundation for the research of narrow gap semiconductor physics. This period was marked mainly by the establishment of narrow gap semiconductor energy band theory.

The second period began in the 1960s, during which a comprehensive investigation was devoted to finding the best narrow gap semiconductor materials for different system applications. Prior to ~1960 people adopted infrared detectors prepared by using PbS, InSb and Ge:Hg (Ge doped with Hg), that were applied in the three "transparent atmospheric windows" with wavelength bands at 1–3 μm (short-wave infrared or SWIR), 3–5 μm (mid-wave or MWIR), and 8–14 μm (long-wave or LWIR), respectively.

According to the spectral distribution rule for blackbody radiation, the thermal radiation of room temperature objects is distributed mainly in the 8–14 μm bands. InSb has a long wavelength cut-off near 5 μm, so its detection efficiency (responsivity) to the radiation emitted by a room temperature object is relatively low. Ge:Hg heavily impurity doped photo-conductive detectors operating in the range of 8–14 μm are suitable for thermal imaging of room temperature objects. However, they only work well when operated at temperatures below ~40 K, which is inconvenient for many practical applications. Moreover, its cut-off wavelength is not optimum. Therefore, people searched for intrinsic photoconductive or photovoltaic detector materials operating at higher temperature while still being responsive in the 8–14 μm range. In order to obtain the best response in the 8–14 μm atmosphere window the detector material must be a semiconductor with its forbidden bandwidth around 0.09 eV. However, there are no elemental semiconductors, or binary compound semiconductors with such forbidden bandwidths. Hence, it was necessary to fabricate an artificially synthesized alloy semiconductor by adjusting the alloy composition to make its forbidden bandwidth about 0.1 eV. The $Hg_{1-x}Cd_xTe$ semiconductor alloy with Cd concentration x, is such an ideal intrinsic infrared radiation detecting material. HgCdTe can be considered a pseudo-binary semiconductor of (HgTe) and (CdTe). Figure 1.1 shows the relationship between the forbidden bandwidth E_g and lattice constant a of some compound semiconductor materials. It can be seen from the figure that the II–VI semimetal compound HgTe ($E_g = -0.3$ eV), and the wide forbidden band semiconductor CdTe ($E_g = 1.6$ eV) both with zinc-blende structures, have lattice constants that are very close, $\Delta a/a = 0.3\%$. This enables HgTe, and CdTe to form continuous solid solutions of $(HgTe)_{1-x}(CdTe)_x$ pseudo binary alloy systems at various concentration ratios x. By varying the concentration of Cd, the alloy can have semimetal behavior like HgTe or semiconductor behavior like CdTe. Since $E_g = E(\Gamma 6) - E(\Gamma 8)$, at 4.2 K, E_g is –0.3 eV when $x = 0$ and 1.6 eV when $x = 1$ (Chen and Sher 1995). The bandwidth varies nearly linearly with x having only a slight bowing between –0.3 and 1.6 eV. At 4.2 K, when the composition $x = 0.161$, $E_g = 0$. Since the band gap of $Hg_{1-x}Cd_xTe$ varies continuously with composition x, this enables it to cover the entire infrared wave band, so it is an important material for preparing infrared detectors. This type of materials can be used not only to replace Ge:Hg for preparing detectors with response wavelength in the range of 8–14 μm with the added advantage of operating at 77 K, but also to substitute for PbS and InSb detectors working at room temperature in the ranges of 1–3 μm and 3–5 μm. By proper composition adjustments this material can also be applied to make PIN-type avalanche photo-diode detectors operating at 1.3 and 1.55 μm for application in optic-fiber communication. This alloy can cover the whole spectral range of 1–30 μm (Long and Schmit 1973; Stelzer et al. 1969).

Besides its many detector applications, this alloy system has proved to be a valuable aid to address fundamental research questions. Because its band gap can be adjusted by changing the composition, it has enabled detailed studies of the band gap variation of transport processes, optical properties, magneto-optic effect, lattice vibration characteristics, etc.

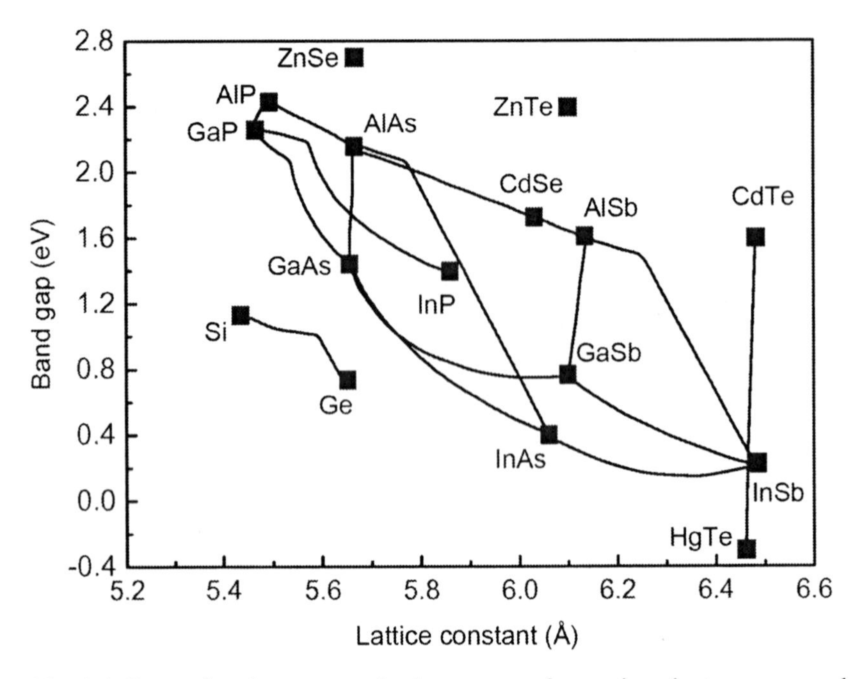

Fig. 1.1. Energy band gap versus lattice constant for semiconductor compounds

In 1959 Lawson and his collaborators first published their research results on HgCdTe (Lawson 1959). Because of the initial difficulty of preparing the material, however, it was not till 1970s, when there was progress of crystal growth capabilities by means of melting and epitaxial techniques, that significant advances on physics research of HgCdTe materials and device fabrication began. In China, Dingyuan Tang initiated in 1967 comprehensive investigations of HgCdTe materials and devices (Dingyuan Tang 1976, 1974). Starting in 1982 the major advances in HgCdTe research were captured in the proceedings of meetings initially entitled "The U.S. Workshop on the Physics and Chemistry of Mercury Cadmium Telluride." In 1992 the meeting's title was changed to add "Other IR Materials," and still later in 1996 it was changed again to become "The U.S. Workshop on the Physics and Chemistry of II–VI Materials." While the term "U.S." appears in the title, in fact it has always been an international meeting.

In the late 1980s HgCdTe materials were applied in the research and development of single-element, multi-element, linear array and two-dimensional focal plane array infrared detector systems. The developments in this period verified that HgCdTe is a relatively ideal infrared detecting material (Long and Schmit 1973). It is a direct band gap semiconductor, so infrared detectors prepared with it are intrinsic. Its corresponding optical process involves a transition between energy band states, which avoids the fundamental disadvantages of impurity-type infrared detectors.

In summary, the main advantages of employing HgCdTe material for preparing infrared detectors are:

- The forbidden bandwidth can be adjusted to cover the entire infrared wave band.
- The material has a large photo-absorption coefficient which enables the inner quantum efficiency generated in a device with a thickness of only 10–15 μm to approach 100%.
- The electron and hole mobilities are high.
- The intrinsic mechanism generates long-lived carriers and a relatively low thermal generation rate, which enables devices to operate at higher temperature.
- There is a good lattice match between CdTe and HgTe, which allows preparation of high quality epitaxial heterostructures on large CdTe substrates.
- The residual impurity concentration can be as low as 10^{14} cm^{-3}.
- It can be doped to make it p-type or n-type.
- Its surface can be passivated.

HgCdTe, however, also has some disadvantages, such as:

- The Hg–Cd bond is weak and at growth temperature Hg vacancies appear if proper controls are not exercised.
- The Hg vacancy is an acceptor, but it also degrades carrier lifetimes so it is not an optimum dopant.
- On a Te site, As is an acceptor, but on a Hg site it is a donor, and it requires stringent growth controls to get it properly positioned.
- Precipitation of Te is also a serious problem.

To realize practical device performance it has been necessary to solve these and other related problems in dopants, point and extended defects, compositional uniformity, etc. In an effort to overcome these materials problems, workers in the field began to adopt epitaxial growth techniques, in addition to conventional bulk crystal growth. HgCdTe based detector systems were advertised by French groups as early as 1967, and by the end of the 1970s the first generation of single-element HgCdTe infrared detectors became mature (Chapman 1979). In 1980s the second generation linear array detectors and small-scale area array detectors (Elliot 1981) were fabricated. Later the third generation of LWIR linear arrays and large-scale focal plane area arrays were successfully developed (Arias et al. 1989). An essential aspect of the ability to grow suitable epitaxial material was the development of large areas of Bridgman technique bulk grown CdZnTe substrate crystals lattice matched to the active HgCdTe devices (Harman 1980; Arias et al. 1993). In order to realize monolithic integration with read out chips, efforts were made to use Si and GaAs crystals as substrates, but their large lattice mismatch to the epitaxial HgCdTe introduced high densities of mismatch and treading dislocations that degrade device performance relative to that achieved on lattice matched CdZnTe substrates (Brill et al. 2001; Varesi et al. 2001). In this period, systematic progress on narrow gap semiconductors was made; on their growth, characterization methods, physics research, and device structures (Dornhaus and Nimtz 1983; Lovett 1977; Dingyuan Tang and Feimin Tong 1991).

The third stage of the development on narrow gap semiconductor physics started in the 1990s. In this stage more and more importance was attached to the research of HgCdTe thin film materials, and to the third generation long linear array and large scale focal plane area array based systems. The study of narrow gap semiconductor physics went deeper and deeper. The quality of HgCdTe thin film materials prepared by means of liquid phase epitaxy (LPE) (Schmit and Bowers 1979), metal-organic chemical vapor deposition (MOCVD) (Irvine and Mullin 1981) and molecular beam epitaxy (MBE) (Faurie et al. 1982) in addition to the bulk material growth methods (Micklethwaite 1981) made substantial advances. The growth techniques were successfully applied in making infrared focal plane arrays. Materials specifications for MWIR devices are more forgiving than those for LWIR devices. The fabrication rules for MWIR applications were refined and those for LWIR were brought to a level capable of supporting practical systems. This involved establishment of greatly improved composition control, electrical parameter control and doping control for large area thin films. In this process the links between material growth, physical characterization, device design, and device fabrication became ever closer. For example it inspired the use of hetero-junctions rather than simple p–n homojunctions: because heterojunctions could be fabricated by epitaxial growth methods. In addition to improving materials features previously mentioned, people now studied the impact on thin films of ion beam etching; this process not only modifies surface impurity states, but can also propagate impurities into existing p–n junctions. The impact of long-term exposure to light on the material and consequent device degradation were determined. Details of the spatial structure of the p–n hetero-junction electrical characteristics and its photoelectric response were extensively modeled, and practical device designs were devised taking advantage of the improved knowledge base. There were also investigations of the band structures at surfaces where two-dimensional electron gases can form to degrade device passivation. On the basis of these studies the technical device specifications were further perfected for the preparation of narrow gap semiconductor infrared focal plane arrays. At present, 512 × 512 and 1,024 × 1,024 element large scale HgCdTe infrared focal plane arrays have been built successfully. At the same time, people are making efforts to explore the new effects of photoelectric conversion, electro-optical conversion and opto-optic conversion and their new applications in infrared focal plane arrays and new types of photoelectric devices (Levine 1993; Levine et al. 1987; Choi and Eglash 1992; Choi and Turner 2000).

Advances of the general semiconductor discipline spilled over into new research ideas for producing narrow gap semiconductors. Many of these centered on research into the photoelectric conversion process in microstructures: superlattices, quantum wires, quantum wells, and quantum dots. Others looked into the advantages that could be derived from sub-band structures of bulk material. Spin electron states were studied in low-dimensional structures along with their optical transition, tunneling and transport rules, in strong magnetic fields and at ultra low temperatures. The research into characterization methods applied to semiconductors involved several optical and electrical experimental means, including infrared fluorescence spectroscopy, infrared magneto-optics spectroscopy, infrared elliptic polarization spectroscopy, micro-regional spectroscopy, flat band capacitance spectroscopy, and quantitative mobility spectroscopy. Additionally, older methods for measuring transport characteristics were extended to ultra-low temperatures and high magnetic fields.

Investigation of narrow gap semiconductors has evolved rapidly as a result of the development of the general semiconductor discipline, and of the ever more stringent requirements placed on infrared devices over the past decade. People continuously discover new phenomena from their investigations, enriching the knowledge base of the semiconductor discipline. Semiconductor science development, at the same time, is closely related with infrared optics and photo-electronic science and with technology, and their applications (including applications in the fields of aeronautical and space infrared remote sensing of the earth environment, military applications, and various ground based civil engineering applications. The accumulation of the research work has spawned the formation of a new discipline; "narrow gap semiconductor physics," with HgCdTe as one of its principal objects.

In addition to HgCdTe and InSb, narrow gap semiconductors also include α-Sn, HgSe, HgCdSe, HgSe, HgCdSe, HgS_xSe_{1-x}, $Hg_{1-x}Mn_xTe$, $Hg_{1-x}Zn_xTe$, PbS, PbSe, PbTe, PbSnSe, PbSnTe, InAs, $InAs_{1-x}Sb_{1-x}$, InAsSb, III–V alloys containing Tl and Bi, and related quaternary system materials. Among all these possible options, in the past 40 years, HgCdTe became the most important semiconductor material for infrared detectors in the mid- and long-wave infrared ranges (3–12 μm). In an effort to overcome the drawbacks of HgCdTe outlined earlier, many alternate materials have been examined, for example, HgZnTe, HgMnTe, PbSnTe, PbSnSe, InAsSb and III–V semiconductors containing Tl or Bi and low dimensional solids. The weak Hg–Te bond requires extreme care to avoid instability and non-uniformity in the bulk of the material, at surfaces, and at interfaces. Nevertheless, HgCdTe still holds the dominant position at present, mainly because it possesses a series of fine material characteristics. An important example is the ease of preparing infrared detectors that function at different wave bands. The lattice constant changes only 0.2% from CdTe to $Hg_{0.8}Cd_{0.2}Te$. Because of this near lattice match between different compositions stacks of layers can be built each of which corresponds to a detector that responds to a different band. Thus two-color and even multi-color pixel arrays have been built, where the different bands from a scene are in spatial register. This greatly simplifies the problem of extracting information about a scene by taking advantage of this multiple band collection ability. Another benefit of the near lattice match among different compositions is the ability to grow heterojunctions in which there is little materials degradation from misfit dislocations. Currently, most of the best detectors are prepared using heterojunction structures.

The lattice constant of $Hg_{1-x}(Cd_{1-y}Zn_y)_xTe$ or $Hg_{1-x}Cd_x(Te_{1-y}Se_y)$ remains almost unchanged when the composition y is adjusted by doping with small amounts of Zn or some Se, and changing x to retain a given band gap. The shorter bond length of ZnTe and CdSe forms stiffer bonds that resist bond angle distortion. This causes the energy of threading dislocations to increase, which tends to suppress their density. Thus the addition of suitably small amounts of Zn or Se can improve the quality of heterostructures, with a corresponding improvement of device performance. From these examples, it is clear that the study of HgCdTe and related alloys enriches our understanding of narrow gap semiconductors and offers opportunities for innovation.

Such research is based on the fundamental physical properties of narrow gap semiconductors. It includes crystal growth, energy band structures, optical constants, lattice vibrations, excitation and transport of carriers, impurity defects, non-linear optical properties, surface–interface states, two dimensional electron gas, super-lattice and quantum wells, device physics and other new phenomena, effects and rules. These are all subjects that will be discussed in this book.

For additional information on narrow gap semiconductor physics, readers can refer to the Semimetals and Narrow-Bandgap Semiconductors

(London) by Lovett (1977), "The Properties and Applications of the HgCdTe Alloy System in Narrow Gap Semiconductors" by Dornhaus and Nimtz (1983), Springer Tracts in Modern Physics Vol. 98, p. 119, "Semiconductors and Semimetals" Vol. 18 by American scientist, collection of review thesis on HgCdTe materials and devices (Willardson and Beer 1981), "Narrow Gap Semiconductor Infrared Detectors" by Chinese scientists Dingyuan Tang and Feimin Tong (1991) ("Semiconductor research and Progress" edited by Shouwu Wang, Beijing: Science Publishing Press, pp. 1–107) and several more papers about physical and chemical properties of HgCdTe narrow gap semiconductors from "Properties of Cd-based narrow-gap compounds" compiled by Capper (1994), as well as the related data and basic formulas of HgCdTe from "Landolt-Börnstein: Numerical Data and Functional Relationships in Science and Technology III/41B Semiconductors: II–VI and I–VII Compounds; Semimagnetic Compounds" republished in 1999, Germany. These references systematically discuss and review the basic physical properties of narrow gap semiconductors, theory and experimental data on these materials and on devices constructed from them.

1.2 Physics of Infrared Photo-Electronics

1.2.1 Infrared Photo-Electronics

Narrow gap semiconductor physics is one of the more important drivers of modern photo-electronics. In the twenty-first century, mankind is entering the photon technology era gradually. Applications requiring rates of hundreds of THz that cannot be accomplished with standard electronics are being done with photons. These include ultra fast computers, and communications as well as advanced imaging applications. On the one hand, people have been establishing a deeper understanding of light, its interaction with matter, and knowledge of how to control them. This background has promoted the rapid advancements of photoelectron physics, and given impetus to the development of further applications. Therefore, the field of narrow gap semiconductors is an important element in the evolution of modern photoelectron technology.

Several major problems have arisen in the research of modern photoelectron physics. As technology advances by leaps and bounds, the time between research on fundamental properties and processes, and high technology applications is becoming shorter and shorter. Moreover, high-tech application requirements have brought about more and more urgent demands on the research into fundamental laws. Even now more demanding application requirements are the major driving force for the development of infrared photoelectron research. From the viewpoint of discipline development in the new century, infrared photoelectron physics will focus on further research on excitations, and IR photon transport and nonlinear interactions in matter. Particularly, expect further research into micro-mechanisms of conversion processes from infrared light to free carriers, from free carriers to infrared light, two infrared light sources into different frequencies, and infrared light and visible light into other visible light beams. Some of these developments will be motivated by optical computing applications.

1.2.2 Thermal Infrared Detector Materials: Uncooled Detectors

Ferroelectric thin films are a type of material that has attracted much attention in recent years. In addition to their use in the development of non-volatile memories and piezoelectric drivers, etc., they can be employed to fabricate infrared broad-bandwidth focal plane arrays (Kruse 1981) working at room temperature. This is one type of detector that falls into the class called "uncooled" detectors. The ferroelectrics function because their effective dielectric constant is temperature dependent. Because of this dependence the voltage across a charged (poled) capacitor made from such materials generates a signal in response to incident radiation as it is absorbed and heats the ferroelectric's surface. Importance is currently attached to the ferroelectric thin films PZT and BST, but it is not certain that these will ultimately prove to be the materials of choice for this application. Generally, the sol–gel method, sputtering, laser-plasma deposition, MOCVD methods and others are employed for their preparation. Research into the physics of ferroelectric thin films, their physical properties related to infrared detectors, in particular, micro mechanism of spontaneous polarization are among the hot issues under investigation in international academic and industrial circles.

A second class materials in competition for uncooled array applications are those that undergo an "insulator-to-metal" phase transition as they are heated. The type most often used at present is VO_{2-x}, where x varies between 0 and 1 (Kruse 1981). At a critical temperature these materials experience a first order phase transition in which the conductivity changes by more than four orders of magnitude. One cannot bias these materials immediately below the critical temperature to take advantage of this huge change, because devices operated in this way are unstable. However, in a range slightly below the critical temperature the material still has a sharp temperature dependence that is useful for thermal detection. VO_{2-x} also can be vapor deposited into thin films on thin support low heat capacity structures to form arrays of detectors with practical detectivities. Their detectivities are not as good as those of cooled photon detector arrays, but they are good enough to satisfy most specifications for some applications where cooling is impractical.

1.2.3 Light Emitting Devices

As we have discussed earlier semiconductor low dimensional structures are alternative infrared photoelectric materials. Structures of III–V semiconductor quantum wells, quantum wires and quantum dots are used in preparation of infrared detectors and focal plane arrays. They were particularly touted for multi-color devices and long wavelength devices, but it is now understood that their operating modes do not allow them to hold a calibration in low light background applications. However, the relatively narrow spectral response feature of the optical transitions between sub-bands of the quantum wells is beneficial for the development of light emitting devices (LED's and Lasers). III–V semiconductor quantum wells have succeeded in emitting light in the mid-infrared wave band.

These low dimensional structures are also being developed as infrared non-linear elements for frequency multiplier and mixer applications. The preparation, control and characterization of semiconductor low dimensional structures form the basis for this direction and are an important aspect of the future development of infrared physics.

1.2.4 Processes of Infrared Physics

The core of infrared photoelectron physics research is focusing on photoelectric conversion, electro-optic conversion and photo–photo conversion processes in infrared responsive matter systems; their mechanisms and methods for controlling them. Photoelectric physics enables us to understand the conversion processes and what must be accomplished in materials processing and design to meet specific goals. These goals are often set by device specifications. Understanding the precise mechanisms also serve as the basis of device design. Thus there is a feedback loop from understanding fundamental mechanisms–fabrication strategies–device design to meet specifications–motivation for research into fundamental mechanisms, all aimed at improving performance. Of course, some research is also driven by the desire to improve man's knowledge of nature. The most important research areas are specified in the previous sections. At the moment, there is an increasing need for various devices, such as, large-scale infrared focal plane arrays, mid-infrared wave band lasers, infrared non-linear devices and infrared mono-photon detectors, etc. These needs have imposed higher and higher requirements on research into mechanisms responsible for infrared photoelectric conversion. For example, while trends in band-to-band and trap-assisted tunneling across p–n junctions have been known for years, no quantitatively reliable theory exists. This knowledge is needed to improve the performance of LWIR focal plane arrays.

Efforts are still underway to discover more quantitative relationships between the photoelectric conversion process and energy band structures, impurity defects and lattice vibrations in order to obtain clearer physical patterns and models. In the past, a relatively deep study of the photoelectric process in three-dimensional systems had been made. Now, it is necessary to study these effects in 2D electron gases at surfaces and interfaces, at impurity defects, and in clusters of concentration fluctuation related quantum dots. New approaches should be explored to advance our knowledge base in each of these areas.

The mechanisms and materials involved in photoelectric emission devices comprise a different set of research areas. Stimulated emission light sources and free electron lasers mostly made from III–V compound alloy based micro-structures, are electro-optic converters covering wide spectral ranges including the entire infrared wave band. These materials and device structures are major subjects of worldwide investigation. The physics of the electro-optic conversion process in these systems is also another hot issue for study. Other major research drivers are the need for high-speed optical switches, electro-optic modulators, mid-infrared lasers, and THz sources. People have worked diligently to discover the physical process responsible limits on high-speed electro-optic modulation and the process of photo-excited emission in mid-infrared. Significant progress has been achieved in the research on electro-optic modulation of GaAlAs based super-lattice structures, on quantum dots lasing of InGaAs, and cascade laser emission of InAsSb based complex super-lattice systems.

Opto-optic mixers require materials transparent over the incident and the mixed frequency bands and having non-linear dielectric constants. Research efforts focus on the mechanisms responsible for the mixing phenomena in new materials, particularly exploration the non-linear optical response of solid low dimensional structures. These are currently active research issues in the infrared field.

Research has always been an essential element in making progress with the interaction between photons and the electrical response of materials. Most future developments will depend on complete accurate physical models based on fundamental precepts. Exploration of new photoelectric testing and measuring methods, are likely paths to unearthing new phenomena, and new device opportunities. Active research in these areas is continuing in various parts of the world without any sign of decline.

1.2.5 Infrared Materials Research, Device, and Applications Status

Infrared photon device physics is the basis of most device applications. Infrared photon devices cover a collection of infrared applications, including large-scale infrared focal plane arrays for cameras, infrared single photon detectors for high speed communications and astronomy, mid-infrared lasers for scientific instruments, etc.

Large-scale infrared focal plane arrays at present are the most advanced third generation infrared scene sensors. They collect information by focusing a scene on the detector array. The IR (infrared) photons striking an element of the array (a pixel) excite carriers in proportion to the intensity of that spot in the scene. In the most sophisticated arrays intensity distributions at several parts of the IR spectrum are separately detected at each pixel. Periodically, at a specified frame rate, the signals generated by the elements of the array are read out by a read-out-integrating-circuit (ROIC). This information is then displayed on a monitor, and processed to extract information about the scene. Arrays that function in this fashion are called “staring arrays”; they have almost completely replaced the older “scanning” systems. The simplest version of a scanning system employs a moveable mirror to scan different regions of a scene sequentially onto a single detector.

Staring arrays have been constructed from narrow gap semiconductors, from quantum well devices operating at low temperature; and from ferroelectrics and insulator-to-metal thin film devices operating at room temperature. The use of staring focal plane arrays has not only greatly simplified the structure of imaging systems, and improved their reliability, but has also increased their detecting performance remarkably. The images and their spectral content obtained from these systems have been used for target tracking, recognition, and even for quantitative analysis of the nature of the materials in the scene. The images have yielded valuable information both on macro objects, such as the health of forests, ground waters, and meteorology data; and on micro objects, such as biological cells.

Although considerable research progress has been made on the fundamental behavior of the materials used in photon detectors, and theories exist that properly predict observed trends and in some cases even quantitative properties, there still are quantitative physical problems that are unresolved. Modern theoretical techniques are gradually eliminating the remaining uncertainties (Chen and Sher 1995). For example, the advent of density functional theory (DFT) has allowed us to calculate, with no adjustable parameters and with remarkable precision, ground state crystal structures, including their bond lengths and elastic constants, as well as the band structures of all the group IV elements and the III–V, and II–VI compound semiconductors. These DFT methods are computationally intensive, but a theory called the hybrid pseudo-potential tight-binding (HPT) theory, while it depends on a few adjustable pseudo-potential form factors for each material, yields equally accurate band structures and wave functions. The form factors can be extracted from the DFT predictions, so it can be argued that HPT results are also parameter free. The HPT theory is much more manageable, and can be coupled with the coherent potential approximation (CPA) method to calculate, in excellent agreement with experiment, the band structures and wave functions of the alloys of all these semiconductors. The quality of the wave functions can be judged from the excellent agreement between the predictions of the temperature and concentration variations of the band gaps for several semiconductors and alloys including HgCdTe (Krishnamurthy et al. 1995). For this result to be reached, the matrix elements of the electron–phonon interactions had to be accurate. Numerous other physical processes can now be predicted in agreement with experiment, e.g. absorption coefficients, carrier transport properties, densities of native point defects as functions of temperature and partial pressures covering the whole stability region for all alloy concentrations, etc. However, there do remain unresolved problems that impact device performance, and as growth processes and device designs improve to overcome old barriers to performance, new challenges constantly arise. Examples include improved control of impurities and extended defects, surface passivation, tunneling through hetero-junctions, and the origin of 1/f noise. An example of a breakthrough in the new era would be LWIR photon based focal plane arrays operating at room temperature. Accomplishing this goal would involve somehow tricking nature to overcome the adverse effects of thermally excited carriers.

A fast infrared single photon detector is another important quantum device with major applications in the giga- to tera-Hertz information transmission field. There are special needs for infrared single photon detectors operating in the 1.3 and 1.55 μm wave bands. These bands are where current optical fiber materials have low absorption coefficients. Development of InGaAs based avalanche photodiodes is an important topic in the development of infrared single photon detectors; and significant progress has been achieved. Other competitors for this application are HgCdTe and HgMnTe with their concentrations chosen so their spin–orbit split-off band and forbidden band gap energies match. This is beneficial to a "resonant ionization by collision" phenomenon in the wave band range of 1.3–1.8 μm, which can yield avalanche photodiodes with high gain and low noise. With a proper composition ratio, a quarternary system HgCdMnTe can be used to prepare an avalanche photodiode with its maximum response at 1.55 μm. Therefore, another important frontier subject at present is research into the absorption and recombination process among conduction bands, valence bands and spin–orbit split bands as well as the theory of the avalanche ionization process. This will help to develop optimum sensitivity HgCdTe, HgMnTe, InGaAs avalanche photodiodes, and to explore other candidate single photon detector mechanisms.

Mid wave band infrared lasers are another important research subject. This activity was initiated at the end of the last century. Now people have started to use III–V semiconductor quantum well structures to prepare mid-infrared cascade lasers, and this work will go forward in the new century. Research on quantum dot lasers, quantum dot infrared detectors and attempts to devise infrared focal plane arrays without attached readout circuits will be carried out vigorously in the new era.

References

Arias JM, Shin SH, Pasko JG, DeWames RE, Gertner ER (1989) Long and middle wave length infrared photodiodes fabricated with $Hg_{1-x}Cd_xTe$ grown by molecular beam epitaxy. J Appl Phys 65:1747–1753

Arias JM, Pasko JG, Zandian M, Shin SH, Williams GM, Bubulac LO, DeWames RE, Tennant WE (1993) MBE HgCdTe heterostructure p-on-n planar infrared photodiodes. J Electron Mater 22:1049–1053

Brill G, Velicus, Boieriu P, Brill G, Velicu S, Boieriu P, Chen Y, Dhar NK, Lee TS, Selamet Y, Sivananthan S (2001) MBE growth and device processing of MWIR HgCdTe on large area Si substrates. J Electron Mater 30:717–722

Capper P (1994) In: Capper P (ed) Properties of Cd-based narrow-gap Compounds. London, INSPEC/IEE

Chapman CW (1979) The state of the art in thermal imaging: Common modules. In: Electro-Optics/Laser 79'Conference and Exposition, Anaheim, California, USA, Oct. 23–25, 1979. Industrial and Scientific Conference Management, Chicago, pp. 49–57

Chen AB, Sher A (1995) Semiconductor Alloys. Plenum Press, New York

Choi HK, Eglash SJ (1992) High-power multiple-quantum-well GaInAsSb/AlGaAsSb diode lasers emitting at 2.1 μm with low threshold current density. Appl Phys Lett 61:1154–1156

Choi HK, Turner GW (2000) Antimonide-based mid-infrared lasers. In: Helm M (ed) Long wavelength infrared emitters based on quantum wells and superlattices. Gordon and Breach Science Publishers, Singapore, Chapter 7

Dingyuan Tang (1974) Properties of ternary system HgCdTe semiconductors. Infrared Physics and Technique, Shanghai Institute of Technical Physics, Chinese Academy of Sciences, 16:345

Dingyuan Tang (1976) HgCdTe material for infrared detectors. Infrared Physics and Technique, Shanghai Institute of Technical Physics, Chinese Academy of Sciences, 4–5:53

Dingyuan Tang, Feiming Tong (1991) Narrow gap semiconductor infrared detectors. In: Shouwu Wang (ed) Development of Semiconductor Devices and Progress. Science Publishing Press, Beijing, pp. 1–107

Dornhaus R, Nimtz G (1983) The properties and applications of the HgCdTe alloy system. In: Narrow Gap Semiconductors, Spring Tracts in Modern Physics, Vol. 98. Springer, p. 119

Elliot CT (1981) New Detector for Thermal Imaging Systems. Electron Lett 17: 312–314

Faurie JP, Million A, Piaguet J (1982) CdTe–HgTe multilayers grown by molecular beam epitaxy. Appl Phys Lett 41:713–715

Harman TC (1980) Liquidus isotherms, solidus lines and LPE growth in the Tellurium-Rich corner of the HgCdTe System. J Electron Mater 9:945–961

Irvine SJC, Mulin JB (1981) The growth by MOVPE and characterization of $Cd_xHg_{1-x}Te$. Thin Solid Films 90:107–102

Kane EO (1957) Band structure of indium antimonide. J Phys Chem Solids 1:249–261

Kane EO (1966) Semiconductors and Semimetals Vol. 1. Academic press, London, p. 75

Kane EO (1981) Narrow Gap Semiconductors: Physics and Applications. Springer-Verlag, Berlin

Krishnamurthy S, Chen AB, Sher A, Van Schilfgaarde M (1995) Temperature dependence of band gaps in HgCdTe and other semiconductors. J Electron Mater 24:1121–1125

Kruse PW (1981) The emergence of $Hg_{1-x}Cd_xTe$ as a modern infrared sensitive material. In: Willardson RK and Beer AC (eds) Semiconductors and Semimetals, Vol. 18. Academic press, London, pp. 1–20

Lawson WD, Nielson S, Putley EH, Young AS (1959) Preparation and properties of HgTe and mixed crystals of HgTe–CdTe. J Phys Chem Solids 19:325–329

Levine BF (1993) Quantum-well infrared photodetectors. J Appl Phys 74:R1–R81

Levine BF, Choi KK, Bethea CG, Walker J, Malik RJ (1987) New 10 μm infrared detector using intersubband absorption in resonant tunneling GaAlAs superlattices. Appl Phys Lett 50:1092–1094

Long D, Schmit JL (1973) Infrared Detectors. Academic Press, New York

Lovett DR (1977) Semimetals and Narrow-bandgap Semiconductors. In: Goldsmid HJ, Ballentyne DWG (eds) Applied Physics Series. Pion Limited, London

Micklethwaite WFH (1981) The crystal growth of Cadmium Mercury telluride. In: Willardson RK, Beer AC (eds) Semiconductors and Semimetals, Vol. 18. Academic Press, London, pp. 70–84

Schmit JL, Bowers JE (1979) LEP growth of $Hg_{0.60}Cd_{0.40}Te$ from Te-rich solution, Appl Phys Lett 35:457–458

Stelzer EL, Schmit JL, Tufte ON (1969) Mercury Cadmium Telluride as an infrared detector material. IEEE Trans. Electron Devices 16:880–884

Varesi JB, Bornfreund RE, Childs AC, Radford WA, Maranowski KD, Peterson JM, Johnson SM, Giegerich LM, de Lyon TJ, Jensen JE (2001) Fabrication of high-performance large-format MWIR focal plane arrays from MBE-grown HgCdTe on 4 inch silicon substrates. J Electron Mater 30:566–573

Willardson RK, Beer AC (1981) Mercury cadmium telluride. In: Willardson RK, New York

2 Crystals

2.1 Theory of Crystal Growth

2.1.1 Introduction

The $Hg_{1-x}Cd_xTe$ Narrow gap semiconductors are nearly random substitutional alloys of the II–VI binary compounds CdTe and HgTe. The fraction (composition) of CdTe is x and of HgTe is $1-x$. The ground states of the HgTe and CdTe compounds as well as the $Hg_{1-x}Cd_xTe$ alloys for all compositions x, have zincblende cubic structures (Daruhaus and Vimts 1983). In this crystal structure there are two interpenetrating face centered cubic sub-lattices, one occupied by the Te anions, and the other shared by the Cd and Hg cations The two interpenetrating face-centered cubic lattices are offset with respect to one another by the displacement $\left(\frac{1}{4}a_0, \frac{1}{4}a_0, \frac{1}{4}a_0\right)$, where a_o is the cube edge length, i.e., the lattice constant. The lattice constant of the HgTe crystal is 6.46 Å, and that of CdTe is 6.48 Å. In Fig. 2.1, cations (Cd or Hg) are listed as type A while Te anions are type B.

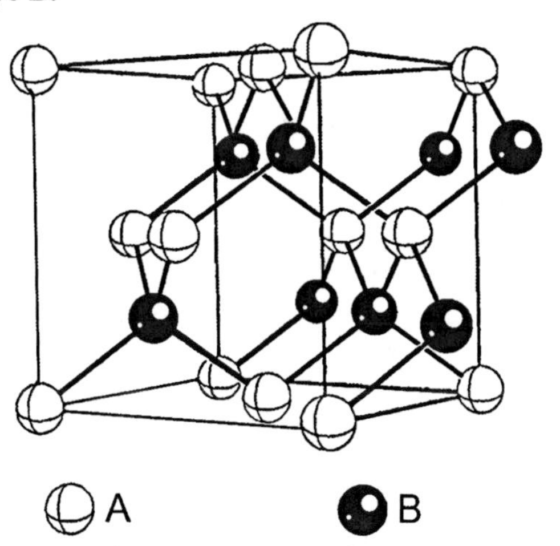

Fig. 2.1. Zincblende crystal structure

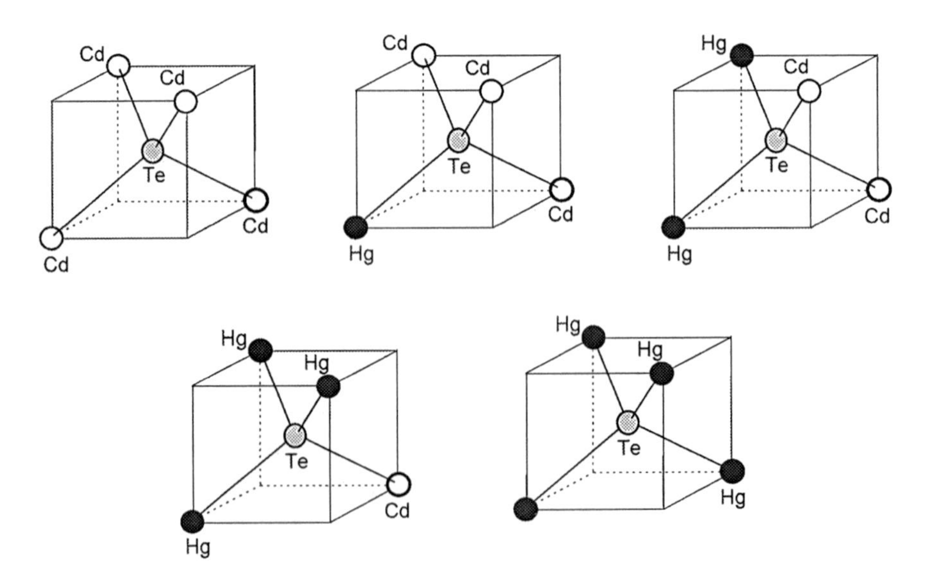

Fig. 2.2. Five kinds of possible unit cells

As shown in Fig. 2.1, the crystallographic unit cell has four cations and four anions, whose coordinates are:

- Cations: $(0,0,0)$, $\left(\frac{1}{2},\frac{1}{2},0\right)$, $\left(\frac{1}{2},0,\frac{1}{2}\right)$, $\left(0,\frac{1}{2},\frac{1}{2}\right)$
- Anions: $\left(\frac{1}{4},\frac{1}{4},\frac{1}{4}\right)$, $\left(\frac{3}{4},\frac{3}{4},\frac{1}{4}\right)$, $\left(\frac{3}{4},\frac{1}{4},\frac{3}{4}\right)$, $\left(\frac{1}{4},\frac{3}{4},\frac{3}{4}\right)$

The distribution of Cd and Hg atoms on sub-lattice A is quasi-random. There are two classes of deviations from strictly random distributions. The first accounts for mesoscopic sized (ten to a few hundred microns) regions where the concentration deviates from its average value. Statistical mechanics requires that such fluctuations exist (Muller and Sher 1999). Moreover, the amplitude and density of such fluctuations is greatest for lattice-matched alloys like HgCdTe. For lattice-mismatched alloys, if there is a region of space where the composition differs from its average value there will be a strain that tends to suppress fluctuations. Fluctuations do affect physical properties that impact the performance of infrared devices.

The second kind of deviation from random alloy arrangements is due to short-range order correlations. The Te atom in lattice B has four nearest-neighbor anions, which are either Cd atoms or Hg atoms. There are five types of these five atom clusters, those with i Cd atoms and 4-i Hg atoms, and the value of i varies from 0 to 4 (Fig. 2.2).

For a random configuration the populations of the different cluster types for a given x form a Bernoulli distribution $P_i(x) = \binom{4}{i} x^i (1-x)^{4-i}$. However, when there are interactions within a cluster among the different types of cations there will be deviations from this simple distribution. In the HgCdTe alloy these cation–cation interactions are small, so short range order deviations are small but not completely negligible.

Another way to look at the correlation problem is to examine only pair interactions between the second neighbor cations, rather than deal with the five atom clusters, or better still 12 atom 16 bond clusters. This leads to a less accurate but simpler formalism often found in the literature. It was developed to treat $GaAs_xP_{1-x}$ crystals (Verleur and Barker 1966). The same formalism applies to cation substituted alloys like HgCdTe. A parameter β, called the short-range order parameter, is used to characterize this deviation from randomness. The probability of finding a Cd atom as a nearest neighbor of another Cd atom is:

$$P_{\mathrm{Cd,Cd}} = x + \beta(1-x)$$

The probability of finding a Hg atom as a nearest neighbor of another Hg atom is:

$$P_{\mathrm{Hg,Hg}} = (1-x) + \beta x$$

The short-range order parameter lies in the range between –1 and 1. Its value depends on the ratio of interaction energies between neighboring cations $2E_{HgCd}/(E_{HgHg} + E_{CdCd})$. If this ratio is unity then $\beta = 0$, and the distribution is completely random. If this ratio is less than 1, then $0 < \beta$, and the populations of the five kinds of tetrahedrons as shown in Fig. 2.2 are modified with the i = 0 and i = 4 populations enhanced at the expense of the others. In the extreme limit where the energy ratio is 0 and $\beta = 1$, there are only two kinds of clusters, i = 0 (HgTe) and i = 4 (CdTe). This leads to spinodal decomposition into regions of HgTe and CdTe with relative volumes determined by x. On the other hand, if the energy ratio is greater than 1, then $\beta < 0$, and the populations of the i = 2 clusters are enhanced at the expense of the others. In the extreme case where the energy ratio is very large then $\beta = -1$, and for special concentrations an ordered crystal is formed. As an example consider the case where $x = 1/2$. Then $P_{HgHg} = P_{CdCd} = 0$, and only the i = 2 cluster has any population. In HgCdTe the tendency is toward ordering, $\beta < 0$. This tendency should be taken into account when calculating alloy properties that are sensitive to short range order arrangements. Outstanding examples are point defect populations and deep state energies.

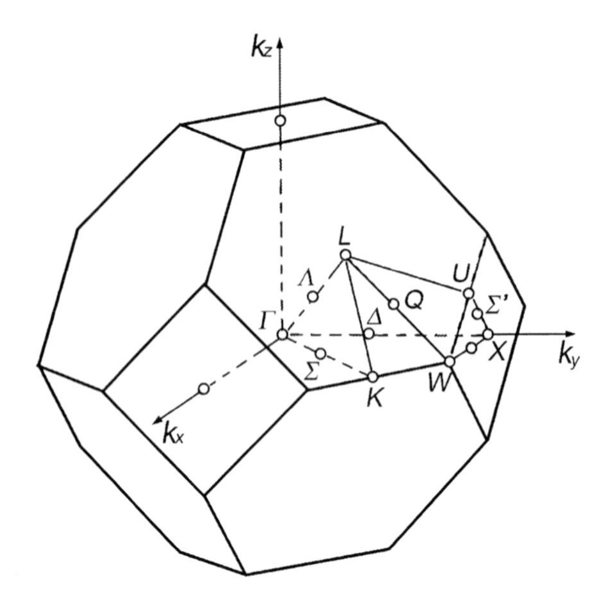

Fig. 2.3. The first Brillouin zone of $Hg_{1-x}Cd_xTe$ alloys

As in all zincblende cubic crystals, each primitive cell of $Hg_{1-x}Cd_xTe$ alloys has two atoms, one anion (Te atom) and one cation (Hg atom or Cd atom). There are six valence electrons outside the core filled shell of a Te atom in a $5S^2$, $5P^4$ configuration; and there are two valence electrons outside the core filled internal-shell in a Hg atom in a: $6S^2$ configuration or a Cd atom in a $5S^2$ configuration. Their chemical bond is mainly covalent, but with a substantial ionic contribution (Harrison 1980). These bonds form along tetrahedral directions. Since the potential of the anions is deeper than that of the cations there is a net electron transfer from the cations to the anions. This is what leads to the ionic component of the bonds. As in all face-center cubic structures, the first Brillouin zone is a truncated octahedron. The crystal point group of a zincblende structure is T_d, as shown in Fig. 2.3.

The lattice constant a_0 of $Hg_{1-x}Cd_xTe$ crystals can be determined by an X-ray technique (Daruhaus and Vimts 1983; Woolley and Ray 1960). The experimental results show that the variation of the constant a_0 with composition x is slightly nonlinear as shown in Fig. 2.4. Using a specific gravity method, the density of $Hg_{1-x}Cd_xTe$ with different compositions can be determined, as shown by the straight line in Fig. 2.4 (Blair and Newnham 1961).

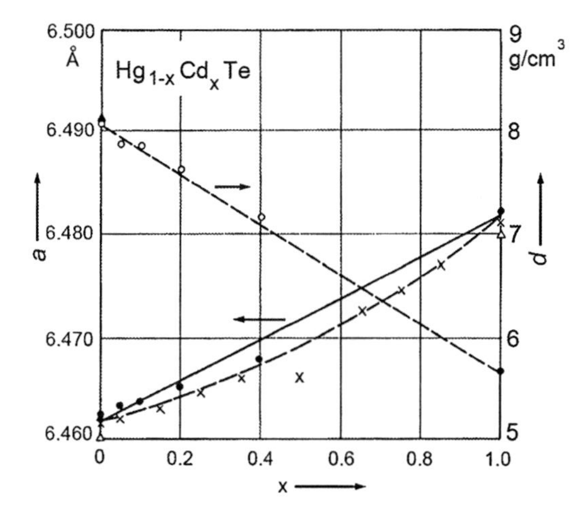

Fig. 2.4. The lattice constant and density versus composition for $Hg_{1-x}Cd_xTe$

The chosen crystal growth procedure must take account of the desired periodic structure at a given composition. While, for example, zincblende is the ground state of HgCdTe it is possible to choose conditions that force the material into the Wertzite structure. Because of its weak bonds and the stability of Hg_2, Cd_2, and Te_4 molecules in the gas phase, the growth of HgCdTe alloys is very difficult. Its growth techniques mainly include: the growth of bulk materials, Liquid Phase Epitaxy (LPE), Metal Organic Chemical Vapor Deposition (MOCVD) and Molecular Beam Epitaxy (MBE). In this chapter, we introduce the common theory of bulk crystal growth at first, and then introduce each of the other techniques used in the growth of narrow gap semiconductors.

2.1.2 The Thermodynamics of Crystal Growth

We begin with a review of the thermodynamics of crystal growth, phase diagrams and their application to the control of growth processes.

The crystal growth procedure involves controlled phase transformations of material under specified thermodynamic conditions. Dealing with a phase variation using thermodynamics, the transformation from the original state to the final state of each phase in the system can be treated without delving into details of mechanisms, such as the binding forces between atoms or how atoms reach their respective final positions. In thermodynamics the state of a system is characterized by some combination of parameters such as temperature T, pressure P, volume V, work W, heat Q, internal energy U, enthalpy H, entropy S, Gibbs free energy G and Heimholz free energy F.

Their interrelations can be obtained from the first and the second laws of thermodynamics.

The first law teaches that the gain of internal energy of a system is equal to the difference between the heat it absorbs from a heat bath and the work it does on an outside entity.

$$\Delta U = Q - W \tag{2.1}$$

Equation (2.1) is the integral expression of the first law, and it's differential expression is:

$$dU = \delta Q - \delta W \tag{2.2}$$

The first law is the energy conservation law of a system as it undergoes a thermal transformation.

There are many seemingly different, but actually equivalent, ways of describing of the second law, of which those of Clausius and Kelvin–Plank are representative (Zemansky 1951). The Clausius statement is: "It is impossible by means of an inanimate material agency to derive a mechanical effect from any portion of matter by cooling it below the temperature of the coldest of the surrounding objects." The Kelvin–Plank statement of the second law is: "It is impossible to construct an engine which working in a complete cycle will produce no effect other than the raising of a weight and cooling of a heat reservoir." One consequence of the Clausius version of the second law is the irreversibility of heat transfer when a system changes phases, while the Kelvin–Plank version asserts the irreversibility of a process transforming work only into heat when systems undergoes phase changes. In general, spontaneous processes in nature are irreversible, which is the more common description of the second law of thermodynamics. The mathematical expression of the second law is:

$$dS \geq \frac{\delta Q}{T} \tag{2.3}$$

Here the inequality holds for irreversible processes, while the equality describes reversible processes where:

$$\delta Q = TdS \tag{2.4}$$

The third law of thermodynamics also has several statements. The earliest statement due to Fowler and Guggenheim is: "It is impossible by any procedure, no matter how idealized, to reduce any system to the absolute zero of temperature in a finite number of operations." A second equivalent statement due to Nernst is: "The entropy change due to any isothermal, reversible process approaches zero as the temperature approaches zero." The third law plays no role in crystal growth, but may have some bearing on device operation at low temperature.

Among the thermodynamic variables there are certain inherent correlations. The state of a closed system can be specified by three independent variables, e.g. for an ideal gas the volume, pressure, and temperature. All the other thermodynamic variables are dependent on those selected to be the independent three. If the system is not closed, so it can exchange matter with a reservoir or there is an external field, then there are extra independent variables. For a closed system at constant pressure in a reversible process, where the only work done is due to a volume change,

$$dU = TdS - PdV \tag{2.5}$$

This expression is a combination of the first and the second law, and is a central statement of thermodynamics. The above formula can be rewritten as:

$$d(U - TS) = -SdT - PdV, \tag{2.6}$$

or

$$dF = -SdT - PdV \tag{2.7}$$

where the Helmholtz free energy F is defined as:

$$F \equiv U - TS \tag{2.8}$$

In an isothermal process, dT = 0, the system does not exchange heat with an external heat bath. If in an isothermal process, work is done on the system by an external agent, its free energy will increase. Or if the system does work on the external agent, its free energy will decrease. When the system is in its equilibrium state no work is exchanged with the outside agent, the volume of system is constant, and the Helmholtz free energy reaches a minimum.

In an isovolumetric process, dV = 0, the work done by pressure P is zero. If the system absorbs heat from an outside bath, the internal energy of system will increase; or if the system gives out heat to an outside bath, the internal energy of the system will decrease:

$$\delta Q = dU + PdV \tag{2.9}$$

In an isobaric process, $dP = 0$, the central statement of thermodynamics can be rewritten as:

$$\delta Q = dU + PdV = d(U + PV) \tag{2.10}$$

Define the enthalpy as:

$$H \equiv U + PV \tag{2.11}$$

Equation (2.10) states that absorbing heat from an outside bath causes an increase of the enthalpy of system. Then if the enthalpy of system decreases, it must supply heat to an outside bath. In an isobaric phase transition, the system's enthalpy is the latent heat of the phase transition.

In an isothermal and isobaric process, $dT = 0$, $dP = 0$, and the system transfers no heat with outside baths. If one generalizes the expression, (2.5), for the central statement of thermodynamics to include both the work done by pressure and the work done by the other generalized forces W′, then it becomes:

$$TdS = dU + PdV + \delta W', \tag{2.12}$$

or

$$\begin{aligned} d(U - TS + PV) &= -SdT + VdP - \delta W' \\ &= dG, \end{aligned} \tag{2.13}$$

where the Gibbs free energy G is now defined as:

$$G \equiv U - TS + PV \tag{2.14}$$

It also can be written as:

$$G = F + PV \tag{2.15}$$

From (2.13) one sees that in an isothermal and isobaric process, only the work done by the other generalized forces changes the Gibbs free energy. In general a system undergoing an isobaric process reaches equilibrium when the Gibbs free energy attains its minimum value. The Gibbs free energy of one mol of matter is named chemical potential μ.

In general, phase transitions can be either first order or second order. A first order phase transition has a latent heat, and it occurs abruptly at a critical temperature. It is characterized by $\Delta H \neq 0$, $\Delta V \neq 0$, $\Delta S \neq 0$, that is, the fist order partial differential of the chemical potential is not zero. In such a phase transition, the relation between the pressure and temperature is given by the Clapeyron equation:

$$\frac{dP}{dT} = \frac{\Delta H}{T\Delta V} \tag{2.16}$$

ΔH is the latent heat of the phase transition, and ΔV is the change of the volume from the initial to the final phase.

A second order phase transition is one in which the specific heat at constant pressure C_P, the coefficient of dilatation α, and the compressibility δ etc. will change, that is, characteristics represented by the quadratic partial differential of the chemical potential will change. These characteristics can be represented as

$$\mu_1 = \mu_2$$

$$V_1 = V_2, \text{ i.e. } \left(\frac{\partial \mu_1}{\partial P}\right)_T = \left(\frac{\partial \mu_2}{\partial P}\right)_T$$

$$S_1 = S_2, \text{ i.e. } \left(\frac{\partial \mu_1}{\partial T}\right)_P = \left(\frac{\partial \mu_2}{\partial T}\right)_P$$

$$C_{p1} \neq C_{p2}, \text{ i.e. } \left(\frac{\partial^2 \mu_1}{\partial T^2}\right)_P \neq \left(\frac{\partial^2 \mu_2}{\partial T^2}\right)_P$$

$$\alpha_1 \neq \alpha_2, \text{ i.e. } \left\{\frac{\partial}{\partial T}\left(\frac{\partial \mu_1}{\partial P}\right)_T\right\} \neq \left\{\frac{\partial}{\partial T}\left(\frac{\partial \mu_2}{\partial P}\right)_T\right\}$$

$$\delta_1 \neq \delta_2, \text{ i.e. } \left(\frac{\partial^2 \mu_1}{\partial P^2}\right) \neq \left(\frac{\partial^2 \mu_2}{\partial P^2}\right) \tag{2.17}$$

In a second order phase transition, the relation between pressure and temperature is given by the Ehrenfest equation:

$$\frac{dP}{dT} = \frac{\alpha_2 - \alpha_1}{\delta_2 - \delta_1} \quad \text{or} \quad \frac{dP}{dT} = \frac{C_{P_2} - C_{P_1}}{Tv(\alpha_2 - \alpha_1)} \tag{2.18}$$

Sublimation of solids to gases, the order-disorder phase transition of alloys, the phase transition from the normal state to the superconducting state of superconductors, are examples of second order phase transitions. In general, first order phase transitions exhibit thermal hysteresis when the temperature is cycled through the critical temperature, and second order phase transitions do not. If a crystal has an allotropic first order phase transition in the cooling process after growth, this usually increases the stress within the crystal that often results in crazing. A phase transition in which the crystal changes from one crystal structure to another, often with a different lattice constant, is allotropic.

2.1.3 The Dynamics of Crystal Growth

The two most important basic elements of crystal growth are the thermal aspects, coupled to the dynamics. From a macroscopic prospective, crystal growth is a transport process of heat, mass and momentum. The force driving crystal growth comes from a super-saturation of the concentration (Δc), or super-cooling (ΔT) behind the growth front growth. The growth process includes the arrival of the source material molecules at the growth interface, their decomposition into the atom species of the unit cell of the crystal, ejection from the surface of the portions of the source molecules that are not to be a part of the crystal, and transport of particles on the growth surface to an appropriate lattice site. The growth surface is never flat; it always consists of an array of mesas. Growth usually proceeds from the base of these mesas. When two mesas meet at their bases, they usually do so to form a perfect crystal. However, if they are mis-aligned they may initiate a dislocation, a twin boundary, or some other type of grain boundary. Some of these defects will anneal out as the growth proceeds. The growth interface transport dynamics is pivotal to the formation of good single crystals with few imperfections. Different crystal faces have different growth speeds, so unless this feature is well controlled it can cause the crystal to have surface roughness, or even pillars and voids. The growth speed must be slow enough so the latent heat given out as the crystal evolves will have time to drive the surface atoms into their proper locations with the desired stoichiometry. Then the crystal growth process will proceed smoothly.

The entities transported in crystal growth are heat, mass, and momentum.

Heat is transported by three methods: radiation, conduction, and convection. In the process of crystal growth, which process is primary depends on the actual technological condition. When the crystal grows at a high temperature, heat radiation transfers the majority of the latent heat of growth deposited at its interface; conduction and convection have subordinate roles. When the crystal grows at a low temperature, the heat transport mainly depends on conduction.

If we neglect the change of the physical constants (such as density, heat capacitance, heat conduction coefficient) with temperature, and do not consider the energy loss in the system, then the net heat delivered in the liquid to the growth surface is governed by the heat exchange equation:

$$\rho c_p \frac{\partial T}{\partial t} + \rho c_p \mathbf{v} \bullet \nabla \boldsymbol{T} = \kappa \nabla^2 \boldsymbol{T} \tag{2.19}$$

Here ∇T is the gradient of temperature, $\nabla^2 = \nabla \bullet \nabla$ is the Laplacian, $\mathbf{v}$ is the velocity of melt, ρ is the density of melt, c_p is the isobaric heat capacitance of melt per unit mass. When the temperature does not change with time, the above equation reduces to the expression:

$$\rho c_p \mathbf{v} \bullet \nabla \boldsymbol{T} = \kappa \nabla^2 \boldsymbol{T} \tag{2.20}$$

If there is no convection in the melt, $\mathbf{v} = 0$, then (2.21) becomes the heat exchange equation:

$$\rho c_p \frac{\partial T}{\partial t} = \kappa \Delta \boldsymbol{T} \tag{2.21}$$

Next we treat mass transport. Mass is transported in crystal growth by two mechanisms: diffusion and convection.

In a solution, the solute density is not always uniform. The spatial distribution of the ith component of the solute density, Ci, is called the ith concentration field of the solute. Concentration gradients are the driving force for diffusion. In three-dimensional space, the concentration gradient ∇C_i can be expressed in Cartesian coordinates as:

$$\nabla C_i = \frac{\partial C_i}{\partial x} \boldsymbol{i} + \frac{\partial C_i}{\partial y} \boldsymbol{j} + \frac{\partial C_i}{\partial z} \boldsymbol{k} \tag{2.22}$$

where **i, j,** and **k** are mutually orthogonal unit vectors. In steady state where the solute concentration at a given position does not change with time, Fick's first law relates the diffusion driven flux current density of the ith solute component $\mathbf{J_i}$ to its concentration gradient ∇C_i,

$$J_i = -D_{li} \nabla C_i, \tag{2.23}$$

or in one dimension $J_{iz} = -D_{li} \dfrac{\partial C_i}{\partial z}$, in which D_{li} is the diffusion coefficient of the ith component with dimensions of length squared per unit time, e.g. cm^2/s. In practical growth processes of crystals, the steady state condition cannot be met, the solute concentration changes with time. Then we must use Fick's second law to describe diffusion:

$$\frac{\partial C_i}{\partial t} = D_{li} \frac{\partial^2 C_i}{\partial z^2} \tag{2.24}$$

In Fick's laws the diffusion coefficient D_{li} is treated as constant, but in fact it is a function of the solute concentration. The quantity appearing in the above expressions is the average of the diffusion coefficient measured from a series of solute concentration. It is called the integral diffusion coefficient. Many recorded diffusion coefficients in the literatures are neither integral nor differential diffusion coefficients, but they are numerical values fitted to experiments supposing that D_{li} is independent of solute concentrations. D_{li} is an important characteristic of materials.

Convective transport of the ith solvent in an incompressible fluid is governed by the expression:

$$\frac{\partial C_i}{\partial t} + \mathbf{v} \cdot \nabla C_i = D_{li} \nabla^2 C_i \tag{2.25}$$

In steady state where $\frac{\partial C_i}{\partial t} = 0$, the above equation simplifies to:

$$\mathbf{v} \cdot \nabla C_i = D_{li} \nabla^2 C_i \tag{2.26}$$

If the solvent is not moving, $\boldsymbol{v} = 0$, the above equation becomes the steady state version of Fick's second law, which can be integrated to yield Fick's first law.

The other process element affecting growth is momentum transport. Momentum transport in a solution stems from convection. Convection has two sources, natural and forced convection. Convection driven entirely by gravity is natural convection. Natural convection includes heat convection and solute convection. The driving forces for heat convection are the temperature gradients of the system. The driving forces for solute convection are solute concentration gradients. The factors influencing heat convection are mostly: the geometrical shape of the container and its boundary properties, the corresponding tropism of heat and melt flow in the gravity field, etc.

The magnitude of the temperature gradient in the melt $|\nabla T|$, the melt's coefficient of thermal expansion α, the melt's viscosity γ, the melt's thermal conductivity κ, a characteristic geometrical parameter of the container l, and the acceleration of gravity g, can be combined to form a dimensionless quantity N_{Ra} called the Raleigh constant, which plays an important role in the in separating different melt state and convection regimes. N_{Ra} is defined as:

$$N_{Ra} \equiv \frac{\alpha g l^3}{\gamma \kappa} |\nabla T| \tag{2.27}$$

N_{Ra} represents the ratio of the buoyancy trending toward instability, to the viscous drag trending toward stability. When the buoyancy and the viscous drag are the same, the melt is stable. The corresponding $(N_{Ra})_c$ is called the critical Raleigh coefficient. When N_{Ra} increases and exceeds the critical value, natural convection increases. The melt's convection becomes unstable, which induces temperature oscillations in the melt, disturbing the stability of the crystal growth interface, and creating growth striations. These striations hurt the optical quality of the crystal. In zero gravity, there is no natural convection, but this case is difficult to realize on earth. Recently, because of the development of space technology, crystals have been grown in microgravity, introducing a new materials science research field.

Forced convection is produced by the rotation of the seed crystal or the crucible. The Reynold's parameter charactering the forced convection state, N_{Re}, is defined as:

$$N_{\mathrm{Re}} \equiv \frac{\pi}{2}\omega d^2 \gamma^{-1} \tag{2.28}$$

where ω is the angular rotational speed of crystal, and d is the diameter of crystal. When N_{Re} approaches its critical value $(N_{Re})_c$, the interface between the solid and liquid is approximately steady, but when N_{Re} exceeds $(N_{Re})_c$, the crystal growth is unstable. From the above equation we can deduce the largest rotational speed permitted for stable crystal growth, which is:

$$\omega_c < 2(N_{\mathrm{Re}})_c \gamma / \pi d^2 \tag{2.29}$$

The $(N_{Re})_c$ values connected with different growth conditions are different.

In bulk crystal growth, besides natural and forced convection, there is still convection induced by non-gravity factors such as volume force, surface tension etc. The volume force is a field effect produced by magnetic polarizability, conductivity, density gradients, and fluid character change with temperature, composition, and pressure. Convection induced by changes of temperature and composition is called Marangoni convection.

Control of fluid convection is essential to the growth of good single crystals. To accomplish this, the design must account for complex macroscopic and microscopic phenomena. The designs of growth chambers, their shapes, the time and spatial arrangement of temperature profiles, solute segregation evolution, etc., were all initially done from empirically gained experience. Now there are elaborate computer aided fluid models that have greatly helped with growth chamber designs.

2.1.4 Applications of Phase Diagrams in Crystal Growth

A phase, as we use the term here, refers to any material that is in a nearly homogeneous state, and for which there are clear boundaries to other different materials. For example, in a condition where ice and water co-exist, no mater how big or how small the ice particle is, there is still a two-phase physical and chemical equilibrium state. A system in phase equilibrium has, at least, the following characteristics:

1. Thermal equilibrium; the temperature is the same for all the phases:

$$T_1 = T_2 = \ldots = T_i = \ldots \tag{2.30}$$

2. Dynamic equilibrium; the pressure is the same for all the phases:

$$p_1 = p_2 = \ldots = p_i = \ldots \tag{2.31}$$

3. Mass-transfer equilibrium; the chemical potential of each chemical constituent j, is the same for all phases:

$$\mu^j{}_1 = \mu^j{}_2 = \ldots = \mu^j{}_i = \ldots \tag{2.32}$$

If $\mu_1 \neq \mu_2$, a phase transformation will happen in the system. A phase transformation usually occurs under isothermal and isobaric conditions, It can be a discontinuous first order process or a continuous second order change, depending on which phases are involved. If there are two phases in a one-component system, and initially the amount of phase A is N_1 moles and phase B is N_2 moles, and suppose during the phase transformation from the first phase A to the second phase B, the amount of phase A decreases by δN_1 moles and phase B increases by δN_2 moles, then $\delta N_1 = -\delta N_2$, and the variation of Gibbs free energy is:

$$\delta G = \delta G_1 + \delta G_2 = \mu_1 \delta N_1 + \mu_2 \delta N_2 = \delta N_1 (\mu_1 - \mu_2) \tag{2.33}$$

At equilibrium, G reaches a minimum and δG vanishes, $\delta G = 0$, so $\mu_1 = \mu_2$. Therefore, a substance always transfers from the higher to the lower chemical potential phase; the relative magnitude of the chemical potentials determines the phase transformation direction. Because chemical potentials are temperature and pressure dependent, external conditions determine which chemical potential is smaller.

The situation is more complicated for a multi-component system (Landau and Lifshitz 1958). If the system is closed exchanging no particles with the outside world, and has k components, and ϕ phases, and if the chemical potential of jth phase of the ith component is μ_i^j , and it contains N_i^j moles, then:

$$\delta G = \sum_j \sum_i \mu_i^j \delta N_i^j \tag{2.34}$$

The multiphase equilibrium condition is:

$$\delta G = 0 \text{, or} \quad \sum_j \sum_i \mu_i^j \delta N_i^j = 0 \tag{2.35}$$

Equation 2.35 can be satisfied if and only if:

$$\begin{cases} \mu_1^1 = \mu_1^2 = \mu_1^3 = \cdots = \mu_1^j = \cdots = \mu_1 \\ \mu_2^1 = \mu_2^2 = \mu_2^3 = \cdots = \mu_2^j = \cdots = \mu_2 \\ \cdots \\ \mu_k^1 = \mu_k^2 = \mu_k^3 = \cdots = \mu_k^j = \cdots = \mu_k \end{cases} \tag{2.36}$$

At fixed temperature and pressure, when the multi-component system is in its phase equilibrium state, the chemical potentials of every component of each phase are equal. Thus, the chemical potential of each component are separate parameters of the equilibrium system. Each of these chemical potentials is a function of k+1 independent variables, P, T and k-1 concentrations of the components differing from the one in question. The conditions, (2.36), constitute a set of k (ϕ-1) equations. The number of unknowns in this set is 2 + ϕ (k-1), which takes account of the system being closed so the total number of moles of each component is constant. To obtain a solution for a given number of unknowns one needs a set of an equal number of independent equations. Here the number of unknowns may exceed the number of equations, which means equilibrium solutions can be obtained with some free parameters. The number of free parameters, denoted f, is called "the thermodynamic degrees of freedom" of the system and is given by [2+ϕ (k-1)] – k (ϕ-1), or

$$f = k + 2 - \varphi \tag{2.37}$$

Thus, the number of degrees of freedom for an equilibrium system equals the number of components plus 2, minus the number of phases. The number 2 in (2.37) arises from the external quantities, pressure and temperature. If there are other environmental conditions, the number 2 should be replaced by the total number n. If the process is isobaric, the then the number 2 will be replaced by 1. Because f is positive semi definite, $f \geq 0$,

$$\varphi \leq k+2 \tag{2.38}$$

The number of phases cannot exceed the sum of the number of components plus 2. This is called the Gibbs phase rule.

The phase rule is useful in setting limits on phase diagrams. It states how many phases can co-exist in an equilibrium system that consists of k components, and how many free parameters there are. For a system of two components in a single phase, the number of degrees freedom is 3, composition x, temperature T, and pressure P. If there are two phases and two components, there are only two degrees of freedom, so there are two fixed relations among the four parameters, temperature, pressure, and concentrations of the two phases. In actual crystal growth applications, it is common practice to find the relationship between temperature and concentration on a liquid–solid phase equilibrium line at fixed pressure, which is a cut along a constant P plane of a three-dimensional phase diagram.

The phase transformation occurring in single component crystal growth always proceeds under isothermal and isobaric conditions; the pressure and temperature can be treated as variables to describe the system's state. Here, the Gibbs function or chemical potential is a function of pressure and temperature $\mu(T,P)$, and in phase equilibrium:

$$\mu_1(T,P)=\mu_2(T,P) \tag{2.39}$$

Therefore, there is a relationship between temperature and pressure in this two phases equilibrium, and the phase equilibrium can only happen at certain specific related temperatures and pressures. From (2.39) one can deduce the functional relationship between P and T that corresponds to the equilibrium phase transformation condition, $P=P(T)$. The curve of the function P(T) is a phase diagram. The phase diagram divides the $P-T$ space into several regions, above the curve is one phase, beneath it is another one; the curve itself identifies conditions for stable two-phase equilibrium states. The points of the $P-T$ curve are the points where equilibrium phase transformations occur. Because crystal growth is an equilibrium transformation process, the phase diagram is an important tool for setting crystal growth conditions.

In order to better understand the phase diagram of narrow gap semiconductor crystal growth, it is helpful to know how to draw the appropriate phase diagram for these two component systems. For a pure single component material, the critical temperature T_c remains constant during growth. For a multi-component material, the crystallization point is

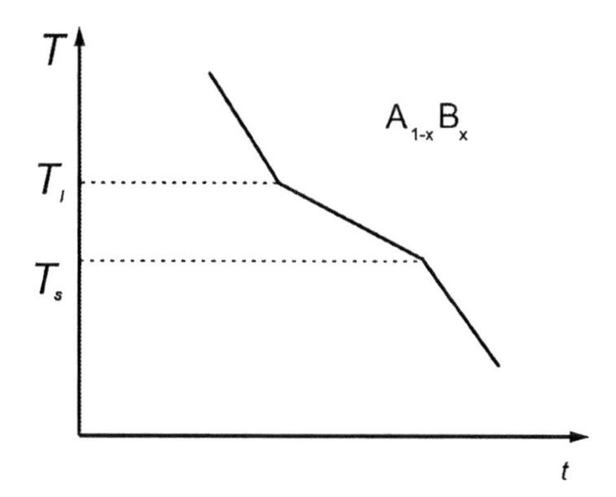

Fig. 2.5. Schematic diagram of phase transition for $A_{1-x}B_x$ two-component system

changeable. We begin by studying a two component systems like $A_{1-x}B_x$ metal alloys, or III–V or II–VI compound alloys where A and B each represents a molecule, e.g. HgTe and CdTe respectively. In such systems the start of growth crystallization temperature is generally higher than the end point crystallization temperature. These temperatures in a two component system, $A_{1-x}B_x$ in the liquid, are functions of the liquid phase composition x at set pressure P. As crystal growth proceeds, if there is no means provided for replenishment, the composition of the melt changes and with it the critical temperature. To obtain a $T-t$ curve in a growth chamber the melt's temperature at first is held above the starting critical temperature, and it is gradually cooled under constant pressure. There is no phase transformation at the beginning, until the temperature, following a specified schedule, drops below the critical temperature corresponding to the initial liquid composition x. When the temperature drops to the initial critical point, the phase transformation begins. As the liquid phase crystallizes and releases latent heat, and the system's radiation rate has not changed, the system temperature drops more slowly, i.e. the $T-t$ slope decreases slightly while growth continues. As the temperature arrives at the lower critical point T_s, the crystallization process stops, no latent heat is released, and then the temperature resumes its more rapid decrease rate. Figure 2.5 indicates the expected trend of such experiments on an A–B two-component system where x of the liquid changes as growth proceeds. A group of T_l points (T_l, x) and T_s points (T_s, x) can be obtained from a $T-x$ phase diagram, T_l points lie on the upper liquidus side of the phase curve and T_s points lie on the lower solidus phase curve. The two curves constitute an A–B two-component alloy phase diagram, like that shown in Fig. 2.6.

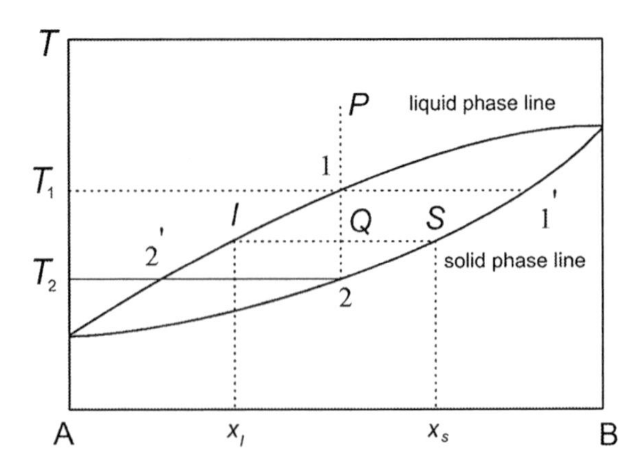

Fig. 2.6. A–B two-component alloy phase diagram

All A–B two-component solid systems may within limits mix. When they mix on a given crystal structure the mixtures are called solid solutions. If they mix in all proportions, that is called an "infinite solid solution." Alloys are solid solutions, some are able to mix in all proportions, and others have solubility limits. The theory of phase diagrams and solubility limits is discussed in detail in Chen and Sher (1995).

In Fig. 2.6 the area above of the liquid phase curve is the area of uniform liquid phases, and the region below the solid phase curve is the solid phase area. The region between the two curves is the two-phase coexistence area. Because the pressure is constant, the Gibbs phase rule becomes:

$$f = k + 1 - \varphi \tag{2.40}$$

here $k = 2$, for example if A and B are the HgTe and CdTe components, in the liquid phase area above the curve, $\varphi = 1$, then $f = 2$. That indicates there are two degrees of freedom, so the temperature T and composition x are independent. In the region where both liquid and solid phases can coexist, $\varphi = 2$, then $f = 1$, and there is only one degree of freedom. This means the composition and temperature are related. At a fixed temperature, as illustrated in Fig. 2.6, the composition of the liquid phase is x_l and the solid phase composition is x_s. Thus if an A–B system starts to cool from state P, approaches state 1, it would begin to grow a solid with concentration x_{s1}. If it continues cooling to the Q point where it remains stationary, a crystal is grown with composition x_s. The composition of the crystalline solid is x_s and the liquid phase composition is x_l. If the

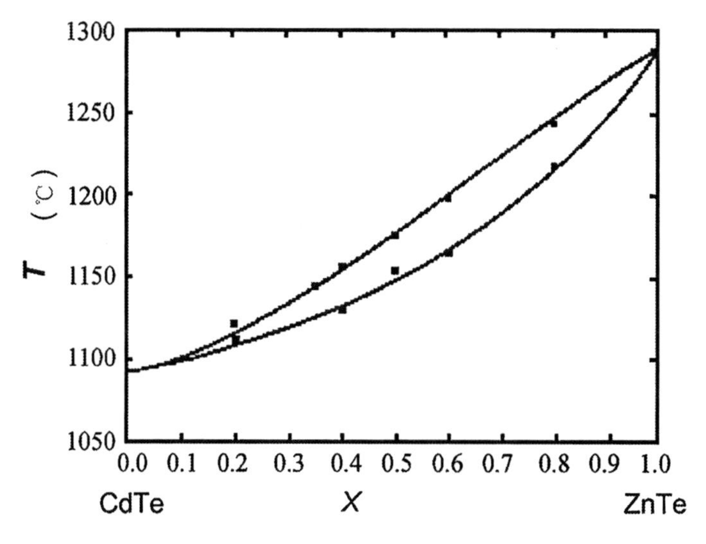

Fig. 2.7. The liquid phase line and solid phase line of CdTe–HgTe pseudo-two component system (Brebrick et al. 1983; Steininger 1970)

temperature decreases below the critical temperature for pure phase A, then growth is finished. The solubility of solid solutions is temperature dependent. If the concentration x_s is below the solubility limit at a temperature where diffusion ceases, then the concentration remains constant as the solid is cooled to room temperature. Practically all alloys are meta-stable at room temperature. The composition of such a solid crystal is $x_{l'}$ to 0. Obviously when a two-component alloy has a solid phase curve close to its liquid phase curve, this favors the growth of uniform composition crystals.

Furthermore, the so-called lever law is often used in the phase diagram to calculate the ratio of the solid to liquid phase masses. At the Q point in the phase diagram, the compositions of liquid and solid phases are x_l and x_s, respectively. The lever rule says that at point Q, if the number of moles in the liquid phase is N_l and in the solid phase is N_s, then:

$$\frac{N_l}{N_s} = \frac{x_l - x_Q}{x_Q - x_s} \tag{2.41}$$

In general in equilibrium, at any temperature in the two-phase region, the liquid phase and the solid phase volumes are different.

Figure 2.7 shows a representative example of a two component phase diagram, that for the CdTe–ZnTe system.

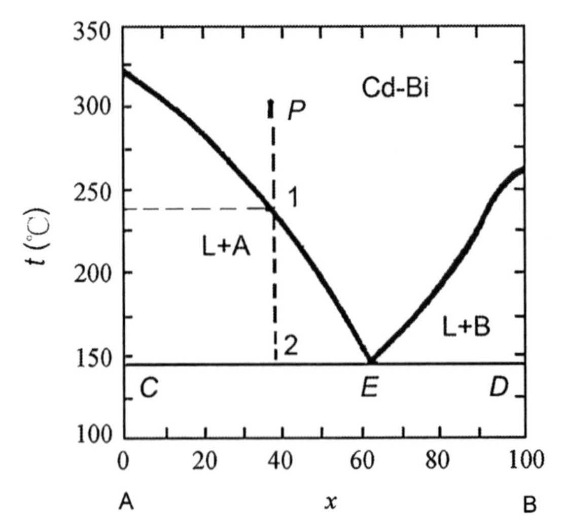

Fig. 2.8. Phase diagram of a eutectic

The eutectic phase diagram, in which the two components are not mutually soluble at all compositions, is completely different. The diagram in Fig. 2.8 for the Cd–Bi two component system is a representative example. The two upper curves in Fig. 2.8 are liquid–solid two phase lines, the regions above the line are liquid phases, solid A phases are below their liquidus line at $x_A = 0$, and the B phases are below their liquidus line at $x_B = 1$. The areas of (L+A) and (L+B) are liquid solid coexistence areas. Line CED is the eutectic line. Once any composition of liquid cools down to the CED line, its liquid composition is x_E. On the CED line, where $k = 2$ and $\varphi = 3$, so $f = 0$, both the temperature and composition of the liquid are fixed. The composition of the liquid phase is x_E, and those of the solid phases are x_C and x_D respectively. Below the CED line, solids co-exist that are mixtures of an almost entirely A compound (typically much less than 1% B), and a primarily B compound. This is a eutectic region.

Between the two extremes, the "infinite solid solution" case and the case where the solid "does not melt" but sublimes, is the limited solid solution case. Figure 2.9 is such a system phase diagram. The concentrations at the C and D points are solubility limits. In Fig. 2.9, the α phase solid solution is one in which B is dissolved in A, conversely, A dissolved in B is the β phase. The area below the temperature at the CED line is the α, and β alloys eutectic area, the $(L+\alpha)$ area is the coexistence area of the liquid phase and the α phase. Similarly, the $L+\beta$ area is the coexistence area of the liquid phase and the β phase. Each area is marked in the figure.

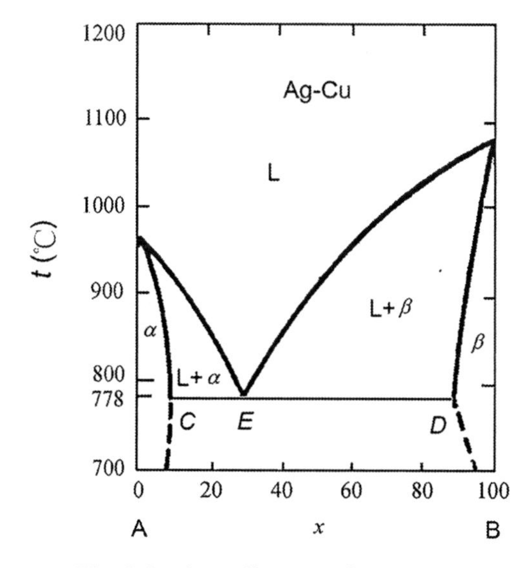

Fig. 2.9. Phase diagram of a eutectic

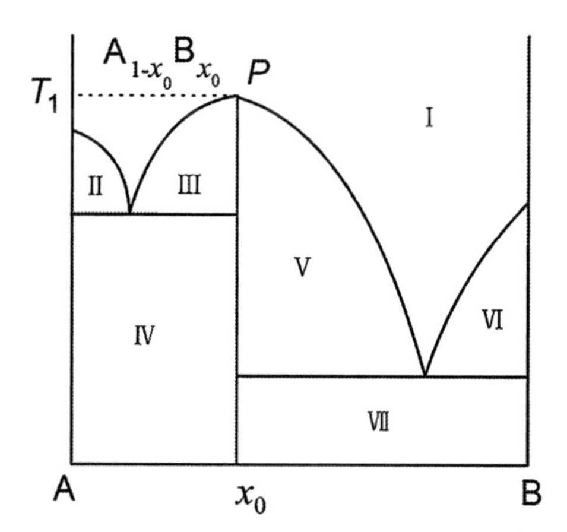

Fig. 2.10. A general phase diagram of systems involving an $A_{1-x_0}B_{x_0}$ compound

A general phase diagram of a compound $A_{1\text{-}x0}B_{x0}$ is shown as in Fig. 2.10. If compound $A_{1\text{-}x0}B_{x0}$ is formed at composition x_0 and melting point T_1, then $A_{1\text{-}x0}B_{x0}$ can be treated as one component of a two component system, with an other component B. Or it can also comprise another two component system with component A. Therefore, there are seven areas in the phase diagram. Area I is a liquid phase L. Area II is a co-existence region of L+A. Area III is L+ $A_{1\text{-}x0}B_{x0}$. Area IV is the A+ $A_{1\text{-}x0}B_{x0}$ eutectic region. Area V is the compound $A_{1\text{-}x0}B_{x0}$. Area VI is L+B. Area VII is B+ $A_{1\text{-}x0}B_{x0}$ eutectic region.

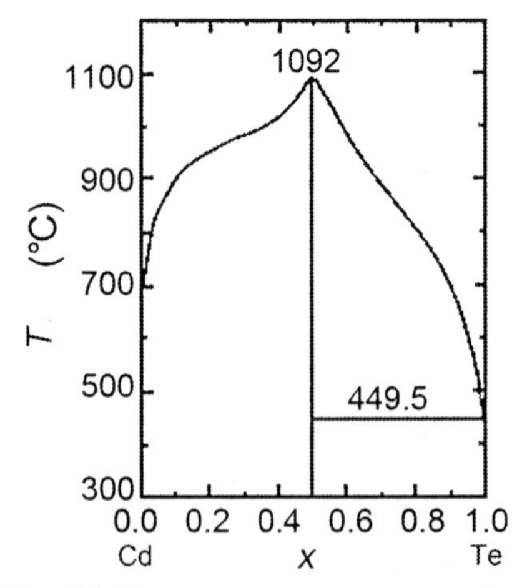

Fig. 2.11. The Cd–Te two component system phase diagram

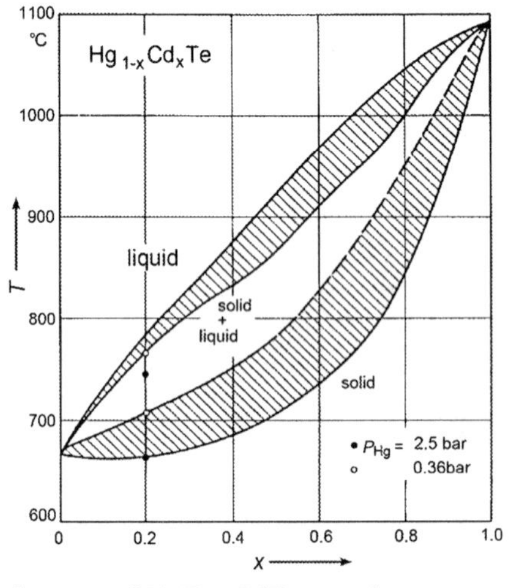

Fig. 2.12. Phase diagram of HgTe–CdTe pseudo two component system

Figure 2.11 displays a Cd–Te two-component phase diagram. A similar phase diagram for InSb appears in the next chapter.

The HgTe–CdTe pseudo-two component system can mutually dissolve in all proportions to form "infinite solid solution" alloy crystals. In fact, measuring and drawing a HgCdTe phase diagram is a complex task. Several authors have presented what look like good results, but their liquidus and solidus phase lines do not agree. The reason can be traced to different Hg partial pressure conditions in their experiments. The liquidus phase line varies with Hg pressure, so the interpretation of a measured T–x

phase diagram must include a specification of the Hg pressure under which it was obtained. Liquidus line and solidus phase lines can be obtained individually, but they can also be deduced by first measuring the liquidus, and calculating the solidus relative to the liquidus based on theory. The range of experimental liquidus and solidus curves for the $Hg_{1-x}Cd_xTe$ alloy is illustrated in Fig. 2.12.

Steininger used a backflow technique under high Hg partial pressure conditions to obtain HgTe–CdTe system's *T–x* phase diagram (Steininger et al. 1970). Data points in Fig. 2.12 are shown for the liquidus, and the solidus curves were obtained from theory. For a simplified case, the HgTe–CdTe *T–x* phase diagram can be also calculated from the empirical formulas listed below:

$$T_l[^0C] = \frac{1000x}{1.37+0.97x} + 668 \qquad (0.1 < x < 1.0)$$

$$T_s[^0C] = \frac{1000x}{5.72-2.92x} + 668 \qquad (0 < x < 0.6) \qquad (2.42)$$

$$T_s[^0C] = \frac{1000x}{5.27-2.92x} + 668 \qquad (0.6 < x < 1.0)$$

These formulas were obtained by fitting average curves in Fig. 2.12. However refined fits are needed take account of the pressure dependence. In experiments a low Hg pressure will result in Hg evaporation from a melting body making it Te rich, with a high density of Hg vacancies, or even Te precipitates. A high Hg pressure will make the material Hg rich, with the excess Hg in interstitial sites or on Te sites.

According to empirical studies, the relation between the equilibrium vapor pressure above liquid Hg (in SI units) and the absolute temperature can be expressed as:

$$\ln P_{Hg} = 11.270 - 7.147/T \qquad (2.43)$$

The same relation above $Hg_{1-x}Cd_xTe$ is:

$$\ln P_{HgCdTe} = 10.206 - 7.147/T \qquad (2.44)$$

Both of these relations are parallel in the $\ln P - T$ phase diagram, therefore,

$$P_{\mathrm{HgCdTe}} = a_{\mathrm{Hg}} P_{\mathrm{Hg}} \qquad (2.45)$$

The parameter $a_{Hg} = 0.345 \pm 0.020$ is the Hg excitation coefficient, and it is independent of x and T. The activity quotient of $Hg_{1-x}Cd_xTe$ is defined as:

$$\gamma_{Hg} = \frac{a_{Hg}}{[Hg]} \tag{2.46}$$

[Hg] is the density of Hg in HgCdTe. While a_{Hg} is independent of x [Hg] is not. Obviously, γ_{Hg} increases, the larger x is, and the smaller [Hg] is. The activity quotient γ_{Hg} must be large enough to allow the excitation coefficient a_{Hg} to remain invariant.

The partial vapor pressures of Hg, Cd and Te above $Hg_{1-x}Cd_xTe$ at a given T can be determined by an atomic absorption line method. In general, the Te vapor pressure is low, in the range of 10^{-5}~10^{-2} atm. The Cd vapor pressure is much lower, in the range of 10^{-8}~10^{-5} atm. The Hg vapor pressure, by contrast, is typically in the range of several atoms. The chemical potential and Gibbs energy of HgTe and CdTe can be deduced once the Hg, Cd and Te partial pressures above the compounds are obtained as functions of temperature.

2.1.5 Segregation Coefficient

In a two-component system composed of solvent A and solute B, according to its T–x phase diagram, the concentration of A and B in the solid and liquid phases will vary with temperature at equilibrium. However, the ratio of the density C_s of solute B in the solid phase to the density C_l in the liquid phase is a constant independent of temperature. This ratio is called the equilibrium segregation coefficient:

$$K_0 = \frac{C_s}{C_l} \tag{2.47}$$

It is the ratio of the solute in the two phases at equilibrium.

At general, solute will exchange between the solid phase and the liquid phase.

In equilibrium, both exchange rates are equal, and according to the "principle of detailed balance" will follow the formula:

$$C_s \exp(-\frac{Q_s}{k_B T}) = C_l \exp(-\frac{Q_l}{k_B T}) \tag{2.48}$$

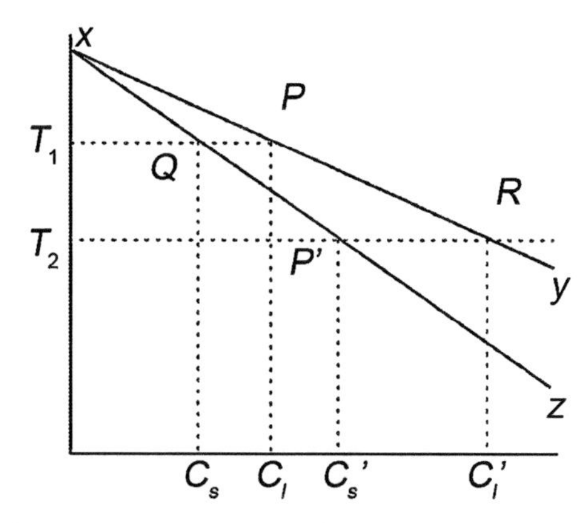

Fig. 2.13. Equilibrium segregation coefficient obtained from the phase diagram

k_B is the Boltzmann constant, Q_s is the activation energy for the transfer of a solute atom from the solid to the liquid phase, and Q_l is the corresponding activation energy of a solute atom from the liquid to the solid phase. Then we find:

$$K_0 = \frac{C_s}{C_l} = \exp(\frac{Q_s - Q_l}{k_B T}) \tag{2.49}$$

If $Q_s < Q_l$, then we have $C_s < C_l$, or $K_0 < 1$. $Q_s < Q_l$ indicates that the probability of solute atom going from the solid into the liquid is large, so the density of C_s must be small to keep both rates equal. The magnitude of K_0 determines the solute density in the solid phase, so it is called the "segregation coefficient." The above formula can also be expressed as:

$$K_0 = \exp(\frac{g_l - g_s}{RT}) \tag{2.50}$$

where R is the gas constant, and g_l and g_s are the chemical potentials of the solute in the liquid state and the solid state, respectively. They are the functions of temperature and pressure. For a situation in which the solute is the same but the solvent is different, the two phase coexistence temperature may be higher or lower than the freezing point of the pure solute. Therefore, segregation coefficients are not only related to the solute but also to the solvent.

The equilibrium segregation coefficient can be obtained directly from a phase diagram, such as that in Fig. 2.13, where *xy* is a liquidus, and *xz* is a solidus line.

Consider the point P on the liquidus, freezing starts from this point at temperature T_1 and the solute density is C_l. Then at the same temperature at the Q point freezing begins at solute density C_s. According to the definition of the segregation coefficient, $C_s = K_0 C_l$. Similarly, if melting starts from point P' on the solidus at solid phase density C_s', when the temperature is T_2, the solute density of the liquid solution will be C_s' / K_0'.

The equilibrium segregation coefficient can be calculated from thermodynamic relationships. Usually it is related to the latent heats of solvents and solutes, as well as their melting points. It is also possible to calculate the maximum solubility in the solid phase (Yu 1984). For an A–B two-component system, in solid–liquid phase equilibrium, the Gibbs free energy is expressed as:

$$\begin{aligned} G &= U + PV - TS = H - TS \\ &= C_A H_A + C_B H_B - T(C_A S_A + C_B S_B) + RT(C_A \ln C_A + C_B \ln C_B) \end{aligned} \tag{2.51}$$

Here, C_A and C_B are the densities of components A and B, respectively, H_A and H_B are the atomic enthalpies of components A and B, and S_A and S_B are the atomic entropies of components A and B. The last term in (2.51) stems from the contribution to the entropy from the number of ways to arrange the A and B atoms randomly on their lattice sites, when their concentration are C_A, and C_B. When solid and liquid are in equilibrium we have:

$$\frac{\partial G}{\partial C_B} = 0, \text{ and } \frac{\partial G}{\partial C_A} = 0.$$

Under dilute solution condition, with $C_A \to 1$, or $\ln C_A \to 0$, the segregation coefficient of solute B can be deduced:

$$\ln K_0 = \ln \frac{C_s}{C_l} = \frac{\Delta H_A}{R}\left(\frac{1}{T} - \frac{1}{T_A}\right) - \frac{\Delta H_B}{R}\left(\frac{1}{T} - \frac{1}{T_B}\right) \tag{2.52}$$

where $\Delta H_A = H_A^S - H_A^l$ is the melting latent heat of solvent A, and $\Delta H_B = H_B^S - H_B^l$ is the latent heat of solute B, and T_A and T_B are the melting points of solute A and solvent B, respectively. T is the melting temperature of an A and B mixed solution. Therefore, the segregation coefficient of solute B in solvent A can be calculated from (2.52). When the two phases are in equilibrium, $T - C_l$ is a liquidus line, while $T - C_s$ is a solidus line. If the liquidus line is measured, then the solidus line can be calculated from the above formula.

At much lower solute density, $C_B \to 0$, the melting points of an A plus B composition solution has $T \cong T_A$. The initial segregation coefficient K_{00} can be obtained from the above formula, which yields the result:

$$\ln K_{00} = -\frac{\Delta H_B}{R}\left(\frac{1}{T_A} - \frac{1}{T_B}\right) \tag{2.53}$$

We may further deduce, when the solute density is C_l, the segregation coefficient of the solvent A in the solution is:

$$\ln K_0 = K_{00} - C_l\left(1 - \frac{\Delta H_B}{\Delta H_A}\right) \tag{2.54}$$

The above two formulas can be used to calculate the segregation coefficient of solute B in solvent A. Equation (2.53) may be used to calculate the maximum solubility in the solid phase. According to (2.53) we find:

$$\ln C_s = K_{00} - C_l\left(1 - \frac{\Delta H_B}{\Delta H_A}\right) + \ln C_l \tag{2.55}$$

The solute B density at the solidus line is a function of that at liquidus line.

Because $\frac{dC_s}{dC_l} = 0$, when $C_l = \left(1 - \frac{\Delta H_B}{\Delta H_A}\right)^{-1}$, C_s has an upper limited maxim.

$$C_{s\max} = \frac{e^{K_{00}-1}}{1 - \frac{\Delta H_B}{\Delta H_A}} \tag{2.56}$$

It is the maximum solubility of solute B.

For the HgTe–CdTe two-component system, with HgTe being the solvent, we have $\Delta H_A =$ 8.7 kCal/mole, and for the CdTe solute, $\Delta H_B = 12$ kCal/mole (Steininger 1976).

The variation of the crystal growth rate has some influence on the solute segregation coefficient. During actual crystal growth, some external factors, such as vibration, heating power fluctuations, environmental fluctuations, melt body dynamical instability, and cooling water flow rate changes, may all effect the crystal growth rate and the segregation coefficient. Also in many crystals, e.g. InSb, growth on different crystal

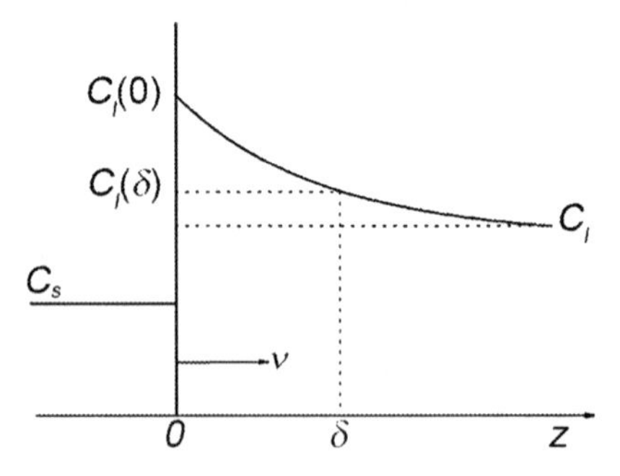

Fig. 2.14. Position dependence of the solute density in a boundary layer

faces has different segregation coefficients, and this causes complicated phenomena to occur during crystal growth. Equilibrium segregation coefficients reflect an idealized situation, in practice treatment of segregation coefficients need special analysis.

The activation energy difference of both solutes across a solid–liquid interface is essentially the difference of their chemical potentials between the solid and liquid phases. In order to make the chemical potentials equal in the liquid and solid phases, account must be taken of the density difference at the interface. In the process of crystal growth, due to the segregation phenomena, there may be a solute boundary layer δ on the liquid side (Yu 1984). Usually the ratio of the solute density in the solid phase, C_s, to that in the liquid at the interface, $C_l(0)$, is an interface segregation coefficient K^*, that is:

$$K^* = \frac{C_s}{C_l(0)} \tag{2.57}$$

If there is no interfacial effect, the interface segregation coefficient is the equilibrium segregation coefficient $K_0 = C_s / C_l$. The variation of solute density with position in the boundary layer can be obtained. It is illustrated in Fig. 2.14, where the z direction is normal to the growth interface, and the solid–liquid interface is at the origin of coordinates $z = 0$.

If one neglects macroscopic convection, then in the boundary layer of the solute satisfies the diffusion equation:

$$D\frac{\partial^2 C_l(z)}{\partial z^2} + v\frac{\partial C_l(z)}{\partial z} = 0 \tag{2.58}$$

The boundary conditions are:

$C_l(0) = C_s / K^*$, and

$C_l(\infty) = C_l$.

Then the solution to (2.58) is:

$$C_l(z) = C_l \left\{ 1 + \frac{1-K}{K} \exp\left(-\frac{v}{D} z \right) \right\} \tag{2.59}$$

D is the diffusion coefficient, v is the interface advancement speed, $K = \frac{K^*}{K_0} = \frac{C_l}{C_l(0)}$. Inserting this expression into (2.59) it becomes:

$$C_l(z) = C_l \left(1 + \frac{C_l(0) - C_l}{C_l} e^{-z/\delta} \right) \tag{2.60}$$

Where δ is the characteristic decay thickness, defined as:

$$\delta \equiv \frac{D}{v} \tag{2.61}$$

Only if $v = 0$, i.e. no growth, is the solute density independent of z.

In this boundary layer derivation of (2.60), the condition as $z \to \infty$, $C_l(z) \to C_l$ was used. If there is forced convection which results from stirring, this condition is modified. Now, for a boundary layer thickness δ_b adjacent to the interface the liquid is stagnant, and beyond that it is moving mixing so the liquid has a uniform density C_l. Thus the boundary conditions change to:

$$\begin{aligned} &C_l(0) \qquad \text{at } z = o, \text{and} \\ &\mathrm{C_l}(\delta_b) = C_l \text{ at } z = \delta_b \end{aligned} \tag{2.62}$$

The solution to (2.58), with boundary conditions (2.62), in the range $0 \le z \le \delta_b$, is:

$$C_l(z) = C_l(0) - \frac{C_l(0) - C_l}{1 - e^{-\delta_b/\delta}} \left(1 - e^{-z/\delta} \right) \tag{2.63}$$

As $\delta_b \Rightarrow \infty$, (2.63) reduces to the no stirring solution (2.60), as expected.

2.1.6 Freezing Process

In the section above we have discussed the solute density distribution in the liquid phase. Next, we will discuss the density distribution in the solid phase in two basic kinds of freezing processes; normal freezing and zone melting. In normal freezing, a material in a liquid state gradually freezes along one direction from beginning to end. Generally, a normally grown crystal has a non-uniform solute distribution $C_s(z)$ along its length. Assuming K^* is constant in the freezing process, and the growth rate is fast enough to justify the neglect of solute atom diffusion in the solid phase, $C_s(z)$ retains the shape frozen in during growth. If g is the volume fraction of the solid phase, L is the solute density in the liquid phase (C_l times the liquid volume fraction), $L = C_l\ (1-g)$, $C_s(z)$ is the solute density in the solid at z near the growth interface, then:

$$C_s(z) = -\frac{dL}{dg} \tag{2.64}$$

Meanwhile, $C_s(z) = K_0 C_l$ so:

$$C_s(z) = \frac{K_0 L}{1-g} \tag{2.65}$$

Substituting (2.65) into (2.64), then integrating from the start of growth to the place where the front reaches z, and therefore g, produces:

$$\int_{L_0}^{L} \frac{dL}{L} = \int_0^g -\frac{K_0}{1-g} dg \tag{2.66}$$

where $L_0 = C_{l0}$ is the value of L at $g = 0$.

Performing the integrals and rearranging terms yields:

$$L = L_0 (1-g)^{K_0} = C_{l0} (1-g)^{K_0} \tag{2.67}$$

Then we have:

$$C_s(z) = -\frac{dL}{dg} = K_0 C_{l0} (1-g)^{K_0 - 1}. \tag{2.68}$$

Given the initial solute density C_{l0}, and the segregation coefficient K_0 this formula predicts the concentration of solute along the length of a normally grown crystal. $C_s(z)$ is a linear function of g, which in turn is a linear function of z if the cross sectional area of the crystal is a constant.

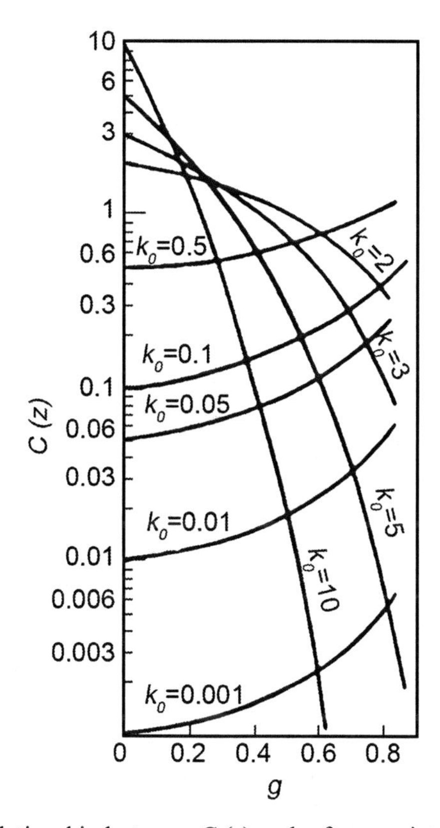

Fig. 2.15. The relationship between $C_S(z)$ and g for a variety of K_0 conditions

Figure 2.15 shows the relationship between $C_s(z)$ and g for a variety of K_0 values. Actually, K_0 varies with solute density; since we have assumed an invariant K_0 in the above discussion, the results only illustrate expected trends.

Another often-used crystal growth method is the zone melting method. In this method one starts with a liquid that has been thoroughly stirred to a uniform concentration C_0 and then quench cooled to form a uniform concentration polycrystalline ingot. This ingot is then raised to a temperature just below its melting point. Then over a short length l at one end of the material the temperature is raised just above the melting point. This melt zone is then caused to pass slowly along the length of the ingot. As the melt zone advances along the z direction of the crystal ingot, the ingot freezes again into large single crystal regions. Now the distribution of solute density in the newly formed solid must be deduced. Suppose that the cross sectional area of the ingot is uniform down its length. If the solute density of the liquid phase at z is $C_l(z)$, that means a number of solute atoms per unit area $K_0C_l(z)dz$ refreezes. Because the crystal ingot's original solute density is C_0, as the front of melt zone advances, it increases the number of solute atoms in the liquid per unit area by C_0dz. Therefore, the variation of solute per unit area in the liquid phase is:

$$dC_l(x) = (C_0 - K_0C_l)dz \tag{2.70}$$

Equation 2.70 can be rewritten in the form:

$$l\frac{dC_l(z)}{dz} + K_0C_l(z) = C_0 \tag{2.71}$$

The solution of this equation is:

$$C_s(z) = K_0C_l(z) = C_0\{1-(1-K_0)\exp(-\frac{K_0}{l}z)\} \tag{2.72}$$

Where, $C_s(z)$ is the solute density at the z position in the solid phase. Note that if $K_0 = 1$, the concentration of the ingot is uniform. It is also nearly uniform beyond a critical point $z_c = l/K_0$. Thus in general the zone melting growth technique produces a larger volume of material with a uniform concentration than the natural growth methods. The problem of growing large volumes of single crystals remains. Ingots grown this way are "mined" for regions that satisfy desired specifications.

How to shape the crystal growth surface as it propagates into the liquid phase is important to the maintenance of a stable surface, and is one important key to high quality crystal growth. If the original growth surface is a smooth surface, this smooth surface will suffer some changes during the growth process. The temperature distribution at the front of the growth surface affects its stability. There are two possible temperature distributions at the growth interface, either with a positive temperature gradient $\frac{dT_l}{dz} > 0$, or with a negative temperature gradient $\frac{dT_l}{dz} < 0$. The former describes a super-heated melting body; and the latter a super-cooled melting body. In a super-heated condition, if there is a bulge at the growth surface, then at the tip of the bulge the growth is slower than at its base. Thus, the bulge will flatten and the growth surface will be smooth. In the super-cooled condition, if there is tiny bulge at the growth surface, then the tip of the bulge grows faster than the base so the bulge tip grows without limit. Obviously, control of the temperature distribution at the growth interface is essential to maintain its stability. This is illustrated in Fig. 2.16a.

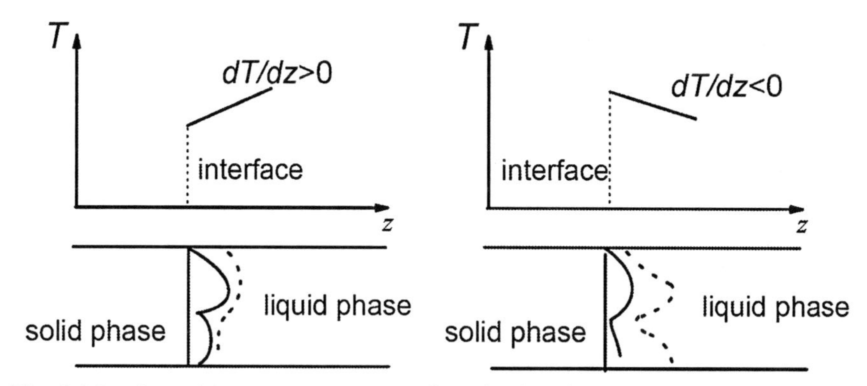

Fig. 2.16a. A positive temperature gradient in the liquid in front of the growth surface keeps the surface stable

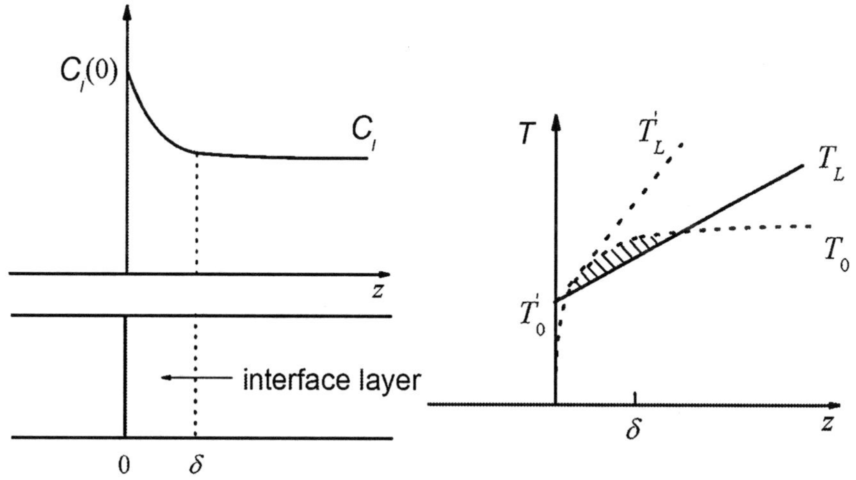

Fig. 2.16b. Increasing the temperature gradient in the liquid ahead of the growth front may overcome compositional super-cooling

There is also a second problem. Because the solute equilibrium segregation coefficient $K_0<1$, as we have seen in (2.60), an enhanced concentration solute boundary layer is formed in front of growth surface with thickness of δ. The solute density increases in this region, causing the freezing point to decrease from T_0 to T_0'. Therefore, without a change in the temperature control, crystal growth would cease. In order to keep the growth going, the heater power must be decreased until the temperature at the growth interface drops down to the new freezing point T_0'. However, if the temperature gradient in the liquid remains constant, it results in the actual temperature in some portions of the δ boundary layer ahead of the growth interface being lower than the freezing point (showed in the hatched area in Fig. 2.16b). If there is a bulge in the growth surface, the bulge will grow rapidly due to its super-cool condition and destroy the growth stability of the surface. This phenomenon that results from the compositional variation of the liquid layer just ahead of the growth front, is called "compositional super-cooling." In order to overcome this growth surface instability, it is useful to increase the temperature gradient in the liquid in front of the growth surface, and arrange for this temperature profile to slide along with the growth front. At any instant the temperature profile required to avoid compositional super-cooling is depicted by the dashed line shown in Fig. 2.16b.

It is useful to examine the parameters that impact the control of compositional super-cooling. The compositional super-cooling critical condition can be established starting from the previous analysis of the solute density distribution relative to the growth interface in the z direction. We repeat (2.59):

$$C_l(z) = C_l\{1 + \frac{1-K}{K}\exp(-\frac{v}{D}z)\} \tag{2.73}$$

where v is the crystal growth rate, and D is the solute diffusion coefficient. In a two-component system the T–x phase diagram, assuming the liquidus line over a concentration range of interest is approximately linear, let its slope be:

$$m = \frac{dT}{dC_l} \le 0 \tag{2.74}$$

Then the freezing point, $T_0(C_l)$, is:

$$T_0(C_l) = T_0(0) + mC_l \tag{2.75}$$

or substituting from (2.73) it becomes:

$$T_0(z) = T_0 + mC_l[1 + \frac{1-K}{K}\exp(-\frac{v}{D}z)] \tag{2.76}$$

where T_0 is freezing point of the pure solvent (or some other specified starting point), then the temperature gradient of the freezing point curve at the growth surface is:

$$G = \left.\frac{dT_0(z)}{dx}\right|_{x=0} = \frac{(-m)C_l(1-K)v}{DK} \ge 0 \tag{2.77}$$

The compositional super-cooling critical condition arises when the temperature gradient G becomes smaller than or equal to the right term of the equation, or $\frac{G}{v} < \frac{mC_l(K-1)}{DK}$. The right term of the above formula is a constant for a given solution system, therefore the smaller the temperature gradient in front of the growth surface, or the larger the growth rate, the easier it is to fall into a compositional super-cooling situation. To avoid compositional super-cooling the condition is:

$$\frac{G}{v} > \frac{mC_l(K-1)}{DK} \tag{2.78}$$

Compositional super-cooling is avoided if the temperature gradient is large and growth rate is slow. In view of this equation, it is easier to avoid compositional super-cooling if the solute density is small, the liquid phase line slope is small, and the segregation coefficient K approaches 1 closely, so the value of the right side of the equation will be small.

2.2 Bulk Crystal Growth Methods

Growth of bulk materials can be grown from a melt and grown from a vapor. Growth from a melt was the traditional method. Both solids and liquids are forms of condensed matter, where the average spacing between constituent atoms is comparable to their atomic sizes. A crystalline solid is characterized by its constituents and its structural symmetry. One or more kinds of atoms arrange regularly to form the crystal lattice. Directional binding forces among groups of atoms make the crystal a rigid solid. In order to turn the crystallized solid into a melt, it is necessary to provide energy to overcome the binding forces, causing the atoms to break away from their average lattice positions and to distribute nearly randomly. The atoms in a melt have only short-range order characterized by radial distribution functions, but no long-range order. Usually heating is used to make a solid undergo a first order phase transition at its melting point. The quantity of heat required to accomplish the transition is the melting latent heat. When the melt freezes, the latent heat is released to reduce the free energy of the system. Only when the free energy decreases, do crystals grow. The ordered lattice arrangement of atoms (or molecules) occurs directly during the course of growth. The structural transition from disorder to symmetrical order is not an volume effect, but is accomplished locally at solid–liquid interfaces. Single crystal growth continues as the interface is caused to propagate into the melt. If there is rapid cooling of a melt so many crystallites nucleate at once, then each will grow independently to form a polycrystalline solid. If the cooling is too rapid then the grains may be so small that the material is effectively amorphous with no long-range order.

There are many methods of crystal growth. Because the melting point of narrow gap semiconductors is not very high, the most common growth techniques from a melt are: Czochralski, Bridgman, Te solvent, half-melting, and solid re-crystallization. The Czochralski technique is one of the principal techniques used for crystal growth. It will be treated next.

2.2.1 Pulling Technique

A pulling technique of crystal growth was proposed by Czochralski in 1917, and now carries his name. It has been improved based on a vast experience in the crystal growth of numerous materials.

Figure 2.17 is a sketch of a pulling apparatus. The constituent materials are melted in the crucible, and the temperature of the melt surface in connection with a seed crystal is brought to the melting point. The seed at the melt interface is neither melting nor growing after being maintained at the constant melt temperature for a period of time. Then the seed is lifted. At the appropriate lift rate and rotation rate, the crystal can grow stably. This requires adjusting the temperature at the solid–liquid interface to maintain the crystal size constant. One can observe the status of the crystal growth through a window placed at the proper position.

The heating methods used in the pulling technique include electric resistance heating and radio-frequency induction heating. Laser beam heating, electron beam heating, plasma heating and arc heating are used when the crystals are grown without a crucible. Resistance heating has the advantages of low cost and easy formation of complex heater shapes, but it also has a temperature response lag. When the temperature is low, using resistance wire and Si–C (Mo) rod (tubes) as heaters, can be done under oxidizing, neutral and reducing atmospheres. When the temperature exceeds 1,400°C, using tungsten and graphite crucibles can be done under neutral and reducing atmospheres with Radio-frequency induction heating. This provides a clean growth condition. It has small temperature lag and can realize precise control of the temperature. Platinum crucibles are used under oxidizing atmospheres when the temperature is below 1,500°C. When the temperature is above 1,500°C, iridium (molybdenum, tungsten, graphite) crucibles are used under reducing and neutral atmospheres.

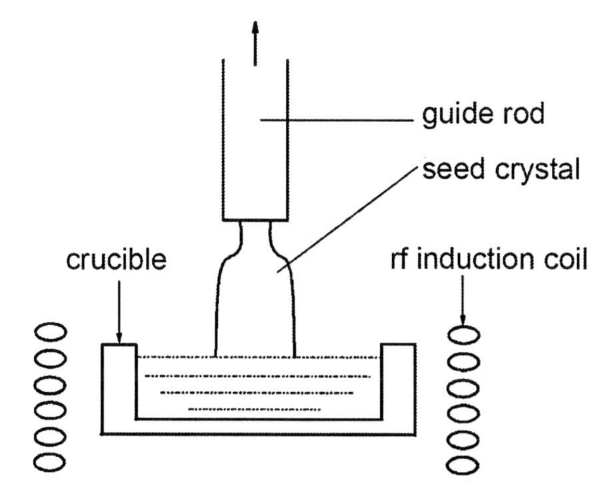

Fig. 2.17. Sketch of a crystal pulling apparatus

Table 2.1. Characteristics of crucible materials in common use

Material	Melting point (°C)	Saturated vapor pressure at the melting point (Pa)	Maximum usable temperature (°C)	Workable atmospheres
SiC			1,500	Unlimited
SiO_2			1,500	Unlimited
Pt	1,774	0.02	1,500	Unlimited
$PtRh_{50}$	1,970		1,800	Unlimited
Ir	2,454	0.47	2,150	Reducing, neutral, weak oxidizing
Mo	2,625	2.93	2,400	Vacuum, reducing, neutral
Ta	2,996	0.66	2,400	Ditto
W	3,410	2.33	3,000	Ditto
C			3,000	Ditto

Commonly, 2,150°C is the upper temperature limit for crystal growth by the Czochralski technique. Table 2.1 lists the characteristics of crucible materials. Good resistance to thermal and mechanical shocks is necessary properties for crucible materials. Crucible materials should be able to endure the working temperature. They should not degrade in the growth atmosphere, and in contact with the melt, and in particular, should not pollute the melt.

Rigorous steady control of the temperature is important to obtain high quality crystals. An ideal temperature control system would constrain temperature fluctuations to less than ±1°C. Using a thermocouple, a pick-up wave coil, and a silicon-light tube, temperatures can be detected with

Table 2.2. Characteristics of typical heat resistance materials

Materials	Melting point (°C)	Maximum useable temperature under an oxidizing atmosphere (°C)	Resistance to thermal shocks	Thermal conductivity
Al_2O_3	2,015	1,950	Good	Middle
BeO	2,550	2,400	Excellent	High
MgO	2,800	2,400	Fine	Middle
ZrO_2	2,600	2,500	Fine	Low
ThO_2	3,300	2,700	Poor	Minimum

sufficient accuracy. After being compared with the specified value, the signals are feed back to a temperature servo control system. This system modulates the power delivered by the heater, and limits the temperature fluctuations to the specified range. Servo systems commonly use a Proportional Integral Derivative (PID) temperature controller. Recently, the integration of multi-segment computer programs exercises the control function by comparing the current reading with a predetermined calibration curve. These systems have improved the precision of temperature control to a level sufficient to yield acceptable crystals.

After-heaters include two types, self-heating and thermal insulation. They are made from heat resistance material. The main function of after-heaters is to modulate the temperature gradient of the growth system. After-heaters can also be used to protect the crystal during the final cool down process. The shapes of after-heaters can be machined to fit growth strategies devised for various materials. The characteristics of heat-resistance materials in common use are listed in Table 2.2.

Crystals are rotated during growth in the pulling technique. The rotation of crystal stirs the melt to drive forced convection. The rotation rate of crystal is one of the most important parameters. It has a direct relationship to the quality of the grown crystals. Rotation enhances the radial symmetry of the temperature field and affects the shape of the interface. With increases of the rotation rate of the crystal, the shape of interface varies from protruding, to plane, and finally to concave. For a given growth system, the rotation rate is set in a predetermined range designed to keep the interface flat. Rotation also changes the temperature gradient near the interface. If natural convection in the melt is dominant, increasing the rotation rate leads to an increase of the radial temperature gradient. If forced convection in the melt is dominant, increasing the rotation rate leads to a decrease of the radial temperature gradient. In general, increasing the rotation rate too far will cause instability in the liquid flow and degrade the resulting crystals.

Rotation of the crystals also will alter the effective segregation coefficient K_0. Increasing the rotation rate, when $K_0<1$, causes K_e to decrease; but when $K_0>1$, K_e will increase. Thus, rotation affects the stability of the interface. When $K_0>1$, increasing the rotation rate leads to a lowering of the interface instability. Even when $K_0<1$, changes of the rotation rate have a complicated influence on the interface stability (Coriell et al. 1976).

The highest growth rate compatible with high crystal quality is desirable. To retain quality, the growth rate of a crystal has a maximum value v_{max}. It is related to growth parameters and properties of type of crystal being grown (Yu 1984).

For pure materials the stability condition of the interface is:

$$v_{\max} = \frac{k_s}{\rho l}\left(\frac{\partial T}{\partial z}\right)_s \tag{2.79}$$

where k_s is the thermal conductivity of the crystal; ρ is its density; l is its length; and $\left(\frac{\partial T}{\partial z}\right)_s$ is the axial temperature gradient.

For doped materials the maximum growth rate becomes:

$$v_{\max} = \frac{D[K_{eff} + (1-K_{eff})\exp(-\frac{v}{D}\delta_c)]}{-mC_{l(B)}(1-K_{eff})}\left(\frac{\partial T}{\partial z}\right)_l \tag{2.80}$$

where D is the diffusion coefficient of the melt; K_{eff} is the effective segregation coefficient; m is the slope of the axial temperature distribution curve; and $C_{l(B)}$ is the concentration of this constituent at the interface. From the above two formulas, the maximum theoretical value of the effective growth rate for a given growth system is obtained.

For some special material and device functions, variable doping is required as crystals grow. If the segregation coefficient of doped compositions is known, the doping concentrations of different parts of crystal can be calculated from the following formula:

$$C_s = K_{eff}C_l = K_{eff}C_0(1-g)^{K_{eff}-1} \tag{2.81}$$

where $g = vt/L$ is the fraction of the melt that has solidified by time t. Doping can change the structural properties and growth characteristics of crystals, which will be discussed in more detail presently.

Diameters of crystals can be auto-controlled by adjusting the pulling technique during growth. One way is to monitor the diameter by light reflection off the meniscus. The meniscus between the edge of the growing crystal and the melt produces a bright ring. A uniform meniscus is assured when the crystal grows with a constant diameter. Thus, when an optical sensor system is focused on a meniscus, its variation can be monitored. Then a feed back servo system adjusts the pulling rate, rotation rate and heater power to maintain the desired diameter.

Crystal imaging techniques can also be used to monitor the diameter. These include images obtained by visible, infrared and X-ray imaging systems.

A weighing technique has also been used to monitor the diameter (Gartner et al. 1972; Okane et al. 1972). During the course of crystal growth, the weight variation of the crystal or crucible can be fed back to the control system. Owing to the complex relation between the weight and diameter, it is necessary pre-calibrate the system from an abundant experimental data set.

There are some other methods of diameter control. Utilizing the Peltier effect, Vojdanl et al. (1974) realized auto-control of the crystal diameter. The principle underlying this kind of diameter control is that when DC current passes through the solid–liquid interface, the heat generated is, $Q = J\alpha T_m$, where J is the current density passing through the interface; T_m is the temperature; and α is a coefficient. If the current flows from the solid into the melt, Q is negative, and the refrigeration effect generated will absorb latent heat as crystallization progresses. If the current flows from the melt into the solid, extra heat is generated. The current density J is independent of the area of the interface, so the net current is proportional to the diameter squared, D^2. For a given diameter, it can be controlled as long as the appropriate current is selected. This Peltier effect occurs at the solid–liquid interface as the crystal grows. There is no thermal hysteresis effect, so temporal temperature fluctuations can still be controlled. But Joule heat will also be generated as current passes. Only a few of materials (e.g. semiconductors like InSb, Ge, and Si), that have high enough conductivities to avoid producing sufficient Joule heating to overwhelm the Peltier effect, can be used this way to control the diameter of crystals during growth. There are other special physical properties of many crystals that have simple relationships with the crystal diameter. Some have been used to control the crystal diameter.

Pulling techniques have been used to prepare the narrow gap semi-conductor InSb. InSb is a III–V compound semiconductor used extensively for making 3–5 μm infrared detector arrays and Hall elements. When preparing an InSb single crystal, it is necessary to purify the components in inert gases and then synthesize a polycrystalline ingot.

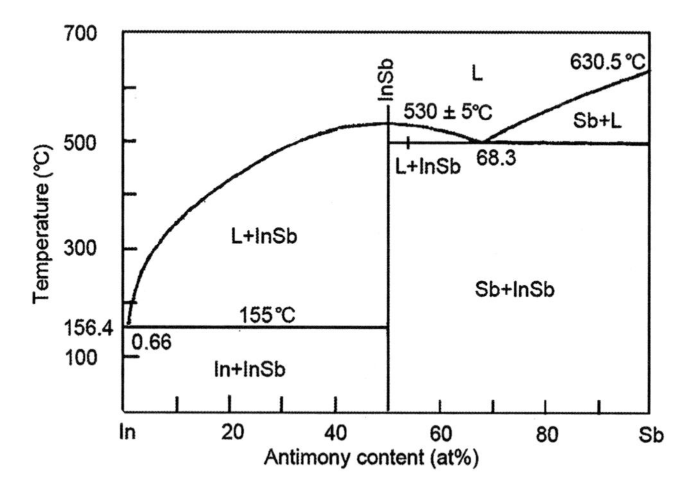

Fig. 2.18. The phase diagram of InSb

A vertical pulling method is used to grow single crystals. The phase diagram is shown in Fig. 2.18.

When the concentration of elementary Sb in the liquid is 50%, freezing will produce a stoichiometric InSb solid phase. The freezing temperature of the solution is a maximum in this situation.

Decreasing the Sb concentration causes the liquid phase line to drop. There exists a broad coexistence zone of a solid InSb + liquid phase. The composition of the liquid and its fractional volume are calculated from the lever rule. When the composition of Sb exceeds 50%, the liquid phase line also drops, and reaches a three-phase point. At this point, solid InSb, liquid Sb, and liquid InSb + Sb coexist. The liquid phase line increases rapidly after the three-phase point. In the phase diagram of InSb, the melting point of In, Sb and InSb are 155, 630, and 530±5°C, respectively, and the triple point is 495°C.

From the InSb *P-T* phase diagram, the vapor pressure of Sb near the melting point is small, about 10^{-5} mmHg. The vapor pressures of In and InSb are smaller still. Therefore, InSb can exit under a protective ambient that is less than an atmosphere. A single crystal is grown after direct heating, synthesizing and purifying.

References (Yu 1984; Yu et al. 1980; Jin 1981) provide details of InSb crystal growth procedure. InSb easily synthesizes in a 1:1 atom ratio. The relative atomic weight is Sb (g) = 1.061×In (g). The vapor pressure of Sb is higher than that of In during the synthesis, purification and single crystal growth processes. The proper state can be reached by the addition of excess Sb (about 0.4%) in the starting mixture. The synthesizing steps are:

a quartz boat with In and Sb is loaded into a clean quartz tube, the tube is evacuated and back filled with pure H_2 gas, the mixture is heated to 800°C to melt it, and the quartz tube is shaken to homogenize the mixture. After holding the temperature stable for 2 h and lowering it to induce a normal freezing process, a polycrystalline material is obtained. A further purification of the polycrystalline material is necessary, because the impurity content is still too high after the initial synthesizing steps. Float zone purification can use either a one-melt zone or a two-melt zone method. The width of the melt zone is generally 1–3 cm. The migration speed of the melt zone is 2–13 cm/h.

InSb expands when it freezes. So as purifying proceeds, the ingot should be kept at a constant obliquity to avoid having material transported to the crystal's tail. After single float zone pass, as we have seen in Sect. 2.1, the concentration of solute at point z in solid is:

$$C_s(z) = C_0[1-(1-K_0)\exp(-\frac{K_0}{l}z)] \tag{2.82}$$

After multiple float zone passes, the ultimate distribution of solute in solid $C_n(z)$ is related to $C_l(z)$ the concentration in the melt zone by:

$$KC_l(z) = C_n(z) \tag{2.83}$$

If the melt zone has length l, $C_l(z)$ can be expressed as:

$$C_l(z) = \frac{1}{l}\int_{z-l}^{z} C_n(z)dz \tag{2.84}$$

or

$$C_n(z) = \frac{K}{l}\int_{z-i}^{z} C_n(z)dz\,.$$

Then a solution is:

$$C_n(z) = Ae^{Bz} \tag{2.85a}$$

where

$$K = \frac{Bl}{1-e^{-Bl}} \tag{2.85b}$$

$$A = \frac{C_0 Bl}{1-e^{-Bl}} \tag{2.85c}$$

and K is the segregation coefficient of the impurity in question. Thus the concentration of an impurity follows an exponential z dependence after multiple float zone passes.

It can be seen from (2.85b) that if $K = 1$, then $B = 0$ and $C_n(z)$ does not vary with z, that is to say purification is ineffective. For $K > 1$ then from (2.85b), B must be positive definite, $B > 0$, and $C_n(z)$ increases with z exponentially. For $K > 1$, B must be negative, $B < 0$, and $C_n(z)$ decreases with z exponentially. For most p-type impurities in InSb, the segregation coefficients exceed 1. These impurities collect at the top of ingots after purifying. The segregation coefficients of most n-type impurity are less than 1. These impurities collect at the tail after purifying. The purest InSb with minimum impurities is in the middle of ingots. In InSb crystals, Zn, Cd and Ge are useful p-type impurities, and Te, Se and S are commonly used n-type impurities. Although impurity densities in InSb are low after many float zone passes troublesome residual donors persist. The donor impurity Te with a segregation coefficient of 0.8, and acceptor impurities, Zn with segregation coefficient of 0.3 and Si with segregation coefficient of 0.1 are the main residual impurities in InSb.

After purification the InSb polycrystalline ingot can be used to grow an InSb single crystal using the vertical pulling method. In order to obtain InSb with required properties, the desired impurities must be doped into InSb during growth. Much attention has been paid to InSb crystalline characteristics, such as growing flat stripes, or eliminating twins. (Yu 1980; Jin 1981)

The advantages of the crystal pulling method are:

- The temperature gradient can be controlled effectively by adjusting the geometrical shapes of the heater, the crucible and the after-heater (or heat preservation cover).
- Temperature, pressure, and image sensors are installed in the apparatus to monitor and control the growth process. The ability to monitor the crystal and control its growth process improves the quality of the resulting material.

Large size single crystals, or even a complete single crystal boule with a prescribed orientation and shape can be grown. Oriented seed crystals, "necking" and "shoulder relaxing" techniques enable these accomplishments. Starting from an oriented seed makes the crystal grow along the desired crystalline axis. The "Necking" technique greatly reduces dislocation densities. The "Shoulder Relaxing" technique increases the size of single crystal regions. It is the most outstanding advantage of the pulling method. Using this method, the crystal is not in contact with the crucible, which avoids both the mechanical stress caused by contact, and nucleation on the crucible wall during the growth process.

The pulling technique is an essential method for fabricating good bulk crystals. Other growth methods often depend on substrates made using pulling methods.

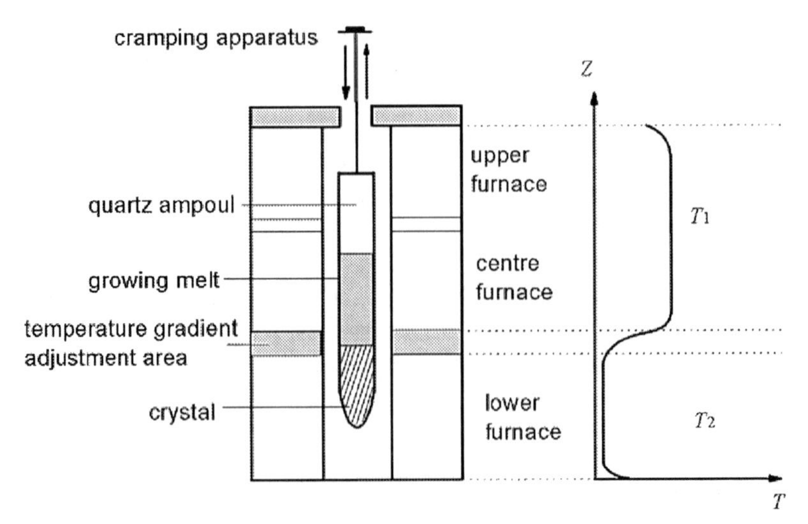

Fig. 2.19. A sketch of a Bridgman vertical growth system and its furnace temperature distribution

2.2.2 Bridgman Method

The main growth methods for fabricating HgCdTe alloy single crystals include the Bridgman technique, the half-melting technique, the Te solvent technique and the solid state re-crystallization technique. Here we focus on the Bridgman technique, which is also called the Bridgman–Stockbarger technique (Bridgman 1925; Stockbarger 1936).

A schematic of a Bridgman apparatus and its temperature distribution is depicted in Fig. 2.19. This method involves a temperature distribution with two zones. An ingot is lowered slowly through a temperature gradient from a high temperature zone to a low temperature zone. The crystal grows at the solid–liquid interface. The rate of lowering must be slow enough, to allow the cross section of the ingot in the quartz tube to equilibrate.

Nucleation is accomplished in one of two ways. The bottom of the quarts crucible terminates in a small cone. A small grain will form there spontaneously to nucleate the growth if nothing else is provided. Often a small oriented seed crystal is inserted into the tip of the crucible to nucleate the growth. This method has the following characteristics. The crystal will grow with the orientation of the seed crystal. The shape of the crystal depends on the shape of the crucible. Both closed and open-ended crucibles have been used. Closed crucibles tend to avoid impurity contamination from the environment, and to reduce volatilization of constituent particles from the melt. Crucible materials are selected that add no pollution to melt, and have a thermal expansion coefficient that matches the crystallizing material. High cleanness of the inner surface of the crucible is necessary to avoid nucleation on its walls. For the sake of promoting the uniformity of the fabricated crystal, rotating crucibles are often used. Automatic system control is programmed. Many crystals of different specifications can be grown simultaneously in one furnace.

The Bridgman technique is a simple and universally adopted method to grow Cd based compound crystals (HgCdTe, CdTe and CdZnTe, etc.) (Shen et al. 1981). In principle, melting the polycrystalline starting material and either moving the quartz crucible or the furnace can crystallize the melt from the top of the quartz crucible down or from its tail up. Bridgman apparatus in common use include both vertical (Muranevich et al. 1983) and horizontal (Lay et al. 1988; Khan et al. 1986) growth techniques. During the course of horizontal growth, the axis of the ampoule is perpendicular to the gravity field. In contrast with vertical growth, the growth interface shape is not affected by the melt's weight. It is possible to control the shape of the growth interface by controlling the transverse temperature gradient. The volume of the quartz ampoule does not stress the crystal ingot, and thermal stresses that occur in the growth process and subsequent cooling are reduced. In the horizontal configuration, the metal vapor pressure can be controlled to adjust the stoichiometry of the melt throughout the growth process. In vertical growth in a closed ampoule the vapor pressures are set by the phase diagram. There is more latitude to adjust the stoichiometry in open ampoule systems (Cheuvart et al. 1990).

As shown in Fig. 2.19, the quartz ampoule loaded with the starting ingot material is placed into the furnace. The furnace is a three-segment resistance heater. The top and middle segments produce temperatures above the melting temperature of the ingot. The bottom segment produces the holding and subsequent annealing temperatures after crystallization. The middle segment is a temperature gradient zone typically of order ~10–20°C/cm. The clamp on the top of furnace supporting the quartz ampoule can move up and down, in a typical case by ~30 cm.

Some preparations for growing crystals by the Bridgman technique are necessary.

- The Quartz ampoule must be soaked in chromic acid and aqua regia for several hours to remove organic and metal impurities.
- Then it is washed in deionized water, dried, and carbonized.

- Once again it is soaked in aqua regia for several hours, washed in deionized water and dried. A slippage line will appear in material if it is grown in an un-carbonized quartz ampoule (Szofran and Lehoczky 1981). The reason for this is that crystals adhere to uncoated quartz ampoules, and strains are induced during cooling because the crystal shrinks faster than quartz. Growing crystal in carbonized quartz ampoules also decreases impurities.
- The high purity component materials are weighed to meet the specified composition.
- They are placed into the quartz ampoule and it is evacuated to 10^{-5} Torr and sealed.
- After being sealed, the ampoule is mounted in a vertical furnace with a three-segment temperature zone, completing the preparation for crystal growth.

Then growth conditions are selected. For example, the temperature gradient at the growth interface is set to 20°C/cm, the lowering rate of the ampoule to 1 mm/h, (after grow is completed), the cooling rate to 20°C/h, and so on. The growth temperature can be obtained from the temperature of the liquid phase line on the HgTe–CdTe psudo-binary phase diagram (Yasumura et al. 1992). This diagram also holds the melting points for different compositions. For stable growth, the heat flow from the interface into the solid should be as small as possible, consistent with criteria derived below. Thus, the melt temperature selected should not be too high; an appropriate temperature is ~10–30°C higher than the melting point. A typical melting point for HgCdTe is ~780°C, and for CdZnTe is ~1,100°C.

Under ideal conditions the Bridgman technique enables the growth of single crystals under quasi-equilibrium conditions. A HgCdTe single crystal is grown via the slow movement of the solid–liquid planar interface. The planar interface shape and its stability are very important. Under this condition, the solute concentration is uniform. Then dislocations and other extended defects do not form easily and thermal stress is minimized. When an instability causes a facet to protrude into the liquid (a convex interface), dislocations and other defects are generated, and tend to migrate toward the outer part of the crystal. Any small grains that nucleate are apt to be annexed into the crystal accompanied by stress. In the worst case, a concave interface forms. Then ringed fringes grow at first followed by polycrystalline growth. The defective volume is apt to concentrate in the interior of the crystal, which produces a large interior stress. Whether the interface becomes convex or concave, it will change the compositional cross section of the ingot.

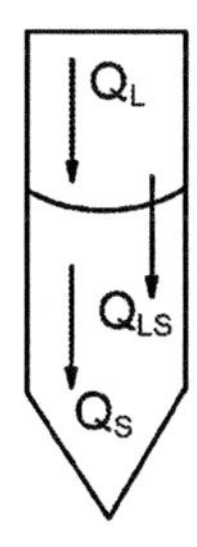

Fig. 2.20. Directions of heat flow near the growth interface

The thermal distribution, the rate of growth, the rates of heating and cooling, the starting composition, and their continuing control during growth are essential issues in the quality of Bridgman grown crystals. The temperature gradient controls the growth rate. Before the initial grain forms, the temperature of the melt is made slight lower than the temperature of the liquid phase line. If the temperature gradient is too small, quite a large a part of the ingot will be in a super-cooled state and the growth rate will run out of control. If the gradient is too large growth will be unacceptably slow. The appropriate temperature gradient can be estimated from the heat equilibrium condition.

Growth dynamics has a definite relationship to the temperature gradient. In general, the growth interface is assumed to be an isothermal plane at an invariant melting point. Let Q_S be the quantity of heat that flows from the interface into the crystal per unit time. Q_L is the corresponding quantity of heat that flows from the melt to the interface per unit time. Q_{LS} is the latent heat per unit time evolved at the interface as the crystal grows (Wood and Hager 1983). The heat flow directions as defined are shown in Fig. 2.20.

Treating the heat flow as one dimensional, as depicted in Fig. 2.20, conservation of energy dictates:

$$Q_L + Q_{LS} = Q_S \tag{2.86}$$

where

$$Q_L = Ak_L(\frac{dT}{dz})_L \tag{2.87a}$$

$$Q_{LS} = L\frac{dm}{dt} = LA\rho_S(\frac{dz}{dt}), \tag{2.87b}$$

and

$$Q_S = Ak_S(\frac{dT}{dz})_S \tag{2.87c}$$

Here k_L and k_S are thermal conductivities of the melt and the crystal, respectively, $(\frac{dT}{dz})_L$ and $(\frac{dT}{dz})_S$ are axial temperature gradients of the melt and the crystal, respectively, $\frac{dz}{dt}$ is the growth rate v of crystal, $\frac{dm}{dt}$ is the crystal mass grown per unit time, L is the latent heat of crystallization per unit mass, ρ_s is the density of crystal, and A is the area of the interface.

Substituting the values of Q_L, Q_{LS}, and Q_S from (2.87) into (2.86), we find:

$$Ak_S(\frac{dT}{dz})_S = Ak_L(\frac{dT}{dz})_L + LA\rho_S(\frac{dz}{dt}) \tag{2.88}$$

or

$$\frac{dz}{dt} = \frac{1}{\rho_S L}\left[k_S(\frac{dT}{dz})_S - k_L(\frac{dT}{dz})_L\right] \tag{2.89}$$

From (2.89), the main factors determining the growth rate are the axial temperature gradients in the melt and the crystal. When $(\frac{dT}{dz})_L = 0$, the growth rate will have its maximum value.

The growth rate is determined not only by characteristics of matter but also by the temperature gradient, the greater the temperature gradient, the slower the growth rate. It is known from the HgTe–CdTe pseudo-binary phase diagram that the segregation coefficient of CdTe in a HgTe solvent is about 3. It has been shown in the discussion of the effective segregation coefficient that increasing the growth rate decreases the effective segregation coefficient. When the growth rate is 1 mm/h, K_{eff} decreases to 1.5. When $v \cong 30$ mm/h, K_{eff} decreases to 1.1. Then the concentration x of a $Hg_{1-x}Cd_xTe$ alloy along the growth direction will approach unity. But the quality of such single crystals is poor. There are composition deviations from unity in cross sections, which are related to the interface shape. Once this effect was discovered many methods to overcome it were proposed, the most obvious being to slow the growth rate.

As we have seen, the flat shape of the solid–liquid interface and supercooling have important influences on Bridgman crystal growth. To maintain a flat interface, it is beneficial to have the term

$$\left[k_S(\frac{dT}{dz})_S - k_L(\frac{dT}{dz})_L\right] \tag{2.90}$$

From (2.89), be large enough so the difference between the radial temperature gradients of the solid and the liquid at the interface, $\left(\frac{dT}{dr}\right)_S - \left(\frac{dT}{dr}\right)_L$, does not have an appreciable effect on growth rates over its area. Then the interface will remain flat. There is an upper limit on $\left[k_S(\frac{dT}{dz})_S - k_L(\frac{dT}{dz})_L\right]$ set by the condition that the growth rate, v, remain small enough to insure the growth of high quality crystals. The maximum diameter of long crystals is constrained by the deleterious influence of radial temperature gradients at the solid–liquid interface.

Too much super-cooling also adversely effects the stability of crystal growth. The super-cooling condition for stable growth is (Harman 1980):

$$\frac{G}{v} < \frac{mC_L(k_0 - 1)}{Dk_0} \tag{2.91}$$

The left side of (2.91) contains externally adjustable parameters, the growth rate v and the temperature gradient in the melt $G_L = \left(\frac{dT}{dz}\right)_L$ just below the interface. The right side contains the average concentration of solvent C_L in the solution (here C_L is not an adjustable parameter, but is determined by the specified concentration x of the crystal), and physical property parameters of the materials. These parameters are the slope m of liquid phase line, the equilibrium segregation coefficient k_0 of the solvent, and the solvent's diffusion coefficient D. From (2.91), we see that in order to maintain a steady growth, a combination of decreasing the growth rate and increasing temperature gradient can be used.

Massive data bases coupled to elaborate three dimensional computer models have provided guidance in the exploration of new growth strategies. Many revised techniques of Bridgman growth have been developed. Lay et al. (1988) devised a ten-segment horizontal temperature gradient furnace in which the temperature profile is moved electrically to avoid deleterious effects from the relative motion of the ampoule and furnace. Controlling the partial pressure of Cd during CdTe growth by having a separate temperature controlled Cd vessel enables precise control of the stoichiometry (Cheuvart 1990; Bell and Sen 1985; Brunet et al. 1993). Lu et al. (1990) grew CdTe crystals by adding a vibration method. Oscillating force fields can be generated by applied electromagnetic fields or by mechanical vibration. A driven force field affects the heat transfer, the shape of growth interface, the defect distribution, and consequently the quality of the resulting crystal. Muhlberg et al. (1990; Pfeiffer and Muhlberg 1992) grew CdTe-based alloys by a small temperature gradient method. They used temperature gradients in the melt of $G_L \approx 4$ K/cm and in the solid of $G_S \approx 8$ K/cm. They obtained CdTe crystals with fewer grain boundaries than those grown with the larger temperature gradient, G $\geq$= 10 K/cm. Evidently in their apparatus the smaller temperature gradient reduced thermal stress and the density of dislocations. Doty et al. (1992; Butler et al. 1993) grew $Cd_{1-x}Zn_xTe$ crystals by a high-pressure technique. During the course of growth, high inert gas partial pressure prevents the evaporation of the intended constituents. Then, gas diffusion is decreased and encapsulation of the ampoule is eliminated. This allows the range of possible crucible materials to be enlarged. The high-pressure method is particularly useful to grow CdZnTe crystals with a high Zn content. Capper et al. (1993) grew CdTe and CdZnTe crystals by an Accelerating Crucible Rotation Technique (ACRT). The ACRT method can make the radial distribution of Zn uniform. The vertical temperature freezing method is similar to the Bridgman method. This method includes: putting polycrystalline material into a quartz crucible, putting the crucible into a vertical resistance furnace, heating to above melting point, then lowering the temperature of the furnace and freezing the solution into a crystal along the temperature gradient. Asahi et al. (1995) used a boron nitride crucible, a separately controlled Cd partial pressure, separate heat units with precise computer controlled temperatures (temperature fluctuation less than 0.1°C), a low temperature gradient, a low cooling rate, and a special crucible shape design to grow large CdZnTe single crystals of 100 mm diameter and 50 mm length.

2.2.3 The Half-Melt and the Te Solvent Methods

The Bridgman method involves quasi-equilibrium growth. Two other quasi-equilibrium growth methods used to prepare HgCdTe single crystals are the half-melt technique and the Te solvent technique. The "half-melt" technique was devised by Harman in 1970 (Harman 1972). This method begins with a quartz crucible filled with an ingot being placed in the high temperature zone in the upper part of the furnace where the ingot melts. Then the quartz crucible is lowered rapidly to a position in the furnace that spans three temperature zones. The states of the material in the crucible at this position can be divided into three segments A, B, and C. Segment C at

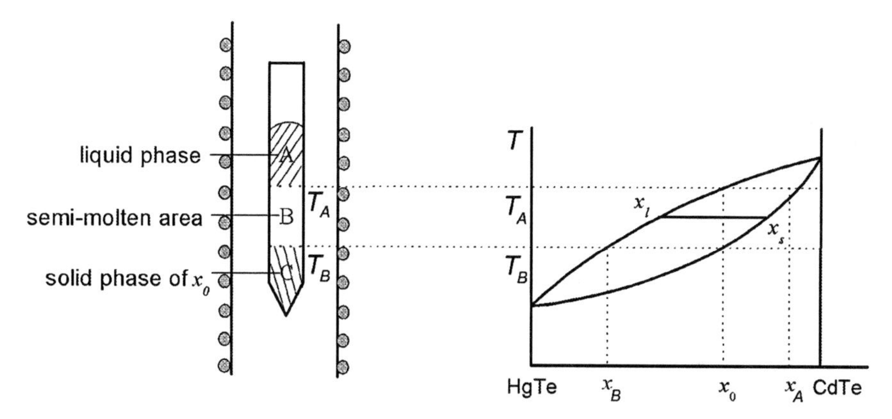

Fig. 2.21. A sketch of the *T*–*x* phase diagram of the HgTe–CdTe pseudo-binary alloy system and the corresponding states of material in the crucible

the bottom is located in the lower temperature zone where the material crystallizes. Segment B in the middle zone is located in a zone with a temperature gradient where the material is in a mixed "half-melt" state. Segment A at the top of the crucible is located in a high temperature zone where the material remains in a molten state. The crucible stays in this position for over 40 days while the crystal evolves. In the half-melt zone, a crystal grows continually until it crystallizes completely. A brief analysis of the physical process follows.

Figure 2.21 is a sketch of the *T*–*x* phase diagram of the HgTe–CdTe pseudo-binary alloy system and the corresponding states of the material in different zones in the crucible. Suppose that T_A is the temperature at the upper edge of the half-melt zone, and T_B is the temperature at the lower edge of this zone. When the crucible is first lowered quickly into the growth position, the initial CdTe concentration of the solid in zone C is x_0, as is the concentration in the molten zone A. The half-melt segment B is the coexistence zone of solid and liquid phases, and their the compositions of the solid and liquid phases are x_S and x_l, respectively. The initial compositions x_S and x_l vary throughout zone B in its temperature gradient. Because T_A is greater than T_B, the material at the upper edge has a higher Gibbs free energy than that at lower edge. A system initially in a non-equilibrium state will evolve toward one in which the chemical potential is a constant. Thus as time passes the chemical potential of upper edge decreases while that of the lower edge increases. This means that the solute CdTe diffuses from the upper edge at temperature T_A to the bottom edge at temperature T_B. This increases the concentration in the liquid phase at the lower edge of zone B leading to supper-cooling there. The super-cooled part freezes into a crystal with composition x_0 or slightly larger than x_0, and the remaining material still complies with the phase diagram. This process continues until the half-melted region crystallizes completely.

The time for this whole process can be calculated from the lever rule in T–x phase diagram and Fick's first law. From the lever rule, the amount of solute required at temperature T_B to transform a thickness dz from liquid into solid in time interval dt is obtained. The lack of CdTe solute in dz at the bottom edge of the half-melt zone is provided by diffusion from the upper edge to thickness dz. From Fick's first law the total time is found from the integral:

$$t = \frac{1}{DG_r^2} \int_{T_B}^{T_A} (x_s - x_0) \frac{dT}{dx_l} \cdot dT \tag{2.92}$$

where D is a Cd and Hg effective inter-diffusion coefficient in the liquid, G_r is the temperature gradient across the half-melt zone, given approximately by:

$$G_r = \frac{T_A - T_B}{l} \tag{2.93}$$

l is length of half-melt zone (the useful section of crystal after growth is completed), and dT/dx_l is the slope of the liquidus phase line which is also a function of T. Using the liquidus and solidus lines and x_0, x_s and x_l can be written as functions of x. Then the value of t can be calculated numerically from the above integral formula. For example, if T_B = 707°C and T_A = 796°C, one finds:

$$\begin{aligned} x_l &= \frac{1.37T - 915}{1690 - 0.97T} && (0 < x < 1) \\ x_s &= \frac{5.72T - 382}{2.92T - 951} && (0 < x < 0.5) \\ x_s &= \frac{5.72T - 3520}{2.92T - 951} && (0.6 < x < 1) \end{aligned} \tag{2.94}$$

Equation 2.94 with $D \approx 5\times10^{-5}$ cm^2/s, $G_r \approx$ 50°C/cm, yields $t \approx 41$ days.

Only the section of crystal in the half-melted zone is useful, the crystal composition difference between the bottom and the top of the crucible is very large. A method has been devised to make the composition more uniform in long crystals (Fiorito et al. 1978). Because the half-melt growth process is controlled mainly by diffusion, the time required to grow crystals is too long for it to be a practical production method.

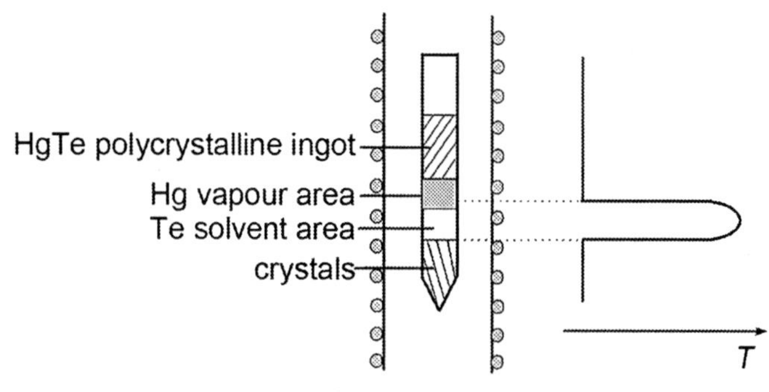

Fig. 2.22. Sketch of the structure of a Te solvent growth system and its temperature distribution

The third quasi-equilibrium method for growing HgCdTe single crystals is the "Te solvent" technique. The Te solvent technique is a vertical melt region method with Te serving as the solvent (Ueda 1972). The method starts with the Hg–Cd–Te components with specified compositions loaded into a quartz ampoule, evacuation of the ampoule and sealing it, placing it in the furnace and heating the material to above its melting point, and then quenching it to room temperature. The result is polycrystalline HgCdTe with uniform composition. Afterward the ampoule is opened, pure Te is added to the bottom of the ampoule, and it is again evacuated and sealed. Then, the quartz ampoule is suspended in the furnace. First, the pure Te at bottom of the ampoule enters the high temperature region and the Te melts. Hg at the interface of the HgCdTe poly-crystal and the Te liquid begins to dissolve. A space is formed between the Te melt and the polycrystalline HgCdTe, which is filled with Hg vapor. As the ampoule continues being lowered, there is excess Te at the surface of the HgCdTe polycrystals. The melting point decreases. Polycrystals begin to melt and dissolve into the Te liquid as solutes. Eventually a solubility limit is reached and the solution becomes saturated. As the ampoule continues being lowered it reaches a place in the furnace where the temperature decreases as shown in Fig. 2.22. The HgTe and CdTe in the solution diffuse and begin to deposit out on the bottom of the ampoule.

Accordingly, a crystal grows from the bottom of the ampoule. In Te, the segregation coefficient of CdTe is greater than that of HgTe, so the composition of $Hg_{1-x}Cd_xTe$ at the beginning is on the high side. But this also decreases the concentration of CdTe in the solution, which reduces its content in the growing crystal. Once a stable condition is reached, the composition becomes close to that of the polycrystalline ingot. As the process continues the lower part of the ampoule contains the growing crystal. Above the crystal there is a zone of a Te liquid solvent with a

HgCdTe solute. Above this zone there is a space that is filled with Hg vapor. The HgCdTe polycrystalline ingot resides at the top. The Te solution zone, as shown in Fig. 2.22, is at the highest temperature.

When growing by the Te solvent technique, there is excess Te is incorporated into the bottom of the polycrystalline ingot, lowering its melting point. For example, a HgCdTe alloy that would melt at 780°C has its melting point decreased to 700°C. Simultaneously, this method has a floating zone aspect. The solubility of impurities in the Te rich segment increases relative to the stoichiometric material, which has the effect of purifying the crystal.

An appropriate temperature gradient is necessary to produce good crystals. If the temperature gradient along the ingot is too large, it causes the interface to be concave and if it is too small the interface is convex. Selecting a proper temperature gradient does not affect the longitudinal heat flow associated with the solidification. Transverse thermal diffusion of the excess Te in the HgCdTe tends to make the interface flat, which is beneficial.

The Te solvent technique is one of the more general class of "traveling hot-zone methods" (THM), also called "float zone methods." As we have noted it has the advantage of growth from a low temperature solution. Furthermore, owing to the impurity segregation effect, the crystal purity improves as it grows. The flaws of this method are:

- The structures of crystals grown by this method are not as good as those grown by the Bridgman technique or by vapor growth methods.
- There are many Te inclusions in these crystals because of the use of Te as the solvent.

In order to improve the purity of these crystals and reduce the density and size of Te inclusions, the growth technique has been revised by the addition of multiple THM passes, a subliming THM coupled with an "accelerated crucible rotation method" (ACRT-THM). Subliming THM reduces the Te inclusion density. ACRT-THM not only reduces the Te inclusions but also increases the growth rate. Using this method, CdZnTe crystals with diameters as large as 50 mm have been grown with growth rates reaching 12 mm per day.

2.2.4 Solid State Re-Crystallization Applied to HgCdTe

Solid state re-crystallizing, as discussed in Sect. 2.2.1 in connection with the growth of InSb, consists of two processes, quenching and re-crystallization (Swink and Brud 1970; Shen et al. 1976). To grow HgCdTe alloys, high purity, 99.9999%, Hg, Cd and Te elements mixed in stoichiometric ratios. Hg vaporizes easily in the closed quartz growth ampoule, so a small surplus of Hg is added. When preparing $x = 0.2$ HgCdTe, an excess of ~40–50 mg/cm^3 of Hg is added. These three materials are placed into a quartz ampoule, which is evacuated and sealed. The ampoule is put into a tube-type furnace that can swag. It is then heated slowly and swaged so the three materials melt and are thoroughly mixed. During the course of heating, when the temperature reaches 700°C a Te–HgTe eutectic is formed. The eutectic has a peak in the effective heat capacity. When the eutectic is reached Hg–Cd–Te begins to form compounds and emit heat. As the process continues the high melting point compounds that are formed separate out from metallic Hg. Metallic Hg has an exponential vapor pressure temperature variation. So it is necessary to control the heating rate rigidly to avoid overlarge Hg pressure induced ampoule explosions. After enough mixing of the Hg–Cd–Te constituents the temperature is raised above the melting point. The temperature is taken above the melting point by ~10°C for x = ~0.17–0.18, ~20°C for x = ~0.20, and ~35°C for x = ~0.25–0.26. Then it is kept constant for 24–48 h and the ampoule is shaken slowly to combine the constituents fully. Finally, the furnace is kept vertical and still for 1–2 h, followed by a rapid quench cooling. This procedure leads to a uniform composition throughout the poly-crystalline ingot.

To understand why the composition of a quenched cooled ingot has a uniform composition it is useful to introduce an effective segregation coefficient for this process. In Sect. 2.1.5, it was shown that the liquid concentration at the growth interface located at z is given by the expression ((2.58) repeated here):

$$C_l(z) = C_l\left[1 + \frac{1-K_0}{K_0}\exp(-\frac{v}{D}z)\right] \tag{2.95}$$

where C_l is the concentration of solute in a part of the solution far away from interface. If the growth rate is very large, viz. $v \to \infty$, then the concentration at any point z is constant:

$$C_l(z) = C_l \tag{2.96}$$

Simultaneously, the thickness of the interface layer, $\delta = \frac{D}{v} \to 0$, vanishes. In general, an effective segregation coefficient can be introduced that is defined by:

$$K_{\text{eff}} \equiv \frac{C_s(z)}{C_l} = \frac{K_0}{K_0 + (1-K_0)\exp(-\frac{v}{D}z)} \tag{2.97}$$

Clearly, $C_s(z) \Rightarrow C_l$, for a fast growth rate as $v \rightarrow \infty$, $K_{eff} \rightarrow 1$ and the concentration is uniform.

While there is no segregation, there are adverse effects associated with rapid quench cooling. Under this condition, the melt cools abruptly, and the crystal's freezing rate cannot keep up with it. Consequently, the temperature of the melt in front of interface is lower than that at the interface, $dT/dz < 0$. So any protruding part of interface will grow more and more rapidly to form dendritic crystallites. Thus, unoriented poly-crystalline HgCdTe is obtained. The boundaries of the crystallites are unstable and therefore the free energy of the system is higher than its equilibrium value. A re-crystallizing process, in which large crystallites grow at the expense of small ones decreases the boundary energy, and drives the system toward a lower free energy more stable state. The re-crystallizing process is carried out at sufficiently high temperatures to permit atomic diffusion through the bulk and along the crystal boundaries.

Most often the re-crystallizing proceeds by first changing the ingot to a new tube, adding a proper quantity of Hg, and evacuating and sealing the tube. Sometimes the old tube is reused. For $x = 0.2$, the Hg pressure at the re-growth temperature is about ten atmospheres. The temperature of furnace has a temperature gradient. The ingot in its tube enters into the high temperature zone slowly. The crystallizing course proceeds gradually from the head to the tail of the ingot. In theory, the free energy of system will decrease if big grains annex little grains. If the whole ingot enters a zone at an elevated temperature but below the melting point, and is maintained there for an extended period, a homogeneous perfect single crystal will grow. The process requires more than 10 days. The re-crystallizing temperature and the Hg pressure is selected from the *P–T* phase diagram. For $x = 0.2$, the re-crystallizing temperature and the Hg pressure should be ~680°C and ten atmospheres respectively. Analyzed in terms of thermodynamics, one can obtain the re-crystallization temperature T_{recryst}, from the melting point T_{melt} and the Debye temperature T_{D}. The relationship is given by the formula:

$$T_{\text{recryst}} \cong \frac{T_{\text{melt}}}{3}(1 + \frac{T_D}{T_{\text{melt}}}) \tag{2.98}$$

The above discussion is based on growing the HgCdTe crystal in a closed quartz tube. But since the Hg pressure involved is high, this mandates a thick walled quartz tube. Because the heat conductivity of quartz is low, and the thick walls have a high heat capacity, the response time to a temperature change is long. This slow response limits the precision of the temperature control system during crystal growth. It presents particular difficulties in fabricating large diameter crystals or alloys with high x values.

The parameter most sensitive to temperature fluctuations is the Hg partial pressure. A high-pressure inert gas re-circulation technique has been proposed to mitigate this problem. In this system the Hg partial pressure above the HgCdTe is controlled by the high-pressure inert gas flow. The choice of the temperature distribution of the whole apparatus, and the pressure of the inert gas, constitute a serious designed problem.

In addition to melt-based growth, vapor-based growth is also an option. The elements of the method for CdTe are:

- Sealing the CdTe-based materials in a quartz ampoule.
- Placing the ampoule in a furnace with a temperature gradient.
- Heating the high temperature zone to a temperature where the CdTe source material sublimes.
- A single crystal of good structure and high purity can be grown by vapor transport and deposition in the colder end of the ampoule.

Vapor growth can be seed free growth (Akutagawa and Zanio 1971; Yellin et al. 1982), seeded growth (Golacki et al. 1982), and moving hot zone growth (Triboulet and Marfaing 1981). Because single crystal vapor growth is slow, CdTe-based crystals of high purity, high resistance, and near perfect structure with dislocation densities of less than 10^4 cm^{-2} are grown. The disadvantages are low growth rates and inability of growing large single crystals. Many attempts have been made to increase the vapor growth rate and crystal volume. Methods that have been tried include: connecting the source with the deposition chamber by a capillary (Buck and Nitsche 1980); growing in an Ar gas atmosphere using a half-closed technique (Durose and Russell 1988; Durose et al. 1993); putting the seed on a sapphire rod through which optical heating is provided (Durose et al. 1993). These improvements did increase the growth rate and produced larger crystals, e.g. single crystals of 5 cm length and 4 cm diameter. But the structures of these crystals are imperfect with dislocation densities ranging up to 10^6 cm^{-2}. Also there is a large amount of sub-structure and inclusions. The vapor-based growth methods still need to be improved to realize practical applications.

2.3 Liquid Phase Epitaxy

The term "epitaxy" is derived from the Greek word meaning "ordered upon." Despite the progress made in bulk crystal growth technology, a heightened interest persists in the growth of HgCdTe as epitaxial layers and layered structures. Among the growth techniques used, LPE, MOCVD and MBE are most important. LPE, MOCVD and MBE offer, in comparison with bulk growth techniques, lower growth temperatures, shorter growth times, and better compositional homogeneity. The growth temperature in MOCVD is around 350°C, which makes the p-doping more difficult owing to vapor phase growth occurring on the Te side of the existence region and the consequent difficulty of getting group five elements to incorporate on the Te sublattice. MBE offers unique capabilities for growing large area epilayers and sophisticated layer structures with abrupt and complex compositions and doping profiles. The growth temperature is less than 200°C. The developments in MBE HgCdTe growth and doping control have led to the demonstration of high-performance infrared focal plane arrays (IRFPAs) for imaging applications over a broad spectral range. The LPE technique yields better compositional uniformity than bulk crystal growth and does not require long annealing times to achieve homogeneity. Moreover, the growth of multiplayer HgCdTe structures with different compositions and doping levels are possible using LPE (Charlton 1982), but they are not as abrupt as those obtained with MBE or MOCVD. However, LPE layers grown from Hg-rich melts can be extrinsically p-doped more easily.

LPE growth of thin HgCdTe layers on CdTe substrates began in the early-to-mid 70s (Konnikov 1975). Large diameter CdTe substrates could be grown because Hg overpressure was not a limiting consideration. Epilayers of HgCdTe may be grown from Te-rich solutions (Wang et al. 1980; Nelson et al. 1980), Hg-rich (Tung 1981a,b; Tung et al. 1987; Herning 1984), and HgTe-rich solutions (Bowers et al. 1980). Diodes fabricated on early LPE material were made by ion implantation into p-type layers that had been annealed to generate Hg-vacancy acceptors. By the mid-80s, LPE growers learned to achieve extrinsic p- and n-type doping, allowing the growth of junctions using p- and n-type doped melts. This resulted in greatly improved device quality and set the stage for the production of both first- and second-generation devices from LPE material. A long wavelength infrared (LWIR) detector with response peak at 17 μm has been demonstrated using a HgCdTe LPE film (Pultz et al. 1991).

LPE growth technology is now very mature for production of both first- and second-generation tactical detectors. Hundreds or thousands of layers are routinely grown each year for a wide variety of production programs as well as for technology demonstration and technology development tasks. Infrared focal plane arrays with sizes of 128×128, 512×512, 480×640 have been realized using HgCdTe LPE films (Amingual 1991; Tennant et al. 1992; Johnson et al. 1990, 1993) in the 1990s. In 2001, 1,024×1,024 and 2,048×2,048 focal plane arrays were realized using HgCdTe LPE films grown on sapphire substrates (Bostrup et al. 2001).

2.3.1 LPE Growth

The LPE process involves the introduction of a carefully prepared substrate platelet into a super-saturated solution, or into a nearly saturated solution in which super-saturation is created after a short period during which substrate etchback may occur. The substrate and the supersaturated solution being in mutual contact represent a thermodynamic system in a non-equilibrium state, which is what causes the growth of the epitaxial layer on the surface of the substrate. Three different approaches have been applied to methods for bringing the liquid solution and the substrate into contact with each other: (a) tipping the melt on and off the substrate (Fig. 2.23, Edwall et al. 1984), (b) dipping the substrate into the melt (Fig. 2.24, Wan et al. 1986; Geibel et al. 1986; Yasumura et al. 1992) and (c) sliding the melt on and off the substrate (Fig. 2.25, Wood and Hager 1983; Tung et al. 1987).

For performing LPE growth it is useful to know the solid–liquid-phase diagram of the system. The phase diagram can be determined experimentally (Tung et al. 1987; Szofran and Lehoczky 1981; Panish 1970) as well as theoretically (Tung et al. 1982; Harman 1980). Figure 2.26 shows the Gibbs triangle for the Hg–Cd–Te ternary system (Herman and Pessa 1985). Isothermal tie lines may be drawn from points *A*, *B*, *C* to a certain composition $Hg_{1\text{-}xD}Cd_{xD}Te$ represented by point *D*. These tie lines show how the solidus composition at point *D* grows from the liquidus of a Te-rich solution $(Hg_{1\text{-}zA}Cd_{zA})_{1\text{-}yA}Te_{yA}$, of a Hg-rich solution $(Hg_{1\text{-}zB}Cd_{zB})_{1\text{-}yB}Te_{yB}$, and from a psudobinnary or HgTe-rich solution $(Hg_{1\text{-}zC}Cd_{zC})_{0.5}Te_{0.5}$. Solidification processes taking place during LPE are shown by arrows on the Hg–Te binary phase diagram (top left), and on the pseudo-binary section of the Hg–Cd–Te ternary phase diagram (top right).

Figure 2.26 shows the relationship between liquidus data for $(Hg_{1\text{-}z}Cd_{z})_{1\text{-}y}Te_{y}$ and solidus composition of LPE layers growth from (a)

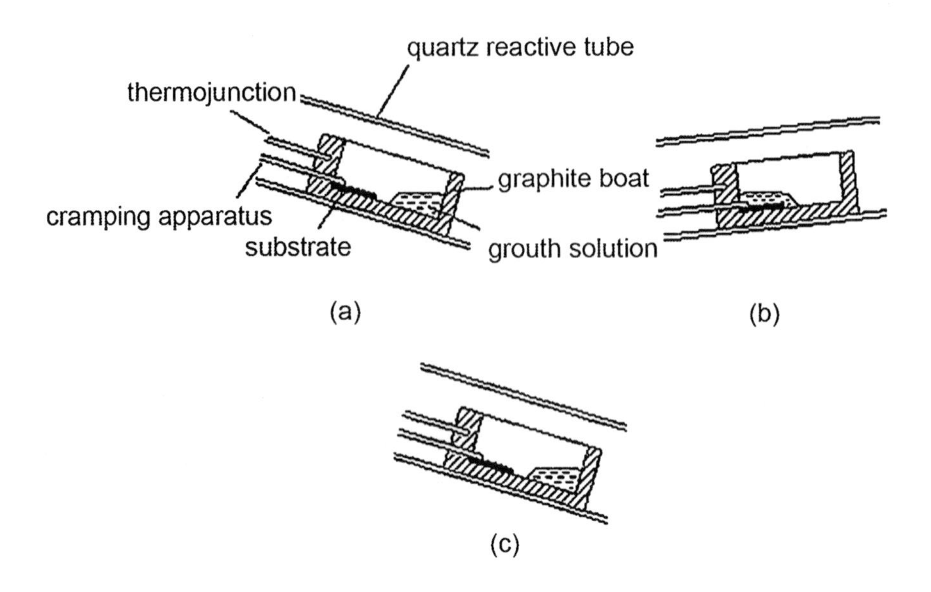

Fig. 2.23. Tipping LPE (**a**) substrate loading, (**b**) epitaxy, and (**c**) substrate removal after growth

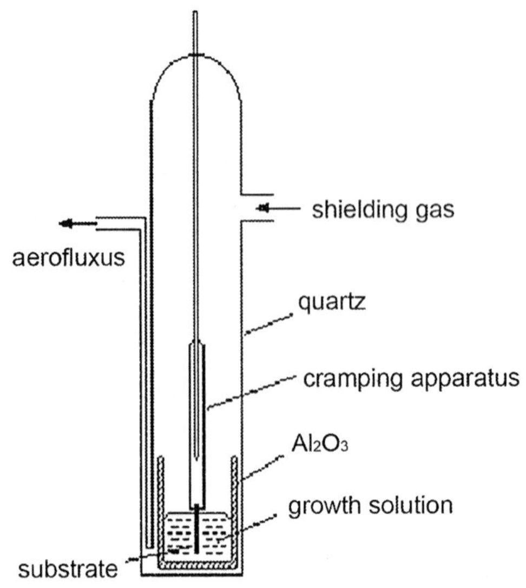

Fig. 2.24. Vertical dipping LPE

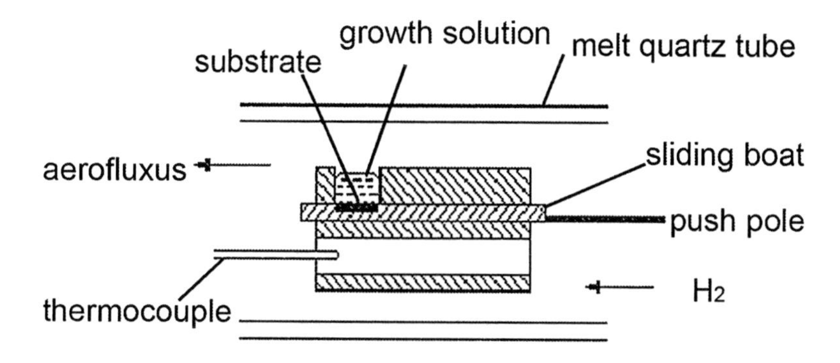

Fig. 2.25. Horizontal sliding LPE

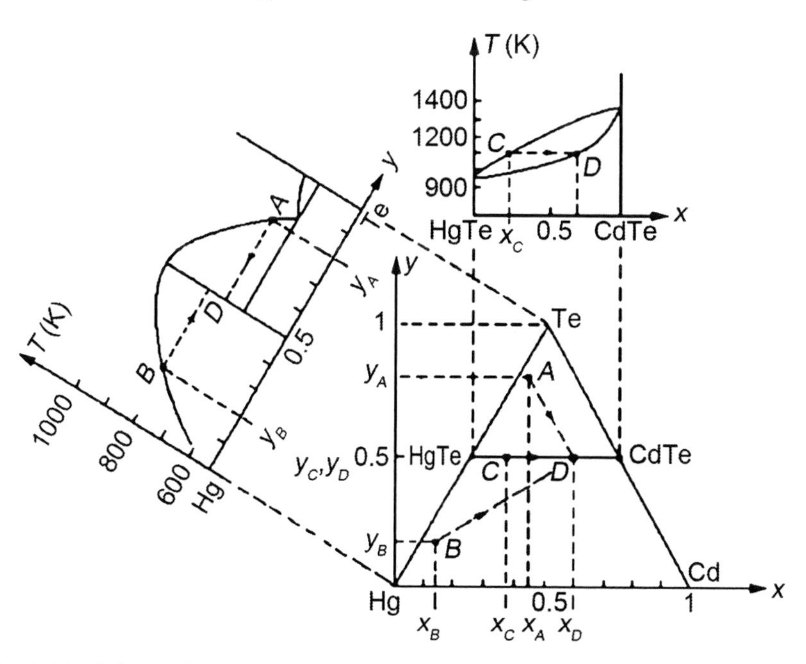

Fig. 2.26. Schematic representation of the Hg–Cd–Te ternary system with the mixed crystal $Hg_{1-x}Cd_xTe$ represented by the line extending from HgTe ($x = 0$) to CdTe ($x = 1$) on the Gibbs composition triangle (*bottom*)

Te-rich and (b) Hg-rich melts. For example, to obtain $x = 0.2$ layers ($Hg_{1-x}Cd_xTe$) from the Te-rich solution $[(Hg_{1-z}Cd_z)_{1-y}Te_y]$ at 500°C would require liquid atomic fractions of Cd and Hg of ~0.01 and ~0.19, respectively (Fig. 2.27). Thus, we require:

$$\begin{aligned} (1-z)(1-y) &= x_{Hg} = 0.19 \\ z(1-y) &= x_{Cd} = 0.01 \end{aligned} \tag{2.99}$$

We have z = 0.05, y = 0.8. To balance the Te would have an atomic fraction of ~0.8 in the LPE solution.

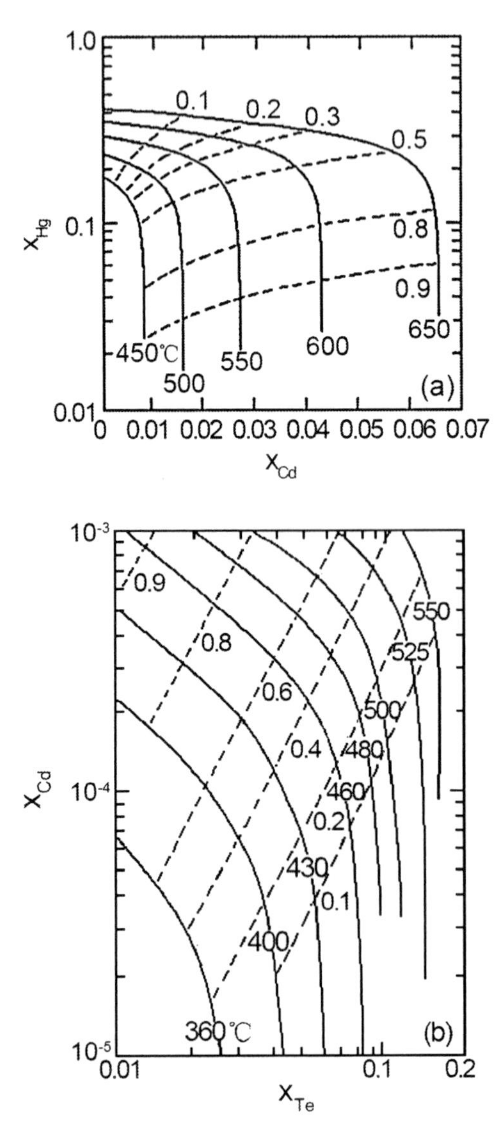

Fig. 2.27. Interrelationship between liquidus data for $(Hg_{1-z}Cd_z)_{1-y}Te_y$ (*solid lines*) and LPE solidus (*dashed lines*) for $Hg_{1-x}Cd_xTe$ for crystal growth from (**a**) Te-rich and (**b**) Hg-rich melts (Tung et al. 1982, 1987)

Generally the mass of Cd, m_{Cd} that is required, is first determined because it is much less than that of Hg and Te. The mass of Hg and Te, m_{Hg} and m_{Te}, are then determined from m_{Cd} according to the following equations:

$$m_{Hg} = \frac{A_{Hg}}{A_{Cd}}(\frac{1-z}{z})m_{Cd} \tag{2.100}$$

$$m_{Te} = \frac{A_{Te}}{A_{Cd}}\frac{y}{(1-y)z}m_{Cd} \tag{2.101}$$

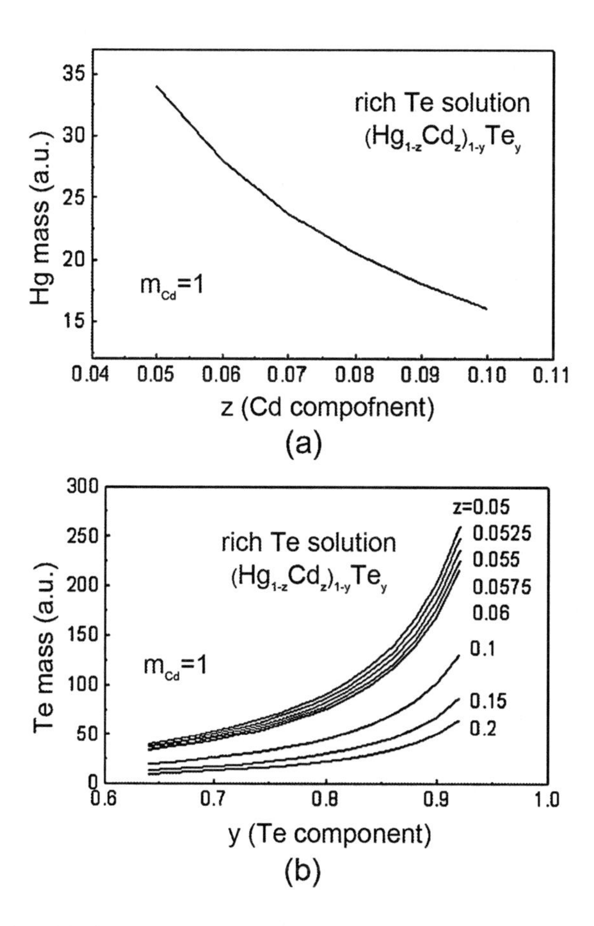

Fig. 2.28. Relationship between **(a)** m_{Hg} and **(b)** m_{Te} and liquidus compositions y and z in $(Hg_{1-z}Cd_z)_{1-y}Te_y$ ($m_{Cd} = 1$)

Here A_{Cd}, A_{Hg} and A_{Te} are the atomic masses of Cd, Hg and Te, respectively. Figure 2.28 shows the relationship between the masses of Hg and Te and the liquidus compositions y and z in $(Hg_{1-z}Cd_z)_{1-y}Te_y$.

Several empirical expressions have been given for the prediction of the approximate composition of the HgCdTe LPE layer based on the experimental data and phase diagrams (Harman 1980). Brice derived an equation for liquidus temperature as a function of liquidus compositions y and z as:

$$T_L = 1102 + 250z + 420yz - 785y \tag{2.102}$$

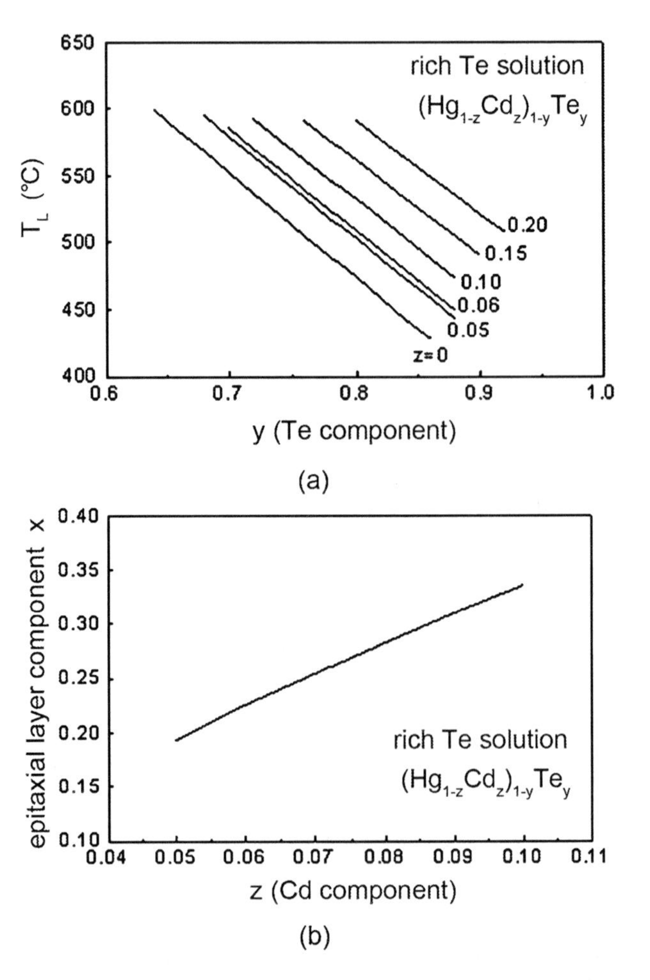

Fig. 2.29. Schematic diagram of (**a**) liquidus temperature and (**b**) solid composition as a function of mole fractions of Cd and Te in the solution

The solid composition is given by:

$$x = z/(0.220 + 0.780z) \tag{2.103}$$

This implies that the composition of the epitaxial layer, x, is mainly determined by the mole fraction of Cd. It is insensitive to the mole fraction of Te. Li et al. derived a quantitative expression for the segregation coefficient of Te-rich Hg–Cd–Te systems as a function of liquid compositions as well as liquid phase epitaxial growth conditions such as supercooling, cooling rate and growth time (Li et al. 1997).

Due to the high diffusivity of the cations in HgCdTe solids, the LPE growth temperature is always set at a relatively low value from 450 to 500°C. The mole fraction of Te is therefore set at ~0.8. Figure 2.29 shows the liquidus temperature and solid composition as a function of mole fractions of Cd and Te in the solution. It is seen that the composition of the epitaxial layer increases only with increasing Cd mole fraction. The liquidus temperature decreases with an increasing Te or a decreasing Cd mole fraction. If the growth temperature is 500°C, the y value will be smaller for LPE growth of long-wavelength HgCdTe and larger for mid- or short-wavelength HgCdTe.

2.3.2 LPE Process

LPE HgCdTe films can be produced through isothermal growth, equilibrium cooling and supersaturated cooling, as shown in Fig. 2.30.

Growth from a supersaturated solution, in the isothermal growth mode, is accomplished by cooling the melt to a temperature below the liquidus temperature (T_l), and subsequently introducing the substrate into the melt. The driving force for epitaxy is provided by super-saturation (supercooling) related to the temperature difference (T_l-T_g), where T_g is the growth temperature. Ideally, the substrate acts as the only region where nucleation takes place.

In equilibrium cooling, the substrate and the melt are brought into contact at or slightly above T_l. Growth is accomplished by decreasing the temperature of the melt and the substrate at a uniform rate. As the melt is cooled, the solution's concentration exceeds its equilibrium value causing epilayer formation on the substrate, which again, acts as the only nucleation site.

Finally, growth from a continuously cooled super-saturated solution can be performed. In this case, growth is obtained by cooling the melt and the substrate (which are not yet in contact) to a temperature below T_l, introducing the substrate into the melt, and further cooling the melt and substrate at a uniform rate.

In a typical growth, appropriate amounts of high purity Hg and Cd are reacted in the Te melt at ~700°C for an hour. For example, a typical ratio of CdTe:HgTe for x = 0.2 HgCdTe growth is 0.004;0.251:0.745. CdTe substrates are first lapped and chemically polished before loading in the growth chamber. Prior to growth, the growth solution is homogenized by stirring at 80 rpm with the temperature of the melt at 40°C above the liquidus temperature, T_l, for 30 min, and cooled at a rate of 0.5°C/min to a temperature of 1–2°C below T_l. The substrate is then inserted in a horizontal orientation into the melt and rotated at a constant rate of from 5

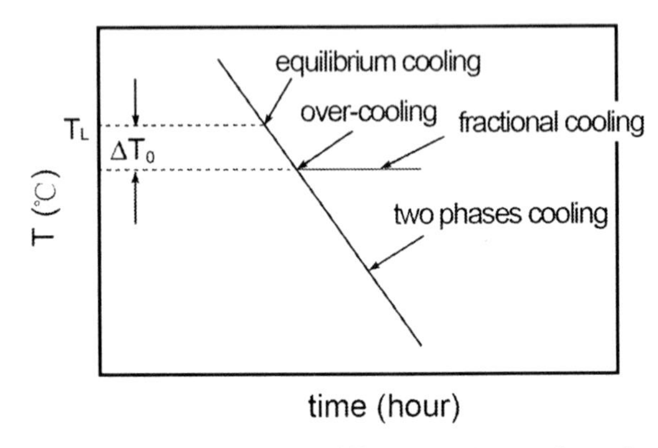

Fig. 2.30. The temperature profiles for different LPE growth methods. An *arrow* indicates the time to introduce the substrate into the melt

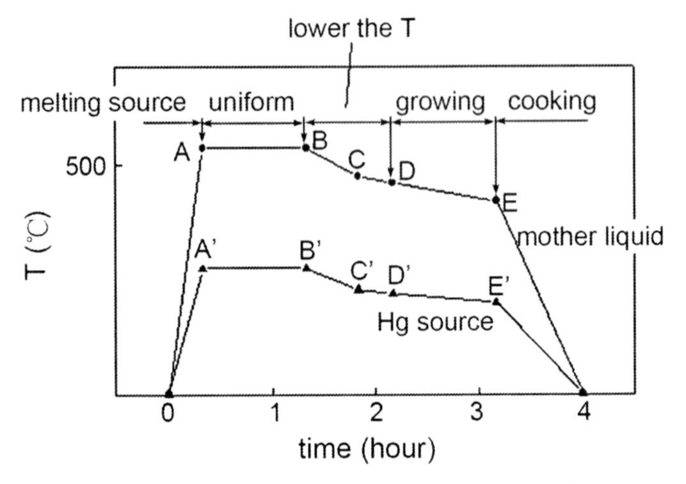

Fig. 2.31. Temperature profile for LPE growth

to 20 rpm, with a cooling rate of 0.05°C/min. At the end of a growth period of 40 min, the rotation is stopped; the substrate is tilted vertically and then removed from the solution. A temperature profile of LPE is shown in Fig. 2.31. This technique usually produces an epilayer free of solidified melt drops with a uniform composition and thickness.

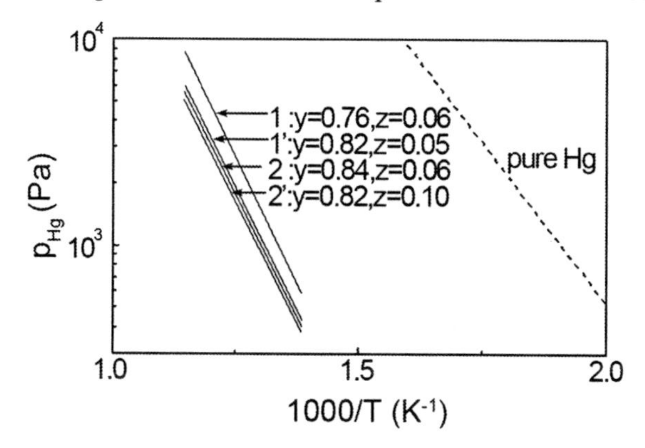

Fig. 2.32. Partial pressure of mercury as a function of temperature for various mole fractions y and z of liquids $(Hg_{1-z}Cd_z)_{1-y}Te_y$. Line 1, $y = 0.76$, $z = 0.06$; line 2, $y = 0.84$, $z = 0.06$; line 1′, $y = 0.82$, $z = 0.05$; line 2′, $y = 0.82$, $z = 0.10$; *dashed line*, vapor pressure of pure mercury

The key factors for LPE growth of HgCdTe films include the following:

(i) Equilibrium Hg Vapor Pressure

LPE from Te-rich melts has the advantage of reducing the Hg partial pressure and increasing the Cd solubility and thereby simplifying the apparatus. However, the difficulty of maintaining the equilibrium mercury vapor pressure during the LPE growth cycle is still a troublesome factor though the Hg pressure over the Te-rich solution is manageably small. This leads to an uncontrolled liquidus temperature and solution composition change that is detrimental to the quality of epilayers and run-to-run reproducibility. Many studies have discussed the effect of non-equilibrium mercury pressures on liquidus temperature and mole fraction of HgCdTe LPE (Brebrick et al. 1983), but no quantitative analytical results were given. Based on the associated solution model and phase diagram theory, Li et al. calculated the variation of the melting point and mole fraction of the growth solution caused by the non-equilibrium Hg pressure. The parameters in the model were determined by fitting to experimental data.

The results are useful in designing LPE growth strategies for $Hg_{1-x}Cd_xTe$ thin films (Li et al. 1995).

In order to maintain a constant solution composition during the Te-rich LPE growth of HgCdTe, consideration of the vapor–liquid relationship is necessary. Figure 2.32 shows the *P–T* phase diagram of the Te-rich corner derived from the associated solution model, i.e. the partial pressure of Hg versus reciprocal temperature for various liquid compositions. The *P–T* relation of pure elemental Hg is also plotted in this figure for comparison. It is seen that there is a change the *P–T* phase diagram slope between the pure Hg elemental case and the alloy melt with various liquid compositions. Therefore the cooling rate of the Hg reservoir must be selected to be different from that of the growth solution for the purpose of properly controlling the balance of Hg pressure above the melt during the LPE process (Li et al. 1995).

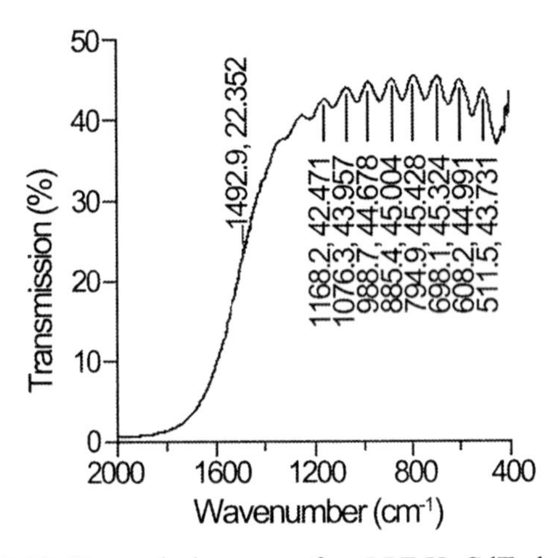

Fig. 2.33. Transmission map of an LPE HgCdTe layer

(ii) Growth Rate and Epilayer Thickness

The epilayer thickness can be derived from a mass transport equation (Hsieh 1974). For a super-cooled solution being further cooled by ramping the temperature, the growth rate and layer thickness relationship is given as (Harman 1980):

$$v(t) = K(\Delta T_s t^{-1/2}/2 + \beta t^{1/2}), \tag{2.104}$$

$$d(t) = K(\Delta T_s t^{1/2} + 2\beta t^{3/2}/3), \tag{2.105}$$

where K is a material constant, ΔT_S is the super-cooling temperature difference, β is the cooling rate during growth, and t is the growth time.

The layer thickness can be measured from its room temperature transmission spectrum. The prominent interference fringes shown in Fig. 2.33 are indicative of a flat CdTe/HgCdTe interface. The film thickness can be derived from the wave-number difference between two adjacent peaks based on the following equation:

$$d = (2n\Delta v)^{-1} \tag{2.106}$$

Here n is the refractive index of the HgCdTe epilayer. In Fig. 2.33 the average wave-number difference is 93.8 cm^{-1}, the epilayer thickness is then calculated as 15.3 μm. The growth parameters for this sample are β = 0.05°C/min, ΔT_S = 1.5°C, and t = 50 min. The material constant K is therefore deduced to be 0.7 μm/(°C $\mathrm{min}^{1/2}$).

(iii) Melt Homogenization

In order to achieve good lateral homogeneity of the composition and a high reproducibility of both the x value and the thickness, the complete liquefaction and homogenization of the source ingot is a necessary prerequisite. It is also important for obtaining good surface morphologies (Wermke et al. 1992).

A criterion for the optimization of the melt homogenization period can be found by calculating the layer thickness in the step-cooling mode (β = 0) of LPE according to (2.105). In order to determine the values of the material constant K and liquidus temperature T_l in (2.105), the duration of the homogenization period, during which the melt is mechanically stirred, has been systematically varied, while the melt composition and temperature of homogenization were kept constant. From at least two experiments for each homogenization time, performed with two different deposition temperatures, while maintaining the growth time constant, the data for K and T_l in (2.105) were determined. From the expressions:

$$d_1 = K(T_L - T_{d1})t^{1/2}, \tag{2.107}$$

and

$$d_2 = K(T_L - T_{d2})t^{1/2}, \tag{2.108}$$

one easily gets:

$$T_L = \frac{T_{d1}d_2 - T_{d2}d_1}{d_2 - d_1}, \tag{2.109}$$

and

$$K = \frac{d_2 - d_1}{(T_{d1} - T_{d2})t^{1/2}}. \tag{2.110}$$

Saturation of the material-specific parameter K can be taken as a criterion for the complete liquefaction and homogenization of the growth melt (Wermke et al. 1992). Following this criterion, conditions for proper source melt homogenization could be determined.

For the vertical dipping LPE system, the homogenization time t_{hom} is related to the stirring frequency f. The following empirical equation was derived by Winkler et al. (1991),

$$t_{\text{hom}} = 3 \times 10^4 \exp(-0.034 f) \tag{2.111}$$

In case where the mercury pressure over the growth melt is not balance during the liquefaction process, the homogenization time can be determined as: (Djuric et al. 1991)

$$t_{\text{hom}} = \frac{4L^2}{\pi^2 D_{Hg}} \ln\left[\frac{4}{(1 - k_N)\pi}\right] \tag{2.112}$$

where L is the height of the growth melt, D_{Hg} is the diffusivity of mercury atoms, and k_N is the ratio of mercury atoms at the bottom to that at the surface. The species in the melt will be balanced when $k_N = 0.9$. If $L =$ 4 mm, and $D_{Hg} = 5\times10^{-5}$ cm^2/s, the time for homogenization is 55 min.

(iv) Supercooling and Cooling Rates

Super-cooling and cooling rates are two important growth parameters for the LPE growth of $Hg_{1-x}Cd_xTe$. It is reported that super-cooling in large Te-rich growth solutions is small, typically 3°C, and that constitutional super-cooling increased with increasing cooling rate (Suh and Stervenon 1988; Wan 1987; Wan et al. 1986). The film grown with larger super-cooling and a faster cooling rate exhibits more terracing and melt pocket inclusions. This is because a high degree of super-cooling in large melts is erratic, which in turn leads to unstable growth with homogeneously precipitated HgCdTe crystals deposited on the film's surface. A fast cooling rate appears to allow more opportunity for melt diffusion in areas of initial surface unevenness or substrate contamination, by lowering the liquidus temperature for those compositions. Both are detrimental to the film's surface morphology (Suh and Stervenon 1988; Mullins and Sekerka 1964).

(v) Stirring Frequency

It is found that stirring during the LPE growth could provide morphological stability by stabilizing the growth plane interface (Li 1996). An LPE study was aimed at the impact of stirring on a phase transition under conditions where the confinement of the sample by its container had a negligible influence. Near a phase transition there is a correlation length over which the properties of the system vary only slowly in space, and this length diverges as the transition temperature is approached. Thus, the boundary layer adjacent to a solid surface acquires a macroscopic thickness. If the fluid phase adjacent to a boundary is of limited spatial extent, then the boundary layer will influence significantly the average macroscopic thermodynamic and transport properties of the system (Min 1982; Burton et al. 1953).

The impact of stirring rate on boundary layer thickness can be described as (Scheel and Elwell 1973):

$$\delta = 1.61 D^{1/3} \gamma^{1/6} \varpi^{-1/2} \tag{2.113}$$

where D is the diffusivity, and γ is the viscosity of the solution. A high stirring rate results in a thinner boundary layer. Since the liquidus temperature is not stable for a very large Hg–Cd–Te solution (Wan 1987; Wan et al. 1986), a thinner boundary layer may lead to easy nucleation during LPE growth.

Li et al. measured the compositional distributions of HgCdTe epilayers grown under different stirring rates by electron scanning probe analysis (Li 1996). It was found that the films grown from stirred melts exhibit more uniform compositions than those from unstirred melts. This occurs because the thin boundary layer of stirred solutions induces stable epitaxial growth (Scheel and Elwell 1973). Further evidence showed that the transverse compositions become non-uniform and the x values have a tendency to decrease from the centre to the periphery of the melt if the stirring frequencies are very large. This is likely to be a simple centrifuge effect, as the stirring rate increases the heavier Hg atoms are preferentially driven by centrifugal force to the outer edge of the melt. The appropriate stirring rates for LPE growth of $Hg_{1-x}Cd_xTe$ are in the range of 15–30 rpm (Li 1996; Wan et al. 1986; Parker et al. 1988; Hurle 1969).

(vi) Substrate

CdTe or CdZnTe are usually employed as substrates for HgCdTe epilayers, because many properties of these compounds match very closely, see Table 2.3. Also it is easier to grow a bulk CdTe crystal than a HgCdTe or HgTe crystal. It has been found that the Cd-terminated CdTe (111) surface is optimal for growing HgCdTe by LPE (Nemirovsky et al. 1982; Harman 1979, 1980) while the Te-terminated surface seems most suitable for the MBE technique.

Table 2.3. Physical parameters of HgCdTe important for heteroepitaxy on a CdTe or CdZnTe substrate

	Crystal structure	Lattice constant (Å)	Melting point (°C)	Coefficient of thermal expansion (K^{-1})	Bandgap (eV)
HgTe	Zincblende	6.461	673	4.0	–0.3
CdTe	Zincblende	6.481	1,092	4.9	1.6
ZnTe	Zincblende	6.100	1,298	8.3	2.26
$Cd_{1-z}Zn_zTe$	Zincblende	$6.481-0.381z$		$(4.9+3.4z)\times10^{-6}$	
$Hg_{1-x}Cd_xTe$	Zincblende	$6.461+0.020x$		$(4.0+0.9x)\times10^{-6}$	

The annealing pretreatment (Astles et al. 1988) or in situ cleaning using a Bi solution (Astles et al. 1987) may improve the quality of the substrate for epitaxial growth. Li (1996) and Parker et al. (1988) investigated the influence of substrate orientation on the growth mechanism by studying LPE layers grown on spherically shaped substrates. The experimental evidence indicated that there is also a critical orientation of the CdTe substrate for smooth HgCdTe film growth, whose value ranges from 1.2° to 2° off the (111) plane. Edwall et al. (1984) reported that flat epilayers were grown on the CdTe substrate misoriented by 1.5–2° off the (111) plane; this may be explained by a surface reconstruction theory that predicts a critical orientation of the substrate for preventing the formation of terraces during epitaxial growth.

2.3.3 Comparison of Different LPE Techniques

Among three LPE systems, tipping may not be suitable for a production-type process. However, it allows performing LPE in a closed evacuated ampoule. It avoids the loss of Hg, and consequently, results in an accurate control of Cd segregation for low x values (Mroczkowski and Vydyanath 1981).

Dipping is particularly well suited to the large-scale production of HgCdTe epilayers. Relatively large melts of 0.1 kg can be used to minimize the melt depletion effects. Therefore, many layers can be grown from the same melt. Melt rotation and agitation are possible, which ensure proper melt homogenization before the growth. The substrate can also be rotated to improve the compositional homogeneity of the layers across the entire substrate (Wan et al. 1986). Multiple slice dipping is also feasible in LPE growth of HgCdTe (Geibel et al. 1986).

The slider boat technique can be used to obtain clean and smooth surfaces and multilayer structures in a closely controlled and reproducible manner. The main disadvantage is the long cycle time of the process to form an epilayer, and thus a limited output, even if multi-substrate slider boats are used. Other problems are small melt volumes and losses of Hg from the melt.

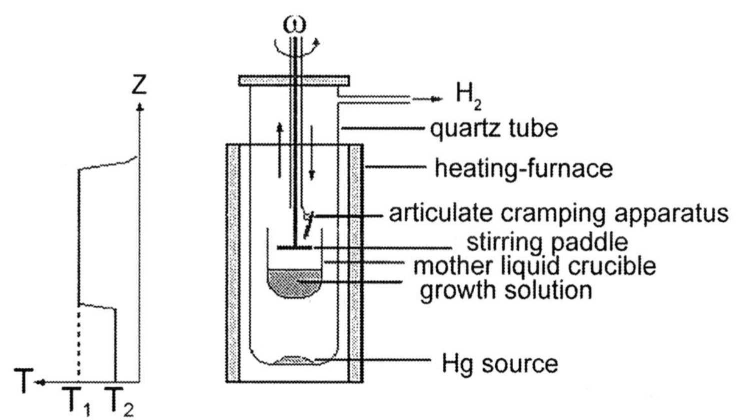

Fig. 2.34. Schematic diagram of the vertical dipping LPE reactor

Growth from Te-rich solutions produces layers with excellent compositional uniformity and a relatively narrow interface zone (~3 μm). Cd has a high solubility in Te (Tung 1981a). This allows small-volume melts to be used with the slider technique without appreciable depletion during a growth run (Wang et al. 1980; Nelson et al. 1980).

Several issues are of concern for Te-rich LPE growth of HgCdTe. The first is control of the equilibrium Hg partial vapor pressure over the growth solution (Bowers et al. 1980). Wang et al. reported a way to avoid Hg loss using a high pressure of Ar and H_2 gas (Wang et al. 1980). The first backside-illuminated HgCdTe photodiode was manufactured using this technique (Lanir et al. 1979). Harman et al. employed an additional Hg source to control the Hg composition in the growth solution (Harman 1981). Chiang et al. designed a TAC setup for their slider boat LPE

(Chiang et al. 1988). A dumbbell-like container with a 0.5 g liquid Hg source at both sides was used to control the equilibrium Hg pressure during LPE growth. A solid HgTe source can also be used to compensate for the Hg loss from the growth solution (Nagahama et al. 1984; Tranchart et al. 1985). The second issue is contamination of the growth solution due to the small surface tension of Te. Te drops residing on the sample result in defects and dislocations in the epilayer because of the thermal expansion mismatch between Te and HgCdTe (Castro and Tregilgas 1988). The solution contamination can be eliminated in a dipping apparatus where the sample is withdrawn in a vertical direction after LPE growth (Wan et al. 1986). Figure 2.34 shows the schematic diagram of a vertical dipping liquid-phase epitaxial growth system (Chen et al. 1993).

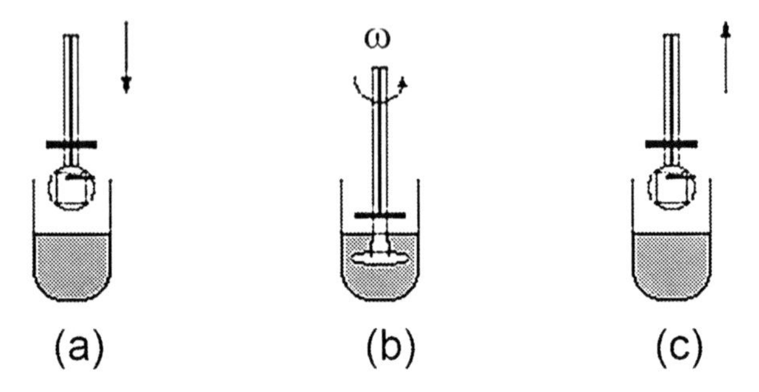

Fig. 2.35. Design of the substrate holder (**a**) dipping into the growth solution in a vertical orientation; (**b**) growth in a horizontal orientation; and (**c**) withdraw in a vertical orientation after LPE

The entire system consists of a container for the growth solution, a semi-transparent gold coated furnace enabling visual observation of the growth chamber, a Hg reservoir for Hg pressure control, and a substrate holder. The substrate holder material was designed to be less dense than the melt so that it floats to a horizontal position when it is dipped into the solution, rotates during growth and is withdrawn in a vertical position due to gravity as illustrated in Fig. 2.35.

Experiments with Hg-rich solvent LPE began in the late 70s, a few years after the initial work with Te-rich solutions. Because of the limited solubility of Cd in Hg, the volume of the Hg melts had to be much larger than that of Te melts in order to minimize melt depletion during layer growth. This precluded the slider growth approach. The mercury pressure is typical 8 atm for a growth temperature of about 500°C which makes open tube growth impossible (Bowers et al. 1980). Growth from Hg-rich solutions results in almost abrupt heterojunctions and the best surface morphology (Tung et al. 1987; Berry et al. 1986). Tung et al. developed an "infinite-melt" vertical LPE method (VLPE) for Hg-melt epitaxy (Tung et al. 1987). Extensive statistics of performance and producibility of the VLPE technology were accumulated to show its maturity and manufacturing readiness. Particularly, double-layer heterojunction (DLHJ) detectors were realized by the VLPE technology for high-performance second-generation focal plane arrays (Tung 1988; Kalisher et al. 1994).

The LPE growth from a HgTe-rich solution leads to uniform composition regions of the epilayer, but results in graded heterojunctions of about 20 μm width. This grading is due to the increased interdiffusion of Hg at the high substrate temperature required for growth. However, growth from a HgTe-rich solution has the advantage that it yields n- or p-type layers easily, with a variety of concentrations (Brown and Willoughby 1982).

2.3.4 Quality Control of HgCdTe LPE Films

In order to prepare large-size and high quality HgCdTe epitaxial films, researchers optimized the LPE parameters based on the phase diagram and their specific setup. Suh et al. investigated the relationship between the morphology of HgCdTe epilayers and cooling rates. The width of growth terraces is inversely proportional to the cooling rate (Suh and Stervenon 1988). Takami et al. studied the dependence of surface flatness on the temperature of melt homogenization and LPE growth. They found that the surface roughness and compositional non-uniformity increase with increasing of the homogenization or growth temperatures (Takami et al. 1992). The best temperature for both melt homogenization and LPE growth ranges from 480 to 500°C.

The epilayers grown from Te-rich solutions are always p-type with a high carrier concentration of 10^{17} cm^{-3}. A defect model where the Hg vacancy native acceptor defects are doubly ionized, and the concentration of native donor defects is negligible, was proposed for bulk and epitaxial $Hg_{0.8}Cd_{0.2}Te$ (Vydyanath 1981a; Vydyanath and Hiner 1989). However, annealing or doping may convert the layers to either weak p-type or n-type.

Vydyanath et al. did Hall-effect measurements on undoped $Hg_{0.8}Cd_{0.2}Te$ films which were grown from Te-rich solutions and were annealed at temperatures varying from 150 to 500°C. Within the phase boundary limits of the material, the hole concentration in the films was found to increase with a decrease in the partial pressure of Hg, in agreement with the behavior of undoped bulk $Hg_{0.8}Cd_{0.2}Te$ crystals at temperature in the range of 400–655°C (Vydyanath and Hiner 1989). The conversion process of sample conductivity that occurs when annealing the sample in a Hg vapor can be understood by assuming that vacancies on the cation sublattice act as ionized acceptors, and that vacancies on the anion sub-lattice act as ionized donors. However this is not the complete picture. While Te vacancies are donors, they are deep donors and because their formation energies are high, their densities are low. The more important native defect donors on the Hg rich side of the existence region are Hg interstitials (Berding et al. 1992), rather than Te vacancies. Annealing of the p-type samples after growth in a Hg-atmosphere reduces the metal vacancy concentration to a level below that of residual extrinsic n-type background impurities in the sample, thus converting the conductivity from p-type to n-type (Capper 1994). Various processes, such as two-step annealing (Chandra et al. 1991), Indium film coating (Koppel et al. 1989), have been developed to improve the electrical properties of annealed samples. In general, the annealing temperature is higher than 350°C for the conductivity conversion from strong p-type to weak n-type. To complete the conversion to n-type, the annealing temperature is typically less than 300°C (Destéfanis 1988; Astles et al. 1988; Chandra et al. 1991; Nouruzi-Khorasani et al. 1989).

Although in the doping of HgCdTe is possible to employ the native mercury vacancies, this has several undesirable traits. These include short Shockley-Read-Hall lifetimes associated with vacancy state mediated recombination (Cheng 1985; Schaaka et al. 1985; Jones et al. 1985). This pointed to the need for an extrinsic dopant. Elements from column I, such as copper and gold, are effective p-type dopants located on interstitial sites, however, they diffuse easily within the crystal lattice which results in the destruction of accurate and abrupt doping profiles. Elements from column VI, such as arsenic and antimony, when located on anion sites are acceptors and do not suffer from the diffusivity problem to the same extent. However, they are not always electrically active because on cation sites they are donors, and often behave as amphoteric dopants (Vydyanath et al. 1987; Harman 1993; Astles et al. 1993). To get them to locate properly they must be introduced in a growth process that has a high Hg partial pressure, e.g. LPE from a Hg melt, to force them onto the anion sublattice.

The electrical properties of the HgCdTe epilayers are also affected by substrate impurities. During the epitaxy and/or annealing processes, impurities such as Li, Na etc. in the substrate may accumulate at the surface of the substrate followed by out-diffusing into the epitaxial films. Techniques to eliminate the unintended impurities from substrates include source purification, in situ cleaning with a Bi solution (Astles et al. 1987), surface modification for enhanced impurity gettering (Parthier et al. 1991), low temperature LPE, and the growth of a buffer layer prior to HgCdTe deposition (Chiang and Wu 1989).

2.4 Molecular Beam Epitaxy Growth of Thin Films

HgCdTe thin films were successfully fabricated for the first time with a MBE method by J.P. Faurie in 1981. Since then, the MBE technique has been improved greatly (Arias et al. 1993, 1994; Brill et al. 2001; He et al. 1998, 2000). Wu et al. (Wu 1993; Lyon et al. 1996) and Faurie's group (Almeida 1995) systematically investigated the growth of HgCdTe thin films on Si substrates by the MBE method. Recently, it was reported by Varesi et al. (2001) that a multilayer structure (ZnTe, CdTe/In-doped HgCdTe/CdTe) on 4 in. Si substrates had been successfully fabricated with an MBE technique. In this structure, the ZnTe and CdTe served as buffer layers, and the last CdTe layer acts as passivation layer. Then 640×480 elements infrared focal plane arrays were fabricated on this structure. In this section, we give a brief introduction to the epitaxial mechanism, growing process, characterization methods, and several other critical issues and key techniques in the process of depositing narrow-gap semiconductor HgCdTe thin films using the MBE technique. In addition, a brief overview of the current research and development activities on the MBE technique will be presented.

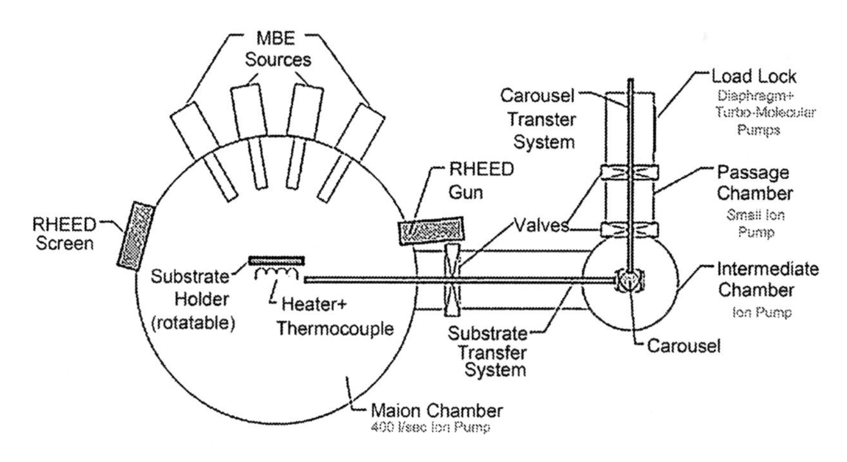

Fig. 2.36. Growth chamber of MBE system

2.4.1 Overview of the Molecular Beam Epitaxy Process

Special equipment is required for growing thin films with MBE techniques. The MBE systems usually consist of three main parts: a growth chamber, a transfer chamber, and a load/lock chamber. The load/lock chamber, used for loading and unloading wafers, is pumped down by an ion pump to pressures as low as 10^{-10} Torr. Gases adsorbed on the substrates are removed in this load/lock chamber before the substrates are transferred to the transfer chamber. Similarly, the transfer chamber is also pumped down to vacuum of 10^{-10} Torr with an ion pump. The transfer chamber is equipped with a four-pole mass spectrometer. The whole system needs to be baked for a long time to reach these low pressures. During this process, the composition of residual gases is analyzed by the four-pole mass spectrometer to ensure that the crucial growth requirements are satisfied. The deposition chamber shown in Fig. 2.36 is the core part of the MBE system. It comprises source ovens, a cooling system, a wafer holder, a substrate heater, monitoring systems and control systems.

An ion pump and a cold bath maintain the ultrahigh vacuum in the growth chamber. The pressure is held at around 10^{-7} Torr during growth. A window facing the wafer holder is installed through which a radiation thermometer can view the wafer. Two strain-free quartz view-ports are located on both sides of the growth chamber, at an angle of 70° with the normal to the substrate. They are used for mounting the optical elements of an ellipsometer. The wafer holder can be rotated and a thermocouple is located at its center. During HgCdTe growth, a specially designed Hg source cell delivers Hg bearing material into the source oven to be heated up during a growth run. The remaining material is retracted back into the cell after completion of a growth run. The starting materials, Hg(7N), Te_2(7N) and CdTe(7N), are of high purity. This ultrahigh vacuum chamber yields low impurity films.

The starting materials used for growing HgCdTe epilayer by MBE are separate CdTe, Te_2, and Hg sources, respectively. The Te source supplies Te_2 gas. The requirement for decomposition of CdTe(s) molecules on the surface determines the lowest temperature for growing films with correct stoichiometry. The decomposition mode of CdTe(s) is:

$$\mathrm{CdTe}(s) \rightarrow \mathrm{Cd}(g) + \frac{1}{2}\mathrm{Te}_2(g) \qquad (2.114)$$

Usually the flux ratio J_{Te_2}/J_{Cd} is 0.5, and is independent of time. The composition x of $Hg_{1-x}Cd_xTe$ is determined by substrate temperature and the flux ratio of J_{CdTe}/J_{Te_2}, where x is given by:

$$x = \frac{K_{\mathrm{CdTe}}(T_m) \times J_{\mathrm{CdTe}}}{[K_{\mathrm{CdTe}}(T_m) \times J_{\mathrm{CdTe}} + K_{Hg}(T_m) \times J_{Hg}]} \tag{2.115}$$

where $K_{\mathrm{CdTe}}(T_m)$, $K_{\mathrm{Te2}}(T_m)$, and $K_{\mathrm{Hg}}(T_m)$ are sticking coefficients of CdTe, Te_2, and Hg respectively, at substrate temperature T_{m}. In most cases, the CdTe and Te_2 sticking coefficients can be treated as 1 when the growth surface is a (111) B face (the anion terminated plane) and the growth temperature is low ($T_{\mathrm{m}} < 200$). The Hg sticking coefficient is small, which necessitates a high Hg flux. During the growth process, the flux ratio J_{Hg}/J_{Te} is maintained at around 110. Instability of Hg source flux often leads to composition gradients normal to the growth surface.

The time sequence of HgCdTe MBE growth consists of three steps: substrate preparation, HgCdTe epilayer growth, and post-growth treatment. Each step is made up of many sub-steps. For instance, substrate preparation includes choosing, polishing, cleaning, etching, and mounting substrate as well as degassing the substrate. The growth steps include substrate temperature control, and source flux control. The post-growth treatment step includes wafer picking, and annealing. All these steps comprise a whole MBE growth process, and poor performance in any of them will lead to a reduction of film quality. Hence, to ensure the reproducibility of the wafers, each step needs to be performed precisely following its specified procedures. These procedures are optimized based on a large number of experiments. For example, to establish the standard procedure for substrate preparation, spectroscopic ellipsometry is employed to monitor the preparation process, and establish a reliable database. Then a microscope is used to observe the surface morphology of the epilayer grown on this substrate. By analyzing the relation between preparation condition parameters and the observed morphology of the grown epilayers, optimized parameters for substrate preparation can be established. Similarly, to establish the surface oxide removal process, from RHEED data collected from dozens of experiments under different conditions, we find the optimized temperature and time for removing oxygen from different substrates. Thereafter, this process can be precisely performed even without RHEED monitoring.

The MBE growth process involves the following issues: the adsorption of constituent atoms or molecules on the surface, their surface migration, the reactions among the surface species and the substrate to nucleate growth. The atomic/molecular species ability to remain on the substrate after hitting it is indexed by sticking coefficients. Usually, the sticking coefficient decreases as the temperature increases in useful temperature regimes. For example, during GaAs binary compound MBE growth, the anion As^{3-} ions stick readily, so the growth rate is mainly determined by the cation (Ga^{3+}) flux. It is much more complicated to determine the adsorption process in a pseudo-binary compound's MBE growth, such as HgCdTe (Smith and Pickhardt 1975).

In MBE growth, the process control parameters include all source oven temperatures, and the substrate temperature. Usually, the temperatures of CdTe, Hg, and Te source ovens are about 500, 100, and 390°C, respectively. The substrate temperature is ~280°C for growing a CdTe buffer layer, whereas the substrate temperature is kept to ~180°C during the growth of the HgCdTe layers. The temperature of source ovens can be accurately controlled, which stabilizes their source flux. It is much more difficult to control the substrate temperature. This is mainly due to the growth being a dynamic process in which the energy lost due to radiation at each moment cannot be compensated instantly.

MBE growth is a non-equilibrium process, which is dominated by molecular fluxes and reactive kinetics. For a given substrate material, crystal axis orientation including any off axis choice, the growth rate of HgCdTe is determined mainly by the substrate temperature, ultrahigh vacuum, and source flux densities.

The commonly used substrates are CdZnTe, or Si and GaAs with ZnTe buffer layers, for example, undoped semi-insulating GaAs (211) B substrates whose dislocation density is less than $5\times10^{4}\ cm^{-2}$. The substrates need to be ultrasonically cleaned in organic solvents to get rid of the organic contamination on the surface, followed by a rinse in running deionized water, and then dehydrated by concentrated H_2SO_4 acid. After that, the substrates are etched in a solution with a composition of $H_2SO_4/H_2O_2/H_2O$ in the ratios 5/1/1 for 3 min, then etched by HCl acid for 1 min. The prepared substrates are mounted on Mo-blocks with indium bonding, and then loaded into the load/lock chamber in a N_2 gas atmosphere. The orientation of the substrate has a great influence on the film growth. The sticking coefficient of Hg is small when growing on smooth (001) plane, therefore a large flux ratio of J_{Hg}/J_{Te2} is needed, and this makes it difficult to obtain intrinsic p-type MCT epilayers. Compared to the (001) plane, there are more steps in a (111) plane. The growth rate on a (111) plane is therefore larger. A disadvantage of growth on (111) planes is it is hard to avoid twin crystal structures. The (211) B face has an intermediate step density. It not only enables relatively high growth rates, but also suppresses twinning and micro defects. So far it is the best substrate crystal orientation for MBE growth of HgCdTe (MCT).

The oxygen removal from the substrate is performed in the growth chamber at an elevated temperature. This process is monitored by an instrument using reflected high-energy electron diffraction (RHEED). The temperature chosen for removing oxygen plays an important role in the quality of the epilayers. A CdTe buffer layer that is 4–6 μm thick is grown first to reduce the lattice mismatched misfit dislocations between HgCdTe and GaAs. Sometimes the buffer layer consists of a thin ZnTe layer followed by a CdZnTe alloy layer that is graded to a composition at its surface that exactly lattice matches the planned HgCdTe layer. RHEED, IR radiation thermometers, and ellipsometry are employed to monitor the compositional variation in real time during the HgCdTe growth. However, even under the best circumstances it has never been possible on GaAs substrates to grow $Hg_{1-x}Cd_xTe$ active layers that have dislocation densities as low as those obtained on properly chosen $Cd_{1-y}Zn_yTe$ substrates.

In the process of MCT (211) growth, it is quite common to achieve MCT films with different stoichiometry by varying the growth temperature and the Hg/Te flux ratio, hence to realize either an n-type or p-type semiconductor compound. Mostly, MCT obtained at low growth temperatures are n-type.

Carrier densities, as low as 5×10^{14} cm^{-3} at 77 K have been achieved. These low carrier densities are due to uncontrolled residual donors. It is also possible to get higher n-type material by in situ doping. For p-type doping, the most commonly used dopant is As. As we discussed before As has two doping behaviors, on the cation sublattice it is a donor while on the anion sublattice it is an acceptor. The likelihood of As taking a cation site is very large when growth proceeds in a Te-rich environment. Gas phase epitaxy of MCT is always done on the Te rich side of the alloy's existence region. If one attempts to increase the Hg flux to reach a Hg rich situation the surface morphology and gross bulk defects degrade to a point were the material is useless. To prevent this tendency, one approach is to grow CdTe/HgTe superlattice structures. One can get As to go onto the anion site in CdTe. Thus a delta doped system can be prepared with heavily p-doped CdTe layers, concentrations ~10^{16}–10^{18} cm^{-3}, and lightly n-doped or intrinsic HgTe layers. Both modulated doping and direct doping obtained through post growth annealing have been done to fabricate p-type material.

The composition of HgCdTe is very difficult to control. Most commonly, control of the composition within tolerances is achieved by constant adjustment of the Cd/Te flux ratio. The substrate temperature is the other parameter difficult to control during growth. By using an IR radiation thermometer for in situ monitoring and calibrating the variation of the HgCdTe surface emissivity with layer thickness, feed back can be adjusted to control the substrate temperature.

The growth temperature range of MBE MCT films is very narrow, only about 10°. The sticking coefficient of Hg increases as the temperature decreases. When the growth temperature is slightly lower than its optimal value, excess Hg sticks on the surface, and the resulting micro twin crystals can be easily seen in RHEED image. The overall defect density on the surface is ~10^6–10^7 cm^{-2} under such growth condition. However, when the growth temperature is slightly higher than its optimal value, a Hg-deficiency occurs on the growth front, which results in voids in the film. Therefore, the growth temperature is a key parameter for MBE MCT growth. Many labs add one more IR radiation thermometer in addition to a thermocouple to enhance their temperature control ability.

Composition control is achieved by adjusting the temperature of the CdTe source oven. However, this method is not an in situ control method. It can only be established by measuring the composition of a number of as-grown films and then adjusting the CdTe source oven temperature from the resulting calibration data set. When growing multilayer heterostructures, it is easy to get accurate compositions for the first layer, but it is difficult to get accurate compositions of other layers. In order to in situ monitor the composition variation during the HgCdTe growth, it is necessary to employ a variety of methods.

Among the different epitaxy techniques, it is only in MBE that many methods can be employed to monitor film growth in real time. All the monitoring set-ups are attached to the ultrahigh vacuum chambers. They are used to study the properties of materials from different perspectives. In modern multi-chamber MBE systems, many monitors are installed. These commonly include:

- Reflection high-energy electron diffraction (RHEED), one of the technique used to study surface structure and morphology;
- Auger electron spectroscopy (AES), used to analyze the chemical composition of both substrate and epilayer as well as their structural depth profile;
- Secondary ion mass spectroscopy (SIMS), used for analyzing the chemical composition of monolayers on the front surface of a film, which provides very high mass resolution for most elements;
- X-ray electron spectroscopy (XPS) and ultraviolet photo-electronic spectroscopy (UPS), used to determine the electron cloud distribution and band structure of an epilayer;
- Ellipsometry has been developed for in situ monitoring of layer thickness and surface morphology.

Among these analysis methods, RHEED, UPS, and ellipsometry can be used in real time, while the other methods can only be employed after film growth.

2.4.2 Reflection High-Energy Electron Diffraction (RHEED)

The reflection high-energy electron diffraction (RHEED) technique serves as a primary characterization tool in monitoring the MBE growth process. Through RHEED, it is possible to study the crystallization kinetics and surface structure as well as morphology: (1) the surface structure morphology of materials can be extracted from the locations and shapes of diffraction streaks in a RHEED image; (2) the intensity oscillation in a RHEED image can help us to understand growth mechanisms and the kinetics of the MBE process, since these oscillations and patterns reflect the film growth kinetics.

The energy of electrons generated by an electron gun used for RHEED analysis is typically in the 10–40 K eV range, corresponding to a de Broglie wavelength of 0.17–0.06 Å, which is much smaller than the lattice parameters. Diffraction patterns, i.e. RHEED images, are formed in special directions when the electron beam is incident on the sample at a grazing angle ($0 < \alpha < 5$). Then the periodic lattice structure of sample acts as an optical grating. The electron diffraction intensity is related to the lattice structure and the diffraction direction. According to diffraction theory, the intensity of a diffracted beam reaches its maximum value when it satisfies the Laue diffraction equation. In the RHEED method because the beam of high-energy electrons is directed onto a sample's surface at grazing incidence, the normal momentum component is very small, which minimizes the penetration of the electron beam to a few atomic monolayers. That enhances the role of the film surface.

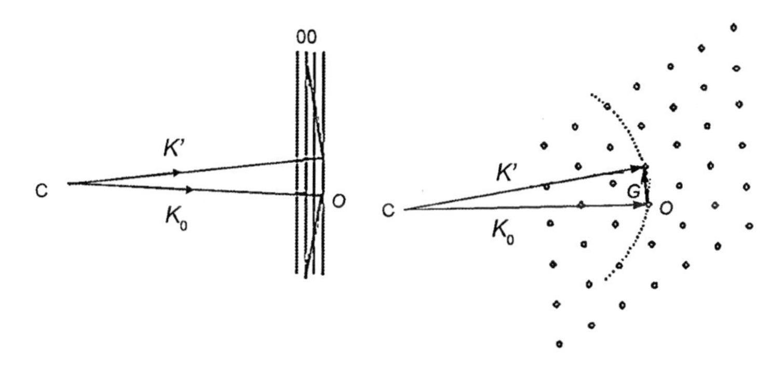

Fig. 2.37. An Ewald Pattern

Usually the Ewald sphere in reciprocal space, shown in Fig. 2.37, is used to describe RHEED images. The wave vector $\mathbf{k}_0$ is drawn in the direction of the incident electron beam, and the origin point C is chosen so that $\mathbf{k}_0$ terminates at some reciprocal lattice point O. The length of CO is the amplitude of $\mathbf{k}_0$, with $[CO] = K_0 = 2\pi/\lambda$. The sphere with radius K_0 centered at point C, is called the Ewald sphere. A diffracted beam will be formed if this sphere intersects any other point in the reciprocal lattice. The surface structure of a flat epilayer is in general a low-dimensional periodic net. This does not necessarily mean that all its atoms lie in a plane, but rather that the structure is periodic only in two dimensions. In the physics of surfaces it is common to speak of a two-dimensional net. The reciprocal net points of a two-dimensional periodic net may be thought of in three dimensions, as rods, because there is no limit along its third axis. Hence the reciprocal lattice points along this axis are moved closer together and in the limit form a rod. The rods are infinite in extent and normal the surface plane, where they pass through the reciprocal net points. Diffraction occurs in the directions from point C to the points where the Ewald sphere intercepts reciprocal net rods, because the Laue diffraction equation is satisfied in these directions.

The back-scattered electron beams are very weak, whereas the forward elastic scattered beams are strong. Diffraction patterns, i.e. RHEED images, are formed on a phosphor screen that is placed across the forward outgoing electron beams. The diffracted electron beam is in the direction $\mathbf{K}' = \mathbf{K} + \mathbf{G}$, where $\mathbf{G}$ is a reciprocal lattice vector.

The radius $\mathbf{K}_0$ of the Ewald sphere for 5–10 K eV electrons will be in the range of 36~105 Å^{-1}. Given the lattice constant of 6 Å, the reciprocal lattice vector $\boldsymbol{2\pi/a}$ is ~1 Å^{-1}. It is obvious that the radius of the Ewald sphere is much longer than the separation between adjacent rods of the reciprocal net. It follows that the Ewald sphere construction will yield a nearly flat surface in the central scattering region. The interception of the rods of the reciprocal net with the nearly flat surface will be a line when the beam is directed at grazing incidence. Considering the thermal vibrations of the lattice, non-collimation and non-monochrome nature of the incident electron beam, the lines become sticks with meaningful diameters. Thus, the observed RHEED image is composed of streaks with widths.

An ideal rod of a reciprocal net is a real one-dimension rod. If the crystal surface deviates from its normal ordering in a certain direction, the average length of the ordering range in this direction will be limited. Hence the rods of the reciprocal net become indefinite in this direction.

Therefore, the RHEED images also provide information about stacking defects on the crystal surface. For example, the RHEED image shows a diploid spot pattern if twins exist on the crystal surface, because in this case the surface structure can be thought of as consisting of two reciprocal net sets. If the crystal surface is polycrystalline, the number of the reciprocal net sets will be very large, resulting in a "polycrystalline ring" pattern in the RHEED image. Thus the RHEED system provides a tool for a qualitative measure of surface morphology and for monitoring the surface structure.

Three processes can be identified from the changes of RHEED images during MBE HgCdTe growth.

- The RHEED image shows a spotty pattern for several seconds after the CdTe source shutter is opened. At this moment, the space lattice of the GaAs (211) B substrate disappears, the film is growing in three-dimensions, and the surface is not flat.
- The RHEED pattern changes into streak lines after 10 min of epitaxial growth. At this stage, the film is beginning to grow in successive two-dimensional layers.
- The streak lines of the RHEED image become narrow and longer after HgCdTe growth is completed.

2.4.3 Monitoring the Growth Temperature

Precise control of the temperatures of the substrate and Hg, CdTe, and Te source ovens is a key HgCdTe MBE growth issue. The stability of these temperatures determines the crystal composition and quality of the final MCT layers. Among these temperatures, the substrate temperature is especially important. However, the substrate temperature is surprisingly difficult to control. By analyzing the IR radiation from the surface of the growing film and comparing it to calibration temperature data, as discussed earlier, the substrate temperature can be deduced. Wang and He (Wang 1996) discovered a rule for the temperature variation, and using it enhanced the precision of the temperature control.

The growth temperature variation also affects crystal quality of MBE HgCdTe. The sticking coefficient of Hg gets smaller and the Hg atoms are easily desorbed from the surface when the growth temperature is higher than its optimal value. As a result at these higher temperatures, there is Hg-deficiency and Te inclusions appear in the structure. When this happens, polycrystalline material is quite often formed. If the growth temperature is slightly lower than the optimal value, Hg enrichment occurs. The excess

Hg atoms will react with the Te atoms to nucleate new mesas, rather than adding to the existing one; this may result in twin crystal structures on the surface. These twin structures degrade the performance of devices. Moreover, the variation of the substrate temperature induces changes of the sticking coefficients of the incident species, resulting in stoichiometry changes. This may also lead to a compositional inhomogeneity through the depth of a layer.

The substrate temperature control usually is realized through a feedback circuit. In one method a signal measured by a thermocouple is fed back to a temperature controller, which responds to modify the applied voltage or current supplied to the heater. In general, the thermocouple is put in the center of the wafer holder. There are two contact modes between the thermocouple and the molybdenum block that houses the heater elements: rotation and non-rotation contact. In the rotation contact mode, there is a gap between thermocouple and molybdenum block to prevent the molybdenum block from touching the thermocouple during rotation. Therefore, there is a small difference between the measured temperature at the back of the substrate holder and the real temperature of the molybdenum block. In a non-rotation contact mode, to ensure good thermal contact between the molybdenum block and thermocouple, a low melting metal gallium layer is used to bond the thermocouple to the molybdenum block, so that the real temperature of the molybdenum is measured directly. Even in the non-rotation mode, these measured temperature values are not the actual temperature of the film's surface or even the substrate's surface.

In order to control the substrate surface temperature accurately, it is necessary to obtain its real temperature. To achieve this goal, non-contacting temperature measurement methods, such as radiation thermometers, are widely used in MBE equipment, especially when rotating the substrate. According to their technical type, radiation thermometers can be classified as: full waveband, single narrow waveband, or ratio thermometers, and thermal imagers. Full waveband and single waveband radiation thermometers are classified according to whether they collect and measure thermal radiation across the full IR spectral band, or in a single narrow spectral region. Ratio thermometers measure the thermal radiation in two or more wavebands simultaneously, and the temperature is extracted from measured radiation intensity ratios. Most MBE systems today include an IR transparent window through which the IR radiation thermometer observes the substrate.

In MBE growth of III–V materials, substrate temperature measurements with IR radiation thermometers were studied in order to improve the accuracy and reliability of the measured temperatures. Neuhaus (1991) worked on reducing the measurement deviation caused by contamination of the optical window. Inoue (1991) used a multiple wavelength technique to automatically monitor and adjust the radiation intensities to resolve this problem. In HgCdTe MBE growth, researchers met the same problems, i.e. window contamination, radiation thermometer reliability, and the oscillation of radiation produced by the growing HgCdTe surface layers. These problems can be resolved by using similar methods to those used in III–V MBE techniques (Bevan et al. 1996).

In general, the IR radiation emitted from a surface is a function of its material species, structure parameters, etc. Because of its complexity, Wang (1996) systematically studied this relationship by computer simulations leading to a thorough investigation of the temperature measurement of epilayers.

Let us do a brief review of the principles of IR radiation underlying their use as thermometers. The radiation spectrum $M_b(T,\lambda)$ of a blackbody, the power per unit volume, at temperature T and wavelength λ, is given by the Planck equation:

$$M_b(T,\lambda)=\frac{2\pi hc^2}{\lambda^5\left[\exp\left(\frac{hc}{\lambda k_B T}\right)-1\right]} \tag{2.116}$$

where c is the velocity of light and h is the Planck constant. The integral of (2.116) over the full wavelength region gives the radiant intensity $M(T)$, the radiation power per unit area. It can be written as:

$$M(T)=\int_0^\infty M_b(T,\lambda)d\lambda \tag{2.117}$$

This integral yields a relation between the temperature and the net radiant intensity, the Stefan–Boltzmann law:

$$M(T)=\sigma T^4 \tag{2.118}$$

where σ, called the Stefan–Boltzmann constant is 5.67×10^{-8} W m^{-2}K^{-4}. The wavelength at the peak intensity of the spectrum shifts with temperature following Wien's Law:

$$\lambda_p T = const. \tag{2.119}$$

where λ_p is the wavelength at the peak of $M(T,\lambda)$. This value moves towards shorter wavelengths as the temperature increases.

The radiant intensity $M_b(\lambda,T)$ of an ideal blackbody is a function of only wavelength λ and temperature T. However, the radiant intensity spectrum of most objects is not only a function of wavelength and temperature, but also other parameters related to their own properties. The radiation intensity spectrum emitted by some objects is close to that of an ideal blackbody, but most objects are far from ideal blackbodies. The emissivity $\varepsilon(\lambda,T)$, defined as the ratio of the radiation intensity $M(\lambda,T)$ of the real object and that $M_b(\lambda,T)$ of an ideal blackbody at the same temperature T, can be used to characterize the emission from real bodies:

$$\varepsilon(\lambda,T)=\frac{M(\lambda,T)}{M_b(\lambda,T)} \tag{2.120}$$

When radiation is incident on a body a fraction of it is reflected, called the reflectance R; a fraction is absorbed, the absorbance A: and a fraction is transmitted, the transmittance T. According to energy conservation, the relationship among A, R, and T is:

$$R+A+T=1 \tag{2.121}$$

When Kirchhoff was studying thermal radiation in 1859 he pointed out that the absorbance $A(\lambda,T)$ of a material must be equal to its emissivity $\varepsilon(\lambda,T)$:

$$\varepsilon(\lambda,T)=A(\lambda,T) \tag{2.122}$$

Therefore, a material with a high absorbance also has a high emissivity. For an ideal blackbody, the absorbance equals to 1, and so does its emissivity $(\varepsilon = A = 1)$, so its reflectance R and transmittance T are zero.

The radiation properties become complicated when a surface layer of different properties coats the object. This is the situation when an epilayer coats a substrate. The radiant intensity of an epilayer is related to its surface emissivity ε_{m}. Assume the surface temperature of the epilayer is T_s, its thickness is d, and the transmittance, reflectance, and absorbance are $T(d,\lambda,T_s)$, $R(d,\lambda,T_s)$, and $A(d,\lambda,T_s)$ at wavelength λ. The quantities $T(d,\lambda,T_s)$, $R(d,\lambda,T_s)$, and $A(d,\lambda,T_s)$ can be measured. Their theory is related ideally to the material's absorption coefficient and dielectric constant; that in turn are calculated from band structure theory and statistical mechanics. Non-ideal effects, like those arising from surface roughness, also have to be included at times.

The surface emissivity of the epilayer, ε_m, can be treated as a function of only wavelength and temperature, $\varepsilon_m = \varepsilon_m(\lambda,T_s)$. Then the radiant intensity generated and emitted from the epilayer is:

$$M_{\mathrm{LPE}}(\lambda,T_{\mathrm{LPE}})=M_b(\lambda,T_{\mathrm{LPE}})\varepsilon_m(\lambda,T_{\mathrm{LPE}})=M_b(\lambda,T_{\mathrm{LPE}})A(d,\lambda,T_{\mathrm{LPE}}) \tag{2.123}$$

Beside the radiation from the epilayer, radiation generated by the substrate is also emitted. If the radiant intensity of the substrate is $M_{sb}(T_{sb},\lambda)$ at temperature T_{sb}, the radiant intensity transmitted through the epilayer is:

$$M'_{sb}(d,\lambda,T_{\mathrm{LPE}},T_{sb}) = T(d,\lambda,T_{\mathrm{LPE}})M(T_{sb},\lambda) \tag{2.124}$$

The integral over the wavelength range $0 \le \lambda \le \infty$ gives the radiant intensity over the whole spectral interval. If there is a full waveband radiation thermometer installed in the direction normal to the epilayer, the detected radiant intensity will be the total radiation from the epilayer, plus the transmitted radiation through the epilayer from the substrate. The display emissivity $E(d,T_s,T)$ from the surface over the full wavelength region can be written:

$$E(d,T_{\mathrm{LPE}},T_{sb}) = \frac{M_{\mathrm{LPE}}(\lambda,T_{\mathrm{LPE}}) + M'_{sb}(d,\lambda,T_{\mathrm{LPE}},T_{sb})}{\sigma T_{\mathrm{LPE}}^4} \tag{2.125}$$

It is also necessary to consider the radiation emitted from the source ovens during MBE growth. This radiation affects the temperature of the wafer surface. These emissions will also be transmitted, reflected, and absorbed by the growing epilayer. All the radiation must be taken into account when calculating the radiant intensity that reaches the radiation

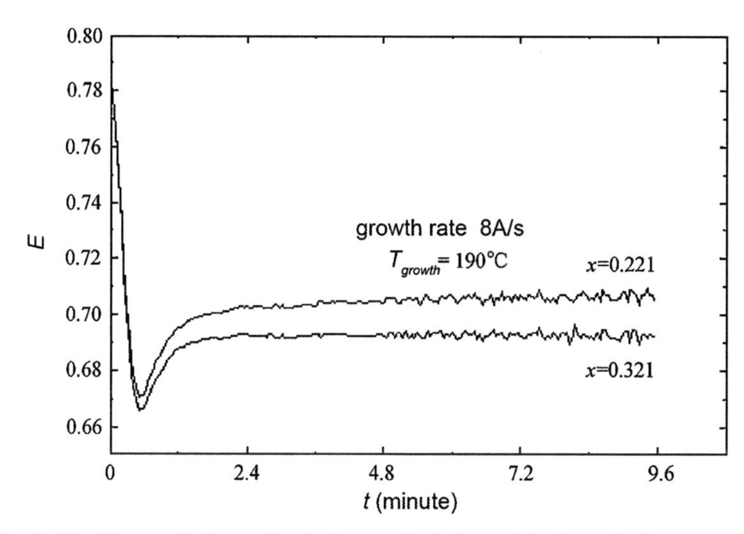

Fig. 2.38. The variation of relative emissivity with time during HgCdTe MBE growing

thermometer. If the source ovens are off axis and the epilayer's reflectivity is specular, the effect of these extra sources can be minimized by narrowing the field of view of the thermometers. However, if the epilayer surface is rough so its reflection is somewhat diffuse, then their effect must be taken into account. In this case, the radiation intensity measured by the radiation thermometers includes more terms.

Now let us take a look at the radiation from the surface when a CdTe buffer layer is grown on a GaAs substrate prior to the HgCdTe nucleation. The radiant intensity from the substrate GaAs stays stable when its temperature is controlled to a fixed value. However after the CdTe buffer layer starts to grow, the radiant intensity from the sample surface begins varying with growth time because of the change of its emissivity. The radiant intensity from the surface of the CdTe layer should be the total of the radiant intensity from the CdTe layer itself and the radiant intensity from the substrate GaAs transmitted through the CdTe layer. Once the CdTe layer is grown, the substrate is in a dynamic equilibrium state before the MCT growth starts. At this moment, the radiant intensity from the sample surface is a constant. Once the MCT grow starts and the MCT layer reaches a finite thickness, the radiant intensity emitted includes the radiation from the MCT layer itself and the transmitted radiation from CdTe/GaAs structure through the MCT layer.

Figure 2.38 shows the time dependence of the relative emissivity of the sample surface during HgCdTe MBE growth measured by Wang and Li He et al. (Wang 1996).

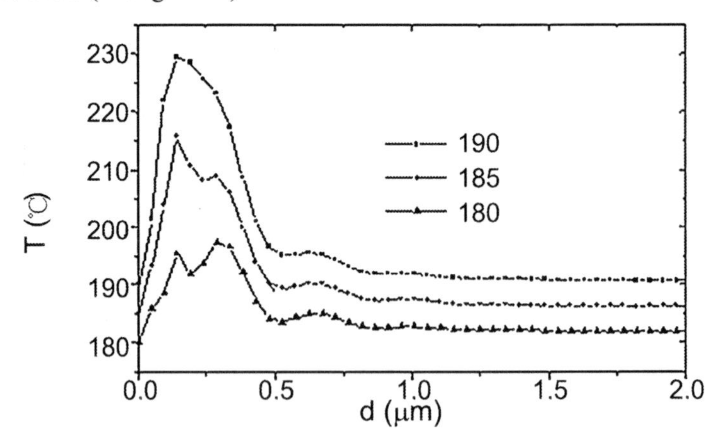

Fig. 2.39. Simulated curve of the reading of a radiation thermometer (in °C) versus epilayer thickness (in μm) for different growth temperatures for MBE HgCdTe films with composition $x = 0.213$

The origin of total radiant intensity from the HgCdTe surface over the full spectral band has been already discussed above. If a single waveband radiation thermometer is used to monitor the MBE growth, it is necessary to specifically calculate both the absorption coefficients of the epilayer and the amount of transmitted radiation from the substrate through the epilayer in this waveband. After considering these cases, Wang and He (Wang 1996) obtained a simulation of the relationship between the reading of a radiation thermometer and the growth time (or epilayer thickness) for MBE HgCdTe with different compositions. Figure 2.39 shows one set of these simulated curves of the expected reading of a radiation thermometer versus the epilayer thickness at different growth temperatures for a HgCdTe film with composition $x = 0.213$. The real temperature of the epilayer surface can be deduced from the reading of the radiation thermometer calibrated via this simulated curve. Hence it can be used for temperature control.

During growth, the substrate temperature is affected by many factors such as substrate size, the surface condition of the heat sink Mo-blocks, the temperatures of the source ovens and so on. As a result, the substrate temperature control is a very complicated process. To control the substrate temperature adequately, it is necessary to take all these factors into account.

Substrate mounting for epitaxy is very important for temperature control during MBE growth. One successful approach that has been used since the early work on MBE, is to bond the semiconductor wafer to a Mo-block using a low melting point metal, such as indium, which provides a liquid thermal contact at the growth temperature. In this case, the surface condition of the Mo-block will affect the substrate temperature. Because the emissivity of a Mo-block is smaller than that of MBE epilayers, more radiant energy will be emitted after the epilayer starts to grow. In order to keep the Mo-block temperature constant, more heater power needs to be supplied. The size of substrate also influences the substrate temperature control. This is because the radiant energy from the surface is related to the area of the sample and the Mo-block. If the substrate is small compared to the area of the Mo-block the surface emissivity of the complex is dominated by the emissivity of the Mo-block. If the substrate is large, the exposed area of Mo-block is small; hence, the surface emissivity of the complex is determined by the emissivity of the sample. These alternatives impact the temperature control mode.

In HgCdTe MBE growth, it has been demonstrated that placing a Si plate over the whole surface of Mo-block prior to mounting a GaAs substrate reduces the variation of the substrate temperature. Although metal bonding is widely used in MBE growth, there are still many problems associated with this method. For instance, if the indium layer is not uniform, the temperature variation across the substrate surface can reach several degrees. This seriously reduces the quality of the grown film. Also, if the wafers are to be processed subsequently in a semiconductor fabrication facility, the indium metal must be thoroughly removed from the back of the wafer before further processing steps to avoid contamination of the process line. To avoid the use of indium backing, the use of sample holders enabling radiative coupling between substrate and heater has become quite popular in recent years. The substrates, which must be of exact dimensions to fit a particular holder, are placed in an open ring, which is then locked into position in close proximity to a ribbon or wire heater wound so as to closely fill the area behind the wafer. Radiation from the heater heats the substrate rapidly because of its low thermal mass. Aside from the advantage of not needing to mount the substrate to the Mo-block with indium metal and then clean the back of the substrate after growth, another advantage of using sample holders is that it is easier to remove the sample after it is processed. However, uneven wafer clamping in this method also results in temperature gradients across the substrate surface.

During MBE growth, the radiation from source ovens also affects the substrate temperature. The radiation from source ovens at different temperatures incident on the substrate surface heats it. This is especially a problem during HgCdTe growth, where the temperature of the CdTe source oven is ~500°C, but the temperature of the substrate is only ~180°C. As a result, the radiation from the CdTe oven reflected from the sample makes the measured temperature slightly higher than its actual value. The good thing is that the part of signal resulting from the source ovens do not change during the growth process, and thus with care it can be subtracted out.

The optimum growth temperature of MBE HgCdTe lies in a very narrow range. A growth temperature that is slightly lower or slightly higher than the optimum value will greatly reduce the quality of the grown film. An understanding of the correlation between the sample surface temperature and its IR radiation provides a theoretical basis for accurate control of the growth temperature, and thus effective improvement the quality of the grown layers.

2.4.4 Composition Control

The band gap of HgCdTe alloys covers a wide IR wavelength range. By adjusting the concentration x of Cd, the band gap of $Hg_{1-x}Cd_xTe$ can be tuned into a specified wavelength range that fits an interesting device application. However, in HgCdTe MBE growth the vapor pressures of sources are relatively high, the substrate surface temperature variation is relatively large, and the sticking coefficients of sources are non-unity. All these constraints make accurate composition control difficult. In recent years, in situ spectroscopic ellipsometry emerged as a promising tool for real time setting growth parameters and determining x values. With a spectroscopic ellipsometer, the optical constants of HgCdTe at growth temperature, and the thickness variation of the compositional distribution can be obtained.

Ellipsometry measurements of surfaces are usually performed under conditions where its surface is in a relatively steady state, so it is an ex situ method. For instance, Alerovitz et al. (1983) studied the surface of a sputtered Si_3N_4 film on a GaAs substrate, and a GaAs-AlGaAs multilayer structure by using a variable incident angle ellipsometer. Merkel et al. (1989) investigated the GaAs/AlGaAs superlattice; Arwin et al. (1983) studied the optical constants of HgCdTe and surface oxide layers derived from different processes by spectroscopic ellipsometry. Orioff et al. (1994) examined the surface of hydrogenated HgCdTe. These are just a few representative instances that have been examined.

Recently, an ellipsometer for in situ real time monitoring has been successfully developed. It can be used for monitoring surface etching, film growth, and other processes. Now, spectroscopic ellipsometry is widely used in III–V and II–VI MBE growth. Maracas et al. (1992) used a spectroscopic ellipsometer capable of operating at multiple wavelengths in the range 250–1,000 nm to monitor the MBE growth of GaAs. The observed variation of ellipsometric parameters with growth time clearly reflected features of the growth process. The optimized growth temperature was determined from the relationship between the observed optical constants and the temperature. The growth rate extracted from the interference curve was close (within 10%) to that obtained from the RHEED oscillations. The advantage of this method is that the spectroscopic ellipsometer can be used for monitoring both uniform epilayer growth and mesa island nucleation and growth, even with a rotating substrate.

It was the French scientists Y. Demay et al. who first used an ellipsometer to monitor the MBE HgCdTe growth. In their experiment, the incident polarized light was directed upon the sample surface at an angle of approximately 70° through a window that was attached to a Riber 2300P chamber. The reflected polarized light was directed back onto the sample surface by a mirror inside the chamber and then was reflected by the sample surface again to the original window. The reflected polarized light was detected. With this method, the inter-diffusion in CdTe/HgTe structures was studied. It was found that the inter-diffusion between CdTe/HgTe is independent of the Hg pressure at 260°C, if the HgTe layer is grown prior to the CdTe layer. However, the density of Hg vacancies has a strong effect on the inter-diffusion between CdTe/HgTe if a HgTe layer covers the CdTe layer. In 1992, Australian researchers Hartley et al. (1992) used a monochromatic polarization modulated ellipsometer to monitor the growth of HgCdTe epilayers and CdTe/HgTe superlattice on ZnCdTe substrates in real time. The compositional deviation they obtained from layer to layer was about ±0.003. The key part of a monochromatic polarization modulated ellipsometer is the polarization modulator, which is made from a quartz block (optical element) attached to a quartz piezoelectric transducer. The quartz block is made to vibrate with the natural resonant frequency of the quartz piezoelectric transducer when an AC field is applied. A mono-axial sinusoidal strain standing wave is formed in the quartz block. Through photo-elasticity, a transparent solid stressed by compression or stretching becomes birefringent. Then different linear polarizations of light travel at slightly different speeds when passing through the material. Therefore, an oscillating strain in the quartz block induces oscillating birefringence; the quartz block causes a time varying linear phase retardation of the wave. The directions of fast and slow axis of the modulation retardation are determined by the strain direction. The amplitude of the sinusoidal retardation is in direct proportion to the strain, and thus direct proportion to the voltage applied on the transducer. In 1996, researchers at Texas Instruments, Bevan et al. (1996) reduced the composition deviations to ±0.0015 by using a polarization-modulated ellipsometer capable of using multiple wavelengths in the range 400–850 nm. Benson et al. (1996) studied the compositional variation, growth rate, and epilayer surface morphology of MBE HgCdTe by using a multiple wavelength rotating analyzer spectroscopic ellipsometer.

An accurate database of optical constants of HgCdTe at growth temperature is necessary for ellipsometer monitoring. Shanli Wang, Li He et al. (Wang 1996) systematically studied the optical constants of HgCdTe for different x values, and applied their results to in situ real time monitoring of HgCdTe MBE growth. In their experiments, a CdTe buffer layer with a thickness of ~3–4 μm was grown on a GaAs (211) B substrate prior to $Hg_{1-x}Cd_xTe$ nucleate. Then a $Hg_{1-x}Cd_xTe$ layer of ~3 μm thickness was grown at ~183°C with a growth rate of 2.5–3μm/h. To monitor the growth process, first the optical path was adjusted so that the incident light fully illuminated the sample. Then, the ellipsometer was calibrated to get the initial values of both polarizer and analyzer angles. During growth, the ellipsometry parameters were continuously obtained in real time. The ex situ ellipsometry measurement was performed on an ellipsometer platform.

In polarization ellipsometry, the polarization state parameter ρ of the elliptically polarized light is defined as the ratio of Fresnel Reflection Coefficients (r_p, r_s), $\rho = r_p / r_s$, where r_p and r_s are for p- and s-polarized light, respectively. According to the Fresnel equation, ρ is related to the optical constants and structure parameters of the material. When a light beam is incident on a HgCdTe/CdTe/GaAs multilayer structure, multiple reflections in the sample, result in an infinite series for the transmitted and reflected light. Using the reflected light as an example, after many reflections and refractions off the interfaces among the layers, the intensity of p-polarized light and that of s-polarized light reflected are obtained. The complex parameter ρ can be written as:

$$\rho = \frac{r_p}{r_s} = \frac{|r_p|}{|r_s|} \exp(i\delta_p - i\delta_s) = \tan\psi e^{i\Delta} \qquad (2.126)$$

Hence, the values of ψ and Δ can be obtained. ψ and Δ are called ellipsometry parameters. For any structure, ψ and Δ are obtained similarly.

A model of the sample's structure needs to be constructed when analyzing the experimental ellipsometric data. The theoretical values of ψ and Δ can be calculated based on this appropriate model, and compared to the experimental results deduced from (2.126). An appropriate structure model can be built by following the following steps:

1. Propose a structure model and calculate the theoretical values of ψ and Δ;
2. Compare the experimental values of ψ and Δ to theoretical values of ψ and Δ;
3. Set an accuracy standard, if the experimental values of ψ and Δ do not match those theoretical values, then adjust the structure model parameters such as the optical constants and thickness and then repeat step 1 to step 3;
4. When the experimental values of ψ and Δ match the theoretical values within the established accuracy standards, then the appropriate structure model is successfully constructed.

Details about how to analyze and fit data will be discussed in Chap. 4 on optical properties.

The composition of HgCdTe is a key issue affecting the sensitivity of measurements of the ellipsometric parameters ψ and Δ. The configuration of the sample also need to be taken into account during calculation of the values of ψ and Δ. For a multiplayer structure, the optical constants and the thickness of each layer all have some influence on the ψ and Δ values.

A change of one of the parameters (an optical constant, or layer thickness) will lead to a change of the ψ and Δ values. In the data fitting procedure, the more unknown parameters there are, the more complicated and less reliable is the analysis. For any single layer case, it is much simpler since there is only one type of material so there are only one or two independent parameters that influence the values of ψ and Δ.

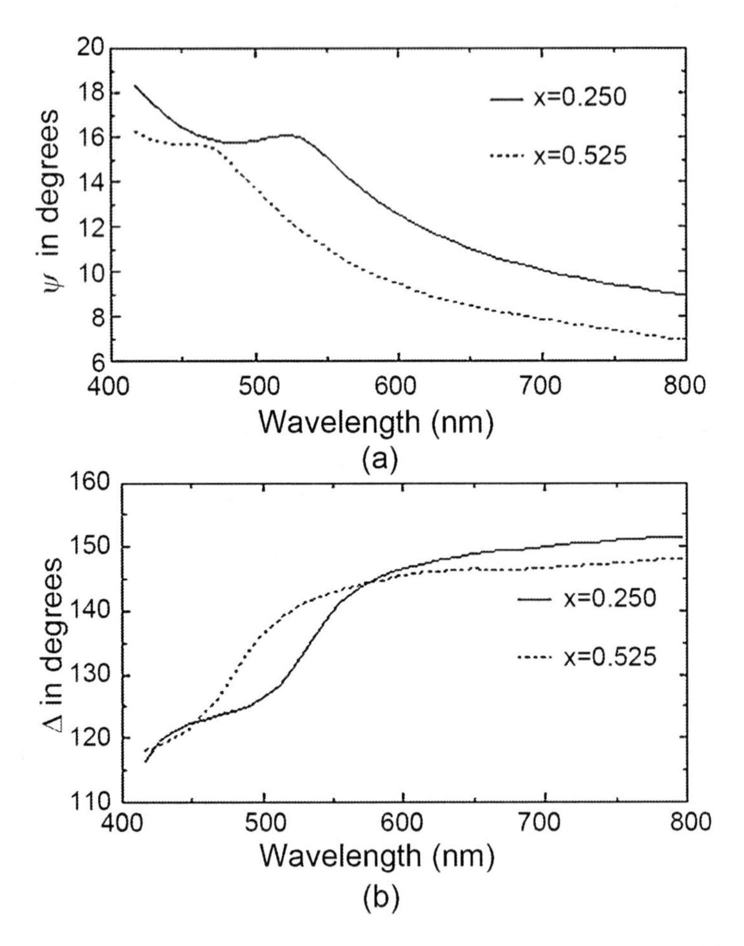

Fig. 2.40. Values of parameters ψ and Δ for HgCdTe with composition $x = 0.25$, 0.525 versus wavelength (**a**) parameter ψ; (**b**) parameter Δ

There is no the problem of oxidation on the surface of a MBE HgCdTe layer, because the epitaxial film growth is carried out in an ultrahigh vacuum environment. In addition, the absorption coefficient of HgCdTe is large in the visible wavelength region, which allows visible light to penetrate into the film to only several hundred angstroms. As a result, the reflected polarized visible light mainly comes from the front layer of the HgCdTe film after the HgCdTe layer reaches a thickness comparable to the absorption depth, and thus the effect of the substrate can be neglected. In this case, only the HgCdTe layer needs to be considered in the model calculation. Since the optical constants of HgCdTe are known functions of its x-value, the relationship between the optical constants and the ellipsometric parameters ψ and Δ can be converted to a relationship between the composition and ellipsometric parameters, ψ and Δ. Figure 2.40 shows the wavelength dependence of the ψ and Δ at an incident angle of 70° for HgCdTe with composition x = 0.25 and x = 0.525, respectively.

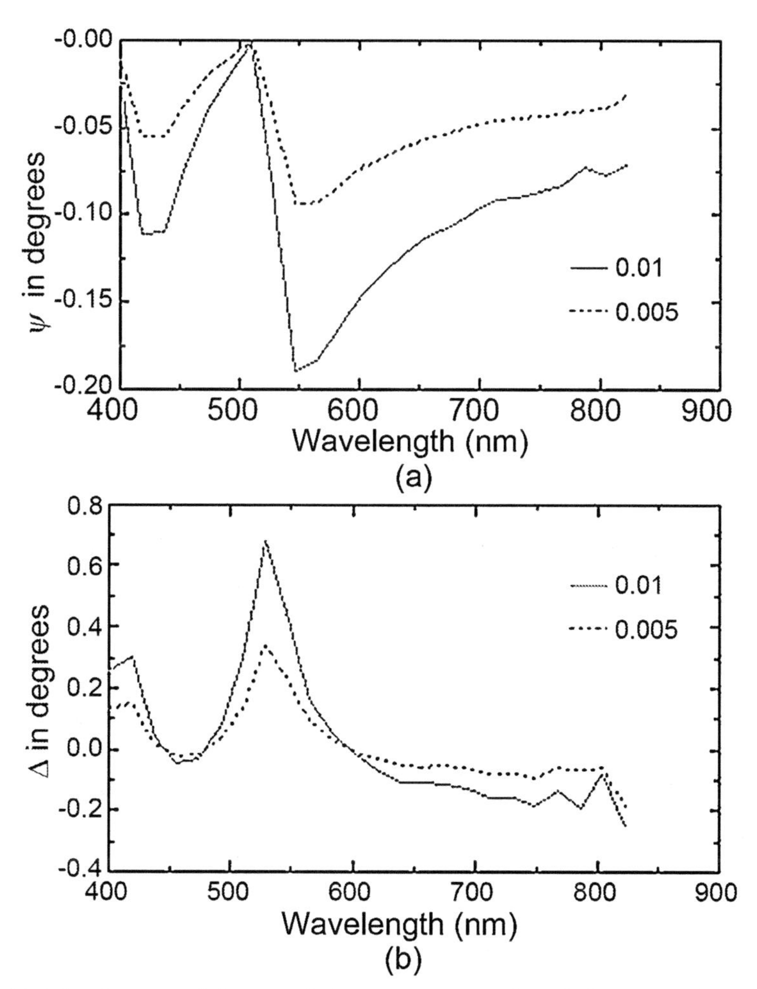

Fig. 2.41. The changes of the ψ and Δ curves vs. wavelength for $Hg_{1-x}Cd_xTe$ with x = 0.25 when the composition deviates from 0.25 by 0.01 and 0.005, respectively

For a sample with composition $x = 0.25$, the maximum changes of ψ and Δ, corresponding to a composition variation of 0.005, are 0.1° and 0.3°, respectively. However, for a sample with composition x = 0.525, the maximum changes of ψ and Δ that correspond to a composition variation of 0.005 are 0.06° and 0.25°, respectively. This clearly indicates that the ellipsometric parameters are very sensitive to the sample composition.

Figure 2.41 shows the changes of the ψ and Δ curves vs. wavelength for $Hg_{1-x}Cd_xTe$ with $x = 0.25$ when the composition deviates from 0.25 by 0.01 and 0.005, respectively.

The temperature also affects the ellipsometric parameters. If we were to plot the wavelength dependence of the optical constants of $Hg_{1-x}Cd_xTe$ with $x = 0.20$ at room temperature and a high temperature, It would be clear that the optical constants at room temperature are different from those at high temperature. Therefore, it is necessary to use the optical constants of HgCdTe at growth temperature, ~180°C, in the fitting data taken during growth. A detailed analysis will be given in Sect. 4.4.

2.5 Perfection of Crystals

2.5.1 X-Ray Double-Crystal Diffraction

X-ray double-crystal diffraction is an important technique to study the quality of crystalline materials. The full width at half maximum (FWHM) of double-crystals rocking curves (DCRC) for materials is the parameter that characterizes the quality of crystals. James (1954) derived the formula for the FWHM W of the reflection curve for a perfect single crystal:

$$W = \frac{2\lambda^2}{\pi V \sin 2\theta_B} |F_{hkl}||C|(\frac{e^2}{mc^2})(\frac{\gamma_0}{|\gamma_{hkl}|})^{1/2} \tag{2.127}$$

where V is the volume of a unit cell, λ is the wavelength of the X-ray being used, F_{hkl} is the structure factor amplitude of reflection from facet (hkl), C is the polarization factor, θ_B is the Bragg angle, e, m, c are fundamental physical constants, and $\left(\frac{\gamma_0}{|\gamma_{hkl}|}\right)^{1/2}$ is a correction factor to account for nonsymmetrical reflection. For a given incidence X-ray wavelength and diffraction facet, F_{hkl} determines the value of the FWHM directly. There exist many kinds of structural defects in practical crystals that for a given facet will increase *W* by different amounts; so the value of the FWHM for a practical crystal is greater than that for a perfect crystal.

$Cd_{1-y}Zn_yTe$ and $Hg_{1-x}Cd_xTe$ all have the sphalerite (or zincblende) structure, which consists of two interpenetrating face-centered cubic lattices displaced relative to one another along the cubic diagonal by a distance of $\frac{\sqrt{3}}{4}a$, where a is the crystal lattice constant; the length of a unit cube edge.

The geometrical structure factor of the unit cell is (Fang and Lu 1980):

$$F_{hkl} = \sum_j f_j e^{2\pi in(hu_j + kv_j + lw_j)} \tag{2.128}$$

where (*hkl*) are the Miller indices of a facet, u_j, v_j and w_j are coordinates of the *j*th ion in the unit cell, n is the index of refraction of the medium, and f_j is the dispersion factor of the *j*th ion. In binary AB zinc blend crystals, (2.128) can be rewritten as:

$$F_{hkl} = F_A + F_B e^{\pi in\frac{h+k+l}{2}} \tag{2.129}$$

In this formula F_A is the structure factor of the A cation face-centered cubic sublattice, defined as:

$$F_A \equiv (1 + e^{\pi in(h+k)} + e^{\pi in(h+l)} + e^{\pi in(k+l)}) f_A \tag{2.130}$$

and f_A is the dispersion factor for a cation of type A. Similarly, F_B is the structure factor of the B anion face-centered sublattice:

$$F_B \equiv (1 + e^{\pi in(h+k)} + e^{\pi in(h+l)} + e^{\pi in(k+l)}) f_B \tag{2.131}$$

and f_B is the dispersion factor for a B type anion.

When (*hkl*) all are odd numbers:

$$F_{hkl} = 4(f_A \pm if_B) \tag{2.132}$$

For the pseudo binary $Hg_{0.80}Cd_{0.20}Te$ alloy, f_A and f_B become (Yu 1979):

$$f_A = 0.80 f_{Hg} + 0.20 f_{Cd} \tag{2.133}$$

$$f_B = f_{Te} \tag{2.134}$$

Similarly for $Cd_{0.96}Zn_{0.04}Te$, we have:

$$f_A = 0.96 f_{Cd} + 0.04 f_{Zn} \tag{2.135}$$

$$f_B = f_{Te} \tag{2.136}$$

According to the values in Table 2.4, the calculated FWHM of ideal single crystal reflection peaks for the (333)-facet of $Cd_{0.96}Zn_{0.04}Te$ and of $Hg_{0.80}Cd_{0.20}Te$ are:

$$W_{CZT} = 5.7(arc.s) \tag{2.137}$$

$$W_{MCT} = 7.6(arc.s) \tag{2.138}$$

In practical DCRC work, the X-rays diffract off two crystals. The FWHM from the first crystal is W_1. The crystal being measured is the second crystal, and the final measured FWHM of DCRC is:

$$W = \sqrt{W_1^2 + W_2^2} \tag{2.139}$$

Table 2.4. Data used to calculate the (333) facet FWHM diffraction peaks of $Cd_{0.96}Zn_{0.04}Te$ and of $Hg_{0.80}Cd_{0.20}Te$

Wavelength of Cuα_1 (λ)	1.5405×10^{-10}m	Correction factor $(\frac{\gamma_0}{\lvert\gamma_{hkl}\rvert})^{1/2}$	1
e^2/mc^2	2.82×10^{-15}m	f_{Te} (Caroline et al. 1962)	31.67
$2\theta_B$	76.49°	f_{Cd} (Caroline et al. 1962)	28.85
Lattice constant of crystal		f_{Hg} (Caroline et al. 1962)	52.05
(a) (MCT)	6.4650×10^{-10}m	f_{Zn} (Caroline et al. 1962)	16.60
(a) (CZT)	6.4658×10^{-10}m		
Polarization factor (C)	1		

A particular double-crystal measurement can serve as an example. In view of the measured FWHM off the (400) facet of a GaAs first crystal, the expected FWHM of the DCRC for perfect $Cd_{0.96}Zn_{0.04}Te$ and $Hg_{0.80}Cd_{0.20}Te$ crystals should be:

$$W_{cal-CZT} = 15(arc.s) \tag{2.140}$$

$$W_{cal-MCT} = 16(arc.s) \tag{2.141}$$

The perfection of crystals can be estimated from their double-crystal diffraction curves. Figure 2.42 is the DCRC off the (111) facet of a CdTe wafer. When $\phi = 0$, the rocking curve is symmetric with a FWHM = 27″.

When $\phi = \pi/2$, the rocking curve broadens and the FWHM ≈ 104″. Also, the multiple peaks that emerge indicate that there exist low-angle grain boundary substructures. Crystals on both sides of sub-grain boundaries have orientation differences that are not larger than 15°. The sub-grain boundaries are composed of dislocation clusters. The sub-grain boundaries are formed during growth when the interface doesn't remain planar. Also impurities tend to collect at these sub-grain boundaries.

The DCRC technique is a point wise measurement that is used to sample crystal perfection at different points, so two-dimensional maps of lattice perfection can be made.

It is necessary to examine the crystal quality of substrates before attempting epitaxial growth on them. Figure 2.43 shows the result of a point wise measurement study on the (111) facet of a $Cd_{0.96}Zn_{0.04}Te$ substrate. FWHM values are distributed in the range of ~16–24″. As the sample is rotated through an angle $\pi/2$, the rocking curve has no clear shape change and remains symmetric without the emergence of multiple peaks. This indicates that the $Cd_{0.96}Zn_{0.04}Te$ substrate crystal has a uniform structure.

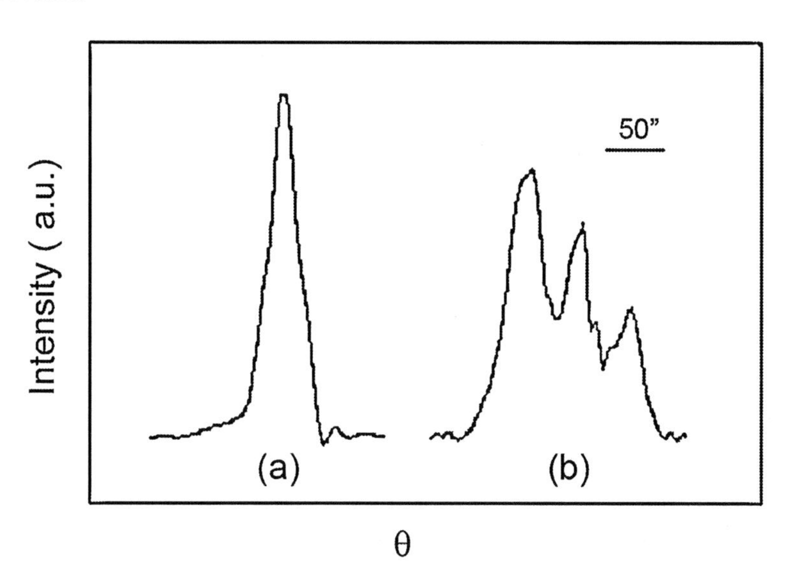

Fig. 2.42. Double-crystals rocking curve of a (111) facet of a CdTe substrate, **(a)** $\phi = 0$ FWHM = 27″, **(b)** $\phi = \pi/2$ FWHM ≈ 104″

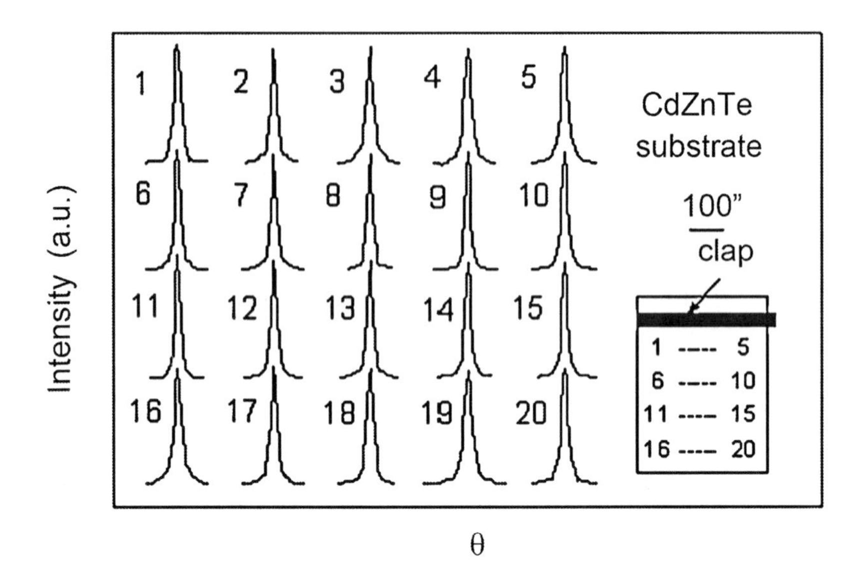

Fig. 2.43. Distribution of X-ray double-crystals rocking curves of a (111) facet of a $Cd_{0.96}Zn_{0.04}Te$ substrate, FWHM: ~16–24″

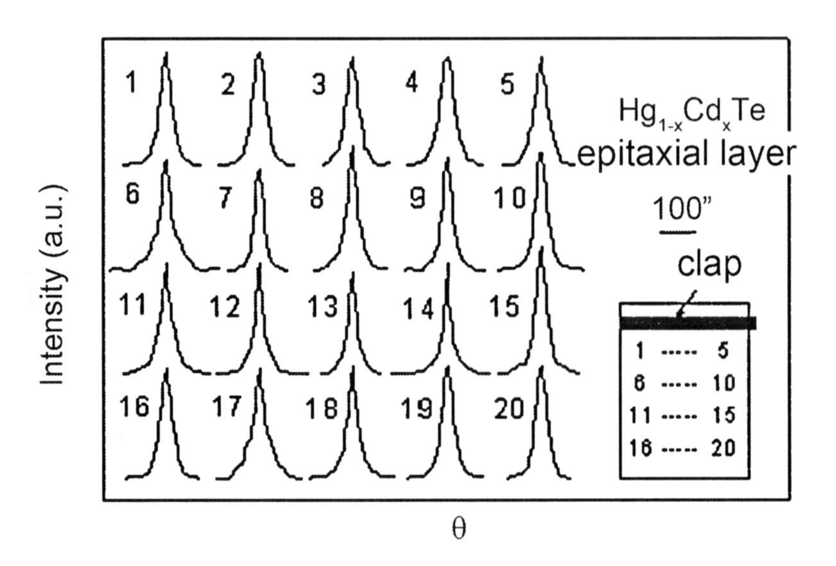

Fig. 2.44. Distribution of X-ray double-crystals rocking curves for an LPE grown $Hg_{1-x}Cd_xTe$ surface, FWHM: ~26–47″

The point wise DCRC measurement technique is also used on $Hg_{1-x}Cd_xTe$ epitaxial films. Figure 2.44 shows such a study on an LPE $Hg_{1-x}Cd_xTe$ layer grown on a substrate (Zhu 1997). It can be seen from Fig. 2.44 that the rocking curve at the edge of the epitaxial layer is broader than that of the central part. The FWHM values vary in the range of ~26–47″. During the course of LPE growth, the quartz clamp holding the substrate acts as a barrier and results in distortions at the substrate edges. Then different parts of the substrate, such as parts between the center and edges, have different super saturation levels that can affect the quality of the epitaxial layer crystal grown. However, the rocking curves still have a high symmetry and multiple peaks do not emerge, which indicates that this $Hg_{1-x}Cd_xTe$ film grown by vertical dipping LPE has a uniform structure and good crystal quality.

The FWHM is related to the dislocation revealing etch pit density on the surface of a sample. A FWHM and etch pits density (EPD) study was done for $Cd_{0.96}Zn_{0.04}Te$ samples, prepared to have different crystal qualities. At least five different zones on each sample were measured and the results averaged to get the data shown in Fig. 2.45. The result shows a strong correlation with EPD's in the range of ~0.4–4×10^5 cm^{-2} corresponding to FWHM values of ~20–60″. Thus the dislocation density of CdZnTe can be estimated from a measurement of the FWHM.

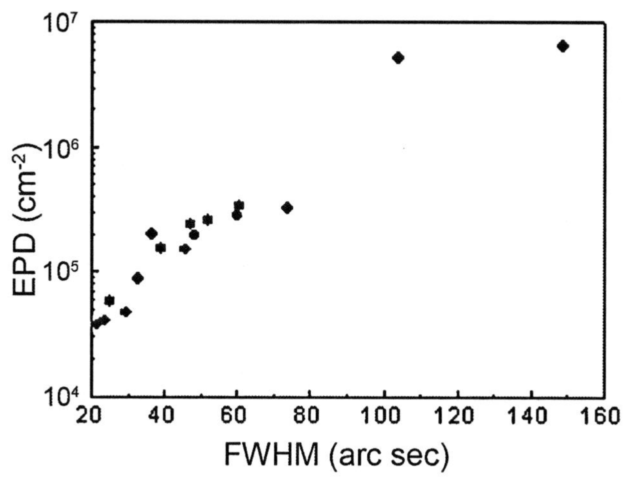

Fig. 2.45. FWHM of double-crystals rocking curves for $Cd_{0.96}Zn_{0.04}Te$ as a function of etch pit density

The dislocation density of a $Hg_{1-x}Cd_xTe$ epitaxial layer also can be estimated from the FWHM of DCRC experiment. Dislocations in epilayers have several sources:

- Structural defects in the substrate, e.g. inclusions, grain boundaries, threading dislocations. Careful preparation of the substrate material can eliminate many of the worst defects, but some threading dislocations are always present. These threading dislocations normally set a minimum on the number of line defects found in an epilayer.
- Lattice constant mismatch between the substrate and the epilayer. This mismatch introduces misfit dislocations near the boundary between the substrate and the epilayer, and also generates new threading dislocations that penetrate through the epilayer.
- Thermal expansion difference between the substrate and the epilayer. Devices function at low temperatures, and the lattice constant changes that occur between the growth temperature and the operating temperature, induce extra stresses that can cause dislocation multiplication.

The thickness of the $Hg_{1-x}Cd_xTe$ epitaxial layer is much less than that of substrate. So mismatches introduce most stress in the epitaxial layer. When the stress exceeds a critical tangential stress of the crystalline material, the lattice will rearrange itself to form stress-relieving dislocations. The lattice constant of an epitaxial alloy layer at a given temperature depends on its composition and its thermal expansion coefficient. Mismatch stress will vary with the compositions of the epitaxial layer and substrate, as well as the difference between the growth and the current temperature. The lattice constant mismatch is defined as:

$$f \equiv \left| \frac{a_0 - a}{a_0} \right| \tag{2.142}$$

In this formula, a_0 and a are the lattice constants of the $Cd_{1-y}Zn_yTe$ and of the $Hg_{1-x}Cd_xTe$ epitaxial layers, respectively. The thermal expansion coefficient of HgTe is 4.0×10^{-6} K^{-1}, of CdTe is 4.9×10^{-6} K^{-1}, and of ZnTe is 8.3×10^{-6}K^{-1} (Basson and Booyens 1983). Supposing that the thermal expansion coefficient varies linearly with composition, the thermal expansion coefficients of $Hg_{1-x}Cd_xTe$ and $Cd_{1-y}Zn_yTe$ can be written as:

$$\alpha(x) = 4.9\times10^{-6} + 9.0\times10^{-7}x(K^{-1}) \tag{2.143}$$

and

$$\alpha(y) = 4.9\times10^{-6} + 3.4\times10^{-6}y(K^{-1}) \tag{2.144}$$

During the course of epitaxial growth, when the layer is just a few atomic layers thick the epitaxial material is strained to match the lattice constant of the substrate. As the layer thickens the stress associated with these strains builds to a point where the material can no longer support it. Once this critical thickness is reached, misfit dislocations form to relieve it.

Dislocations form in a variety of complex patterns (Hirth and Lothe 1982), but for this discussion we will only consider misfit and threading dislocations. A primitive picture of a misfit dislocation is an extra closed fragment of a plane of atoms (or a missing plane fragment) that is nearly parallel to the growth interface. The misfit dislocation lies at the perimeter of the enclosed plane fragment. A threading dislocation is formed by an extra half atom plane (or missing half plane) in the crystal. The threading dislocation lies at the edge of the half plane. The threading dislocations most studied in this field are those that cross the interface and/or terminate at the surface of the epilayer.

Matthews (1975) derived an expression for the critical thickness (h_c) for the onset of stress relaxation of an epitaxial layer:

$$h_c = \frac{b(1-\nu\cos^2\beta)}{8\pi f(1+\nu)\sin\beta\cos\zeta}\ln(\frac{\alpha h_c}{b}) \tag{2.145}$$

Here b is the Burgers vector; ν is Poisson's ratio; β is the angle between the Burgers vector and dislocation line; ζ is the angle between the growth interface and the slip plane, and α is a factor related to the formation energy of a dislocation.

Because HgCdTe crystals have the sphalerite structure, their slip system is {111}<111> (Long and Schmit 1970). The dominant dislocations in this slip system are the so-called 60° dislocations, i.e. the dislocation line lies along the <110> direction, the Burgers vector is in the <101> direction, and β = 60°. Often $Hg_{1-x}Cd_xTe$ epitaxial layers are grown on a (111) surface of CdTe-based substrates. Then we have ζ = 0. Accordingly, (2.145) can be written:

$$h_c = \frac{b(1-0.25\nu)}{6.93\pi f(1+\nu)}\ln(\frac{\alpha h_c}{b}) \tag{2.146}$$

For 60° dislocations in MCT alloys, with Burgers vector $b = a/\sqrt{2}$, a typical value of α is 4 (Basson and Booyens 1983). From the given elastic constants of CdTe and HgTe, the Poisson ratio of MCT alloys is ~0.32 (Basson and Booyens 1983). For a $Hg_{1-x}Cd_xTe$ epitaxial layer with x = 0.2, the critical thickness is less than 0.8 μm, which is less than typical practical epitaxial layer thicknesses of ~10 μm.

In general, the FHWM of rocking curves can be affected by a variety of structural defects in crystals, such as composition variation induced stress, lattice bending, and dislocations. Their contributions to the rocking curves add nearly incoherently. Thus their contributions to the FHWM (W_m) add in quadrature and can be expressed as (Qadri and Dinan 1985):

$$W_m^{\ 2} = w_s^{\ 2} + w_b^{\ 2} + w_d^{\ 2} + \ldots \qquad (2.147)$$

Where w_s, w_b and w_d are the contributions to the rocking curve FHWM from stress, lattice bending and dislocations, respectively. Owing to the excellent uniformity of the transverse variation of x in current material (Δ x≤0.002), this contribution to the FHWM now can be ignored. In addition, the thicknesses of epitaxial layers are generally much greater than the critical thickness for stress relaxation (Long and Schmit 1970), so the residual misfit induced stress in the upper layers is approximately zero. In principle the combined epilayer plus substrate will bend due to the thermal expansion coefficient mismatch between them. However, the thickness of $Hg_{1-x}Cd_xTe$ epitaxial layers is much less than that of substrates, so the bending is small and can be ignored to first approximation. There is no obvious evidence to indicate the existence of significant lattice bending in good epitaxial layers (Qadri and Dinan 1985). Thus in layers having no inclusions or voids, the remaining broadening of the FHWM of the $Hg_{1-x}Cd_xTe$ epitaxial layer rocking curves result mainly from the density ρ of threading dislocations through the relation (Dunn 1957):

$$\rho = w_m^{\ 2} / 4.35 b^2 \qquad (2.148)$$

Table 2.5. Fitted values of dislocation densities of different points of a $Hg_{1-x}Cd_xTe$ LPE surface

Point	FWHM (arcsec)	Dislocation density ($\times 10^6$ cm^{-2})	Point	FWHM (arcsec)	Dislocation density ($\times 10^6$ cm^{-2})
1	43.3	4.85	11	34.7	3.11
2	43.3	4.85	12	26.0	1.75
3	43.3	4.85	13	27.7	1.98
4	46.8	5.66	14	27.7	1.98
5	45.5	5.35	15	35.5	3.26
6	34.7	3.11	16	40.7	4.28
7	32.1	2.66	17	36.4	3.43
8	30.3	2.37	18	34.7	3.11
9	33.9	2.97	19	39.0	3.93
10	39.0	3.93	20	39.0	3.93

where b is the Burgers vector. For $Hg_{1-x}Cd_xTe$ crystals, $b = a/\sqrt{2}$, with a being the lattice constant. From (2.148), the dislocation density for the upper "bulk like" layer of $Hg_{1-x}Cd_xTe$ epitaxial films is obtained (see Table 2.5). The average "bulk" dislocation density of a typical $Hg_{1-x}Cd_xTe$ epitaxial film is 3.57×10^6 cm^{-2}.

Further analyses reveal the dislocation density near the surface. Figure 2.46 shows measured values of the FWHM as a function of the thickness of a $Hg_{1-x}Cd_xTe$ epitaxial layer on a CdZnTe substrate. For each thickness, the FWHM in the figure is the value averaged over at least three different points. The results show that the FWHM has its largest values for thin films and decreases monotonically as the thickness increases. For thin epitaxial layers ($\leq\sim1$ μm), the diffraction peaks coming from substrates also can be observed. It can be seen from Fig. 2.46 that the crystal quality of thick epilayer surfaces is higher than it is at their interface. When the thickness of an epitaxial layer exceeds ~4 μm, the variation of the FWHM slows, saturating at about 40″. This saturation indicates that beyond ~4 μm, the longitudinal structure of this epitaxial layer is uniform. From Fig. 2.46, the estimated dislocation density of the interfacial layers is $\sim2.42\times10^7$ cm^{-2}, while that of the near surface layers, is $\sim3.57\times10^6$ cm^{-2}. The difference is the misfit dislocation density.

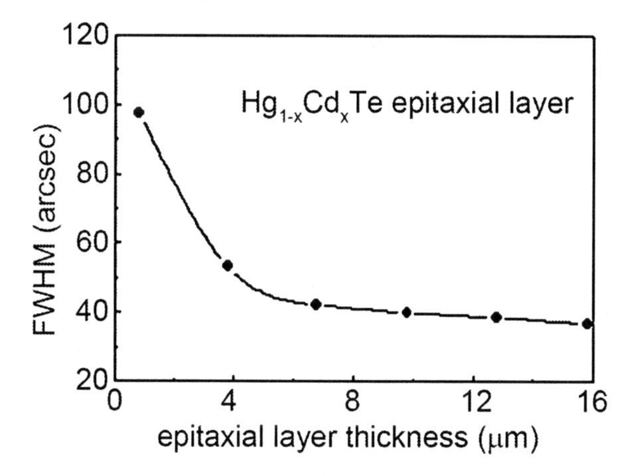

Fig. 2.46. FWHM values as a function of the thickness of $Hg_{1-x}Cd_xTe$ LPE layers

The bond length of CdTe is slightly smaller than that of HgTe. As a consequence when a MCT layer is grown on a CdTe substrate there is a small but non-negligible lattice constant mismatch. The addition of a small amount of Zn to CdTe, forming a CdZnTe alloy, can bring the lattice constants of a MCT epilayer and the CZT substrate into an exact lattice match. The amount of Zn needed in the alloy depends on the Cd concentration in the MCT. Thermal expansion effects can still introduce misfit dislocations, but fewer than those due to compositional lattice constant differences. Wood and co-workers (Wood et al. 1985) conducted a transmission electron microscopy (TEM) measurement on a $Hg_{0.8}Cd_{0.2}Te$/ CdTe sample and found an interface dislocation density of 3×10^{8} cm^{-2}. Bernardi and co-workers (1991) studied the crystal quality of a $Hg_{0.78}Cd_{0.22}Te$/CdTe interface. They found a FWHM that became ~150–170″ as the thickness of the epitaxial layer was reduced. The dislocation density of the interfacial layer reported in Fig. 2.46 is lower than that of the above referenced work (Wood et al. 1985) and (Bernardi et al. 1991). Mainly because the misfit dislocation density of a $Hg_{1-x}Cd_xTe$/CdZnTe system is lower than that of a $Hg_{1-x}Cd_xTe$/CdTe system.

If $x = 0.2$, where the $Hg_{1-x}Cd_xTe$ epitaxial layer contacts the substrate, with an LPE growth temperature of 500°C, the mismatch fraction in the $Hg_{0.8}Cd_{0.2}Te/Cd_{0.96}Zn_{0.04}Te$ system obtained from (4.19) is $f = 5.4\times10^{-4}$, which is smaller than that of the $Hg_{1-x}Cd_xTe$/CdTe system where $f = 2.8\times10^{-3}$. If all of this mismatch ($f = 5.4\times10^{-4}$) is relaxed by misfit dislocations, then at the growth interface there will be one dislocation every ~1,900 atm. Any voids or inclusions near the interface, on either side, (Nouruzi-Khorasani et al. 1989) also give rise to dislocations at the interface (Bernardi et al. 1991). It is worth pointing out that the crystal quality of the $Hg_{1-x}Cd_xTe$/CdZnTe interfacial layer is the determining factor of the crystal quality of the surface layer.

Scanning ion-beam mass spectroscopy (SIMS) is often used to measure the spatial variation of the composition of alloys. Figure 2.47 is a profile measured by allowing an Ar ion beam to sputter through the epilayer, while measuring the masses of the material coming off the surface (Zhu 1997). After ~2 μm the alloy composition of the epilayer becomes almost constant. X-ray double-crystal rocking curves also can be used to study the effect of processing technique changes on the perfection of crystals. For example, Fig. 2.48 is a typical set of double-crystal rocking curves of a CdZnTe wafer before and after annealing. The surface of the annealed sample was untreated. As seen from the figure, the sample's rocking curve is nearly symmetric before annealing. However, the rocking curve loses its symmetry after annealing for 1 h. The right part of the rocking curve has a slope, which becomes less steep as the annealing time is increased. Results of this kind of measurement are valuable aids for adjusting processing strategies.

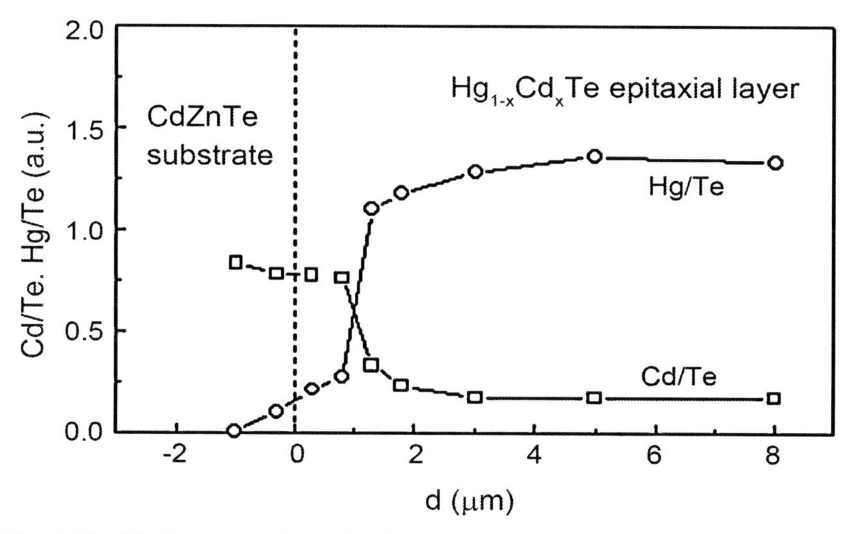

Fig. 2.47. SIMS measured longitudinal distribution curve of Cd/Te and Hg/Te concentrations near the interface of a $Hg_{1-x}Cd_xTe$/CdZnTe system

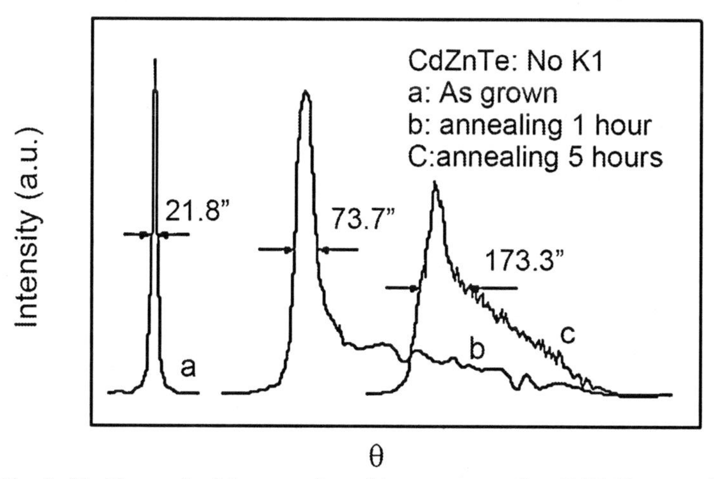

Fig. 2.48. X-ray double-crystal rocking curves of a CdZnTe sample before and after annealing

2.5.2 Morphology

The surface morphology of HgCdTe epitaxial films is mainly determined by the growth technique, but also has some relationship to the substrate quality. Surface morphologies can be observed by many methods, such as X-ray topological morphology (XRT), TEM, scanning electron microscopy

(SEM), metallographic microscopy, energy-dispersive X-ray spectroscopy (EDX) and so on.

XRT techniques have several forms with different characteristics and application ranges. The scanning reflection morphology technique is a useful method to study the effect of the surface morphology of substrates on $Hg_{1-x}Cd_xTe$ LPE films. The principles underlying scanning reflection morphology will be considered first. For a perfect crystal, a monochromatic X-ray beam of wavelength λ is diffract off lattice planes with spacing interval d. There is a strong diffraction ray at an angle of $2\theta_B$ from the incident beam of radiation. The Bragg formula determines θ_B:

$$2d\sin\theta_B = n\lambda \tag{2.149}$$

where n is index of refraction. If the orientation of lattice planes or their separation experiences a local change (e.g. around a dislocation), a perfect region and a distorted region cannot both satisfy the Bragg condition. Hence, in the presence of distortions the intensity of the Bragg peak decreases by amounts depending on the relative area of distorted material.

Scanning reflection morphology not only enables morphology studies of surface layers but also of those below the surface. By changing the angle of incidence and the X-ray wave length, which changes the absorption characteristics encountered in a study, the depth examined can be varied in the range of ~1–15 μm.

The surface morphology of a $Hg_{1-x}Cd_xTe$ LPE film depends on the orientation of the substrate, the growth temperature, the growth rate, and the nucleation mode. The typical morphology of such layers is a collection of "steps" caused by inclusions, grain boundaries, surface pits and so on. Typical steps have amplitudes that can be clearly resolved. The surface structure consists of a series of roughly parallel and periodic step facets. The steps are micro facets that all have about the same orientation, close to that of the substrate. Small protrusions alternate between their neighboring facets. The width of typical protrusions is about 20 μm, with heights in the range of 0.1–1.0 μm. The angle between the surface and substrate planes varies continuously in one- and two-dimensional orientations. These rippled surfaces are caused by a variety of imperfections to be examined in more detail next. When all the major imperfections are eliminated, often there is still a regular "cross-hatch" pattern of strained material on the surface in the <110> directions (Rhiger et al. 1997). The cross-hatch period in each direction is typically ~100 μm. This strain distribution at the epilayer surface is caused by even a very small lattice constant mismatch between the substrate and the epilayer material (Berding et al. 2000). When focal plane arrays are built on such material, the crosshatch pattern results in a similar pattern of pixels with degraded noise characteristics (Rhiger et al. 1997).

Inclusions, mainly those of Te, are often found in both MCT epilayers as well as CZT substrates. On epilayer surfaces they can be observed in a metallurgical microscope as small black spots. Their morphology can be seen in more detail with SEM and EDX spectra measurements. Typically their diameters fall in the range of ~10–20 μm. The inclusions have the lattice structure of metallic Te with small concentrations of Hg, and Cd impurities. It is not surprising that Te inclusions are found unless care is taken to prevent them. The Te corner of the Hg–Cd–Te phase diagram has a HgCdTe–Te coexistence region, so when $Hg_{1-x}Cd_xTe$ layers are grown from a Te-rich solution by LPE Te inclusions can form. The growth temperature and melt concentrations must be selected to minimize these inclusions. While the as grown $Hg_{1-x}Cd_xTe$ LPE samples may still contain some inclusions their number and size can be decreased further by post growth annealing in a Hg atmosphere. This will convert the metallic Te regions to HgCdTe. These high Hg partial pressure anneals are also done to reduce the density of Hg vacancies, changing the material into weakly p-type or even n-type semiconductors required in device applications. Observation of annealed $Hg_{1-x}Cd_xTe$ samples with an SEM shows that the density of black inclusions decreases greatly. This occurs because the Hg overpressure shifts the system into a single-phase region of the phase diagram. Then Hg atoms diffusing into the epilayer during annealing dissolve the Te inclusions into the alloy.

The next issue is "grain boundaries." If there are defects in the substrates, such as polycrystalline grain boundaries, twin crystal boundaries, or threading dislocations, they will extend to the epitaxial layer as LPE growth proceeds. The resulting imperfections of the crystal structure will affect the performance of devices fabricated on such material. If part of a CdTe substrate is polycrystalline, this structure will reappear in the epitaxial film. These structures degrade surface morphology, and with it carrier mobility, and junction uniformity.

For HgCdTe prepared by vertical dipping LPE without proper care, epilayer surface pits or voids arise from micro-grain impurities that adhere to the substrates. LPE growth is preceded by a high temperature process designed to make the mother liquid uniform. The temperature in this molten stage is high and the time it lasts is long. The partial pressures of Hg, Cd, and Te are large. After the molten source homogenizing process, the substrate surface can become rough, where many grain-like adherents and pits appear. Spectral analysis indicates that Hg metal peaks appear on the CdTe substrates. This is due to the large partial pressure of Hg at the substrate surface. Even if the surfaces of substrates are covered, Hg atoms still infiltrate them and react with them.

HgCdTe epitaxial layers grown on these rough substrates are also rough with pitted surfaces. HgCdTe epitaxial layers grown on smooth substrates have good morphologies. The pits in epitaxial layers appear above sites of adherents. Nucleation and growth above the adherents are interrupted, so little or no growth occurs there leaving pits.

In order to mitigate the pollution phenomenon, the surface of the substrate can be covered by a quartz cover glass during the molten source stage. While this reduces the problem it does not eliminate it completely. Addition of a melt-back process is employed for further improvement. The melt-back technique is:

- First, the substrate is dipped into an under-saturated solution to absorb the Hg metal adherents, and to melt a few atomic layers from the substrate.
- Then the substrate is brought into contact with the growth solution for the required epitaxial layer.

Using this melt-back technique, LPE layers that are free of pits can be grown.

A second melt-back procedure is also used at the end of the epilayer growth, to avoid material from the mother solution adhering to the epitaxial layer surface as it is removed from the melt. In this procedure, just at the end of growth for a short time the temperature is raised to melt-back a few atomic layers of the epilayer, and reduce the adhesion of mother solution while the sample is being removed from it. Then the sample is lifted vertically to leave the mother solution, and the power is shut off to allow the growth system to cool rapidly. The under-saturated solution is sometimes used to wash any residual mother solution off the epilayer surface and melt back another few atomic layers. Morphologies of $Hg_{1-x}Cd_xTe$ films grown by this melt-back LPE method are smooth, as long as the degree of under-saturation of the second solution is not too large. It is observed in the experiments that the degree of under-saturation should be in the range of ~0.5–1.5.

2.5.3 Precipitated Phase in $Hg_{1-x}Cd_xTe$ Epitaxial Films

CdZnTe crystals are not only important substrate materials for growing $Hg_{1-x}Cd_xTe$ films by LPE, but are also used to prepare X-ray and γ-ray detectors, photoelectric modulators, solar cells, laser windows etc. But as discussed earlier, Te metal precipitates form easily during the course of $Hg_{1-x}Cd_xTe$ LPE film growth from Te-rich solutions. Also, in the process of growing CdTe-based bulk crystals, due to the difficulty of controlling the thermo-kinetics Te and Cd second phase inclusions are formed in crystals. The existence of such precipitated phases destroys the uniformity of the bulk material and its surface morphology and degrades devices fabricated from it. Microscopic Te inclusions in epitaxial layers have been shown to affect their infrared transmission. Li et al. (1995) analyzed the infrared absorption spectrum of $Hg_{1-x}Cd_xTe$ LPE films to verify this point, identifying free carrier absorption and scattering from the Te metal inclusions. Zhu et al. (1997) determined the Te micro inclusion characteristics in $Hg_{1-x}Cd_xTe$ epitaxial films using TEM and Raman spectra measurements. The sizes of these micro-inclusions are in the range of ~40–60 nm. The inclusions are too small to measure their compositions by EDX. But their infrared absorption spectrum still identifies them as principally Te metal micro-inclusions.

Usually some Te inclusions are found in CdZnTe substrates. Te inclusions are due to excess Te atom concentrations in the melt caused by the evaporation of Cd. At a given temperature different partial pressures of Te, Cd and Zn are in the vapor phase. A typical situation has 96 at% Cd above the melt. From the ideal gas law, the excess Te atom concentration in a melt caused by the evaporation of Cd is:

$$N_{T_e} = \frac{P_{Cd} N_A}{RT} \frac{h_V}{h_L} \tag{2.150}$$

where p_{Cd} is the partial pressure of Cd, N_A is the Avogadro constant, R is the gas constant, T is the temperature, h_V/h_L is the ratio of free height h_V above the melt to the melt height h_L in the ampoule. When the temperature is 1,100°C, the partial pressure of Cd is about 10^5 Pa (Lorentz 1962), and the excess Te caused by Cd loss is $N_{T_e} = 5 \times 10^{18} \frac{h_V}{h_L} \mathrm{cm}^{-3}$. When the excess concentration of Te is in the range of ~10^{18}–10^{19} cm^{-3}, this variation of the melt composition has to be considered. The excess Te is incorporated in the CdZnTe in equilibrium as it grows at high temperature. When the crystal is cooled, the solid solubility of Te decreases, and then the supersaturated Te precipitates.

Te deposits are different from Te inclusions. In general, we do not distinguish them. Until now all Te metal micro-grains incorporated into CdTe based alloys have been called inclusions. Now we will distinguish two types of Te micro-grains that differ by their average size, and the mechanism through which they form. Zanio (1978) first introduced the concept of a distinction between inclusions and deposits. Many micro-grains observed have been classified as deposits incorrectly. Rudolph et al.

(1993) established criteria for identifying them more accurately. According to their research, inclusions originate from micro-grains generated at unstable regions of the growth front. These unstable regions occur in the nooks of grain boundaries, at twin crystals projecting through the growth surface, and at contacts of the growth front with the vessel's walls. The concentration of inclusions is determined by the growth rate (Rudolph et al. 1995). By contrast, the formation of deposits is due to conservation of components as the crystal is cooled. Initially the excess Te is incorporated as native point defects, Cd vacancies, Te interstitials, and Te antisites. For most growth conditions Cd vacancies dominate. In any one crystal it is difficult to tell which mechanism is responsible for a given micro-grain because they both contribute in every matrix. The size of deposits is in the range of ~10–30 nm, which requires an HRTEM measurement to see them. Inclusions tend to be larger ~1 μm and can be observed using an infrared microscope. Assuming that Te deposits/inclusions in CdTe-based materials are spherical and excluding point defects, the total excess concentration of Te in a crystal is (Rudolph et al. 1993):

$$N_{Te} = \frac{4\pi\rho_{Te}N_A}{3A_{Te}}\sum_i r_i^3\rho_i \tag{2.151}$$

where N_A is the Avogadro constant, r_i and ρ_i are the radius and the density of depositions/inclusions, respectively, A_{Te} is the atomic mass of Te, ρ_{Te} is the mass density of Te micro-grains; the subscript i labels different grains with different diameters.

Te deposits can be studied using a differential scanning calorimeter (DSC). The operating principle of a DSC is:

- Heat is supplied to a sample under test and to a reference sample at a constant ramp rate.
- The temperature difference between the test sample and the reference is measured.
- In some instruments this temperature difference versus time is used to extract information about the properties of the sample.
- In other instruments there are two heat sources, one for the sample and one for the reference. Their difference is controlled to maintain a zero temperature difference between the sample and reference. Then it is the extra power versus either time or temperature that is used to extract information.

In the second power difference method there are some endothermic peaks and exothermic valleys in the curve. The position of sharp features on the temperature (or time) axes; their shapes and numbers are related quantitatively to the characteristic properties of the sample being investigated. The area of a peak (or valley) is related to the enthalpy of the reaction producing it. Given a calibration to extract an enthalpy per particle of the reaction, the area can be used to estimate the quantity of material participating it that reaction. Burger and Morgan (1990) first used this technique in studies of the evolution of inhomogeneous regions in HgI_2 single crystals, and found that DSC could determine the iodine deposition density. Jayatirtha et al. (1993) examined undoped and Cl doped CdTe crystals by DSC. They not only determined the existence of Te depositions but also measured them quantitatively.

From DSC curves of CdZnTe samples, if the endothermic peaks appear at the known melting point temperature of pure Te, it is determined qualitatively that the Te depositions exist. At the same time, knowing the latent heat of melting for pure Te per unit mass ΔH_f^0, the area of the endothermic peak ΔH_f can be used to extract the Te content in CdZnTe. The relation is:

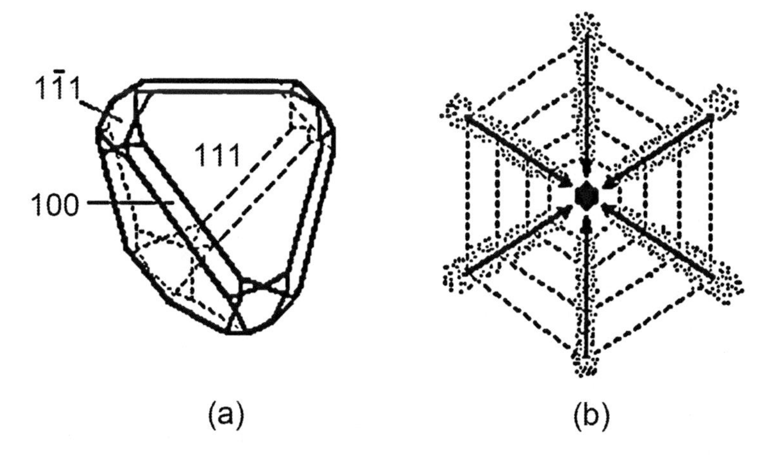

Fig. 2.49. (**a**) The shape of a tetrahedral matrix around an inclusion, and (**b**) the crystallization course of a drop in a supersaturated solution

$$w_{Te} = \frac{\Delta H_f}{\Delta H_f^0} \text{ (wt\%)} \tag{2.152}$$

Usually, inclusions are distributed inhomogeneously, and concentrate in small-angle grain boundaries and twin boundaries. Inclusions produced by super-cooling behind the growth front result in cellular growth at small-angle boundaries. Solution drops easily aggregate in these regions. These aggregates behind the growth interface produce sub-grain boundaries in the material grown above them. These large size inclusions have no clear orientation with the surrounding matrix but have resolvable boundaries. During growth, they are usually located in one of the two adjacent {111} facets of the CdZnTe sphalerite structure (Rudolph et al. 1995) (Fig. 2.49). When the ingot cools, the Te crystallites are located in an approximately uniform temperature field (in a low temperature gradient). The supersaturation of the Te-rich melt will result in a quasi-symmetrical crystallization at the center of the Te metal regions. At around 450°C, the crystals obtained include some formed at the eutectic, whose composition is almost entirely Te. Of course, when the growth front is concave on the {111} facets, the growth rate decreases, but the alloy still grows. As a result, a large number of little solution drops are formed, and a defuse outline is built around each Te inclusion.

Deposits and inclusions in crystals can be observed with infrared microscopes and transmission electron microscopes. Analyzing the variation of precipitated phase sizes and densities before and after annealing is the best way to determine proper annealing conditions. Because of scattering of infrared radiation caused by Te deposits, the infrared transmission spectrum measures the quantity of Te deposits in a sample. These measurements offer guidance to the development of improved annealing techniques. Figure 2.50 shows the infrared transmission curves of a CdZnTe wafer before (curve a) and after annealing under a high Cd partial pressure at a temperature of 700°C for 1 h (curve b), 3 h (curve c) and 5 h (curve d), respectively. The transmittance curve d shows the annealing has improved the quality of the sample.

Rutherford backscattering spectroscopy (RBS) can be used to study deposits or inclusions in wafers. The RBS technique is a method of studying surface characteristics (Winton et al. 1994; Strong et al. 1986), which has been in use since 1970. An energetic ion beam that can penetrate the material bombards the sample's surface. If the sample is a single crystal and the beam is aligned with a channel between the atom columns, many emerge from the back of the crystal and few are backscattered along the channel direction. If there is an imperfection along the channel the ions are scattered, and the number reaching the back through the channel is greatly reduced. If the sample is not a perfect crystal, then numerous ions are scattered into random directions and some are backscattered along the channel. If the number of ions in a given time detected over a wide solid angle in the backward direction is N_R, and the number of ions backscattered in the channel direction is N_C, then the ratio

$\chi = \frac{N_C}{N_R}$ can be used to characterize the crystal quality. A smaller χ corresponds to a higher crystal quality. Figure 2.51 shows the Rutherford backscattering curves of a CdZnTe sample before and after annealing. The channel spectrum of the sample is obtained on a portion of a sample thinned to 100 μm. The results show that the crystal quality after annealing is better than that before annealing; χ_{annealed} (≈ 0.06) < $\chi_{\text{as-grown}}$ (≈ 0.14).

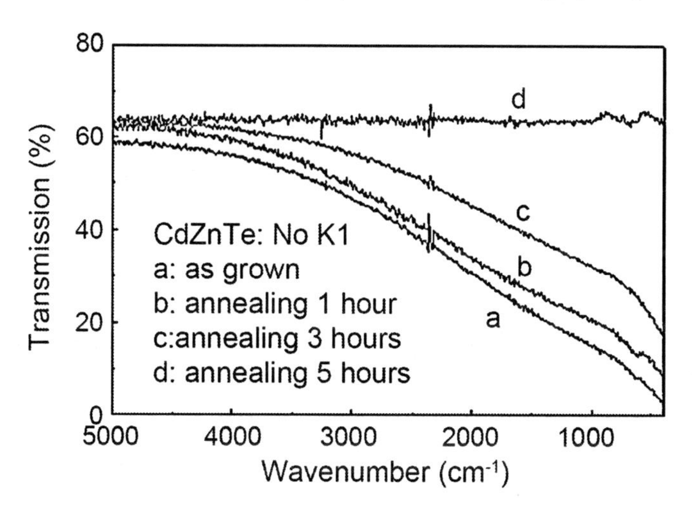

Fig. 2.50. Infrared transmission coefficient curve of a CdZnTe wafer before and after annealing

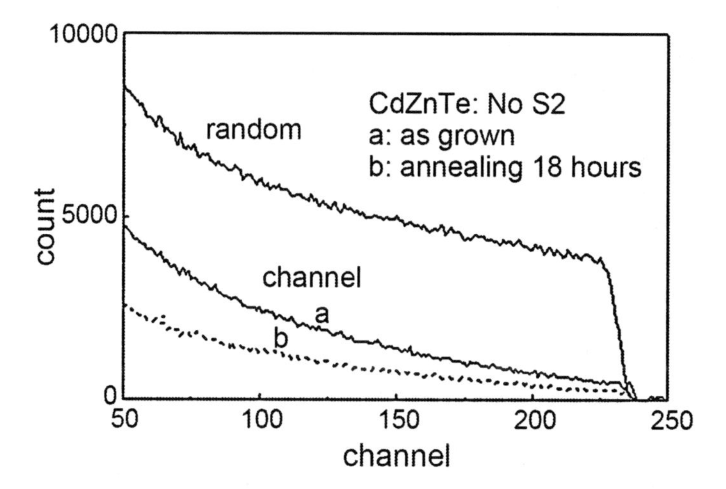

Fig. 2.51. Rutherford backscattering curves of CdZnTe sample before and after annealing

Annealing in a high Cd partial pressure can lower the Te inclusion density, and decrease the size of Te deposits, but does not completely eliminate them. The annealing mechanism responsible for decreasing excess Te is not settled. Vydyanath et al. (1992) put forward two mechanisms, Te thermal migration to surfaces in a temperature gradient, and diffusion of Cd from the vapor to combine with the excess Te adding to the CdTe based solid. Berding et al. (1995) based on accurate first principles density functional calculations to obtain defect formation enthalpies and a generalized quasi-chemical statistical theory, demonstrated that the densities of Te interstitials and vacancies in CdTe are low at the usual annealing temperatures. These two native defect types provide the means for Te diffusion, and because they are scarce Te diffusion is slow. This prediction agrees with experiments. Thus the principal mechanism for the reduction of excess Te when a high Cd partial pressure is present is Cd diffusing in to combine with the Te. If higher temperatures are used and annealing times are extended then Te thermal migration can contribute to reducing Te inclusions (Vydyanath et al. 1993; Bronner and Plummer 1987; Wang 1989). The experimental results show that deleting Te depositions is more difficult than deleting Te inclusions. Anthony and Cline (1971, 1972) proved this point when they studied the thermal migration of a solution drop in KCl.

2.5.4 Native Point Defects

There are six classes of native point defects in a CA (cation and anion) binary compound semiconductor:

- Cation and anion vacancies, empty lattice sites, denoted respectively V_C and V_A.
- Cation and anion interstitials, atoms in interstitial positions in the lattice, denoted respectively C_I and A_I. In fact there are a number of possible interstitial sites, so the number of native point defects could be greater than six in some materials. Normally one type of site is dominant.
- Cations on anion lattice sites, and anions on cation sites called antisite defects, denoted respectively A_C, and C_A.

Native point defects account for any lack of stoichiometry of compounds. In $Hg_{1-x}Cd_xTe$ crystals, because they are dopants, native point defects also have important effects on their electrical properties. The HgTe bond is weak; consequently the volatility of Hg is larger than Cd and even Te. So unless care is taken to control the Hg partial pressure properly during growth or annealing, the concentration $[V_{Hg}]$ of Hg vacancies is large. Hg interstitials have much smaller densities but can be present in high enough numbers to impact Hg diffusion rates. The second most concentrated native point defect in MCT is Te_{Hg} antisites. Since Hg vacancies are acceptors, and Te antisites are donors, the electrical properties of these crystals can be adjusted to some extent by controlling their relative concentrations, as Fig. 2.52 shows (Vydyanath et al. 1992, 1981a; Berding et al. 1995). Of course impurities also play key roles in the carrier concentrations of these alloys.

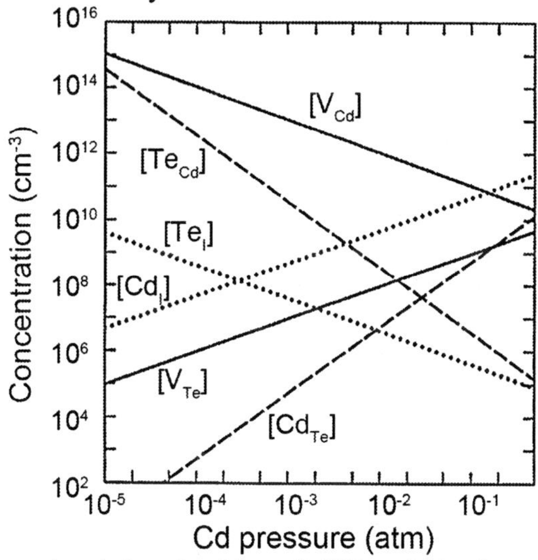

Fig. 2.52. Native point defect densities of CdTe within the stability region of HgTe at 700°C

When a $Hg_{1-x}Cd_xTe$ alloy is heat-treated, its native point defect concentrations vary depending on the temperature and the Hg partial pressure. It is impossible experimentally to measure separately the densities of each of these native point defect types; only their net effect on a variety of physical properties can be determined. Because these defects interact with one another and some of them may be neutral or charged, any attempt to calculate their densities must be done in a quasi-chemical formalism that properly accounts for the interactions among them. Since the densities of these defects cannot be measured separately, the free energies that enter into the reaction constants also cannot be properly assigned. Fortunately, modern theory has advanced to a point where trust worthy trends, and even quantitative predictions of these densities and their ionization states can be calculated. Next we will outline the elements of this theory and compare its predictions for MCT alloys with experiments.

The interactions among the native point defects can be written as six chemical reactions:

$$\mathrm{HgTe} \Rightarrow \mathrm{V_{Hg}Te} + \mathrm{Hg_g} \tag{2.153}$$

$$\mathrm{Hg_g} \Rightarrow \mathrm{HgV_{Te}} \tag{2.154}$$

$$2\mathrm{Hg_g} \Rightarrow \mathrm{HgHg_{Te}} \tag{2.155}$$

$$2\mathrm{HgTe} \Rightarrow \mathrm{TeHg_{Te}} + 2\mathrm{Hg_g} \tag{2.156}$$

$$\mathrm{Hg_g} \Rightarrow \mathrm{Hg_I} \tag{2.157}$$

$$\mathrm{HgTe} \Rightarrow \mathrm{Te_I} + \mathrm{Hg_g} \tag{2.158}$$

The reference states for these reactions are Hg_I atoms in the gas phase, and HgTe pairs on lattice sites in the crystal. This set of reactions is complete, but not unique. For example, if (2.157) is substituted into (2.153) one gets the Frankel defect formation reaction:

$$\mathrm{HgTe} \Rightarrow \mathrm{V_{Hg}Te} + \mathrm{Hg_I} \tag{2.159}$$

in which a Hg atom goes from a lattice site into an interstitial position.

The concentrations of the defects in (2.153)–(2.158) can be obtained from the evaluation of the corresponding reaction constants.

$$K_{V_{Hg}} = \left[V_{Hg}\right] p_{Hg} \tag{2.160}$$

$$K_{V_{Te}} = \left[V_{Te}\right] / p_{Hg} \tag{2.161}$$

$$K_{Hg_{Te}} = \left[Hg_{Te}\right] / p_{Hg}{}^{2} \tag{2.162}$$

$$K_{Te_{Hg}} = \left[Te_{Hg}\right] p_{Hg}{}^{2} \tag{2.163}$$

$$K_{Hg_I} = \left[Hg_I\right] / p_{Hg} \tag{2.164}$$

$$K_{Te_I} = \left[Te_I\right] p_{Hg} \tag{2.165}$$

These reaction constants can be found in standard texts (Reif 1965; Chen and Sher 1995), but more specific results for this problem are in Berding et al. (1994). The enthalpies and entropies that enter are calculated from solutions of the density functional form of Schroedinger equation in the linearized muffin tin approximation (LMTO) using the full-potential (FP) Harris Foulkes approximation (Anderson et al. 1985). This method allows accurate calculations of atomic forces in solids, enabling determination of equilibrium lattice constants, elastic constants and normal mode frequencies, and relaxation around defect sites. In doing these relaxation calculations the presence of Cd was not properly accounted for. This is probably the largest source of error in this procedure. Then formation enthalpies and vibrational contributions to the entropies of the reactions listed in (2.160)–(2.165) can be determined (Berding et al. 1993). The enthalpy results along with the cohesive energies of the compounds with Te anions are listed in Table 2.6. While occupied electron state energies were given accurately by this formalism, and so total energies could be found, excited state energies were still not precise. In particular band gaps were too small. Thus only semi-quantitative values of donor and acceptor state energies could be determined. That is why only qualitative comments are given in Table 2.6 for the ionization state energies of the native state defects. States that will be highly ionized at 77 K are designated as shallow, and those that will not are called deep.

Table 2.6. Formation energies and ionization states of the native point defects and some cohesive energies

Formation energy and ionization state		
Defect	(eV)	State
V_{Hg}	1.93	shallow acceptor
V_{Te}	2.39	donor
Te_{Hg}	2.68	shallow donor
Hg_{Te}	0.75	deep acceptor
Te_{I}	4.47	shallow donor
Hg_{I}	1.75	shallow donor

Cohesive energies	
ZnTe	4.66(4.8)[a]
CdTe	4.17(4.4)[a]
HgTe	3.37(3.3)[a]

Notes: Formation energies refer to the neutral defect reactions in (2.160)–(2.165) in the text. See text for discussion of shallow and deep levels
[a]Experimentally determined cohesive energies

To evaluate the densities of these defects account must be taken of the number of them that are ionized. Thus there are additional interactions to include. These are the ionization of donors, acceptors and creation of electron–hole pairs. The electron–hole pair interaction must be included because the Fermi energy enters into the reaction constants for the ionization reactions, and the equilibrium densities are often determined at high temperature where the samples are nearly intrinsic. Also one needs to calculate the carrier concentrations at low temperatures for situations where the defect densities are in non-equilibrium states that were frozen in when samples were quench cooled. Even in annealing studies during cool down a temperature is reached where for the time allotted, the defect populations are frozen.

A measure of the accuracy of this method can be found by comparing the predicted and measured cohesive energies in Table 2.6 of ZnTe, CdTe, and HgTe. The differences range from ~2 to 5%. This is actually a hard test, because the correlation energies play a significant part in the total energy of free atoms, and the cohesive energies are differences between total energies of the assembled solid and free atoms. Gradient corrections to the correlation energy terms in the density functional equation have to be included to reach this degree of accuracy.

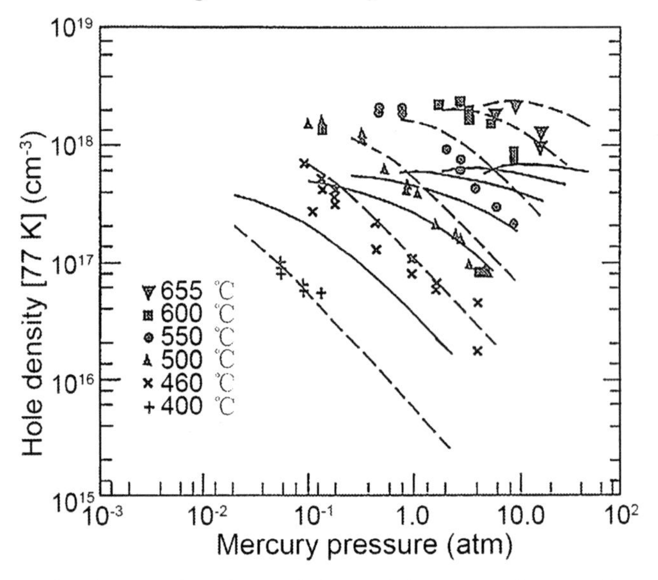

Fig. 2.53. Hole densities as functions of Hg partial pressure for samples quenched from various temperatures. The *solid curves* are direct predicted values, while the *dashed curves* where obtained by increasing all defect formation energies by 10%, and then rigidly shifting the curves up by a factor of 5.5

Figure 2.53 is a comparison of the predicted hole concentration with those measured as a function of Hg partial pressures on samples quench cooled from various high temperatures between 400 and 650°C (Vydyanath 1981b). The dashed theory curves are obtained by increasing all defect formation energies by 10% relative to those calculated from first principles and then a rigid shift of the resulting curves upward by a factor of 5.5. The theory (solid lines) predicts the behavior only semi-quantitatively. Even this is remarkable in view of the exponential dependences in the theory, and that there were no parameters adjusted to fit the experiments. The dashed lines are intended to test the sensitivity of the predictions to small changes in a few parameters.

Other aspects of the predictions can be seen from Figs. 2.54 and 2.55. The total densities of neutral plus ionized V_{Hg} (solid lines) and V_{Te} (dashed lines) defects are plotted in the existence region of HgTe as a function of T^{-1}. The upper boundary corresponds to Te saturation, and the lower boundary to Hg saturation. Lines of constant partial pressures in atmospheres are also shown. The Hg vacancy density exceeds that of Te vacancies except near the melting point, so in equilibrium undoped samples are always p-type.

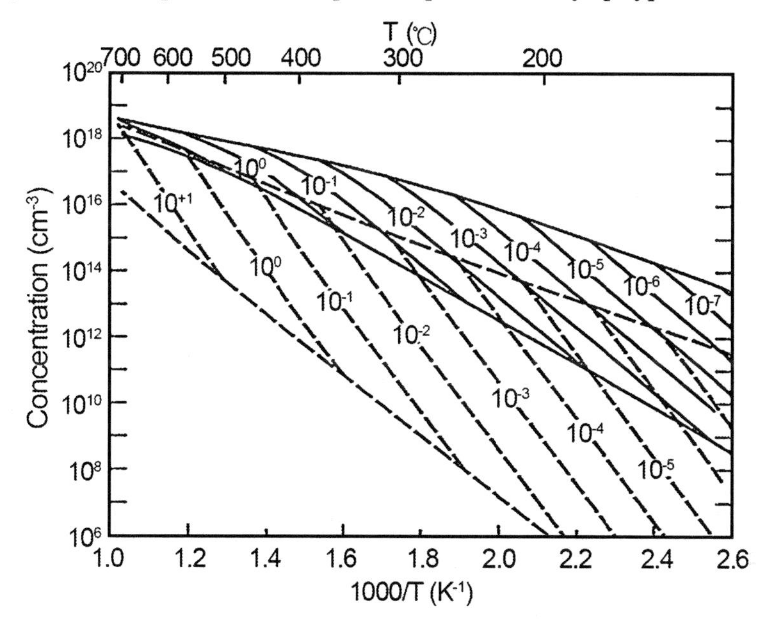

Fig. 2.54. The Hg vacancy (*solid line*) and Te vacancy (*dashed line*) total concentrations as functions of T^{-1} throughout the existence region of HgTe. Curves of constant partial pressure in atmospheres are also shown

Figure 2.55 are plots of the concentrations of native point defects within the HgTe existence region for samples equilibrated at three different temperatures. The curve at 185°C is typical of MBE growth temperatures, another at 220°C is typical of the condition used in Hg saturated anneals, and the last at 550°C is typical of LPE growth temperatures. Also shown is a range for the possible concentrations of a Hg vacancy-Te antisite complex. This complex is predicted to have a binding energy of 1.1 eV, which accounts for its high density. The uncertainty in the density arises because we do not know its ionization energy.

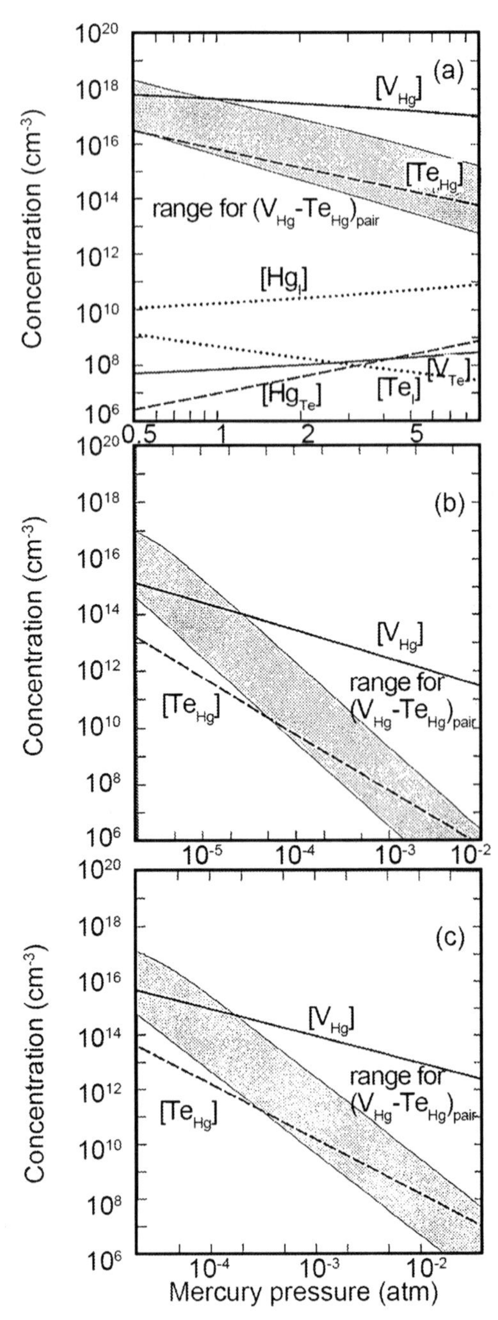

Fig. 2.55. The equilibrium concentrations of the native point defects in HgTe as functions of Hg partial pressure within the phase stability region at (**a**) 500°C, (**b**) 185°C, and (**c**) 220°C

In the lower part of the range the pair complex is neutral, and in the upper end it is a donor whose ionization state resonates with the conduction band edge.

According to Fig. 2.55, independent of the annealing condition used, undoped material in equilibrium is always p-type. However, bulk, LPE, and MBE grown high purity material following a 220°C, Hg rich anneal are all converted to n-type with carrier concentrations measured at 77 K of a few times 10^{14} cm^{-3}. An outstanding issue in MCT alloys is: what is the residual donor? No one has been able to identify an impurity that behaves this way. Berding et al. have argued that it may be a non-equilibrium concentration of either Te vacancies or $V_{Hg}Te_{Hg}$ pair complexes. These defects, it is argued diffuse very slowly and their high temperature concentration may be frozen in. MBE grown material is always grown from the Te edge of the phase diagram, and surface effects can enter to make the V_{Te} concentration higher than its equilibrium value at the ~185°C growth temperature. This remains an unresolved issue.

In practice, the typical heat treatment of n-typed HgCdTe LPE films is conducted in a closed quartz tube for 48 h at 250°C under a Hg saturated vapor pressure. For LWIR n-type HgCdTe at 77 K, the minimum electron concentration that can be obtained is ~2–10×10^{14}cm^{-3} and the mobility is greater than 6×10^{4} cm^{2}/V s. A typical p-type heat treatment for MBE grown material can be conducted in situ under ultra-vacuum conditions. At 77 K, the concentration of holes is ~0.5–2×10^{14} cm^{-3} and their mobility, at best, is ~600 cm^{2}/V s. However when the acceptors are V_{Hg} it is well established that the hole lifetimes are shorter than those from impurity acceptors, e.g. As_{Te}. Thus impurity acceptor doping, while presenting fabrication problems, is preferred.

$Hg_{1-x}Cd_xTe$ LPE films are grown to prepare high performance n–p photovoltaic infrared focal plane detector arrays. One class of devices requires that the active absorption region of the $Hg_{1-x}Cd_xTe$ be weakly p-type. The desired hole concentration is ~10^{16} cm^{-3}, and the mobility should be as high as possible. But most $Hg_{1-x}Cd_xTe$ films grown by LPE are strongly p-type, with vacancy concentrations leading to $p > 10^{17}$ cm^{-3}. Therefore, the as grown LPE films have to be annealed. Following the Hg vacancy reduction annealing heat treatment prescription outlined above, an in situ annealing (anneal in the growth furnace) is done. Because the residual donor concentration range is ~5× 10^{14}–10^{15} cm^{-3}, if samples' are exposed to an excessive Hg pressure they will become n-type. When p-type behavior with a hole concentration of ~10–16 cm^{-3} is the goal a different prescription is followed. A reasonable set of annealing conditions are: for $x = 0.23$ the temperature of the sample is held at 380°C, and a Hg source is held at 360°C, for 6 h. Such an annealing treatment produces a p-type concentration of ~$0.5–6\times10^{16}$ cm^{-3}, and a hole mobility of up to 300 cm^2/V s. The reason for the low hole mobility is extra ionized impurity scattering from compensation of the residual donors.

Additional information on heat treatment techniques can be obtained by consulting references (Xiao 1981, 1995; Yang et al. 1990). Techniques for improving the control of the Hg vacancy concentration was reported by Yang et al. (1985) who deduced experimentally a relationship among the concentration of Hg vacancies, the heat treatment temperature, and the temperature of a Hg source. They also deduced theoretically an analytic expression for this relationship.

For epitaxial materials, because the traditional heat treatment techniques will affect the surface of the narrow gap material, often a thin layer of wide gap material (CdTe or ZnS) is deposited to cover the surface prior to the p-type heat treatment. The concentration of Hg vacancies of epitaxial materials still can be controlled effectively (Yang et al. 1998, 2002). The disadvantage of the method is that misfit dislocations are introduced in the heat treatment process because of lattice constant and thermal expansion coefficient differences between the two materials. Another method conducts the heat treatment in an open-tube, which uses heated HgTe powder as its source. The ability of this technique to control Hg vacancy concentrations without surface damage is as effective as the wide gap thin cover layer technique (Yang et al. 2001).

For n-type conversion, it is necessary to use an open-tube heat treatment technique under a Hg saturated condition (Chen et al. 1997). A high temperature technique has also been devised to decrease the dislocation density and to reduce the surface degradation of the material (Yu et al. 1999; Yang 1997).

References

Akutagawa W, Zanio K (1971) Vapor growth of cadmium telluride. J Cryst Growth 11:191–196

Alerovitz SA, Bu-Abbud GH, Woollam JA, Liu DC (1983) An enhanced sensitivity null ellipsometry technique for studying films on substrates: Application to silicon nitride on gallium arsenide. J Appl Phys 54:1559–1569

Almeida J (1995) Finite semigroups and universal algebra. World Scientific, Singapore

Amingual D (1991) SPIE 1512:40–51

Anderson OK, Jepsen O, Glotzel D (1985) In: F Bassani et al. (eds.) Highlights of condensed matter theory. North Holland, Amsterdam The Netherlands, p. 59
Anthony TR, Cline HE (1971) Thermal migration of liquid droplets through solids. J Appl Phys 42:3380–3387
Anthony TR, Cline HE (1972) Thermomigration of gold-rich droplets in silicon. J Appl Phys 43:2473–2476
Arias JM, et al. (1993) J Electron Mater 22:1049
Arias JM, Pasko JG, Zandian M, Kozlowski LJ, et al. (1994) J Opt Eng 33:1422
Arwin H, Aspnes DE, Rhiger DR, et al. (1983) Properties of $Hg_{0.71}Cd_{0.29}Te$ and some native oxides by spectroscopic ellipsometry. J Appl Phys 54:7132–7138
Asahi T, Oda O, Taniguchi Y, Koyama A (1995) Characterization of 100 mm diameter CdZnTe single crystals grown by the vertical gradient freezing method. J Cryst Growth 149:23–29
Astles MG, Blackmore G, Steward V, Rodway DC, Kirton P (1987) The use of in-situ wash melts in the LPE growth of (CdHg)Te. J Cryst Growth 80:1–8
Astles MG, Hill H, Blackmore G, Courtney S, Shaw N (1988) The sources and behaviour of impurities in LPE-grown (Cd,Hg)Te layers on CdTe(111) substrates. J Cryst Growth 91:1–10
Astles MG, Shaw N, Blackmore G (1993) Semicond Sci Technol 8:S211
Basson JH, Booyens H (1983) Phys Stat Solid (a) 80:663
Bell SL, Sen S (1985) J Vac Sci Technol A 3:112
Benson JD, Cornfeld AB, Martinka M, et al. (1996) J Electr Mater 25:1406
Berding MA, van Schilfgaarde M, Sher A (1992) J Vac Sci Technol B10:1471
Berding MA, van Schilfgaarde M, Sher A (1993) J Electron Mater 22:1005
Berding MA, van Schilfgaarde M, Sher A (1994) Phys Rev B 50:1519
Berding MA, van Schilfgaarde M, Sher A (1995) J Electron Mater 24:1127
Berding MA, Nix WD, Ghiger DR, Sen S, Sher A (2000) J Electron Mater 29:676
Bernardi S, Bocchi C, Ferrari C, Franzosi P, Lazzarini L (1991) Structural characterization of $Hg_{0.78}Cd_{0.22}Te$/CdTe LPE heterostructures grown from Te solutions. J Cryst Growth 113:53–60
Berry JA, Sangha SPS, Hyliands MJ (1986) SPIE 659:115
Bevan MJ, Duncan WM, Weatphal GN, et al. (1996) J Electron Mater 125:1371
Blair J, Newnham R (1961) In: Metallurgy of elemental and compound semiconductors, Vol. 12. Inter Science, New York, p. 393
Bostrup G, Hess KL, Ellsworth J, et al. (2001) J Electron Mater 30:560
Bowers JE, Schmit JL, Speerschneicder CJ, et al. (1980) IEEE, Trans Electron Devices ED-27:24
Brebrick RF, Su Ching-Hua, Liao Pok-Kai (1983) Associated solution model for Ga–In–Sb and Hg–Cd–Te. In: Willardson RK, Beer AC (ed.) Semiconductors and semimetals Vol. 19. Academi Press, London, p. 171
Brill G, Velicu S, Boieriu P, et al. (2001) J Electron Mater 30:717
Bronner GB, Plummer JD (1987) J Appl Phys 61:5286
Brown M, Willoughby AFW (1982) Diffusion in $Cd_xHg_{1-x}Te$ and related materials. J Cryst Growth 59:27–39

Brunet P, Katty A, Schneider D, et al. (1993) Mater Sci Eng B 16:44
Buck P, Nitsche R (1980) Sublimation growth and X-ray topographic characterization of CdTe single crystals. J Cryst Growth 48:29–33
Burger A, Morgan S (1990) J Cryst Growth 99:988
Burton JA, Prim RC, Slichter WP (1953) J Chem Phys 21:1987
Butler JF, Doty FP, Apotovsky B, et al. (1993) Mater Sci Eng B 16:290
Capper P (1994) Annealing of epitaxial HgCdTe. In: Properties of narrow gap cadmium-based compounds. INSPEC, the Institution of Electrical Engineers, London, p. 152
Capper P, Harris JH, O'Keefe E, et al. (1993) Mater Sci Eng B 16:29
Caroline H, Macgillavry, Rieck GD (1962) International tables for X-ray crystallography Vol. III physical and chemical tables. The Kynoch Press, Birmingham, England, p. 210
Castro CA, Tregilgas JH (1988) Recent developments in HgCdTe and HgZnTe growth from Te solutions. J Cryst Growth 86:138–145
Chandra D, Tregilgas JH, Goodwin MW (1991) Dislocation density variations in HgCdTe films grown by dipping liquid phase epitaxy: Effects on metal-insulator-semiconductor properties. J Vac Sci Technol B 9:1852
Charlton DE (1982) Recent developments in cadmium mercury telluride infrared detectors. J Crystal Growth 59:98–110
Chen AB, Sher A (1995) Semiconductor alloys: physics and materials engineering. Springer, New York
Chen XQ, Chu JH, Zhang SM, et al. (1993) J Funct Mater (Chinese) 24:231
Chen XQ, et al. (1997) Patent. ZL 97 106610.8 (Chinese)
Cheng DT (1985) J Vac Sci Technol A 3:128
Cheuvart P, El-Hanani U, Schneider D, Triboulet R (1990) CdTe and CdZnTe crystal growth by horizontal bridgman technique. J Cryst Growth 101:270–274
Chiang CD, Wu TB (1989) Low temperature LPE of $Hg_{1-x}Cd_xTe$ from Te-rich solution and its effects. J Cryst Growth 94:499–506
Chiang CD, Wu TB, Ghung WC, Yang J, Pang YM (1988) A new attachment for stable control of mercury pressure in the slider LPE of $Hg_{1-x}Cd_xTe$. J Cryst Growth 87:161–168
Coriell SR, Hurle DTJ, Sekerka RF (1976) Interface stability during crystal growth: The effect of stirring. J Crystal Growth 32:1–7
Daruhaus R, Vimts G (1983) The Properties and applications of the $Hg_{1-x}Cd_xTe$ alloy system, in narrow gap semiconductors. In: Springer Tracts in modern physics Vol. 98. Springer, New York, p. 119
Destéfanis GL (1988) Electrical doping of HgCdTe by ion implantation and heat treatment. J Cryst Growth 86:700–722
Djuric Z, Jovic V, Djinovic Z, et al. (1991) J Mater Sci: Mater Electron 2:63
Doty FP, Buther JF, Schetzina JF, et al. (1992) J Vac Sci Technol B 10:1418
Dunn CG, Koch EF (1957) Acta Metal 5:548
Durose K, Russell GJ (1988) J Cryst Growth 86:471
Durose K, Turnbull A, Brown P (1993) Mater Sci Eng B 16:96
Edwall DD, Gertner ER, Tennant WE (1984) J Appl Phys 55:1453

Fang JX, Lu D (1980) Solid state physics. Shanghai Science and Technology Press, Shanghai, p. 61

Fiorito G, Gaspavvini G, Passoni D (1978) J Electrochem Soc 125:315

Gartner KJ, et al. (1972) J Crystal Growth 13/14:619

Geibel C, Maier H, Ziegler J (1986) SPIE Mater Technol IR Detectors 659:110

Golacki Z, Gorska M, Makowski J, et al. (1982) J Cryst Growth 56:213

Harman TC (1972) J Electron Mater 2:230

Harman TC (1979) J Electron Mater 8:191

Harman TC (1980) J Electron Mater 9:945

Harman TC (1981) J Electron Mater 10:1069

Harman TC (1993) J Electron Mater 22:1165

Harrison WA (1980) Electronic structure and properties of solids. WH Freeman and Company, San Francisco

Hartley RH, Folkard MA, et al. (1992) J Crystal Growth 117:166

He L, Yang JR, Wang SL, et al. (1998) SPIE 3553:13

He L, Wu Y, Wang S, et al. (2000) SPIE 4086:311

Herman MA, Pessa M (1985) J Appl Phys 57:2671

Herning PE (1984) J Electron Mater 13:1

Hirth JP, Lothe J (1982) Theory of dislocations, 2nd Ed. Wiley, New York

Hsieh JJ (1974) J Cryst Growth 27:49

Hurle DTJ (1969) J Cryst Growth 5:162

Inoue N (1991) MBE monolayer growth control by in-situ electron microscopy. J Cryst Growth 111:75–82

James RW (1954) The optical principles of the diffraction in crystals. Wiley, New York, p. 59

Jayatirtha HN, Henderson DO, Burger A (1993) Appl Phys Lett 62:573

Jin G (1981) Infrared Phys Technol (Chinese) 5:80

Johnson SM, Kalisher MH, Ahlgren WL, et al. (1990) Appl Phys Lett 56:946

Johnson SM, Avigil J, et al. (1993) J Electron Mater 22:83

Jones CE, James K, Merz J, et al. (1985) J Vac Sci Technol A 3:131

Kalisher MH, Herning PE, Tung T (1994) Prog Crystal Growth 29:41

Khan AA, Allred WP, Dean B, et al. (1986) J Electron Mater 15:181

Konnikov SG (1975) US patent 3,902,924

Koppel P, Owens KE, Longshore RE (1989) SPIE 1106:70

Landau LD, Lifshitz EM (1958) Statistical physics. Addison-Wesley Publishing Company, Reding, MA

Lanir M, Wang CC, Vanderwyck AHB (1979) Appl Phys Lett 34:50

Lay KY, Nichols D, McDevitt S, et al. (1988) J Cryst Growth 86:118

Li B (1996) Ph.D. Thesis, Shanghai Institute of Technical Physics, CAS

Li B, Chu JH, Chen XQ, et al. (1995) J Cryst Growth 148:41

Li B, Chu JH, Zhu JQ, Chen XQ, Cao JY, Tang DY (1997) Characterization of Hg(1-x)Cd(x)Te epitaxial films on various vicinal planes. Acta Physica Sinica 46:1168–1173

Long D, Schmit JL (1970) Mercury–cadmium telluride and closely related alloys. In: Willardson RK, Beer AC (eds.) Semiconductors and semimetal Vol. 5. Academic Press, London, p. 175

Lorentz MR (1962) J Phys Chem Solids 23:939
Lu YC, Shiau JJ, Fiegelson RS, et al. (1990) J Cryst Growth 102:807
Lyon de TJ, Rajavel RD, Jensen JE, et al. (1996) J Electron Mater 25:1341
Maracas GN, Edwards JL, Shiralagi K, et al. (1992) J Vac Sci Technol A 10:1832
Matthews JW (1975) Coherent interfaces and misfit dislocations. In: Matthews JW (ed.) Epitaxial growth Part B. Academiac Press, London, p. 559
Merkel KG, Snyder PG, Woollam JA, et al. (1989) Jpn J Appl Phys 28:133
Min NB (1982) Physical base of crystal growth. Shanghai Science and Technology Press, Shanghai, p. 17
Mroczkowski JA, Vydyanath HR (1981) J Electrochem Soc 128:655
Muhlberg M, Rudolph P, Genzel C, et al. (1990) J Cryst Growth 101:275
Muller MW, Sher A (1999) Phys Rev Lett 74:2343
Mullins WW, Sekerka RF (1964) J Appl Phys 35:444
Muranevich A, Roitberg M, Finkman E (1983) J Cryst Growth 64:285
Nagahama K, Ohkata R, Nishitani K, et al. (1984) J Electron Mater 13:67
Nelson DA, et al. (1980) SPIE 225:48
Nemirovsky Y, Margalit S, Finkman E, et al. (1982) J Electron Mater 11:133
Neuhaus D (1991) Device for radiometric or high temperature radiance measurement in a vacuum. Rev Sci Instrum 62:2291–2292
Nouruzi-Khorasani A, Jones IP, Dobson PS, et al. (1989) J Cryst Growth 96:348
Okane DF, et al. (1972) J Crystal Growth 13/14:624
Orioff GJ, et al. (1994) J Vac Sci Technol A 12:1252
Panish MB (1970) J Electrochem Soc 117:1202
Parker SG, Weirauch DF, Chandra D (1988) J Cryst Growth 86:173
Parthier L, Boeck T, Winkler M, et al. (1991) Behaviour of impurities in (Hg,Cd) Te layers grown by LPE. Crystal Properties Preparation 32–34:294
Pfeiffer M, Muhlberg M (1992) J Cryst Growth 118:269
Pultz GN, Norton PW, Krueger EE, et al. (1991) Growth and characterization of p-on-n HgCdTe liquid-phase epitaxy heterojunction material for 11–18 μm applications. J Vac Sci Technol B 9:1724
Qadri SB, Dinan JH (1985) Appl Phys Lett 47:1066
Reif F (1965) Fundamentals of statistical and thermal physics. McGraw-Hill, New York
Rhiger DR, Sen S, Peterson JM, Chung H, Dudley M (1997) J Electron Mater 26:515
Rudolph P, Neubert M, Muhlberg M (1993) J Cryst Growth 128:582
Rudolph P, Engel A, Schentke A (1995) J Cryst Growth 147:297
Schaaka HF, Tregilgas JH, Beck JD, et al. (1985) J Vac Sci Technol A 3:143
Scheel HJ, Elwell D (1973) J Electrochem Soc 120:818
Shen J, Li HD, Chen JZ, et al. (1976) Infrared Phys Technol (Chinese) 4–5:11
Shen J, Chen JZ, Tang DY, et al. (1981) Chinese Sci Bull 10:593
Smith DL, Pickhardt VY (1975) J Appl Phys 46:2366
Steininger J (1970) J Appl Phys 41:21
Steininger J (1976) J Electron Mater 5:299
Steininger J, Strauss AJ, Brebrick RF (1970) J Electrochem Soc 117:1305
Stockbarger DC (1936) Rev Sci Instrum 7:133

Strong RL, Anthony JM, Gnade BE, et al. (1986) J Vac Sci Technol A 4:1992
Suh SH, Stervenon DA (1988) J Vac Sci Technol A 6:1
Swink LN, Brud MJ (1970) Matall Trans 1:629
Szofran FR, Lehoczky SL (1981) J Electron Mater 10:1131
Takami A, Kawazu Z, Takiguchi T, et al. (1992) J Cryst Growth 117:16
Tennant WE, et al. (1992) J Vac Sci Technol B 10:1359
Tranchart JC, Latorre B, Foucher C, et al. (1985) J Cryst Growth 72:468
Triboulet R, Marfaing Y (1981) J Cryst Growth 51:89–96
Tung T (1988) J Cryst Growth 86:161
Tung T, Golonka L, Brebrick RF (1981a) J Electrochem Soc 128:1601
Tung T, Golonka L, Brebrick RF (1981b) J Electrochem Soc 128:451
Tung T, Su Ching-Hua, Liao Pok-Kai, et al. (1982) J Vac Sci Technol 21:117
Tung T, Kalisher MH, Stevens AP, et al. (1987) In: Farrow RFC, Schetzina JF, Cheung JT (eds.) Materials for infrared detectors and sources: Mater Res Soc Symp Proc Vol. 90. Mater Res Soc, Pittsburgh, p. 321
Ueda R, et al. (1972) J Crystal Growth 13/14:668
Varesi JB, Bornfreund RE, Childs AC, et al. (2001) J Electron Mater 30:566
Verleur HW, Barker Jr AS (1966) Phys Rev B 149:71
Vojdanl S, et al. (1974) J Crystal Growth 24/25:374
Vydyanath HR (1981a) J Electrochem Soc 128:2609
Vydyanath HR (1981b) J Electrochem Soc 128:2619
Vydyanath HR, Hiner CH (1989) J Appl Phys 65:3080
Vydyanath HR, Ellsworth JA, Devaney CM (1987) J Electron Mater 16:13
Vydyanath HR, Ellsworth JA, Kennedy JJ, et al. (1992) J Vac Sci Technol B 10:1476
Vydyanath HR, Ellsworth JA, Parkinson JB, et al. (1993) J Electron Mater 22:1073
Wan CF (1987) J Cryst Growth 80:270
Wan CF, Weirauch DF, Korenstein R, et al. (1986) J Electron Mater 15:151
Wang J (1989) Ph.D. thesis, Shanghai Institute of Technical Physic, CAS
Wang SL (1996) Ph.D. Thesis, Shanghai Institute of Technical Physic, CAS
Wang CC, Shin SH, Chu M, et al. (1980) J Electrochem Soc 127:175
Wermke A, Boeck T, Gobel T, et al. (1992) J Cryst Growth 121:571
Winkler M, Teubner T, Jacobs K (1991) Crystal Properties Preparation 36–38:218
Winton GH, Faraone L, Lamb R (1994) J Vac Sci Technol A 12:35
Wood RA, Hager RJ (1983) Horizontal slider LPE of HgCdTe. J Vac Sci Technol A 1:1608
Wood RA, Schmit JL, Chung HK, et al. (1985) J Vac Sci Technol A 3:93
Woolley JC, Ray B (1960) J Phys Chem Solids 13:151
Wu OK (1993) Mat Res Soc Symp Proc 302:423
Xiao JR (1981) Research Report, Shanghai Institute of Technical Physic
Xiao JR (1995) Survey report of CAS
Yang JR (1997) Chinese Patent ZL 97 10663.9
Yang JR, Yu ZZ, Tang Dingyuan (1985) J Cryst Growth 72:275
Yang JR, Yu ZZ, Liu JM, et al. (1990) Infrared Res (Chinese) 9:351
Yang JR, et al. (1998) Chinese Patent ZL 98 111054.1

Yang JR, et al. (2001) Chinese Patent ZL 01 131924.0
Yang JR, Chen XQ, He L (2002) SPIE 4795:76
Yasumura K, Murakami T, Suita M, et al. (1992) J Cryst Growth 117:20
Yellin N, Eger D, Shachna A (1982) J Cryst Growth 60:343
Yu FJ (1979) Infrared Phys Technol (Chinese) 5:45
Yu ZZ (1984) Narrow-gap semiconductor materials. Shanghai Institute of Physics
Yu ZZ, Jin G, Chen XQ, et al. (1980) Acta Phys Sin 29:1
Yu MF, et al. (1999) Chinese J Semicond 20:378
Zanio K (1978) Semiconductors and semimetals Vol. 13. Academic Press, New York, p. 125
Zemansky MW (1951) Heat and thermodynamics: an intermediate textbook for students of physics, chemistry, and engineering. McGraw-Hill, New York
Zhu JQ (1997) Ph.D. Thesis, Shanghai Institute of Technical Physics, CAS
Zhu J, Chu JH, Li B, et al. (1997) J Cryst Growth 117:61

3 Band Structures

3.1 General Description of Band Structures

3.1.1 Band Structure Theory Methods

Semiconductor crystals are composed of a large number of atoms with nuclei and electrons. A solution of the Schrödinger equation of this many-body problem would yield the electron states and their characteristics for a given arrangement of the atoms. The equilibrium arrangement would be found by varying it until the minimum energy configuration is located. Semiconductor band theory simplifies this many-body problem to a single electron problem. The Born–Oppenheimer adiabatic approximation takes account of the fact that electron masses are small compared to those of nuclei and therefore electrons respond more rapidly than nuclei to changes in local potentials. The electrons are divided into two classes, core electrons that are localized on their nucleons to form ions, and valence electrons that move through the solid lattice. The adiabatic approximation is the basis of a theory in which the movements of the valence electrons can be decoupled from those of ions. In this way, the many-body problem is reduced to a many-electron problem with the electrons moving in the localized potentials of the ions and the other valence electrons. Then the Hartree-Fock approximation, in which each electron is assumed to be in the fixed potential field of all the lattice ions, and in the Coulomb potential of an average distribution of all other electrons, reduces the many electron problem to a single electron problem. Because the potential distribution from both the lattice ions and the averaged Coulomb interaction from all other electrons is periodic, the problem is reduced to the movement of an electron in a periodic potential.

There are many approximation methods to deal with the problem of a single electron in a three-dimensional periodic potential. Normally the electron wavefunction in a crystal is assumed to be a superposition of Bloch functions. A Bloch function is a product of a plane wave and a pre-exponential factor that has the periodicity of the Bravais lattice. These wave functions form a complete set of basis functions in momentum space. The corresponding set in real space, Fourier transforms of the Bloch functions, are Wannier functions, which of course also are a complete set. If the superposition of Bloch functions (or Wannier functions) is substituted into the Schrödinger equation the result is a secular determinant.

The solution of the secular equation yields the coefficients of the eigenvectors and the corresponding eigenvalues. This is one of the basic theoretical approaches to semiconductor energy band theory. The differences among computational methods lie in the selection of basis-functions (e.g. Bloch or Wannier functions) and approximations to the potentials. The two most elementary treatments are the "nearly free electron" and the "tight binding" methods.

There is a whole collection of methods to deal with the energy band structure of solids, such as the "orthogonalized plane-wave" (OPW), the "tight-binding," the "augmented plane-wave" (APW), the Green's function method of Korringa, Kohn, and Rostoker (KKR), the empirical pseudopotential, and the $\boldsymbol{k}\cdot\boldsymbol{p}$ perturbation methods. Some of these are used in combination with others.

The orthogonalized plane-wave method (OPW), proposed by Herring (1940), starts with plane waves and unmodified atomic states for the core electrons. Then the plane waves are orthogonalized to all the core states of all the ions in the solid (Harrison 1970). The OPW wave functions then are used as the basis functions for solutions to the Schrodinger equation for the valence electrons. This method overcomes the difficulty associated with the rapid oscillation of valence electron wave functions in the regions near the atom cores. Moreover if the wave functions expressed as a superposition of OPW wave functions are substituted into the Schrodinger equation and terms are rearranged, the new equation has a smooth but non-local pseudopotential with eigenfunctions expressed in terms of plane waves, and eigenvalues identical to those of the original equation. Using a similar method, one can describe the symmetry points in the Brillouin zone through a combination of normalized OPW functions, and produce an irreducible representation for the basis functions of the crystal space group.

The APW method computes the bands using a muffin-tin potential (Fig. 3.1) approximation to the real potential distribution of a solid. The muffin-tin potential is defined to be a constant potential between spheres surrounding the atoms, and a spherically symmetrical potential within the spheres. The wave functions of electron states inside the spherically symmetrical potential satisfy the Schrödinger equation that is solved in spherical coordinates. The wave functions outside the spheres are superpositions of plane waves. The coefficients of these superpositions are chosen

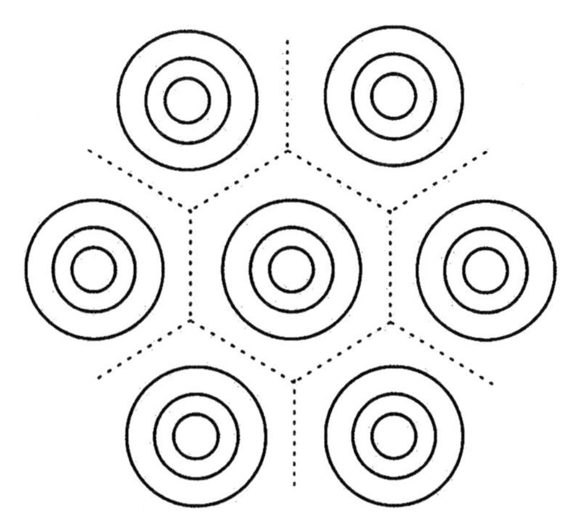

Fig. 3.1. Schematic diagram of a muffin-tin potential

to match the interior wave functions at the boundaries of the spheres. This is the origin of the name "augmented plane wave" (APW) method. The method was proposed first by Slater in 1937 (Slater 1937). With the development of computer techniques, it has become straightforward to perform such a calculation. The number of APW functions required depends on the crystal structure as well as the nature of the states contributing to the bands. Normally there is a faster convergence with bands originating from *s* and *p* atomic states, and a slower convergence for those of *d* bands.

The KKR method is also aimed at solving the problem of a muffin-tin potential distribution. The Schrödinger equation can be viewed as describing the motion of a particle scattered by the potential. Korringa (1947) separated the wavefunctions into two components, an incident wave and a scattered wave. Kohn and Rostoker (1954) solved the problem starting from the integral form of the Schrödinger equation. This integration is written as the sum over all the spherical terms of the muffin-tin potential, and then using the periodicity of the lattice it is reduced into an integration surrounding the center of one muffin-tin sphere. It is in fact very similar to the APW method. In the KKR method, however, the sum is over all the reciprocal lattice vectors. In this way, the secular equation contains the contribution from different spherical harmonic functions.

The pseudo-potential derived from OPW's is still quite complicated for some crystals. An alternative replaces the core potential with a parameterized weak potential. The parameters are selected to fit a few of the observed band structure symmetry points in the Brillouin zone. This is the so-called empirical pseudo-potential method.

In the ***k·p*** method, perturbation theory is used to study wavefunctions according to the crystal symmetry, so that band structures around some points in k space can be obtained. The coefficients in the energy expressions can be found through some limited, experimentally determined parameters, such as E_g and m^*, so that the energy dispersion can be determined.

The ***k·p*** method has become very important in calculating regions of band structures near extreme points at tops and bottoms of bands. Here ***k*** is the electron wave vector and ***p*** is its momentum. The extra ***k·p*** term in the Hamiltonian arises from the kinetic energy of the electron, taking into account that is moving through a periodic lattice. It was devised by Seitz; Shockley; Dresselhans, Dip, and Kiffel (DDK); and especially by Kane (1940). Seitz proposed an expression for effective masses based on the ***k·p*** method. In 1950, Shockley applied the effective mass equation to some more complicated degenerate band structures. In a 1955 paper about cyclotron resonance, Dresselhans set the foundation for the extension of the ***k·p*** method by adding a spin–orbit interaction term to the equation (Dresselhaus 1955). The spin–orbit interaction arises from a relativistic correction to the Schrodinger equation. Kane dealt with the band structures of p-type Si and Ge in 1956, and InSb in 1957 using this method (Kane 1957). In 1966, Kane published a systematic review of the ***k·p*** method, which has become known as the Kane theory (Kane 1966). Through application of second order perturbation theory to InSb, we can understand how the spin–orbit interaction acts to split the threefold orbital degenerate valence band (six fold including the spin degeneracy) into two bands; the upper valance band (the band edge) is still two fold degenerate and the split off band is non-degenerate. The ***k·p*** interaction further splits the upper band into a heavy hole and light hole band. At the center of the band structure where $k = 0$, the valence band remains two fold degenerate, but away from that point the degeneracy is split. The ***k·p*** interaction also tends to sharpen the bottom of the conduction band which has the effect of decreasing the effective mass relative to the free electron mass. As we will see the conduction band and light hole effective masses are directly proportional to the band gap. Thus in narrow gap semiconductors one always finds small effective masses.

In Sect. 2.5.4 in connection with the treatment of native point defects one version of self-consistent density functional methods was introduced. In general this method starts with an assumed electron density distribution, from which the electron–electron interaction and screening is calculated to obtain an effective one electron periodic potential. Then the Schrödinger equation is solved for the one-electron wave functions. This yields a different electron density and periodic potential. The procedure is repeated until a self-consistent result is obtained. The method is computationally intensive, and a variety of approximation methods have been devised to produce results. The accuracy of the predictions is remarkable, and steadily improving. While this true, because the method is computational, it is difficult to gain insight into the underlying mechanisms responsible for effects. As a consequence, in the following presentations the older methods will be used.

3.1.2 A Brief Treatment of Band Structures of Narrow Gap Semiconductors

II–VI compound semiconductors such as HgTe and CdTe have zincblende structures (Dornhaus and Nimtz 1983; Long and Schmit 1973). The lattice constant of HgTe is 6.46 Å and that of CdTe is 6.48 Å. They can form HgCdTe solid solutions with any composition.

HgCdTe alloys also have the zincblende cubic structure. It consists of two interpenetrating face-centered cubic (fcc) lattices, displaced along the body diagonal of the cubic cell. In this configuration each ion has four near neighbors located in a tetrahedral arrangement. The positive anions, Cd or Hg, share the sites on one sublattice, while the negative cations, Te, occupy the sites on the other sublattice. Every unit cell includes four atoms, two cations and two anions. There are six valence electrons outside the core of Te in a $5s^2$, $5p^4$ configuration, while there are two valence electrons outside the cores of Hg or Cd, in $6s^2$, and $5s^2$ configurations respectively.

The main contribution to bond energies in these crystals is covalent binding but there is a substantial ionic contribution (Chen and Sher 1995). Because the lowest unfilled Te states lie below the energy of the filled Hg and Cd states there is a net electron transfer from the Hg and Cd cations to the Te anions. This transfer direction occurs despite the fact that Te has more electrons than Hg or Cd.

The first Brillouin zone of this structure is a truncated octahedron. The zincblende crystal belongs to the T_d point group; its Brillouin zone is shown in Fig. 3.2.

The lattice constant a_0 (the length of the edge of a unit cube) of HgCdTe can be measured by an X-ray diffraction method (Woolley and Ray 1960; Dornhaus and Nimtz 1983). It was found that the variation of a_0 with composition x is slightly non-linear, bowing below a linear dependence, as is shown in Fig. 3.3. The density dependence of the HgCdTe on composition, which can be determined by the specific gravity method, is linear as also shown in Fig. 3.3 (Blair and Newnham 1961).

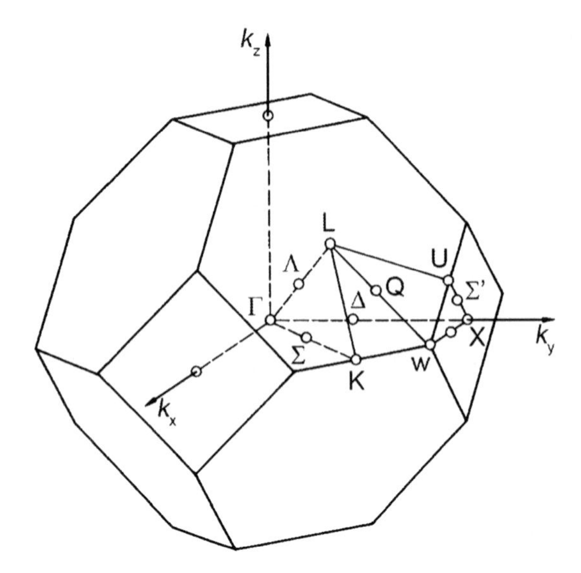

Fig. 3.2. The first Brillouin zone in a zincblende structure. The labels on the symmetry points are in the standard notation adapted in this field

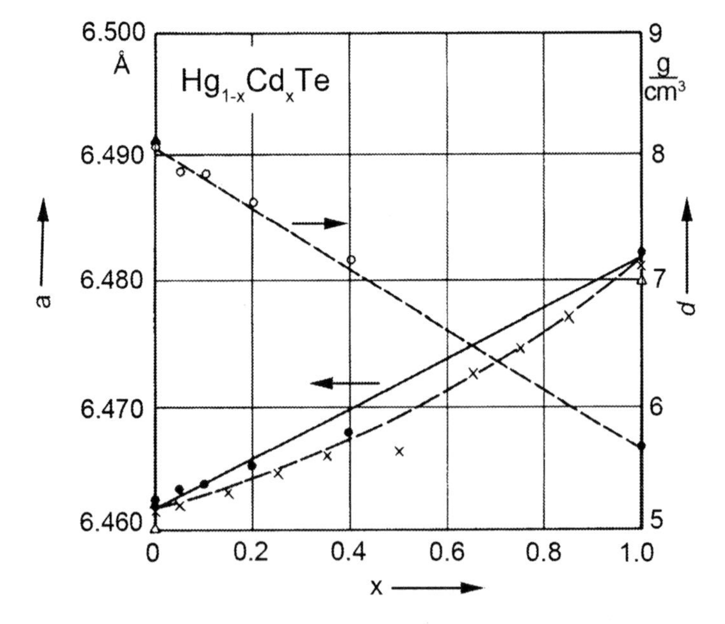

Fig. 3.3. Dependence of lattice constant and density of HgCdTe on composition

Dresselhaus (1955) and Parmenter (1955) have studied the band structures of zincblende compounds in detail. They concluded that the extreme values of both the conduction band and the valence band are located at Γ, the center of the first Brillouin zone and therefore they are direct gap semiconductors. While this is true for II–VI compounds, it is not the case for all III–V compounds. Better calculations, in agreement with experiments, find that some of these compounds have indirect gaps, with valence band edges at the Γ point but with conduction band edges at the x or L points (Chen and Sher 1978, 1995). As seen in Fig. 3.2 the x points are in (100) directions and the L points are in (111) directions. The top of the valence band is made up of p-symmetry states, mostly Te p-states and therefore, including spin, is six fold degenerate. The bottom of the conduction band is made up of s-symmetry states, mostly derived from Cd or Hg s-states, and is two fold degenerate. The effect of the $\boldsymbol{k \cdot p}$ and the spin–orbit interaction terms which break the symmetry of the Hamiltonian, splits the degeneracy of the valence band into what is called Γ_8 bands ($m_J = \pm 3/2$), a heavy hole and a light hole band, and a split-off band, called Γ_7 ($m_J=1/2$). At $k=0$, the Γ_8 band is four-fold degenerate and Γ_7 is two-fold degenerate. The split-off energy difference between Γ_8 and Γ_7, is denoted Δ. The bottom of the conduction band is denoted Γ_6.

For the band structure in CdTe and HgTe, the relativistic corrections leading to the spin–orbit interaction must be considered. These corrections become progressively larger with atomic number, because in the presence of their high charges electrons move faster in their orbits. Thus in Cd, Hg and Te the effect is large. This effect is included in the Dirac relativistic formula. According to Herman et al. (1963), the Hamiltonian for a single electron includes, besides the normal non-relativistic terms H_1, the correction terms H_D, H_{mv} and H_{so}, which represent the Darwin interaction, the mass–velocity interaction, and the spin–orbit interaction, respectively. As shown in Fig. 3.4, the energy levels in the two compounds derived from the non-relativistic terms H_i is the same. The large difference between the atomic masses of Hg and Cd (M_{Hg}=200.6, M_{Cd}=112.4) leads to very different corrections induced by H_D and H_{mv} respectively, in the two compounds. Because the Te dominates the spin–orbit interaction, the correction coming from it for the two compounds is the same. Therefore, the contribution from the relativistic-interaction that decreases the energy of the s-state, in HgTe interchanges the positions of the Γ_6 state with the Γ_8 state. This is the underlying reason that HgTe is a semimetal.

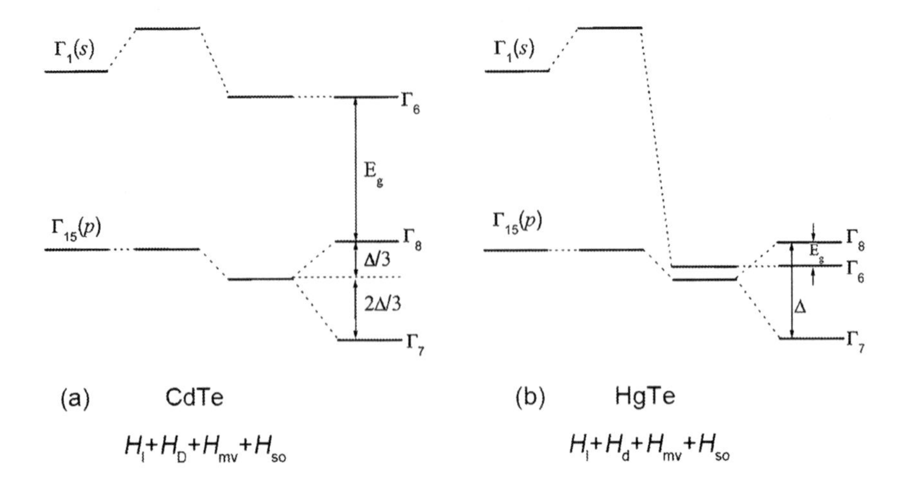

Fig. 3.4. The formation of energy levels at Γ in CdTe and HgTe

Figure 3.5a shows the band structure near the Γ point in CdTe. Γ_6 forms the conduction band and Γ_8 forms the valence band. There is an energy difference of 1.6 eV between them. Γ_7 is the spin–orbit split-off band, which lies below the Γ_8 band with an energy distance of $\Delta = 1\text{eV}$. Compared with CdTe, HgTe has an inverted band structure. Because of the relativistic effect, the Γ_8 state lies above the Γ_6 state, so what is normally thought of as the band gap $E_g = E(\Gamma_6) - E(\Gamma_8)$ is smaller than zero, and is about –0.3 eV at 4 K.

In HgCdTe alloys, Hg and Cd are located on cation sublattice sites nearly randomly. There are short range correlations (Chen and Sher 1985) and mesoscopic size regions with concentrations that differ from the average value (Muller and Sher 1999), but these effects can be neglected in first order. Nevertheless, there is no translational periodicity and Bloch functions do not exist. This difficulty was overcome initially by using a virtual-crystal approximation (VCA) method to solve the problem (Nordheim 1931; Phillips 1973). This method employs a concentration weighted average potential to replace the real crystal potential V induced by Hg and Cd atoms:

$$\bar{V} = \mathrm{x} V_{\mathrm{Cd}} + (1 - \mathrm{x}) V_{\mathrm{Hg}} \tag{3.1}$$

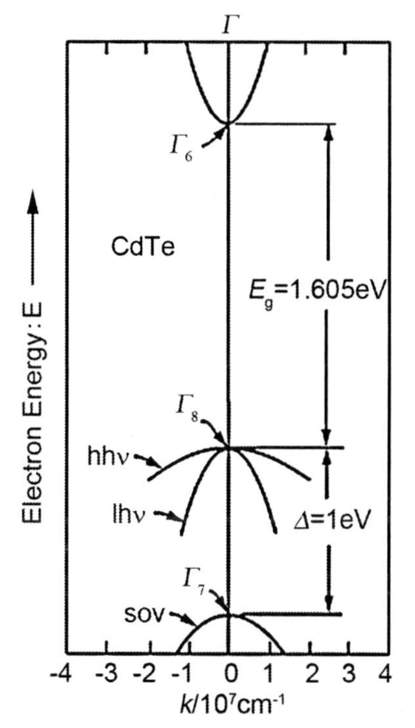

Fig. 3.5a. Band structures of CdTe at Γ

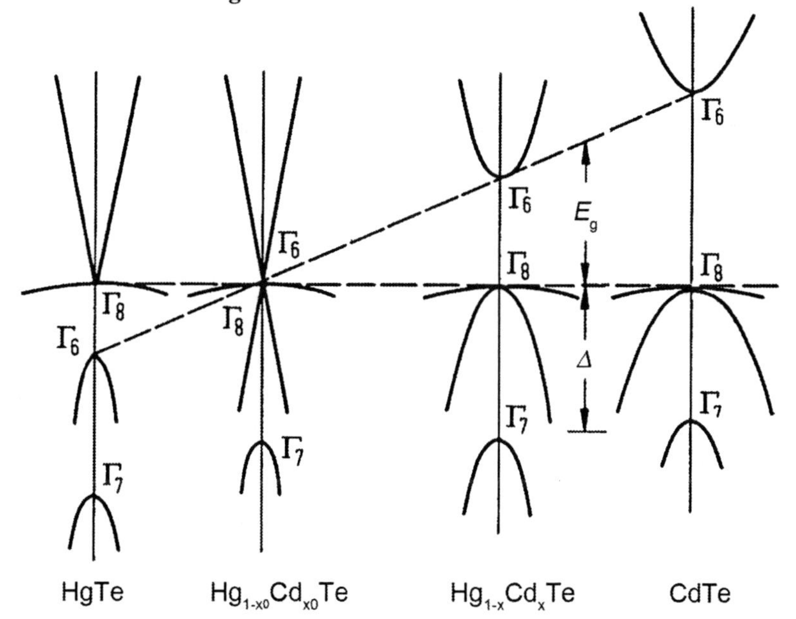

Fig. 3.5b. Change of band structure of $Hg_{1-x}Cd_xTe$ with composition x near Γ

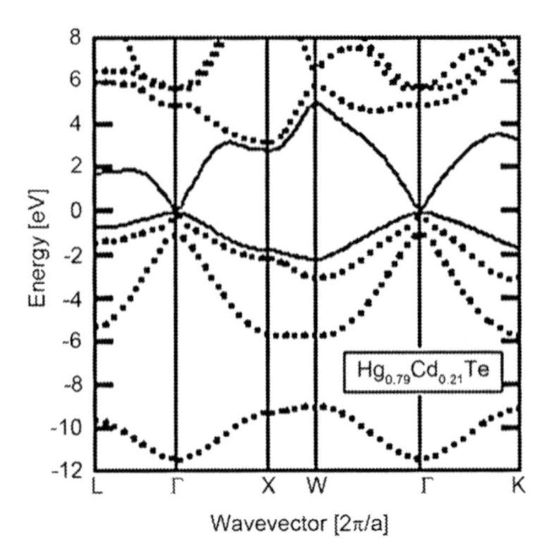

Fig. 3.6. Electronic band structure of $Hg_{0.79}Cd_{0.21}Te$ at $T = 0$. The *solid lines* denote the conduction and valence band states connected to those at the fundamental gap at Γ

In which V_{Cd} and V_{Hg} are the crystal potentials produced by the sublattice of Cd or Hg, respectively. If the average $\overline{V}$ of the cation potentials is taken, translational periodicity of primitive cells is restored, and Bloch functions can be found. Then the dispersion relation around the Γ point can be determined. Figure 3.5b shows the dependence on the composition x of the band structure of HgCdTe around Γ. In the VCA the band gap of the inverted structure of HgTe varies nearly linearly with composition to that of CdTe. With a decrease of the Hg concentration in the crystal, the relativistic effect also decreases, together with the value of $E(\Gamma_8) - E(\Gamma_6)$, which becomes 0 at $x = x_0$. For $x < x_0$, the alloy has the same structure as the semi-metal HgTe. For $x > x_0$, the state Γ_6 is above the state Γ_8, and at the same time the Γ_6 band and the Γ_8 band change the signs of their curvatures to those of semiconductors as the gap opens.

While the VCA model is appealing and we will use it along with $\boldsymbol{k \cdot p}$ theory to lend insight into trends in narrow gap $Hg_{1-x}Cd_xTe$ band structures, it is not valid. In fact this is the first alloy in which was found experimentally that VCA fails (Spicer et al. 1982). The major part of the failure occurs well away from the fundamental band edges. However there are small but important effects there as well. The coherent potential approximation (CPA) theory solves this dilemma, properly predicting the experimental results (Chen and Sher 1995). An example of a full band structure of the $Hg_{0.79}Cd_{0.21}Te$ alloy at T = 0 is in Fig. 3.6 (Krishnamurthy et al. 2006). The main differences in the behavior of states near the gap edges are that the concentration variation of the gap bows slightly below a straight line, and the shape of the bottom of the conduction band more closely resembles a hyperbola than a parabola. The same consideration applies to the light hole effective mass. CPA applied to this system must be done in a formalism that ends in a computer intensive numerical analysis. Thus, we will continue to develop the VCA analysis to gain insight into the mechanisms responsible for the behavior of these materials in device applications, but remember to rely on more accurate results when doing device design. An outline of the VCA/***k·p*** procedure is in Appendix 3B.

3.2 The *k·p* Perturbation Method and Secular Equations

The ***k·p*** perturbation method is one effective way to deal with band structures of narrow gap semiconductors. All of the characteristic trends of InSb and $Hg_{1-x}Cd_xTe$ at Γ can be deduced. In this section the ***k·p*** formalism and the resulting secular equation is introduced. In the following sections, calculated energy band structures and some often-used formulas are deduced.

3.2.1 The *k·p* Formalism

Where does the ***k·p*** term originate? When the kinetic energy operator in the Schrödinger equation for a single electron in a periodic potential is applied to a Bloch-function a term with ***k·p*** is produced. The ***k·p*** method treats this term as a perturbation.

In a periodic potential, the Schrödinger equation for a single electron is:

$$H\Psi = \left(\frac{p^2}{2m} + V(\mathbf{r})\right)\Psi = E\Psi \tag{3.2}$$

where Ψ is a Bloch-function (see Appendix 3A):

$$\Psi = u_{n\mathbf{k}}(\mathbf{r})e^{i\mathbf{k}\cdot\mathbf{r}} \tag{3.3}$$

The factor $u_{n\mathbf{k}}(\mathbf{r})$ has the same period as $V(\mathbf{r})$. $\mathbf{k}$ is the wave vector in the first Brillouin zone (see Appendix 3A), and n is the band index. Placing (3.3) into (3.2), we have:

$$\left(\frac{p^2}{2m} + \frac{\hbar}{m}\mathbf{k}\cdot\mathbf{p} + \frac{\hbar^2 k^2}{2m} + V \right) u_{n\mathbf{k}} = E_n(k) u_{n\mathbf{k}} \tag{3.4}$$

or

$$\left(\frac{p^2}{2m} + \frac{\hbar}{m}\mathbf{k}\cdot\mathbf{p} + V \right) u_{n\mathbf{k}} = \left(E_n - \frac{\hbar^2 k^2}{2m} \right) u_{n\mathbf{k}}(\mathbf{r}) \tag{3.5}$$

Define:

$$H_{\mathbf{k}_0} = \frac{p^2}{2m} + \frac{\hbar}{m}\mathbf{k}_0\cdot\mathbf{p} + \frac{\hbar^2 k_0^2}{2m} + V \tag{3.6}$$

Then we find:

$$\left\{ \mathrm{H}_{\mathbf{k}_0} + \frac{\hbar}{m}(\mathbf{k}-\mathbf{k}_0)\bullet\mathbf{p} + \frac{\hbar^2}{2m}\left(k^2 - k_0^2\right) \right\} u_{n\mathbf{k}} = \mathrm{E}_n(\mathbf{k}) u_{n\mathbf{k}} \tag{3.7}$$

If the wave function at $\mathbf{k} = \mathbf{k}_0$ is known, then the wave function at any wave vector $\mathbf{k}$ can be expressed as:

$$u_{n\mathbf{k}}(\mathbf{r}) = \sum_{n'} C_{n'n} \,|(\mathbf{k}-\mathbf{k}_0)|\, u_{n'\mathbf{k}_0}(\mathbf{r}) \tag{3.8}$$

Substituting (3.8) into (3.7), multiplying the both sides of the equation by $u^*_{nk_0}(\mathbf{r})$, and integrating over the volume of a unit cell, a secular equation for the eigenvalues is obtained:

$$\sum_{n'} \left\{ \left[E_n(\mathbf{k}_0) + \frac{\hbar^2}{2m}\left(k^2 - k_0^2\right) \right] \delta_{nn'} + \frac{\hbar}{m}(\mathbf{k}-\mathbf{k}_0)\cdot\mathbf{p}_{nn'} \right\} C_{n'n} = E_n(\mathbf{k}) C_{nn} \tag{3.9}$$

where

$$\mathbf{p}_{nn'} \equiv \int u^*_{n\mathbf{k}_0}(\mathbf{r})\,\mathbf{p}\,u_{n'\mathbf{k}_0}(\mathbf{r})\,d\mathbf{r} \tag{3.10}$$

In the secular determinant arising from (3.9), the terms in the bracket are diagonal, and the terms of the form $\frac{\hbar}{m}(\mathbf{k}-\mathbf{k}_0)\bullet\mathbf{p}_{nn'}$ are off diagonal. When $\mathbf{k}$ is near $\mathbf{k}_0$, this off-diagonal term can be treated as a perturbation. Using perturbation theory the eigenvalues and eigen functions can be calculated.

In the calculations above, there are many $u_{n'\mathbf{k}}(\mathbf{r})$ wave functions. It would become very complicated if all the states were considered. Kane divided these states into two categories: A and B. There are strong interactions among the states in category A. But the interactions between the states in A and those in B are very weak. In such a first order approximation, the index n'in $u_{n'\mathbf{k}_0}(\mathbf{r})$ includes only several basis functions. In a higher order approximation, the interaction between the states in A and B can be added as a second order perturbation. In Kane's calculation, the states near the bottom of the conduction band and near the top of the valence band are included in A, while the states from remote bands are included in B.

Taking $h_{nn'}(\mathbf{k})$ defined as:

$$h_{nn'}\left(\mathbf{k}\right) \equiv E_n\left(\mathbf{k}_0\right)\delta_{nn'} + \frac{\hbar}{m}\left(\mathbf{k} - \mathbf{k}_0\right)\mathbf{p}_{nn'} \tag{3.11}$$

and $h'_{nn'}$ as the interaction matrix elements after renormalization by inclusion of the of states in the B manifold as a second order perturbation we have:

$$h'_{nn'} = h_{nn'} + \sum_m^B \frac{h_{nm}h_{mn'}}{\mathrm{E}_n - h_{mm}} \tag{3.12}$$

where n, n' belongs to A, while m belongs to B. E_n is the eigenvalue of state n. Substituting (3.11) and (3.12) into (3.9) and dropping the term in $k^2 – k_0^2$ by limiting the treatment to k values close to k_0 yields a new secular equation:

$$\sum_n^A (h'_{nn'} - E_n\delta_{nn'})C_{nn'} = 0 \tag{3.13}$$

If m belongs to B and n belongs to A, then

$$C_{mn} = \sum_{n'}^A \frac{h_{nn'}C_{n'm}}{\mathrm{E}_n - h_{n'n'}}, \quad \begin{pmatrix} m \subset B \\ n, n' \subset A \end{pmatrix} \tag{3.14}$$

and finely:

$$u_{n\mathbf{k}} = \sum_{n'}^A C_{nn'}u_{n'\mathbf{k}_0} + \sum_m^B C_{nm}u_{m\mathbf{k}_0} \tag{3.15}$$

The way that changes the band structures will be discussed in the following sections.

3.2.2 Complete Secular Equation

Because of the high charge number of Cd, Hg and Te atoms, the Schrödinger equation for electrons in the crystal potential of HgCdTe must include relativistic terms. These terms arise from relativistic modifications of the magnetic field generated by the rapid electron orbital motion in these high atomic number atoms with their large charge. In an electron's coordinate system, besides the interaction with the crystal potential, there is an extra magnetic field interaction. The interaction between the magnetic field generated by the electrons' orbital motion and its spin gives rise to the spin–orbit coupling term, which influences the electron's energy levels. Starting from the Dirac-relativistic equation, the Hamiltonian for an electron in a solid becomes:

$$H = H_1 + H_D + H_{mv} + H_{so} \tag{3.16}$$

where $H_D = -\frac{\hbar^2}{4m^2c^2}\nabla V \cdot \nabla$ is the Darwin interaction term, $H_{mv} = \frac{E' - V}{4m^2c^2} p^2$ is the mass–velocity interaction term $\left(E' = E - mc^2\right)$, $H_{so} = \frac{\hbar}{4m^2c^2}(\nabla V \times \mathbf{p}) \cdot \sigma$ is the spin–orbit interaction term; σ is the spin operator, and $H_1 = \frac{p^2}{2m} + V(\mathbf{r})$ is the non-relativistic term.

The derivation of the above equations will be presented presently, and their application analyzed.

For an electron that moves in a crystal potential, in the coordinate system moving with the electron, besides the interaction from the coulomb field, there is an additional interaction from the magnetic field induced by the orbital motion of the electron. This magnetic field interacts with the spin of the electron and leads to a spin–orbit, or $S - L$, coupling term. The coupling can be deduced from the Dirac equation. The Dirac equation starts with the relativistic relation between energy and momentum:

$$E^2 = c^2 p^2 + m^2 c^4 \tag{3.17}$$

If one substitutes operators for the E and **p** terms:

$$E \rightarrow i\hbar\frac{\delta}{\delta t} \text{ and } \mathbf{p} = -i\hbar\nabla \tag{3.18}$$

then (3.17) becomes the Schrödinger form of a relativistic equation:

$$-\hbar^2 \frac{\delta^2 \Psi}{\delta t^2} = -\hbar^2 c^2 \nabla^2 \Psi + m^2 c^4 \Psi \tag{3.19}$$

This relativistic equation only applies to spin-less particles.

Dirac began from the linearized version of (3.17) written in the form:

$$E = \sqrt{c^2 p^2 + m^2 c^4} \cong \vec{\alpha} \cdot c\mathbf{p} + \beta \cdot mc^2 \tag{3.20}$$

and replacing E and $\mathbf{p}$ by their operators obtained the general form of the Dirac Hamiltonian $\hat{H}$ and the Dirac equation:

$$\begin{aligned} &\hat{H} = \hat{\vec{\alpha}} \cdot c\mathbf{p} + \hat{\beta} mc^2, \text{ or} \\ &\left(\hat{\vec{\alpha}} \cdot c\mathbf{p} + \hat{\beta} mc^2\right)\hat{\psi} = E\hat{\psi} \end{aligned} \tag{3.21}$$

To make (3.21) agree with (3.19) Dirac showed that the components of $\hat{\vec{\alpha}} : \hat{\alpha}_x, \hat{\alpha}_y, \hat{\alpha}_z$ and $\hat{\beta}$ had to anticommute in pairs and that their squares are unity. The lowest order matrices that satisfy these conditions are 4×4. They describe spin 1/2 particles (electrons). Higher order matrices that satisfy the anticommute conditions are for particles with higher spins. The forms of these matrices for a spin 1/2 particle are:

$$\begin{aligned} &\beta = \begin{pmatrix} I & 0 \\ 0 & I \end{pmatrix}, \\ &\hat{\alpha}_x = \begin{pmatrix} 0 & \hat{\sigma}_x \\ \hat{\sigma}_x & 0 \end{pmatrix}, \hat{\alpha}_y = \begin{pmatrix} 0 & \hat{\sigma}_y \\ \hat{\sigma}_y & 0 \end{pmatrix}, \text{and } \hat{\alpha}_z = \begin{pmatrix} 0 & \hat{\sigma}_z \\ \hat{\sigma}_z & 0 \end{pmatrix} \end{aligned} \tag{3.22}$$

where the $\hat{\sigma}_j, j = x, y, z$ are 2×2 matrices, called "spinors" and I is the unit matrix:

$$\hat{\sigma}_x = \begin{pmatrix} 0 & 1 \\ 1 & 0 \end{pmatrix}, \hat{\sigma}_y = \begin{pmatrix} 0 & -i \\ i & 0 \end{pmatrix}, \hat{\sigma}_z = \begin{pmatrix} 1 & 0 \\ 0 & -1 \end{pmatrix}, \text{and } I = \begin{pmatrix} 1 & 0 \\ 0 & 1 \end{pmatrix} \tag{3.23}$$

In a crystal, the periodic potential V has to be added to the Hamiltonian to yield:

$$\left(\hat{\vec{\alpha}} \cdot c\mathbf{p} + \hat{\beta} mc^2 + V(\mathbf{r})\right)\hat{\psi} = E\hat{\psi} \tag{3.24}$$

where $\hat{\Psi}$ is a 4×1 matrix, $\hat{\Psi}_1$ and $\hat{\Psi}_2$ are 2×1 matrices:

$$\hat{\psi} = \begin{pmatrix} \psi_1 \\ \psi_2 \\ \psi_3 \\ \psi_4 \end{pmatrix} = \begin{pmatrix} \hat{\Psi}_1 \\ \hat{\Psi}_2 \end{pmatrix} \tag{3.25}$$

Then the Dirac equation becomes:

$$\begin{cases} c\hat{\vec{\sigma}}\mathbf{p}\hat{\Psi}_2 + mc^2 I\hat{\Psi}_1 + V\hat{\Psi}_1 = E\hat{\Psi}_1 \\ c\hat{\vec{\sigma}}\mathbf{p}\hat{\Psi}_1 - mc^2 I\hat{\Psi}_2 + V\hat{\Psi}_2 = E\hat{\Psi}_2 \end{cases} \tag{3.26}$$

The second equation can be solved for $\hat{\Psi}_2$ and substituted into the first one to find:

$$\begin{aligned} &\frac{p^2}{2m}\hat{\Psi}_1 - \frac{E'-V}{4m^2c^4}p^2\hat{\Psi}_1\nabla V\nabla\hat{\Psi}_1 + \frac{\hbar^2}{4m^2c^4}\hat{\vec{\sigma}}\left[\nabla V \times \mathbf{p}\right]\hat{\Psi}_1 + V\hat{\Psi}_1 \\ &= E\hat{\Psi}_1 \end{aligned} \tag{3.27}$$

where $E' = E - mc^2$, the first and the fifth terms on the left side of the equation are non-relativistic; the second term H_{mv} is the mass–velocity interaction; the third term appears only in the Dirac-equation, and is the Darwin interaction; the fourth term H_{so} is the spin–orbit interaction. If the potential V(**r**) is expanded about the sites located at $\mathbf{R}_\ell$:

$$V(\mathbf{r}) = \sum_\ell V_\ell(r - \mathbf{R}_\ell) \tag{3.28}$$

then the spin–orbit term can further be written as:

$$\begin{aligned} &\frac{\hbar}{4m^2c^2}\hat{\vec{\sigma}} \cdot \sum_\ell \frac{\partial V_\ell(\mathbf{r}-\mathbf{R}_\ell)}{\partial|\mathbf{r}-\mathbf{R}_\ell|}\frac{(\mathbf{r}-\mathbf{R}_\ell)\times\mathbf{p}}{|\mathbf{r}-\mathbf{R}_\ell|} \\ &= \frac{1}{2m^2c^2}\sum_\ell \frac{1}{|\mathbf{r}-\mathbf{R}_\ell|}\frac{\partial V_\ell(\mathbf{r}-\mathbf{R}_\ell)}{\partial|\mathbf{r}-\mathbf{R}_\ell|}\cdot(\boldsymbol{S}\cdot\boldsymbol{L}_\ell) \end{aligned} \tag{3.29}$$

In (3.29) the second equality is for a spherical approximation to the V_ℓ potential and the relations $\mathbf{r} - \mathbf{R}_\ell \times \mathbf{p} \equiv \boldsymbol{L}_\ell$ and $\frac{\hbar}{2}\hat{\vec{\sigma}} \equiv \boldsymbol{S}$ have been used. The fourth term is also known as the $S-L$ coupling. Remember that $\boldsymbol{S}$ is a complicated vector, sometimes called a super-vector, because every component is a matrix.

Applying the fourth term H_{so} to a Bloch function, we have:

$$\begin{aligned} H_{so}\psi &= \frac{\hbar}{4m^2c^2}\vec{\sigma}\cdot(\nabla V\times \mathbf{p})u_{n\mathbf{k}}(\mathbf{r})e^{i\mathbf{k}\cdot\mathbf{r}} \\ &= \frac{\hbar}{4m^2c^2}\left[\vec{\sigma}\cdot\nabla V\times\left(\hbar\mathbf{k}e^{i\mathbf{k}\cdot\mathbf{r}}\right)u_{n\mathbf{k}}(\mathbf{r})+e^{i\mathbf{k}\cdot\mathbf{r}}\vec{\sigma}\cdot\nabla V\times \mathbf{p}u_{n\mathbf{k}}(\mathbf{r})\right] \end{aligned} \tag{3.30}$$

In (3.30) the 2×2 matrix notation has been deleted, but the $\boldsymbol{\sigma}$ vectors are still 2×2 matrices and so are the $u_{n\mathbf{k}}$ functions. This compressed notation will be used in the rest of this section, unless otherwise noted.

Equation (3.29) can be simplified to become:

$$\left[\frac{\hbar}{4m^2c^2}\vec{\sigma}\cdot(\nabla V\times \mathbf{p})+\frac{\hbar}{4m^2c^2}\vec{\sigma}\cdot(\nabla V\times \boldsymbol{k})\right]u_{n\mathbf{k}}(\mathbf{r})=E_{so}u_{n\mathbf{k}}(\mathbf{r}) \tag{3.31}$$

In the $u_{n\mathbf{k}}$ representation, $H_{so}=H_{sol}+H_{sok}$ is defined where H_{sol} and H_{sok} are the first and second terms in the above equation. H_{sok} is often referred to as the $\boldsymbol{k}$ -dependent spin–orbit term.

In Kane's $\boldsymbol{k{\cdot}p}$ method, the two terms H_D and H_{mv} are not discussed, because they introduce constant shifts to all the bands at Γ. The energy splitting induced by the interactions H_D and H_{mv} at Γ contributes little to E_g. Therefore Kane concentrated on the spin–orbit term H_{so}, and including it the secular equation becomes:

$$\left[\frac{p^2}{2m}+V(\mathbf{r})+\frac{\hbar}{4m^2c^2}(\nabla V\times \mathbf{p})\cdot\vec{\sigma}\right]e^{i\mathbf{k}\cdot\mathbf{r}}u_{n\mathbf{k}}(\mathbf{r})=E_{n\mathbf{k}}e^{i\mathbf{k}\cdot\mathbf{r}}u_{n\mathbf{k}}(\boldsymbol{r}) \tag{3.32}$$

In the $u_{n\mathbf{k}}$ representation the Schrödinger equation for the periodic function $u_{n\mathbf{k}}$ of a primitive cell is:

$$\left\{\begin{aligned} &\frac{p^2}{2m}+V(\mathbf{r})+\frac{\hbar}{m}\boldsymbol{k}\cdot\boldsymbol{p}+\frac{\hbar}{4m^2c^2}(\nabla V(\mathbf{r})\times \boldsymbol{p})\cdot\vec{\sigma} \\ &+\frac{\hbar^2}{4m^2c^2}\left[\nabla V(\mathbf{r})\times \boldsymbol{k}\right]\cdot\vec{\sigma} \end{aligned}\right\}u_{n\mathbf{k}}(\mathbf{r})=E'_{n\mathbf{k}}u_{n\mathbf{k}}(\mathbf{r}). \tag{3.33}$$

The quantity $E_{n\mathbf{k}}$ on the right side of (3.33) is the energy shift of an electron in state n, $\mathbf{k}$:

$$E'_{n\mathbf{k}}\equiv E_{n\mathbf{k}}-\frac{\hbar^2k^2}{2m} \tag{3.34}$$

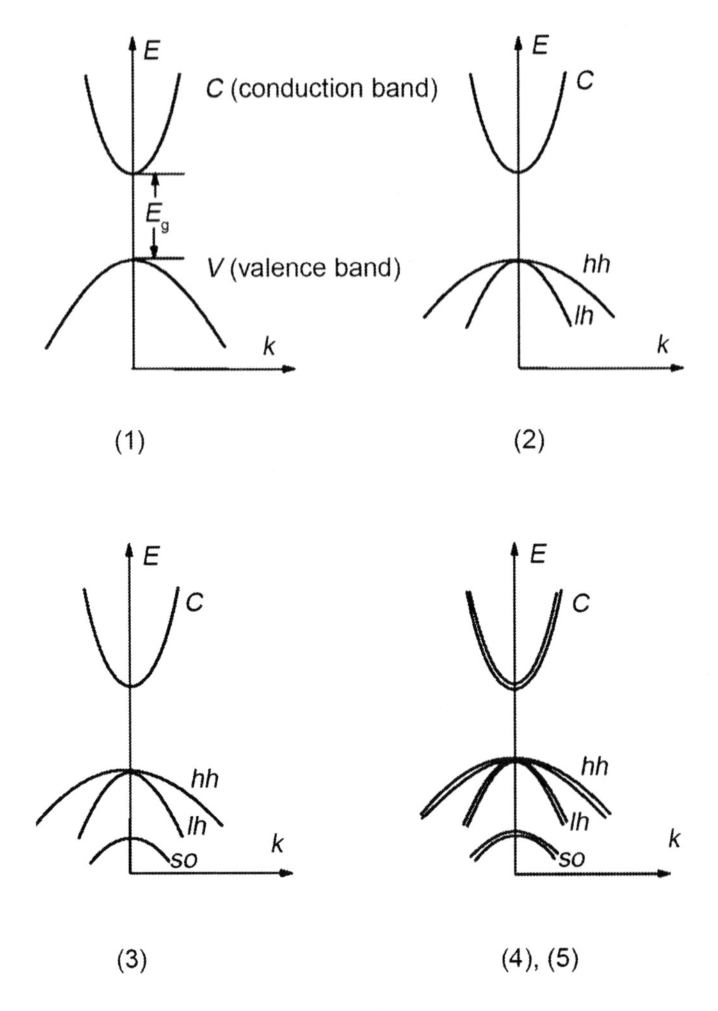

Fig. 3.7. Step by step analysis of the different term contributions to energy bands. The designations *hh*, *lh*, and *so*, label the heavy-hole, light-hole and split-off bands

The third term at the left side of (3.32) is the $\boldsymbol{k}\cdot\boldsymbol{p}$ interaction. The forth and fifth terms come from the spin–orbit coupling and the fifth term is the $\boldsymbol{k}$-dependent one. From (3.33) we can examine step by step the contributions of the different terms to the energy band structures as shown schematically in Fig. 3.7.

In general if the correction terms are small; perturbation theory can be used to find the deviations they introduce into the band structure. First, according to (3.33), when only the first and second terms are considered and the perturbation expansion is to be around the Γ point at $\boldsymbol{k}_0 = 0$ the equation for the basis set is:

$$\left(\frac{p^2}{2m}+V\right)u^0_{n\mathbf{k}_0}=E^0_{n\mathbf{k}}u^0_{n\mathbf{k}_0} \tag{3.35}$$

Adjacent to the band gap at Γ, the solution to this equation yields a two-fold degenerate conduction band and a six-fold degenerate valence band.

Once the correction terms to the Hamiltonian are added the energy $E'_{n0}=E^0_{n0}+E^1_{n0}+E^2_{n0}+\cdots$, is determined to various orders of perturbation theory using the complete set $\{u^0_{n0}\}$ as the basis.

Second, when only the $k{\bullet}p$ interaction is included, in first order perturbation the conduction band is unchanged because the spin degeneracy is not lifted, and since the $\boldsymbol{p}$ matrix element is multiplied by $\boldsymbol{k}_0 = 0$ this contribution vanishes. The situation at the valence band edge is somewhat different, since $\boldsymbol{p}$ acts on the orbital terms and degenerate perturbation theory must be used. The net result is that the heavy hole band and the light hole band, away from $\boldsymbol{k} = 0$, are split as shown in Fig. 3.7(2). This only happens in crystals like zincblende that do not have a center of symmetry.

Third, when the $k{\bullet}p$ and the $(\nabla V \times \mathbf{p}){\bullet}\hat{\vec{\sigma}}$ terms are both included as first order perturbations, as illustrated in Fig. 3.7, the band gap narrows, the conduction band (labeled C) and light hole band (labeled lh) masses are decreased, the heavy hole band (labeled hh) and the light hole band remain degenerate at $k = 0$, and the split-off band (labeled so) energy at $k = 0$ is decreased. Each of these bands is still two-fold (spin) degenerate.

Fourth, when the $k{\bullet}p$ and $(\nabla V \times \mathbf{p}){\bullet}\hat{\vec{\sigma}}$ interactions are treated in second order perturbation theory even the two-fold spin degeneracies are lifted. In addition the curvatures of the conduction and light hole bands are increased still further. Because the effective mass of a band is inversely proportional to its curvature, the effective masses of these bands are reduced.

Fifth, if the $\mathbf{k}$ -dependent spin–orbit term H_{sok}, is treated as a first order perturbation, the two fold degeneracy of the heavy hole band is further lifted and its top is not at Γ, rather it is displaced for InSb by ~3% along the $\langle 111 \rangle$ direction. Its energy is also lifted by $\sim 10^{-3}$ eV for InSb, and $\sim 10^{-5}$ eV for HgCdTe. While the shift of the top of the valence band away from $k = 0$ is predicted to this level of perturbation it is not observed. More complete theories place the top of the valence band at Γ (Chen 1995).

In the perturbation calculations, accurate evaluations of the matrix elements are essential. They depend on the crystal symmetry and resulting selection rules.

3.2.3 Selection Rules

The symmetry of a crystal determines the gross characteristics of its band structure. Group theory is an effective way to deal with crystal symmetry. Many crystals, in particular those of interest in this book, fall into the fcc Bravais lattice system, with two atoms per primitive cell on each fcc site; diamond, and zincblende crystal structures are examples. The first Brillouin zone in reciprocal space for all of these structures is a truncated octahedron. Symmetry points Γ, X, K and W are shown in Fig. 3.2 (see Appendix 3A). In the figure, Δ, Λ, Σ, Q and Z label axes. The zero of the periodic potential is taken to be at its peaks between lattice sites; at the base potential in the free electron limit. Then the symmetry of the wave functions remains unchanged, and the energy expression is:

$$E = \frac{\hbar^2}{2m}(k + k_n)^2 \tag{3.36}$$

where k_n is a reciprocal lattice vector (see Appendix 3A). The symmetry of the points in the Brillouin zone determines the symmetry of the wave functions, and also the degeneracy of the electron states. The higher the symmetry, the larger is the degeneracy number. Much of the degeneracy is lifted when a periodic potential is included.

The influences of symmetry on band structures are summarized in character tables. The symmetry operations of a group can be collected in matrices. The trace of every matrix is the sum of its diagonal elements and is invariant to unitary transformations. Transformations that diagonalize matrices are unitary, thus the sum of the diagonal elements before and after it is diagonalized are the same. A group can be represented as linear transformations, which can be applied to a vector in a specified space. There are an unlimited number of matrixes to represent a group. But there is an irreducible representation A, which is defined such that A = M^{-1}AM (where M is a unitary matrix). The character of an irreducible representation can be derived without knowing its matrix. A group has limited irreducible representations and their number is equal to the rank of the group (conjugated elements are equivalent, and the class of equivalent elements is named the "rank"). The sum of squares of the representation dimension is equal to the element number in the group.

Table 3.1. Table of symmetry operation for a cubic group

E	Rank	Description	Influence on point (x, y, z)
E	1	Identity	X,Y,Z
C_4^2	3	Rotate 180° about an axis	$X\bar{Y}\bar{Z}, \bar{X}Y\bar{Z}, \bar{X}\bar{Y}Z$
C_4	6	Rotate 90° about an axis	$\bar{Y}XZ, \bar{X}ZY, \bar{Z}YX, Y\bar{X}Z, XZ\bar{Y}, ZY\bar{X}$
C_2	6	Rotate 120° about X Y Z 0	$YX\bar{Z}, Z\bar{Y}X, \bar{X}ZY, \bar{Y}\bar{X}\bar{Z}, \bar{Z}\bar{Y}\bar{X}, \bar{X}\bar{Y}\bar{Z}$
C_3	8	Rotate 120° about a cube diagonal	$ZXY, YZX, Z\bar{X}\bar{Y}, \bar{Y}\bar{Z}X, \bar{Z}\bar{X}Y, \bar{Y}Z\bar{X}, \bar{Z}X\bar{Y}, Y\bar{Z}\bar{X}$
J	1	Inversion through the center	$\bar{X}\bar{Y}\bar{Z}$
JC_4^2	3	J, C_4^2	$XY\bar{Z}, \bar{X}YZ, X\bar{Y}Z$
JC_4	6	J, C_4	$Y\bar{X}\bar{Z}, \bar{Y}X\bar{Z}, \bar{X}Z\bar{Y}, \bar{X}\bar{Z}Y, \bar{Z}\bar{Y}X, Z\bar{Y}\bar{X}$
JC_2	6	J, C_2	$\bar{Y}\bar{X}Z, \bar{Z}Y\bar{X}, X\bar{Z}\bar{Y}, YXZ, ZYX, XZY$
JC_3	8	J, C_3	$\bar{Z}\bar{X}\bar{Y}, \bar{Y}\bar{Z}\bar{X}, \bar{Z}XY, YZ\bar{X}, ZX\bar{Y}, Y\bar{Z}X, Z\bar{X}Y, \bar{Y}ZX$

For a zincblende structure, the Brillouin zone center point Γ has 24 symmetry operations. These operations can be divided into many ranks. The rank of E, the identity operation is unity, and its operation transforms a matrix into itself. The rank of C_4, the rotation by $\pm\pi/2$ about each fourfold axis, is 6 because it constitutes six operations. Similarly the rank C_4^2, the rotation by π about each fourfold axis is 3. The rank of C_3, the rotation by π/3 or 2π/3 about each threefold axis is 8. J is the inversion operation with rank 1. There are also products like JC_4^2. JC_4^2 and JC_2 are reflections in the symmetry planes perpendicular to the fourfold and twofold axes, respectively.

A group is a set of operations A, B…C. These elements satisfy the relations in a group product table. For a crystal with cubic symmetry, there are 48 elements. The elements of this group operate as shown in Table 3.1.

It has been mentioned before that all wave vectors of states $\mathbf{k}+\mathbf{k}_\mathbf{n}$, where $\mathbf{k}_\mathbf{n}$ is a reciprocal lattice vector, are equivalent to that at point $\mathbf{k}$ in the Brillouin zone. There is a more normal expression for the symmetric points:

$$\beta\mathbf{k} = \mathbf{k} + \mathbf{k}_n \tag{3.37}$$

Table 3.2. Character table χ of the cubic symmetry group

XYZ	E	$3C_4^2$	$6C_4$	$6C_2$	$8C_3$	J	$3JC_4^2$	$6JC_4$	$6JC_2$	$8JC_3$
Γ_1	1	1	1	1	1	1	1	1	1	1
Γ_2	1	−1	−1	−1	1	1	1	−1	−1	1
Γ_{12}	2	2	0	0	−1	2	2	0	0	−1
Γ_{15}	3	−1	1	−1	0	−3	1	−1	1	0
Γ_{25}	3	−1	−1	1	0	−3	1	1	−1	0
Γ_1'	1	1	1	1	1	−1	−1	−1	−1	−1
Γ_2'	1	−1	−1	1	1	−1	−1	1	1	−1
Γ_{12}'	2	2	0	0	−1	−2	−2	0	0	1
Γ_{15}'	3	−1	1	−1	0	3	−1	1	−1	0
Γ_{25}'	3	−1	−1	1	0	3	−1	−1	1	0

where β is an operator, a group operation of a crystal space group. In order to study it in detail, it is necessary to construct a character table for every symmetry point in the Brillouin zone. As an example, consider the Brillouin zone center point Γ, and apply the Γ group. The states characterized by wave vectors that belong to the same irreducible representation must have the same energy, because their wave functions can be transformed into each other through group operations. Therefore, the band structure has the same symmetry as the reciprocal lattice. If the state in a point group is degenerate, its band is also degenerate.

Every elemental operation restores a cubic lattice into its original structure. For InSb, which has the zincblende *fcc* structure, its first Brillouin zone is a truncated octahedron. The point has cubic symmetry. In a group's operations, every element is expressed through a different matrix. The trace, the sum of the diagonal elements of each matrix, is its character. Every character table has a different set of representations. But any complicated representation can be reduced to ten, as is shown in Table 3.2.

How can we determine that a point x Y Z belongs to the Γ'_{15} symmetry? We will show first for point XYZ, which matrixes we need to realize the 48 operations, which character system they satisfy, and then we know how the corresponding representations can be found from the character table.

First apply the operation E following Table 3.1:

$$\begin{pmatrix} X \\ Y \\ Z \end{pmatrix} = \begin{pmatrix} 1 & 0 & 0 \\ 0 & 1 & 0 \\ 0 & 0 & 1 \end{pmatrix} \begin{pmatrix} X \\ Y \\ Z \end{pmatrix} \tag{3.38}$$

and find its character is $\chi(E) = 3$.

$$\begin{pmatrix} X \\ \bar{Y} \\ \bar{Z} \end{pmatrix} = \begin{pmatrix} 1 & 0 & 0 \\ 0 & -1 & 0 \\ 0 & 0 & -1 \end{pmatrix} \begin{pmatrix} X \\ Y \\ Z \end{pmatrix} \tag{3.39}$$

and find its character is $\chi(C_4^2) = -1$.

Apply the operation C_4:

$$\begin{pmatrix} \bar{Y} \\ X \\ Z \end{pmatrix} = \begin{pmatrix} 0 & -1 & 0 \\ 1 & 0 & 0 \\ 0 & 0 & 1 \end{pmatrix} \begin{pmatrix} X \\ Y \\ Z \end{pmatrix} \tag{3.40}$$

and find its character is $\chi(C_4) = 1$.

Apply operation C_2:

$$\begin{pmatrix} Y \\ X \\ \bar{Z} \end{pmatrix} = \begin{pmatrix} 0 & 1 & 0 \\ 1 & 0 & 0 \\ 0 & 0 & -1 \end{pmatrix} \begin{pmatrix} X \\ Y \\ Z \end{pmatrix} \tag{3.41}$$

and find its character is $\chi(C_2) = -1$.

Apply the operation C_3:

$$\begin{pmatrix} Z \\ X \\ Y \end{pmatrix} = \begin{pmatrix} 0 & 0 & 1 \\ 1 & 0 & 0 \\ 0 & 1 & 0 \end{pmatrix} \begin{pmatrix} X \\ Y \\ Z \end{pmatrix} \tag{3.42}$$

and find its character is $\chi(C_3) = 0$. The characters are 3, –1, 1, –1 and 0, which belongs to the Γ_{15} representation. Because the $|x\rangle, |y\rangle$ and $|z\rangle$ state functions of ***p***, are proportional to *x*, *y* and *z*, respectively, the state functions of ***p*** also belong to the Γ_{15} representation.

Let's look at the representation of $z^2, x^2 - y^2$, a subset of the d state functions.

Applying the operation E:

$$\begin{pmatrix} Z^2 \\ X^2 - Y^2 \end{pmatrix} = \begin{pmatrix} 1 & 0 \\ 0 & 1 \end{pmatrix} \begin{pmatrix} Z^2 \\ X^2 - Y^2 \end{pmatrix} \tag{3.43}$$

yields the character $\chi(E) = 2$.

Applying the operation C_4^2:

$$\begin{pmatrix} Z^2 \\ X^2 - Y^2 \end{pmatrix} = \begin{pmatrix} 1 & 0 \\ 0 & 1 \end{pmatrix} \begin{pmatrix} X^2 - Y^2 \\ Z^2 \end{pmatrix} \tag{3.44}$$

yields the character $\chi(C_4^2) = 2$.

Applying the operation C_4:

$$\begin{pmatrix} Y^2 - X^2 \\ Z^2 \end{pmatrix} = \begin{pmatrix} -1 & 0 \\ 0 & 1 \end{pmatrix} \begin{pmatrix} X^2 - Y^2 \\ Z^2 \end{pmatrix} \tag{3.45}$$

yields the character $\chi(C_4) = 0$.

Applying the operation C_2:

$$\begin{pmatrix} Y^2 - X^2 \\ Z^2 \end{pmatrix} = \begin{pmatrix} -1 & 0 \\ 0 & 1 \end{pmatrix} \begin{pmatrix} X^2 - Y^2 \\ Z^2 \end{pmatrix} \tag{3.46}$$

we have the character $\chi(C_2) = 0$.

Applying the operation C_3:

$$\begin{pmatrix} Y^2 - X^2 \\ Z^2 \end{pmatrix} = \begin{pmatrix} -1 & 0 \\ 0 & 1 \end{pmatrix} \begin{pmatrix} X^2 - Y^2 \\ Z^2 \end{pmatrix} \tag{3.47}$$

yields the character $\chi(C_3) = -1$. Therefore, the character of $z^2, x^2 - y^2$ is 2, 2, 0, 0 and -1, which belongs to the Γ_{12} representation. More specifically, $d_{z^2}, d_{x^2-y^2}$ belongs to the Γ_{12} representation. Similarly d_{xy}, d_{yz}, d_{zx} belongs to the Γ'_{25} representation, and s belongs to the Γ_1 representation.

The representations, to which the wave functions belong, are very important for simplifying matrixes, and this will be demonstrated in the following calculations.

Table 3.3. Character table of the Γ representation of the T_d single group

Representation	Basis	Rank/1E	$3C_4^2$	$8C_3$	$6JC_4$	$6JC_2$
Γ_1	1	1	1	1	1	1
Γ_2	$x^4(y^2-z^2)+y^4(z^2-x^2)$ $+z^4(x^2-y^2)$	1	1	1	−1	−1
Γ_{12}	$x^2-y^2, z^2-\frac{1}{2}(x^2+y^2)$	2	2	−1	0	0
Γ_{15}	x, y, z	3	−1	0	−1	1
Γ_{25}	$z(x^2-y^2)$	3	−1	0	1	−1

Table 3.3 is the character table of the Γ representation in the single group T_d. From the table we can see that, the character of E is the dimension of the representation. It is also the degree of the degeneracy. Therefore, Γ_1 and Γ_2 are one-dimensional representations, Γ_{12} is a two-dimensional representation, and Γ_{15} and Γ_{25} are three-dimensional representations. The total number of elements in the group is 1+3+8+6+6=24. The sum of the squares of the number of the dimensions for the representations is $1^2+1^2+2^2+3^2+3^2=24$. The basis shows the symmetry of the wave functions. Therefore, the wave function of type Γ_1 has the total symmetry of the point group; Γ_{15} is three-fold degenerate, and it has the symmetry of x, y and z. Γ_{12} is two-fold degenerate, and has the symmetry of type x^2-y^2 and type $z^2-\frac{1}{2}(x^2+y^2)$, which have a two-fold degeneracy and two knot-surfaces. In fact, the four orbits of the tetrahedron in the zincblende structure are related to the irreducible reduced Γ representation. State Γ_{15} is for a p electron state. The gross features of the band structures of a zincblende crystal can be obtained from its symmetry. The characteristics of the points Δ, Λ and Σ, X in the Brillouin zone can be found in Parmenter (1955).

When the spin–orbit coupling is included, a double group has to be applied. Due to time inversion symmetry, the electron spin leads to an additional degeneracy. For most points in the Brillouin zone, when the spin is included, the representation becomes two-fold degenerate with a very small energy change. But for those points with a high degree of symmetry, the consideration of spin can lead to energy splittings. To solve this problem, one needs to consider the product of the Γ representation and the spin-function s, which leads to the so-called double group. The inclusion of the spin into the point group doubles the number of elements (Elliott 1954).

Table 3.4 is a character table for the Γ double group. The direct product can be reduced. For example, $S^{-1} \times \Gamma_{15}$ gives $\Gamma_7 + \Gamma_8$. Therefore, when the spin is considered, one energy level is split into two levels and each one has its own degeneracy degree.

Table 3.4. Complete character table of Γ in the double group T_d^2

	E	E	$6C_4^2$	$8C_3$	$8\bar{C}_3$	$6JC_4$	$6J\bar{C}_4$	$12JC_2$
Γ_1	1	1	1	1	1	1	1	1
Γ_2	1	1	1	1	1	–1	–1	–1
Γ_3	2	2	2	–1	–1	0	0	0
Γ_4	3	3	–1	0	0	–1	–1	0
Γ_5	3	3	–1	0	0	1	1	1
Γ_6	2	–2	0	1	–1	$\sqrt{2}$	$-\sqrt{2}$	0
Γ_7	2	–2	0	1	–1	$-\sqrt{2}$	$\sqrt{2}$	0
Γ_8	4	–4	0	–1	1	0	0	0

Table 3.5. Relations between the single and double Γ groups

Single-group	Γ_1	Γ_2	Γ_{12}	Γ_{15}	Γ_{25}
Double-group	Γ_6	Γ_7	Γ_8	$\Gamma_7 + \Gamma_8$	$\Gamma_6 + \Gamma_8$

The relations between the single Γ group and double Γ group are shown in Table 3.5.

Similar discussions can be made for other points in the Brillouin zone of the zincblende structure. It can also be applied to all other of crystal types (Parmenter 1955; Dresselhaus 1955).

The narrow band gap semiconductors HgCdTe and InSb have zincblende structures. Their first Brillouin zones are truncated octahedrons, which has cubic symmetry at the Γ point and belongs to the fcc group O_h. When the problem needs to be solved precisely, the tetrahedron, T_d group of the zincblende structure, must be used. When the spin is included, the T_d^2 group must be used. The irreducible representation of the O_h group is Γ_1, Γ_2, Γ_{12}, Γ'_{15}, Γ'_{25}, Γ'_1, Γ'_2, Γ'_{12}, Γ_{15} and Γ_{25}; The irreducible representation of the n group is Γ_1, Γ_2, Γ_{12}, Γ_{15} and Γ_{25}; The irreducible representation of the T_d^2 group is Γ_1, Γ_2, Γ_3, Γ_4, Γ_5, Γ_6, Γ_7 and Γ_8. The former five irreducible representations belong to both group T_d and group T_d^2. In some of the literature for calculating the matrix elements of the terms $\frac{\hbar}{m}\boldsymbol{k}{\bullet}\boldsymbol{p}$, $\frac{\hbar}{4m^2c^2}(\nabla V\times\boldsymbol{p}){\bullet}\sigma$ and $(\nabla V\times\boldsymbol{p}){\bullet}\sigma$, the following selection rules are often used. Using these selection rules, one can determine which elements are zero and which are non-zero, thereby simplifying the matrix.

- Rule 1: For a matrix element $\langle\Psi_i|R|\Psi_j\rangle$, if Ψ_i is orthogonal to $R\Psi_j$, then it is zero; otherwise, it is non-zero. The necessary condition for a non-zero element is that $\Gamma_i\times\Gamma_R$ contains Γ_j. Namely the direct product of the irreducible representation of Ψ_i and that of R contains the irreducible representation of Ψ_j. This condition is expressed as:

$$\Gamma_i\times\Gamma_R\times\Gamma_j \text{ contains } \Gamma_1. \tag{3.48}$$

  This is a necessary condition for a non-zero element.
- Rule 2: Even though $\langle\Psi_i|R|\Psi_j\rangle$ satisfies the above general non-zero condition, it is still zero under following special conditions:
  (a) R can be applied to Ψ_j orΨ_i directly. Its first application gives an odd function, which leads the integration over one period to be zero.
  (b) R cannot be applied to Ψ_j or Ψ_i directly. Then we can reflect Ψ_iR to the basis axis or basis plane of Ψ_j. If the result changes its sign, we can further reflect the character of Γ_j on the basis plane to see if the sign changes. If one application changes the sign and the other one does not, then the integration is zero, so the matrix element is zero.

To implement selection Rule 1, the often used direct product equations are:

$$\begin{aligned}&\Gamma_1\times\Gamma_{15}=\Gamma_{15}\\&\Gamma_{15}\times\Gamma_{15}=\Gamma_1+\Gamma_{12}+\Gamma'_{15}+\Gamma'_{25}\qquad(\text{group O}_\text{h})\\&\Gamma_{15}\times\Gamma_{15}=\Gamma_1+\Gamma_{12}+\Gamma_{15}+\Gamma_{25}\qquad(\text{group T}_\text{d})\end{aligned}\tag{3.49}$$

In group O_h, the s state belongs to the Γ_1 representation, whereas the x, y and z states as well as the $\frac{\partial}{\partial x},\frac{\partial}{\partial y},\frac{\partial}{\partial z}$ belong to the Γ_{15} representation. States d_z, $d_{(y^2-z^2)}$ belong to the Γ_{12} representation and states d_{xy}, d_{yz} and

d_{zx} belong to the Γ_{25}' representation. According to the above-mentioned selection rule, obviously we have $\langle s|\mathbf{k}\bullet\mathbf{p}|s\rangle = 0$, because the s state belongs to Γ_1, the p state belongs to Γ_{15}. Their direct product $\Gamma_1\times\Gamma_{15}$ does not contain Γ_1 and this leads to a zero matrix element. Similarly, $\langle x|\mathbf{k}\bullet\mathbf{p}|x\rangle = 0$ because $\Gamma_{15} \times \Gamma_{15}$ does not contain Γ_{15}. However, $\langle s|\frac{\hbar}{m}\mathbf{k}\bullet\mathbf{p}|x\rangle = k_x\langle s|\frac{\hbar}{m}p_x|x\rangle = ik_xP$ is non-zero, because $\Gamma_1\times\Gamma_{15}$ contains Γ_{15}.

Table 3.6. Some useful relations of direct products

T_i	Γ_1	Γ_2	Γ_3	Γ_4	Γ_5	Γ_6	Γ_7	Γ_8
$\Gamma_i\times\Gamma_4$	Γ_4	Γ_5	$\Gamma_4+\Gamma_5$	$\Gamma_1+\Gamma_3+\Gamma_4+\Gamma_5$	$\Gamma_2+\Gamma_3+\Gamma_4+\Gamma_5$	$\Gamma_7+\Gamma_8$	$\Gamma_6+\Gamma_8$	$\Gamma_1+\Gamma_7+2\Gamma_8$

For another example: $\langle is|(\nabla V\times\mathbf{p})\bullet\sigma|(x-iy)/\sqrt{2}\rangle = 0$, because $\Gamma_1\times\Gamma_{15}\times\Gamma_{15} = \Gamma_{15}\times\Gamma_{15}$ does not contain Γ_{15}, while among the states on the right, both x and y belong to Γ_{15}. Therefore the matrix is 0. In the same way, we can obtain

$$\begin{gathered}\frac{\hbar}{4m^2c^2}\left\langle\frac{x-iy}{\sqrt{2}}\uparrow\left|(\nabla V\times\mathbf{p})\cdot\sigma\right|\frac{x-iy}{\sqrt{2}}\uparrow\right\rangle = \frac{\Delta}{3}\\ \Delta \equiv \frac{3\hbar i}{4m^2c^2}\langle x|(\nabla V\times\mathbf{p})_z|y\rangle\end{gathered} \tag{3.50}$$

In second-order perturbation theory, the T_d group or T_d^2 group are often encountered. The relations for the direct products encountered in T_d^2 or T_d groups are shown in Table 3.6.

p_x, p_y, p_z and $\frac{\partial}{\partial x}, \frac{\partial}{\partial y}, \frac{\partial}{\partial z}$ belong to the Γ_{15} representation in the T_d group, and belong to the Γ_4 representation in the T_d^2 group.

For example, we have:

$$\left\langle x\left|\frac{\hbar}{m}\boldsymbol{k}\cdot\boldsymbol{p}\right|x\right\rangle_{\text{second order}} = \frac{\hbar^2}{m^2}\sum_{j\alpha l}\frac{\langle x|\boldsymbol{k}\cdot\boldsymbol{p}|u_{j\alpha l}\rangle\langle u_{j\alpha l}|\boldsymbol{k}\cdot\boldsymbol{p}|x\rangle}{E_v - E_j} = \frac{\hbar^2}{m^2}\sum_{j\alpha l}\frac{\left|\langle x|\boldsymbol{k}\cdot\boldsymbol{p}|u_{j\alpha l}\rangle\right|^2}{E_v - E_j} \tag{3.51}$$

where j,α and l refer the j band, the α state and the l representation, respectively. Because x belongs to the Γ_{15} representation, and $\boldsymbol{p}$ belongs to the Γ_{15} representation, we have:

$$\Gamma_{15}\times\Gamma_{15} = \Gamma_1 + \Gamma_{12} + \Gamma_{15} + \Gamma_{25} \qquad (T_d\, group) \tag{3.52}$$

Therefore, $u_{j\alpha l}$ may belong to the representations $\Gamma_1, \Gamma_{12}, \Gamma_{15}$ and Γ_{25}.

$$\begin{aligned}
\text{Above equaiton} &= \frac{\hbar^2}{m^2}\left[\sum_j^{\Gamma_1}\frac{\left|\langle x|\boldsymbol{k}\cdot\boldsymbol{p}|u_{j\alpha l}\rangle\right|^2}{E_v - E_{j\alpha}} + \sum_j^{\Gamma_{12}}\frac{\left|\langle x|\boldsymbol{k}\cdot\boldsymbol{p}|u_{j\alpha l}\rangle\right|^2}{E_v - E_{j\alpha}}\right. \\
&+ \left.\sum_j^{\Gamma_{15}}\frac{\left|\langle x|\boldsymbol{k}\cdot\boldsymbol{p}|u_{j\alpha l}\rangle\right|^2}{E_v - E_{j\alpha}} + \sum_j^{\Gamma_{25}}\frac{\left|\langle x|\boldsymbol{k}\cdot\boldsymbol{p}|u_{j\alpha l}\rangle\right|^2}{E_v - E_{j\alpha}}\right] \\
&= \frac{\hbar^2}{m^2}\left[\sum_j^{\Gamma_1}\frac{\left|\langle x|k_x p_x|u_j\rangle\right|^2}{E_v - E_j} + \sum_j^{\Gamma_{12}}\frac{\left|\langle x|k_x p_x|u_j\rangle\right|^2}{E_v - E_j}\right. \\
&+ \sum_j^{\Gamma_{15}}\frac{\left|\langle x|k_z p_z|u_{j\alpha}\rangle\right|^2}{E_v - E_j} + \sum_j^{\Gamma_{15}}\frac{\left|\langle x|k_y p_y|u_{j\alpha}\rangle\right|^2}{E_v - E_j} \\
&+ \left.\sum_j^{\Gamma_{25}}\frac{\left|\langle x|k_y p_y|u_{j\alpha l}\rangle\right|^2}{E_v - E_j} + \sum_j^{\Gamma_{25}}\frac{\left|\langle x|k_z p_z|u_{j\alpha l}\rangle\right|^2}{E_v - E_j}\right] \\
&= k_y^2\left(F' + 2G\right) + \left(k_y^2 + k_z^2\right)\left(H_1 + H_2\right)
\end{aligned} \tag{3.53}$$

where F', G, H_1, and H_2 are defined as:

$$
\begin{aligned}
F' &\equiv \frac{\hbar^2}{m^2}\sum_j^{\Gamma_1}\frac{\left|\langle x|p_x|u_j\rangle\right|^2}{E_v - E_j} \\
G &\equiv \frac{1}{2}\frac{\hbar^2}{m^2}\sum_j^{\Gamma_{12}}\frac{\left|\langle x|p_x|u_j\rangle\right|^2}{E_v - E_j} \\
H_1 &\equiv \frac{\hbar^2}{m^2}\sum_j^{\Gamma_{15}}\frac{\left|\langle x|p_y|u_j\rangle\right|^2}{E_v - E_j} \\
H_2 &\equiv \frac{\hbar^2}{m^2}\sum_j^{\Gamma_{25}}\frac{\left|\langle x|p_y|u_j\rangle\right|^2}{E_v - E_j}
\end{aligned}
\tag{3.54}
$$

3.3 Calculation of Band Structures

3.3.1 Solution at k_0=0

If ***k·p*** term and spin–orbit coupling terms are not considered, (3.7) at k_0=0 takes the following simple form:

$$
\left(\frac{p^2}{2m}+V\right)u_i = E_i' u_i \tag{3.55}
$$

In this case the general wave function $u_\mathrm{n}(\mathbf{k})$ in (3.7), where i represents a state with a band index n and a spin orientation index, reduces to just the band index because there is no spin dependent term in this zeroth order Hamiltonian. The resulting band states are all two-fold degenerate. The energy E_i' is given by $E_i' = E_i - \dfrac{\hbar^2 k^2}{2m}$ to agree with (3.7). The wave function solution $\{u_i\}$ set to this equation at $k_0 = 0$ serves as the basis for the ***k·p*** perturbation.

At $\mathbf{k}=0$, from group theory the electron wave functions describing the conduction band are two-fold degenerate with the symmetry properties of s-state functions, $S\uparrow$ and $S\downarrow$, when spin is included. They have a Γ_1 representation, with corresponding energy E_c. The electron wave functions of the valence band are p-state functions with a six-fold degeneracy including spins. This representation is a tight binding like basis. For many-electron atoms, the solution is the same as the solution of the Schrödinger equation for a hydrogen-like atom, i.e.

$$\Psi_{nlm}(r,\theta,\varphi) = R_{nl}(r) Y_{lm}(\theta,\varphi) \tag{3.56}$$

The angular part is the same as $Y_{lm}(\theta,\varphi)$; for its S is constant, while for p it is:

$$\begin{aligned} Y_{11} &\sim p_1 \sim \sin\theta e^{i\Phi} = \sin\theta(\cos\Phi + i\sin\Phi) \sim x + iy \\ Y_{10} &\sim p_0 \sim \cos\theta \sim z \\ Y_{1,-1} &\sim p_{-1} \sim \sin\theta e^{-i\theta} \sim x - iy \end{aligned} \tag{3.57}$$

These p -state functions are usually expressed in terms of their normalized linear combinations as:

$$\begin{aligned} p_x &= (p_1 + p_{-1})/\sqrt{2} &\sim \sin\theta\cos\Phi &\sim x \\ p_y &= -i(p_1 - p_{-1})/\sqrt{2} &\sim \sin\theta\sin\Phi &\sim y \\ p_z &= p_0 &\sim \cos\theta &\sim z \end{aligned} \tag{3.58}$$

When the spin is included, the three p -state functions are represented as: $x\uparrow, x\downarrow, y\uparrow, y\downarrow, z\uparrow, z\downarrow$, and their degenerate energy is E_v. Under symmetry operations p -state functions, p_x, p_y, and p_z, transform like x, y, and z functions, and fall into the Γ_{15} representation. At $\mathbf{k}=0$, $E_c^{'}(k)$ and $E_v^{'}(k)$ are E_s^0, and E_p^0 respectively, and from (3.34) with energies at finite k:

$$\begin{aligned} E_c^{'}(\mathbf{k}) &= E_s^0 + \frac{\hbar^2 k^2}{2m} \\ E_v^{'}(\mathbf{k}) &= E_p^0 + \frac{\hbar^2 k^2}{2m} \\ E_s^0 - E_p^0 &= E_g^0 \end{aligned} \tag{3.59}$$

where m is the free electron mass and E_g^0 the band gap arising from the periodic potential. The bands' masses and the band gap will be modified once the $\boldsymbol{k} \bullet \boldsymbol{p}$ and the spin–orbit interactions are added.

3.3.2 First Order Perturbation Correction Due to the *k·p* Term

If the ***k·p*** term is included, but the spin–orbit term is not, then (3.7) around $\mathbf{k}_0=0$ takes the form:

$$\left(\frac{p^2}{2m}+V(r)+\frac{\hbar}{m}\boldsymbol{k}\cdot\boldsymbol{p}\right)u_i(\mathbf{k})=E_i'u_i(\mathbf{k}) \tag{3.60}$$

where i = c,v1,v2,v3. In first order the $u_i(\boldsymbol{k})$ is replaced by $u_i(0)$. The secular equation obtained taking only states arising from the two fold degenerate conduction band and six fold degenerate upper valence band (including spin) introduced by the ***k·p*** term in the representation $|is\rangle,|x\rangle,|y\rangle$and$|z\rangle$ is:

$$\begin{vmatrix} E_s^0-E_i' & k_xP & k_yP & k_zP \\ k_xP & E_p^0-E_i' & 0 & 0 \\ k_yP & 0 & E_p^0-E_i' & 0 \\ k_zP & 0 & 0 & E_p^0-E_i' \end{vmatrix}=0 \tag{3.61}$$

where by symmetry:

$$P\equiv -i\frac{\hbar}{m}\langle s|p_x|x\rangle=-i\frac{\hbar}{m}\langle s|p_y|y\rangle=-i\frac{\hbar}{m}\langle s|p_z|z\rangle \tag{3.62}$$

P is real because $|is\rangle$ rather than $|s\rangle$ was selected as a basis state. Note that (3.61) is the complete first order secular equation for the limited subset of four bands consisting of the conduction and the top three valence bands (including spin there are eight bands).

Taking the top of the unperturbed valence band as the zero energy point, the solution of (3.61) is:

$$\begin{cases} E_c(k)=\dfrac{\hbar^2k^2}{2m}+\dfrac{1}{2}E_g+\dfrac{1}{2}\sqrt{(E_g)^2+4k^2P^2} \\ E_{v1,3}=\dfrac{\hbar^2k^2}{2m} \\ E_{v2}=\dfrac{\hbar^2k^2}{2m}+\dfrac{1}{2}E_g-\dfrac{1}{2}\sqrt{(E_g)^2+4k^2P^2} \end{cases} \tag{3.63}$$

The contribution of the first order ***k·p*** perturbation term at small k, the has decreased conduction band mass, and while the top of the valence band is still triply degenerate, the light hole mass has also changed. The uppermost valence band (the heavy hole band) has the wrong curvature at this level of approximation. At small k the energy of the state *v2* (the light hole band) is:

$$E_{v2} = \left(1 - \frac{mP^2}{2\hbar^2 E_g}\right)\frac{\hbar^2 k^2}{2m} \tag{3.64}$$

For this to have the observed downward curvature, one must have $\frac{mP^2}{2\hbar^2 E_g} > 1$. Whether or not this condition is satisfied for a given semiconductor depends on details. It is satisfied for most semiconductors. For larger k, the conduction and light hole bands are not parabolic at this level of perturbation.

3.3.3 Perturbation with Both *k·p* and $(\nabla V \times \mathbf{p})\bullet\sigma$ Included

In the above discussion the spin–orbit coupling term was omitted. If perturbations from both ***k·p*** and $(\nabla V \times \mathbf{p})\bullet\sigma$ terms are taken into account, then the heavy hole valence band and the spin–orbit band are split. Taking the eight basis functions from left to right to be:

$$\begin{array}{cccc} \left|is\downarrow\right\rangle & \left|(x-iy)\uparrow/\sqrt{2}\right\rangle & \left|z\downarrow\right\rangle & \left|(x+iy)\uparrow/\sqrt{2}\right\rangle \\ \left|is\uparrow\right\rangle & \left|-(x+iy)\downarrow/\sqrt{2}\right\rangle & \left|z\uparrow\right\rangle & \left|(x-iy)\downarrow/\sqrt{2}\right\rangle \end{array} \tag{3.65}$$

The top four functions are orthogonal respectively to one below it because of the spin orientation. We note that:

$$\begin{aligned} &\langle\downarrow|\boldsymbol{\sigma}_z|\downarrow\rangle = 1,\ \langle\downarrow|\boldsymbol{\sigma}_x|\downarrow\rangle = 0,\ \langle\downarrow|\boldsymbol{\sigma}_y|\downarrow\rangle = 0 \\ &\langle\uparrow|\boldsymbol{\sigma}_z|\uparrow\rangle = 1,\ \langle\uparrow|\boldsymbol{\sigma}_x|\uparrow\rangle = 0,\ \langle\uparrow|\boldsymbol{\sigma}_y|\uparrow\rangle = 0 \\ &\langle\downarrow|\boldsymbol{\sigma}_z|\uparrow\rangle = 0,\ \langle\downarrow|\boldsymbol{\sigma}_x|\uparrow\rangle = 1,\ \langle\downarrow|\boldsymbol{\sigma}_y|\uparrow\rangle = i \\ &\langle\uparrow|\boldsymbol{\sigma}_z|\downarrow\rangle = 0,\ \langle\uparrow|\boldsymbol{\sigma}_x|\downarrow\rangle = 1,\ \langle\uparrow|\boldsymbol{\sigma}_y|\downarrow\rangle = -i \end{aligned} \tag{3.66}$$

In the representation (3.65) the 8 × 8 interaction matrix is block diagonal and can be written as $\begin{pmatrix} H_D & 0 \\ 0 & H_D \end{pmatrix}$ where for simplicity we have taken $\mathbf{k} = k\hat{z}$:

$$H_D = \begin{pmatrix} E_s^0 & 0 & kP & 0 \\ 0 & E_p^0 - \frac{\Delta}{3} & \frac{\sqrt{2}\Delta}{3} & 0 \\ kP & \frac{\sqrt{2}\Delta}{3} & E_p^0 & 0 \\ 0 & 0 & 0 & E_p^0 + \frac{\Delta}{3} \end{pmatrix} + \frac{\hbar^2 k^2}{2m} I \tag{3.67}$$

and as before $P = -i\frac{\hbar}{m}\langle s|p_z|z\rangle$, E_s^0 and E_p^0 are eigenvalues of the unperturbed Hamiltonian at k=0, I is the unit matrix, and Δ is defined as:

$$\begin{aligned} \Delta &\equiv \frac{3\hbar i}{4m^2c^2}\left\langle x\left|\frac{\partial V}{\partial x}p_y - \frac{\partial V}{\partial y}p_x\right|y\right\rangle \\ &= \frac{3\hbar i}{4m^2c^2}\left\langle y\left|\frac{\partial V}{\partial y}p_z - \frac{\partial V}{\partial z}p_y\right|z\right\rangle \\ &= \frac{3\hbar i}{4m^2c^2}\left\langle z\left|\frac{\partial V}{\partial z}p_x - \frac{\partial V}{\partial x}p_z\right|x\right\rangle \end{aligned} \tag{3.68}$$

All matrix elements of H_{so} in which x, y, or z enter an odd number of times vanish because the matrix element integrals are over symmetric intervals in each of these parameters. Equation (3.67) was written assuming ***k***//**z**, if ***k*** is along another principal axis H_D can be also described by an equation like (3.67) obtained by a rotation of basis functions. Expanding the secular equation (with as before $E_i^{'} = E_i - \frac{\hbar^2 k^2}{2m}$):

$$\det\left|H_{lm} - E_i^{'} I\right| = 0 \tag{3.69}$$

in minors we find:

$$\begin{cases} E_p^0 + \frac{\Delta}{3} - E_{v1}^{'} = 0 \\ \left(E_s^0 - E\right)\left(E_p^0 - \frac{\Delta}{3} - E_i^{'}\right)\left(E_p^0 - E_i^{'}\right) - \left(E_s^0 - E_i^{'}\right)\frac{2}{9}\Delta^2 \\ -k^2P^2\left(E_p^0 - \frac{\Delta}{3} - E_i^{'}\right) = 0 \end{cases} \tag{3.70}$$

The solutions to (3.70) at k = 0 are:

$$\begin{aligned} E_s(0) &= E_s^0 = E_g \\ E_{v1}(0) &= E_p^0 + \frac{\Delta}{3} = 0 \\ E_{v2}(0) &= 0 \\ E_{v3}(0) &= -\Delta \end{aligned} \tag{3.71}$$

The state *v1* can be identified with the heavy hole band, *v2* with the light hole band, and *v3* with the split-off band. The top of the valence band at k = 0 has been set to zero energy, and the band gap is E_s^0 referenced to the zero of energy at $E_p^0 + \frac{\Delta}{3}$. Notice that the spin–orbit term has narrowed the band gap by $\Delta/3$ relative to its unperturbed value. If only terms to order k^2 are retained then the solutions to (3.70) are:

$$\begin{aligned} E_s(k) &= E_g + \frac{\hbar^2 k^2}{2m_c}; & \frac{1}{m_c} &\equiv \frac{1}{m} + \frac{2P^2\left(1 + \frac{2\Delta}{3E_g}\right)}{\hbar^2\left(E_g + \Delta\right)} \\ E_{hh}(k) &= \frac{\hbar^2 k^2}{2m}; & & \\ E_{lh}(k) &= -\frac{\hbar^2 k^2}{2m_{lh}}; & \frac{1}{m_{lh}} &\equiv \frac{4P^2}{3\hbar^2 E_g} - \frac{1}{m} \\ E_{so}(k) &= -\Delta - \frac{\hbar^2 k^2}{2m_{so}}; & \frac{1}{m_{so}} &\equiv \frac{2\ {}^2P}{3\hbar^2\left(E_g + \Delta\right)} - \frac{1}{m} \end{aligned} \tag{3.72}$$

The spin–orbit split band is separated by Δ from the heavy hole band. At this level of approximation the heavy hole band still has the wrong curvature, and the light hole and split-off bands can have the right curvatures depending on details. The solutions of the cubic equation, 3.70, yield non-parabolic bands, which mean the effective masses become **k** dependent at larger **k** values. All the masses and energies will be modified once more when second order terms, the B type states, and terms that lift the spin degeneracy are added.

A specific example of this non-linearity of the bands is seen if (3.70) is solved in Kane's narrow gap limit $\Delta >> k^2P^2, E_g$ (Kane 1957) where the solutions are:

$$\begin{cases} E_c &= \dfrac{\hbar^2 k^2}{2m} + \dfrac{E_g}{2} + \dfrac{1}{2}\sqrt{E_g^2 + \dfrac{8k^2P^2}{3}} \\ E_{v1} &= \dfrac{\hbar^2 k^2}{2m} \\ E_{v2} &= \dfrac{\hbar^2 k^2}{2m} + \dfrac{E_g}{2} - \dfrac{1}{2}\sqrt{E_g^2 + \dfrac{8k^2P^2}{3}} \\ E_{v3} &= -\varDelta + \dfrac{\hbar^2 k^2}{2m} - \dfrac{P^2k^2}{3\left(E_g + \varDelta\right)} \end{cases} \tag{3.73}$$

The impact of these solutions will be discussed in the next section where effective masses are treated in more detail.

The perturbation terms contributing to second order corrections that are most important are those that couple the conduction band to the top of the valence band. Woolley and Ray (1960), devised an efficient way to deduce the appropriate wave functions. Based on the first order energy values, modified wave functions can be derived for the conduction and valence bands. Let the index i now refer to the conduction band, the light hole valence band and the spin–orbit split band.

At $\mathbf{k} = 0$ the new wave functions are linear combinations of the basis functions:

$$\begin{aligned} \varPhi_{i\alpha} &= a_i\left[is\downarrow\right] + b_i\left[(x-iy)\uparrow/\sqrt{2}\right] + c_i\left[z\downarrow\right] \\ \varPhi_{i\beta} &= a_i\left[is\uparrow\right] + b_i\left[-(x+iy)\downarrow/\sqrt{2}\right] + c_i\left[z\uparrow\right] \\ \varPhi_{v_1\alpha} &= \left|(x+iy)\uparrow/\sqrt{2}\right\rangle \\ \varPhi_{v_1\beta} &= \left|(x-iy)\downarrow/\sqrt{2}\right\rangle \end{aligned} \tag{3.74}$$

hence

$$\begin{pmatrix} E_g & 0 & Pk \\ 0 & -\dfrac{2}{3}\varDelta & \dfrac{\sqrt{2}\varDelta}{3} \\ Pk & \dfrac{\sqrt{2}\varDelta}{3} & -\dfrac{\varDelta}{3} \end{pmatrix} \begin{pmatrix} a_i \\ b_i \\ c_i \end{pmatrix} = E_i' \begin{pmatrix} a_i \\ b_i \\ c_i \end{pmatrix} \tag{3.75}$$

where now Kane's approximation is used so $E_i^{'} \equiv E_i - \dfrac{\hbar^2 k^2}{2m}$ for all i. For the heavy hole valence band, we have $a = c = 0$, $b = 1$. Then we find for the others:

$$\begin{cases} \left(E_g - E_i'\right)a_i + Pkc_i & = & 0 \\ \left(-\dfrac{2}{3}\Delta - E_i'\right)b_i + \dfrac{\sqrt{2}}{3}\Delta c_i & = & 0 \\ Pka_i + \dfrac{\sqrt{2}}{3}\Delta b_i - \dfrac{\Delta}{3}c_i - E_i'c_i & = & 0 \end{cases} \tag{3.76}$$

Here the third equation in (3.76) is dependent. From the first and second equations we have:

$$\begin{aligned} Pkc_i &= a_i\left(E_i' - E_g\right) \\ \frac{\sqrt{2}}{3}\Delta c_i &= b_i\left(E_i' + \frac{2}{3}\Delta\right) \end{aligned}$$

therefore

$$\begin{cases} a_i & = & Pk\left(E_i' + \dfrac{2}{3}\Delta\right)/N \\ b_i & = & \dfrac{\sqrt{2}}{3}\Delta\left(E_i' - E_g\right)/N \\ c_i & = & \left(E_i' - E_g\right)\left(E_i' + \dfrac{2}{3}\Delta\right)/N \end{cases} \tag{3.77}$$

N is the normalization constant, $E_i^{'}$ is the solution of (3.72). In (3.77) the index, i, refers only to the conduction band, the light hole valence band and the spin–orbit split band. For the heavy hole valence band, $a = c = 0$, and $b = 1$ are used.

In the above expressions in the small k approximation, we have:

$$\begin{array}{lll} a_c = 1 & b_c = c_c = 0 & \\ a_{v2} = 0 & b_{v\ 2} = \sqrt{\dfrac{1}{3}} & c_{v2} = \sqrt{\dfrac{2}{3}} \\ a_{v3} = 0 & b_{v3} = \sqrt{\dfrac{2}{3}} & c_{v3} = -\sqrt{\dfrac{1}{3}} \end{array} \tag{3.78}$$

Substituting these values into (3.74) yields the modified wave functions (the ***J***, ***m****j* representation):

$$\left.\begin{array}{lll}
\left|J,m_j\right\rangle & & \\
\left|\frac{1}{2},-\frac{1}{2}\right\rangle & \Phi_{c\alpha} = & \left|is\downarrow\right\rangle \\
\left|\frac{1}{2},\frac{1}{2}\right\rangle & \Phi_{c\beta} = & \left|is\uparrow\right\rangle \\
\left|\frac{3}{2},-\frac{1}{2}\right\rangle & \Phi_{v2\alpha} = & \sqrt{\frac{1}{3}}\left|(x-iy)\uparrow/\sqrt{2}\right\rangle+\sqrt{\frac{2}{3}}\left|z\downarrow\right\rangle \\
\left|\frac{3}{2},+\frac{1}{2}\right\rangle & \Phi_{v2\beta} = & -\sqrt{\frac{1}{3}}\left|(x+iy)\downarrow/\sqrt{2}\right\rangle+\sqrt{\frac{2}{3}}\left|z\uparrow\right\rangle \\
\left|\frac{1}{2},-\frac{1}{2}\right\rangle & \Phi_{v3\alpha} = & \sqrt{\frac{2}{3}}\left|(x-iy)\uparrow/\sqrt{2}\right\rangle-\sqrt{\frac{1}{3}}\left|z\downarrow\right\rangle \\
\left|\frac{1}{2},+\frac{1}{2}\right\rangle & \Phi_{v3\beta} = & -\sqrt{\frac{2}{3}}\left|(x+iy)\downarrow/\sqrt{2}\right\rangle+\sqrt{\frac{1}{3}}\left|z\uparrow\right\rangle \\
\left|\frac{3}{2},\frac{3}{2}\right\rangle & \Phi_{v1\alpha} = & \left|\frac{x-iy}{\sqrt{2}}\uparrow\right\rangle \\
\left|\frac{3}{2},-\frac{3}{2}\right\rangle & \Phi_{v1\beta} = & \left|-\frac{x+iy}{\sqrt{2}}\downarrow\right\rangle
\end{array}\right\} \quad (3.79)$$

This set of wave functions is the new basis for the second order perturbation calculation and the linear k term perturbation.

3.3.4 Second Order Perturbation Between $\Phi_{i\alpha}$ and $\Phi_{i\beta}$

Some of the degeneracy among Φ_i and Φ_j states has been removed by the first order perturbation calculation. Second order perturbation of the ***k·p*** term must be performed to remove the remaining degeneracy between states $\Phi_{i\alpha}$ and $\Phi_{i\beta}$.

For each two-fold state the second order perturbation calculations of ***k·p*** and spin–orbit interactions are performed. First the coordinate system x, y, z is rotated into a system x', y', z' to make the k vector parallel to the z'

direction. The corresponding change of the wave function Φ into Φ' is shown in the table Eq. (3.80) below.

	$\Phi'_{i\alpha}$	$\Phi'_{i\beta}$
$\Phi'_{i\alpha}$	$E'_i + H_{\alpha\alpha}$	$H_{\alpha\beta}$
$\Phi'_{i\beta}$	$H_{\beta\alpha}$	$E'_i + H_{\beta\beta}$

(3.80)

The secular equation can be written as:

$$\begin{vmatrix} E'_i + H_{\alpha\alpha} - \lambda & H_{\alpha\beta} \\ H_{\beta\alpha} & H'_i + H_{\beta\beta} - \lambda \end{vmatrix} = 0$$

and

$$H_{\alpha\alpha}(i) = H_{\beta\beta}(i)$$
$$H_{\alpha\beta} = H_{\beta\alpha}$$

therefore

$$\lambda = E'_i + H_{\alpha\alpha} \pm H_{\alpha\beta} \tag{3.81}$$

is the energy modification due to the second order perturbation. $2H_{\alpha\beta}$ is the energy interval between two otherwise degenerate states. For the conduction band, the light hole valence band, and the spin–orbit split band, the results are as follows:

$$\begin{aligned} H_{\alpha\alpha} &= a_i^2 A' k^2 \\ &+ b_i^2 \left\{ Mk^2 + \frac{(L-M-N)\left(k_x^2 k_y^2 + k_y^2 k_z^2 + k_z^2 k_x^2\right)}{k^2} \right\} \\ &+ c_i^2 \left\{ L'k^2 + 2\frac{(L-M-N)\left(k_x^2 k_y^2 + k_y^2 k_z^2 + k_z^2 k_x^2\right)}{k^2} \right\} \\ H_{\alpha\beta} &= \pm\sqrt{2}\,\frac{a_i b_i B}{k} \left\{ k^2 \left(k_x^2 k_y^2 + k_y^2 k_z^2 + k_z^2 k_x^2\right) - 9k_x^2 k_y^2 k_z^2 \right\}^{1/2} \end{aligned} \tag{3.82}$$

For the heavy hole valence band, since a=c=0 and b = 1, the results are:

$$\begin{aligned} H_{\alpha\alpha} &= Mk^2 + \frac{(L-M-N)\left(k_x^2 k_y^2 + k_y^2 k_z^2 + k_z^2 k_x^2\right)}{k^2} \\ H^{\alpha\beta} &= 0 \end{aligned} \tag{3.83}$$

It can be seen from (3.82) and (3.83) that the existence of $\pm H_{\alpha\beta}$ lifts the two-fold degeneracy for the conduction band, the light hole valence band and the spin–orbit split band, but not for the heavy hole valence band. The energy modification for the heavy hole valence band is:

$$E_{hh} = \frac{\hbar^2 k^2}{2m} + Mk^2 + (L - M - N)\frac{\left(k_x^2 k_y^2 + k_y^2 k_z^2 + k_z^2 k_x^2\right)}{k^2}$$

usually also written as

$$E_{hh} = -\frac{\hbar^2 k^2}{2m_{hh}} \tag{3.84}$$

The expressions, (3.72)–(3.74) are widely used *E-k* relations for narrow gap semiconductors. In the above formula the parameters *A′, M, L, N, L′, B* all are interaction matrix elements, their expressions are:

$$\begin{aligned}
L' &= F' + 2G, \qquad & L &= F + 2G \\
M &= H_1 + H_2 \\
N' &= F' - G + H_1 - H_2, \qquad & N &= F - G + H_1 - H_2
\end{aligned}$$

$$\begin{aligned}
F' &= \frac{\hbar^2}{m^2}\sum_j^{\Gamma_1}\frac{\left|\langle x|p_x|u_j\rangle\right|^2}{E_v - E_j}, \quad G = \frac{\hbar^2}{2m^2}\sum_j^{\Gamma_{12}}\frac{\left|\langle x|p_x|u_j\rangle\right|^2}{E_v - E_j} \\
H_1 &= \frac{\hbar^2}{m^2}\sum_j^{\Gamma_{15}}\frac{\left|\langle x|p_y|u_j\rangle\right|^2}{E_v - E_j}, \quad H_2 = \frac{\hbar^2}{m^2}\sum_j^{\Gamma_{25}}\frac{\left|\langle x|p_y|u_j\rangle\right|^2}{E_v - E_j} \\
A' &= \frac{\hbar^2}{m^2}\sum_j^{\Gamma_{15}}\frac{\left|\langle s|p_y|u_j\rangle\right|^2}{E_c - E_j} \\
B' &= \frac{2\hbar^2}{m^2}\sum_j^{\Gamma_{15}}\frac{\langle s|p_x|u_j\rangle\langle u_j|p_y|z\rangle}{\frac{1}{2}(E_c + E_v) - E_j} \\
F &= F' + P^2/(E_v - E_c)
\end{aligned} \tag{3.85}$$

Depending on details m_{hh} may be positive or negative. To agree with experiment it must be positive.

3.3.5 Contribution from the Linear *K* Term H_{kso}

The contribution to the Hamiltonian from the linear *k* term has the form:

$$H_{kso} = \frac{\hbar^2}{4m^2c^2}(\nabla V \times \mathbf{k}) \cdot \sigma \tag{3.86}$$

Since the contribution from the H_{kso} term is small, and the conduction and the spin–orbit split bands' degeneracy has been removed already, its contribution to these bands can be ignored. However the degeneracy for the heavy hole valence band remains, and at the Γ point Φ_{v1} and Φ_{v2} are also degenerate. Therefore only the contribution from the linear *k* term that splits the heavy hole and light hole valence bands need be considered.

By using the selection rules in Sect. 3.2.3, the following interaction matrix elements of H_{kso} can be obtained.

$$\begin{array}{cc} & \begin{array}{cccc} \Phi_{v1\alpha} & \Phi_{v1\beta} & \Phi_{v2\alpha} & \Phi_{v2\beta} \end{array} \\ H_{ij} = -\dfrac{c}{2} & \begin{pmatrix} 0 & \sqrt{3}(k_x - ik_y) & k_x + ik_y & -2k_x \\ \sqrt{3}(k_x + ik_y) & 0 & -2k_z & -(k_x + ik_y) \\ k_x - ik_y & -2k_z & 0 & \sqrt{3}(k_x + ik_y) \\ -2k_x & -(k_x + ik_y) & \sqrt{3}(k_x - ik_y) & 0 \end{pmatrix} \end{array} \tag{3.87}$$

Here the coefficient c in H_{ij} takes the values c_a, c_b, or c_c

$$\begin{aligned} c_a &= -\frac{\hbar^2}{2\sqrt{3}m^2c^2}\left\langle x \left| \frac{\partial V}{\partial y} \right| z \right\rangle \\ c_b &= -\frac{\hbar^2}{2\sqrt{3}m^2c^2}\sum_j^{\Gamma_{12}} \frac{\langle x|p_x|\Psi_j\rangle \left\langle \Psi_j \left| \frac{\partial \Psi}{\partial z} p_x - \frac{\partial V}{\partial x} p_z \right| y \right\rangle}{E_v - E_j} \\ c_c &= \frac{\hbar^2}{2\sqrt{3}m^2c^2}\sum_j^{\Gamma_{25}} \frac{\langle x|p_y|\Psi_j\rangle \left\langle \Psi_j \left| \frac{\partial V}{\partial z} p_x - \frac{\partial V}{\partial x} p_z \right| x \right\rangle}{E_v - E_j} \end{aligned} \tag{3.88}$$

where c_a is for the H_{kso} term, c_b and c_c are cross terms of the results from the second order perturbation calculation of both the ***k·p*** and the $(\nabla V \times \mathbf{p}) \cdot \sigma$ terms. The numerical estimate by Dresselhaus (1955) shows

$c_a\left(\frac{k_{\max}}{2}\right)\sim 0.02\text{eV}$ for InSb, where $k_{\max}$ is the wave vector at the boundary of Brillouin zone. c_b, c_c are very small.

The modified energy values can be obtained by diagonalizing (3.88):

$$\lambda_{\substack{1,2\\3,4}} = \pm c\left\{k^2 \pm \sqrt{3}\left(k_x^2k_y^2 + k_y^2k_z^2 + k_z^2k_x^2\right)^{1/2}\right\}^{1/2} \tag{3.89}$$

Then the energies for the heavy hole and the light hole valence bands are modified:

$$E = E_{v1,2} \pm c_a\left[k^2 \pm \sqrt{3}\left(k_x^2k_y^2 + k_y^2k_z^2 + k_z^2k_x^2\right)^{1/2}\right]^{1/2} \tag{3.90}$$

Thus, the twofold degeneracy of the heavy hole valence band is removed. Equation (3.90) also raises a new issue. If both signs "$\pm$" in the equation take the "+" sign, and since

$$\begin{aligned} k_x &= k\sin\theta\cos\phi \\ k_y &= k\sin\theta\sin\phi \\ k_z &= k\cos\theta \end{aligned}$$

Equation (3.90) becomes a term linear in k:

$$\lambda = +ck\left[1+\sqrt{3}\sin\theta\left(\cos^2\theta+\frac{1}{4}\sin^2\theta\sin^2 2\phi\right)^{1/2}\right]^{1/2} \tag{3.91}$$

Therefore the energy is:

$$E_{v1} = Ak^2 + Bk \tag{3.92}$$

From $\frac{\partial E_{v1}}{\partial k} = 2Ak + B = 0$, one can determine the maximum point of the $E-k$ curve, and according to this equation it is not at the point $k=0$. This prediction is never observed. The problem may be corrected by the inclusion of the B type terms.

For the E_{v1} band, the k term is two times larger than the k^2 term. However for the E_{v2} band, the contribution from the k^2 term is much larger than that from k term. Therefore at $|k|>|k_m|$ (k_m is the wave vector at the predicted, but not observed, maximum of valence band), the v_1 band is well separated from the v_2 band, that is $E_{v1}-E_{v2}>h_{12}$. It is reasonable to omit the term $\langle\Phi_{v1}|(\nabla V\times\mathbf{k})\cdot\sigma|\Phi_{v2}\rangle$ simplify the matrix in (3.87).

The matrix in (3.67) was obtained for $k // z$. The energy expression corresponding to (3.73) and (3.74) for different directions are obtained by rotating the x, y, z basis system into a x', y', z' system where $k // z'$. Then the modified energy can be calculated.

Since the term $\left\langle \Phi_{v1} \middle| (\nabla V \times \mathbf{k}) \cdot \sigma \middle| \Phi_{v2} \right\rangle$ is small and can be omitted, the matrix in (3.87) can be divided into two 2×2 matrices,

For the heavy hole band it becomes:

	$\Phi'_{v1\alpha}$	$\Phi'_{v1\beta}$
$\Phi'_{v1\beta}$	$H_{1\alpha\alpha}$	$H_{1\alpha\beta}$
$\Phi'_{i\beta}$	$H_{1\beta\alpha}$	$H_{1\beta\beta}$

and for the light hole band it becomes:

	$\Phi'_{v2\alpha}$	$\Phi'_{v2\beta}$
$\Phi'_{v2\alpha}$	$H_{2\alpha\alpha}$	$H_{2\alpha\beta}$
$\Phi'_{v2\beta}$	$H_{2\beta\alpha}$	$H_{2\beta\beta}$

(3.93)

It can be shown that:

$$H_{1\alpha\alpha} = H_{v1\beta\beta} = 0$$

$$H_{v1\alpha\beta} = 3\sqrt{3}c\left[\frac{kk_z\left(k_x^2 - k_y^2\right) + ik_x k_y\left(k^2 + k_z^2\right)}{k^2\left(k_x^2 + k_y^2\right)^{1/2}}\right]$$

$$H_{v1\beta\alpha} = 3\sqrt{3}c\left[\frac{kk_z\left(k_x^2 - k_y^2\right) - ik_x k_y\left(k^2 + k_z^2\right)}{k^2\left(k_x^2 + k_y^2\right)^{1/2}}\right]$$

The modified energy can be obtained from the secular equation,

$$\begin{vmatrix} -\lambda & H_{\alpha\beta} \\ H_{\beta\alpha} & -\lambda \end{vmatrix} = 0$$

to be:

$$\delta E_{v1} = \lambda = \pm\sqrt{\mathrm{H}_{\alpha\beta}\mathrm{H}_{\beta\alpha}} = \pm\frac{3\sqrt{3}c}{k^2}\left[\left(k_x^2 + k_y^2\right)\left(k_y^2 + k_z^2\right)\left(k_z^2 + k_x^2\right)\right]^{1/2} \quad (3.94)$$

for the heavy hole band, and

$$\delta E_{v2} = \pm \frac{\sqrt{3}c}{k^2}\left[\left(k_x^2+k_y^2\right)\left(k_y^2+k_z^2\right)\left(k_z^2+k_x^2\right)-8k_x^2k_y^2k_z^2\right]^{1/2} \tag{3.95}$$

for the light hole band. Equation (3.94) shows that the band E_{v1} will be split except in the (100) direction, and band E_{v2} will be split except in the (111) and (100) directions. Since the degeneracy of E_{v2} was already removed in the former discussion, the contribution of the linear k term H_{kso} to the removal of degeneracy is more meaningful for E_{v1}.

To sum up the results from step 4 and step 5, the energy E_{v1} of the heavy hole band can be expressed as:

$$\begin{aligned} E_{v1} \quad = \quad & \frac{\hbar^2k^2}{2m_0} + Mk^2 + \left(L-M-N\right)\frac{k_x^2k_y^2+k_y^2k_z^2+k_z^2k_x^2}{k^2} \\ \pm \quad & \frac{3\sqrt{3}c}{k^2}\left[\left(k_x^2+k_y^2\right)\left(k_y^2+k_z^2\right)\left(k_z^2+k_x^2\right)\right]^{1/2} \end{aligned} \tag{3.96}$$

The maximum of the heavy hole band in the (111) direction is found from:

$$\frac{\partial E_{v1}}{\partial k} = 0$$

which yields:

$$\begin{cases} k_* = \left|\dfrac{2\sqrt{2}c}{\dfrac{\hbar^2}{m_0}+\dfrac{2}{3}\left(L+2M-N\right)}\right| \\ E_0 = \left|\dfrac{4c^2}{\dfrac{\hbar^2}{m_0}+\left(L+2M-N\right)\dfrac{2}{3}}\right| \end{cases} \tag{3.97}$$

It can be seen that when $c \neq 0$ the maximum point k_* in this formalism departs from the Γ point. For InSb and HgCdTe k_* is predicted to be ~0.3 times the distance to the edge of the Brillouin Zone, and E_0~10^{-5}–10^{-4} eV. Once again this prediction does not agree with experiments.

In Kane's ***k·p*** perturbation method, the energy band parameters E_g, P, M, L, N, etc. must be deduced from experiments. Through the measurements of the energy gap, the split off energy gap, the conduction band electron effective mass, the heavy hole and light hole effective masses and the Luttinger parameters (Luttinger 1956; Mavroides 1972), the values of E_g, and the other parameters can be deduced.

The ***k·p*** perturbation method is a very useful method to obtain insight into the origin of band structure features of narrow gap semiconductors. However it does not predict quantitative results, since its parameters are obtained by fitting numerous parameters to experimental results.

3.4 Parameters of the Energy Bands

3.4.1 The Energy Gap

The energy band structures derived in Sect. 3.3 are analytic expressions for the energy-wave vector relation near the origin of the Brillouin Zone in narrow gap semiconductors. However there are many undetermined parameters involved in the expressions. These energy band parameters include the following quantities: the energy gap E_g, the momentum matrix element P, the heavy hole effective mass m_{hh}, the spin–orbit splitting Δ, the conduction band electron effective mass m_c, and combination parameters like the intrinsic carrier concentration n_i. These energy band parameters can also be obtained by theoretical calculations but until recently with large errors. Katsuki and Kunimune (1971), Chadi and Cohen (1973) calculated the HgCdTe energy bands using a virtual crystal pseudopotential method. Figure 3.8 shows their variation of the energy gap E_g and the spin–orbit splitting Δ with composition x. It is evident from the figure that the variation of E_g with composition x is nearly linear and can find the composition x for which the energy gap is zero $(E_g = 0)$. However, the accuracy of this theoretical calculation is about 0.05 eV that corresponding to a 400 cm^{-1} wave number. This error is obviously too large to guide infrared detector research. A more recent calculation Chen and Sher (1995) using the hybrid pseudopotential method coupled to the CPA yields much better results, see Fig. 3.9. Their results agree with experiment to within a few millivolts over the concentration range $0 < x < 0.4$. Still it is necessary to obtain the applicable values of the energy band parameters by experiments.

The most important parameter is the energy gap E_g. We need determine E_g from experiment and obtain its values at different temperatures and for $Hg_{1-x}Cd_xTe$ compositions.

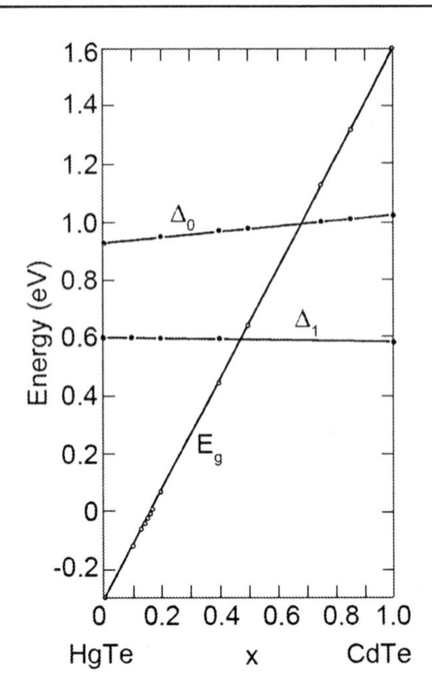

Fig. 3.8. E_g and Δ versus composition x from theoretical calculations (Chadi and Cohen 1973)

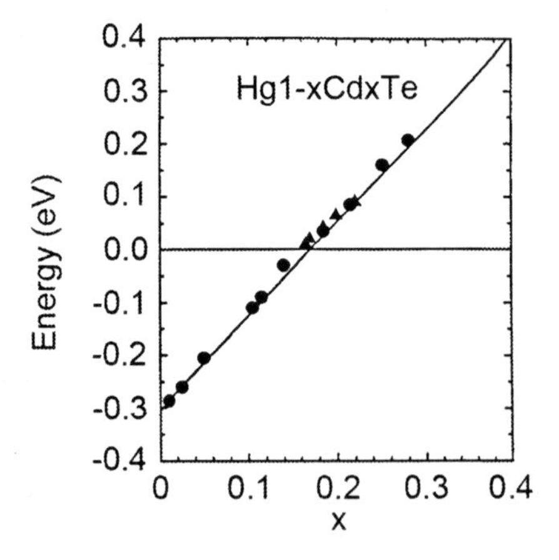

Fig. 3.9. Band gap of $Hg_{1-x}Cd_xTe$ as a function of the concentration x. The *solid line* is theory (Chen 1995) and the *dots* (Guldner et al. 1977) and *triangles* (Dornhaus and Nimtz 1977) are experimental values at 4.2 K

There are a variety of experiments used to determine E_g (Dornhaus and Nimtz 1983), such as the Subnikov-de Hass effect (Antcliffe 1970), magnetic plasma reflection (Galazka and Sosnowski 1967), interband magnetic reflection and magnetic absorption (Harman et al. 1961), cyclotron resonance and non-resonant cyclotron absorption (Wiley and Dexter 1969; Ellis and Moss 1971), temperature dependence of carrier concentrations (Mallon et al. 1973; Finkman 1983), optical absorption (Scott 1969; Blue 1964), and photoluminescence (Elliott et al. 1972). The Subnikov-de Hass effect and magnetic plasma reflection can be used to obtain carrier concentrations and to deduce the energy gap. In interband magneto-reflection and magneto-absorption experiments, the characteristic spectrum depends on the energy gap and magnetic field. At different magnetic fields, the characteristic spectrum changes. Extrapolating the observed spectrum to zero magnetic, we can obtain a characteristic energy spectrum that depends on the energy gap, and from which it can be deduced. We can also find the energy gap by measuring the cyclotron resonance frequency that depends on the conduction band electron effective mass, and therefore the energy gap. In photo-luminescence experiments by measuring the interband recombination rate of optically excited carriers, the energy gap can be also obtained. Intrinsic optical interband absorption measurements yield the energy gap directly.

The intrinsic optical absorption measurement is the most direct method to determine the energy gap. The absorption spectrum has a steep absorption edge as a function of wave number followed by a slowly increasing band-to-band transition absorption band. For a thick sample only the absorption edge can be measured, where the absorption coefficient is small enough to permit enough light transmission to be measured. However, the absorption coefficient of the interband transition is very high for thick samples. The light in this wave band cannot penetrate the sample very far, and therefore cannot be detected in thick samples. So if one wants to measure the interband transition, a very thin sample must be adopted for the measurement. Another option is to use a reflection measurement. After obtaining the intrinsic absorption spectrum, the location of band gap can be determined from the spectrum.

Scott (1969) had measured the absorption spectrum of HgCdTe, over a wide range of compositions, shown in Fig. 3.10. Since the samples he used were rather thick, he obtained only the absorption edge. He took the photon energy, where the absorption coefficient $\alpha = 500\ \text{cm}^{-1}$, as the value of energy gap and then found a relation between the energy gap and

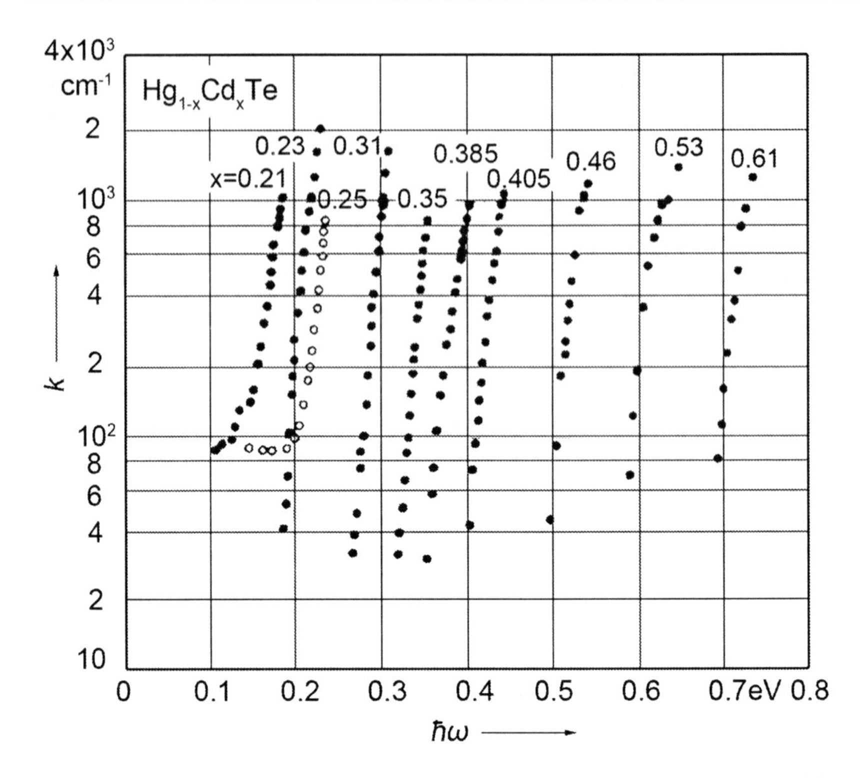

Fig. 3.10. The absorption spectrum of $Hg_{1-x}Cd_xTe$ for different compositions at room temperature (Scott 1969)

composition, at different temperatures. He deduced an empirical formula for the energy gap:

$$E_g(\mathrm{eV}) = 0.303 + 1.37x + 5.6 \times 10^{-4} \times (1 - 2x)T + 0.25x^4 \tag{3.98}$$

Measurements of the photoconductive and the photovoltaic device spectral response also can be used to determine the energy gap. The spectral response curve of a device is the electrical current produced by the optical excitation of electrons from the valance band to conduction band. So it reflects essentially the optical absorption spectrum. Schmit and Stelzer (1969) adopted the method of measuring photoconductive and photo-voltaic device spectral response and took as E_g the photon energy corresponding to an arbitrary cut-off wavelength λ_{co}, defined as one half of the peak of the long wavelength response. Using this technique to deduce the energy gap, at different temperatures, T, and compositions, x, an empirical formula $E_g(\mathrm{x}, T)$ was obtained:

$$E_g(\mathrm{eV}) = -0.25 + 1.59x + 5.233(10^{-4}) \cdot (1 - 2.08x)T + 0.327x^3 \tag{3.99}$$

Finkman and Nemirovsky (1979) performed an optical absorption experiment on thin $Hg_{1-x}Cd_xTe$ samples with compositions x = 0.205 to 0.220. The absorption coefficients in the range ~1,000–2,000 cm^{-1} were measured. In the measured absorption coefficient vs. photon energy curve, they took the photon energy corresponding to the absorption coefficient of 1,000 cm^{-1} as E_g, and also obtained a formula for $E_g(x,T)$:

$$E_g(\text{eV}) = -0.337 + 1.948x + 6.006(10^{-4}) \cdot (1-1.89x)T \tag{3.100}$$

Weiler (1981) based on magneto-optical experimental data, deduced still another formula:

$$\begin{aligned} &E_g(\text{eV}) \\ &= -0.304 + 6.3(10^{-4}) \cdot^{2} (1-2x)T/(11+T) + 1.858x + 0.054x^2 \end{aligned} \tag{3.101}$$

This formula agrees well with the other experimental values when the temperature is below ~100 K and $x \le 0.3$.

In 1982, Hansen et al. (1982) analyzed systematically the data accumulated at the Honewell laboratory, as well as that issued by other laboratories. Most of this data was collected from magneto-optical experiments at low temperature. Their empirical ex-pression for $E_g(x,T)$ is:

$$\begin{aligned} &E_g(\text{eV}) \\ &= -0.302 + 1.93x + (1-2x)5.35(10^{-4})T - 0.810x^2 + 0.832x^3 \end{aligned} \tag{3.102}$$

It is applicable to the regimes $0 \le x \le 0.6$, and $4.2\text{ K} \le x \le 300\text{ K}$.

Of the various experimental methods used to determine E_g, the values obtained by measuring interband magneto-absorption at low temperature is acknowledged to be the most accurate. But magneto-optic experimental data is usually restricted to be collected below 77°K. Measuring the intrinsic absorption spectrum, we can obtain absorption data at temperatures ranging from 4.2 to 300°K. Identifying the band gap with a photon energy corresponding to a particular absorption coefficient, e.g. 500 or 1,000 cm^{-1}, or to a cut off wave length in a photoconductivity experiment is inherently inaccurate. As Tang pointed out (Tang et al. 1958), the shape of the long wavelength spectral response is sample dependent. It changes with surface preparation, and impurity content. Scattering induced band tails of both the conduction and valence band edges are common. Also mesoscopic concentration fluctuations, electron–phonon scattering, and strain induced piezoelectric fields, all contribute to the band tails. Thus only data collected on carefully prepared surfaces and samples with low impurity densities can be trusted to yield band gap properties of the intrinsic material.

A better criterion for determining the energy gap from an absorption spectrum is the location of the turning point where the steep absorption edge ends and the flat intrinsic absorption band begins (Finkman and Nemirovsky 1979; Chu 1984; Chu et al. 1982, 1983). The reason is that the steep absorption edge is dominated by transitions between the band tails of conduction and valence bands, while onset of the flat intrinsic absorption band is due to band-to-band transitions. This method for determining E_g has also been applied to other narrow gap semiconductors, such as $Hg_{1-x}Zn_xTe$ (Wu et al. 1995a,b). In order to determine the location of the onset of band-to-band absorption the complete absorption spectrum must be measured. In transmission spectroscopy, the transmittance is: $T \propto \exp(-\alpha d)$, where α is the absorption coefficient and d the thickness of the sample. To get a reasonable accuracy of spectral measurements a transmittance of at least ~10^{-3}–10^{-4} is needed. This can be attained for large absorption coefficients only with very thin samples, those having a thickness d of ~10 μm. After obtaining the complete absorption spectrum, one can fit the data to a theoretical curve to deuced m_c, m_{hh}, Δ, and then calculate n_i.

To measure the complete absorption spectrum, ranging from its onset into the intrinsic absorption band, absorption coefficients as high as 10^3~10^4 cm^{-1} must be measured. Therefore the samples must be very thin. As the photon energy decreases gradually from a value above the energy gap to the energy gap, the absorption coefficient decreases gradually. The feature of the transmission spectrum of thin samples as functions of wave number is that the transmittance increases slowly at first from zero to a critical value, then increasing more rapidly to form the transmission edge. The energy range, over which the transmittance increases rapidly, corresponds to the band tail dominated absorption edge, and the energy range over which the transmittance increases slowly from zero to a critical value, corresponds to the intrinsic absorption band. Figure 3.11 shows a transmission spectrum of a thin $Hg_{1-x}Cd_xTe$ sample (x=0.443; d=2.5 μm). The experimental temperatures were 300, 250, 200, 50, 100, 77 and 4.2 K. At each temperature a transmission spectrum was measured as shown. From the figure we can see that when ν = 5,000 cm^{-1}, (λ = 2 μm), the sample already has about a 10% transmittance. As the wave number decreases the transmittance increases slowly until at T = 4.2 K the critical wave number reaches ~4,000–4,150 cm^{-1}. The critical wave number increases as the temperature increases. By measuring this change the variation of the energy gap with temperature can be deduced (Chu 1984).

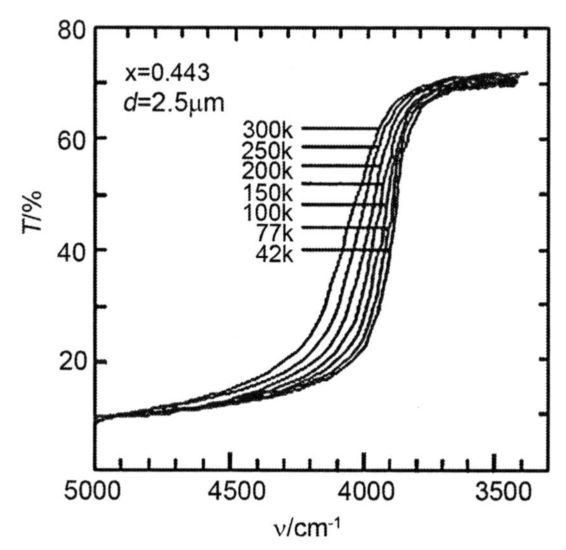

Fig. 3.11. Transmission spectra of a $Hg_{1-x}Cd_xTe$ ($x = 0.443$, $d = 2.5$ μm) thin sample at 4.2–300°K

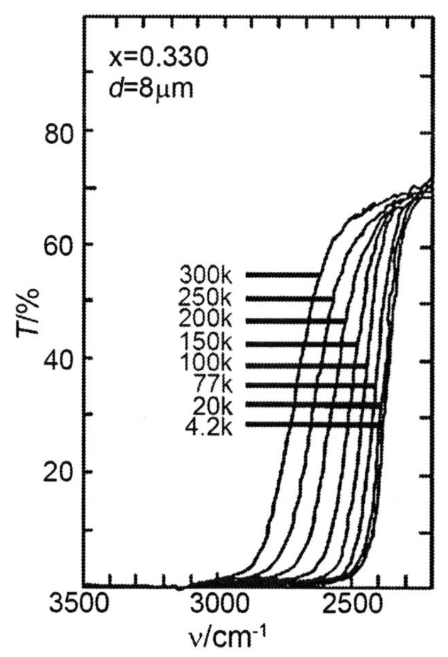

Fig. 3.12. Transmission spectra of a $Hg_{1-x}Cd_xTe$ ($x = 0.330$, $d = 8$ μm) sample at temperatures ranging from 4.2 to 300 K

Figures 3.12 and 3.13 show transmission spectra of HgCdTe samples with $x = 0.330$ ($d = 8$ μm), $x = 0.276$ ($d = 6$ μm), respectively, at different temperatures.

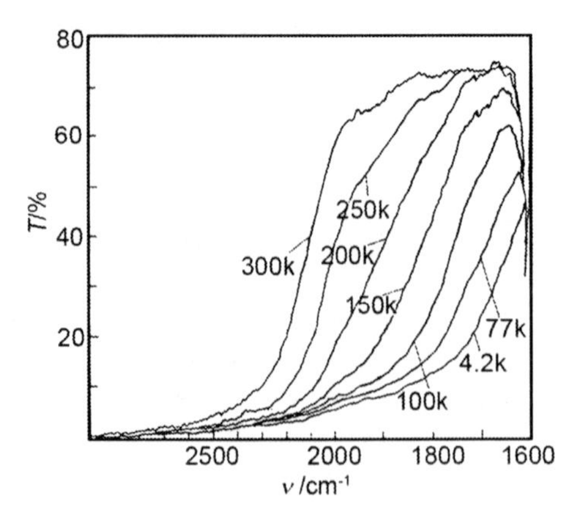

Fig. 3.13. Transmission spectra of a $Hg_{1-x}Cd_xTe$ ($x = 0.276$, $d = 6$ μm) sample measured at temperatures ranging from 4.2 to 300 K

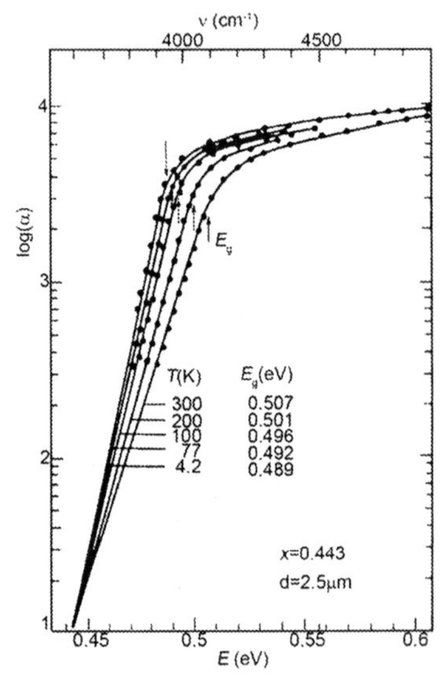

Fig. 3.14. Absorption spectra of a $Hg_{1-x}Cd_xTe$ ($x = 0.443$, $d = 2.5$ μm) sample at different temperatures

Based on these measured transmission spectra, the absorption spectra with different compositions at different temperatures can be obtained. The specific data reduction method is presented in Sect. 4.2. Figures 3.14–3.16 show the absorption spectra of HgCdTe thin samples with $x = 0.443$, 0.276 and 0.200 at different labeled temperatures. At the wave number, where the absorption edge ends and the intrinsic absorption band begins, the corresponding phonon energy is taken to be the energy gap E_g. In Figs. 3.14–3.16, the location of the critical wave number for each temperature is identified with an arrow.

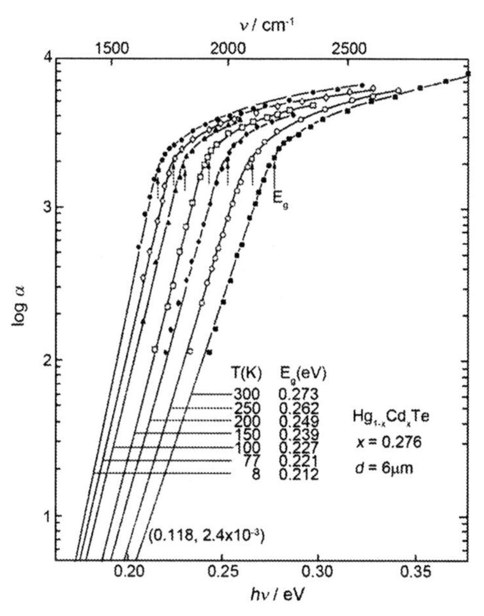

Fig. 3.15. Absorption spectra of a $Hg_{1-x}Cd_xTe$ (x = 0.276, d = 6 μm) sample at different temperatures

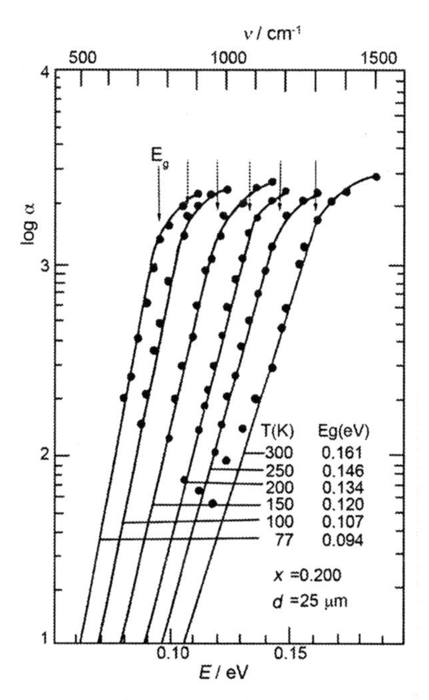

Fig. 3.16. Absorption spectra of a $Hg_{1-x}Cd_xTe$ (x = 0.2, d = 25 μm) sample at different temperatures

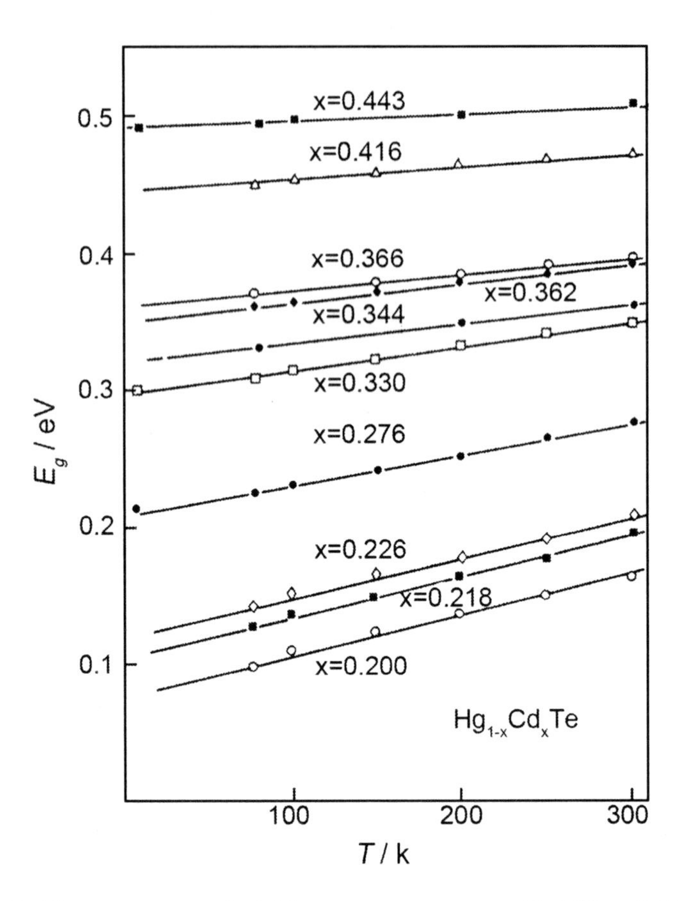

Fig. 3.17. The energy gap E_g (eV) versus temperature T of $Hg_{1-x}Cd_xTe$ samples with different compositions

Table 3.7. $\partial E_g/\partial T$ of HgCdTe with different compositions

X	0.200	0.218	0.226	0.276	0.330	0.344	0.362	0.366	0.416	0.443
$\partial E_g/\partial T$ (10^{-4} eV/K)	3.1	3.14	2.9	2.34	1.7	1.4	1.37	1.2	0.8	0.5

Based on these measured absorption spectra, the energy gap E_g(eV) at different temperatures can be obtained for the different compositions (Fig. 3.17). From these curves the temperature coefficient of the energy gap $\partial E_g/\partial T$ can be deduced for the samples with different compositions (Table 3.7). Further the $\partial E_g/\partial T$ verses x relation curve can be obtained (Fig. 3.18).

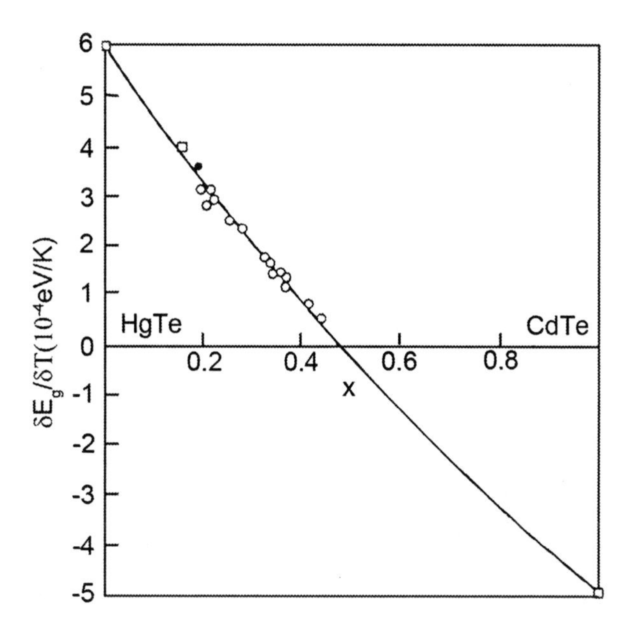

Fig. 3.18. The temperature coefficient of the energy gap $\partial E_g / \partial T$ versus composition x

In Fig. 3.18 the black dots are taken from the experimental results of Chu et al. (1982), the circular dots are deduced from Fig. 3.18 (Chu et al. 1983), and the square dots are taken from the results of other authors. The other values taken from the literature are, $\frac{\partial E_g}{\partial T}$=$4\times10^{-4}$ eV/K for x=0.16 (Bajaj et al. 1982), $\frac{\partial E_g}{\partial T}$=$6\times10^{-4}$ eV/K for HgTe (Pidgenon and Groves 1967), and $\frac{\partial E_g}{\partial T}= 5\times10^{-4}$ eV/K for CdTe (Pidgenon and Groves 1967; Tsay et al. 1973). From these experimental results, it is clear that the relation between $\partial E_g / \partial T$ and composition x is slightly non-linear, bowing below a straight line. The temperature coefficient of energy gap can be expressed as:

$$\frac{\partial E_g}{\partial T} = (6 - 14x + 3x^2) \times 10^{-4}\,\text{eV/K} \tag{3.103}$$

The curve in Fig. 3.18 is calculated from (3.103). Note when the composition is about $x = 0.48$, the temperature coefficient is zero.

The temperature coefficient of the energy gap E_g of HgCdTe is a non-linear function of composition, which accords with result of theoretical analysis (Pawlikowski and Popko 1977; Pawlikowski et al. 1976). The temperature coefficient of energy gap generally comes from two contributions, thermal dilation (expansion) of the crystal, and electron–phonon interaction induced band structure modifications:

$$\frac{\partial E_g}{\partial T} = \left(\frac{\partial E_g}{\partial T}\right)_{Di} + \left(\frac{\partial E_g}{\partial T}\right)_{ph} \tag{3.104}$$

in which

$$\left(\frac{\partial E_g}{\partial T}\right)_{Di} = -3ac\left(\frac{\partial E_g}{\partial T}\right)_{T} = ax + b \tag{3.105}$$

where a and b are constants. Equation (3.105) is a linear function of x, but

$$\left(\frac{\partial E_g}{\partial T}\right)_{ph} = \frac{\left[m_e \varepsilon_c^2 + \left(m_e^{3/2} + m_h^{3/2}\right)^{3/2} \varepsilon_v^2\right] G}{1 + F\left[\varepsilon_c^2 + \varepsilon_v^2 \left(m_e^{3/2} + m_h^{3/2}\right)^{-1/3} m_e^{-1/2}\right]} \tag{3.106}$$

in which F, and G are both slowly varying functions of the temperature, the composition, the average mass of atoms in the unit cell, the spin–orbit splitting and various physical constants (Vasilff 1957). Equation (3.103) was derived based on parabolic bands, and as we will see later in this section, this is a poor approximation for narrow gap materials. ε_c, and ε_v are deformation potential constants of the conduction band and the valence band respectively, and they are also x dependent. m_e is the conduction band electron effective mass and depends on x. In general (3.106) is a non-linear function of x, and numerous adjustable parameters that can be selected to cause it to coincide with the experimental results.

A more rigorous approach was taken to the question of the temperature variation of the band gaps of semiconductors, including pseudo-binary alloys by Krishnamurthy et al. (1995). They started from complete band structures calculated from the HPT method and in the case of alloys CPA was added. The electron–phonon interactions were added and all matrix elements were calculated throughout the Brillouin zone, that couple all of the bands to the conduction and valence band edges. These couplings always lead in second order to repulsions between the unperturbed states. For semiconductors the conduction band edge is driven down by the states

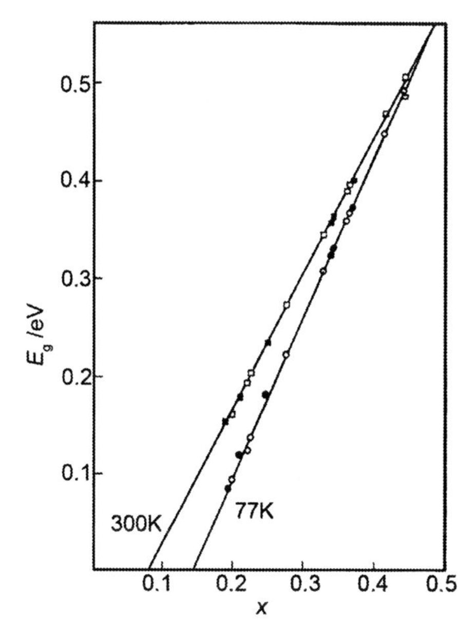

Fig. 3.19. The energy gap E_g (eV) versus composition x at two different temperatures. The experimental *dots* are taken from absorption data and the two *solid lines* are calculated from (3.106) at 300 and 77 K

above it and up by those in the valence band that lie below it. The same is true for the valence band edge. Because of selection rules, contrary to expectations, both band edges decrease in energy as the temperature increases. Moreover, the transverse acoustic phonons dominate the behavior. For wide gap materials the conduction band edge decreases faster than the valence band edge so the gaps close as the temperature increases. While for narrow gap materials the reverse is true and the gap opens. With no parameters fitted to the temperature variation data this theory predicts results in fair agreement with experiments on a wide variety of semiconductors. For example this theory predicts $\mathrm{d}E_g/\mathrm{d}T = 0$ for HgCdTe with $x_0 = 0.50$ compared to the experimental concentration of $x_0 = 0.48$ (see Fig. 3.18).

Based on the experimental values of energy gap for HgCdTe samples with different compositions and at different temperatures, we can obtain a relation between E_g and x, T. Figure 3.19 shows the variation of E_g (eV) and x, square dots are the experimental results at 300°K; circular dots are the experimental results at 77°K. To sum up the above experiment results, the empirical formula is (Chu et al. 1983):

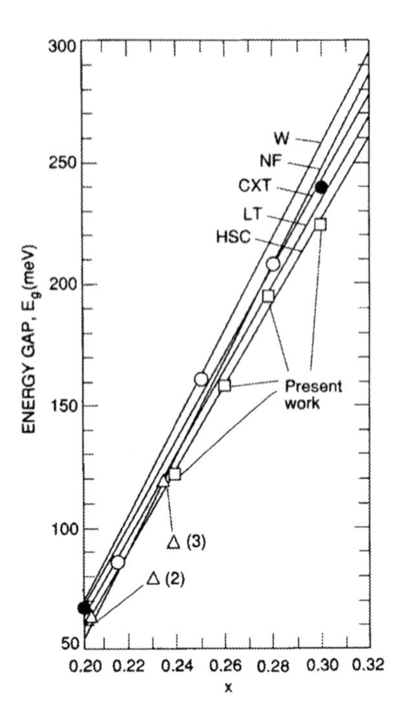

Fig. 3.20. Comparison of the energy gap of HgCdTe calculated from different formulas (Seiler et al. 1990)

$$
\begin{aligned}
&E_g(\text{eV}) \\
&= -0.295 + 1.87x - 0.28x^2 + (6 - 14x + 3x^3)(10^{-4})T + 0.35x^4
\end{aligned}
\tag{3.107}
$$

This formula is applicable for $0.19 \le x \le 0.433$, $4.2\text{K} \le \text{T} \le 300\text{K}$. In Fig. 3.20, two solid lines are fit using (3.107) at 300 and 77 K, respectively, experimental dots are taken from the absorption data. Equation (3.107) was named by Seiler et al. and other authors as the CXT formula (Seiler 1989; Seiler et al. 1990). Because (3.107) was obtained from the energy gap data determined from intrinsic absorption spectra, it has the best obvious physical meaning.

An obvious characteristic of (3.107) is the introduction of the x^2 term. Introducing this term is reasonable from theoretical considerations (Cadorna 1963). Based on virtual crystal approximation (VCA), virtual crystal Hamiltonian can be written as (Zhong 1982):

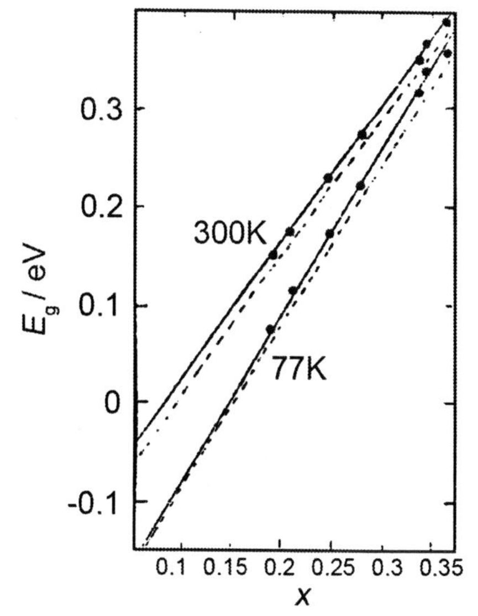

Fig. 3.21. Comparison of some experimental E_g data and results calculated from (3.107) (*solid line*) and (3.102) (*dashed line*)

$$\begin{aligned} H(\text{HgCdTe}) &= -\frac{\hbar^2}{2m}\nabla^2 + xV_{\text{CdTe}}(r) + (1-x)V_{\text{HgTe}}(r) \\ &= xH(\text{CdTe}) + (1-x)H(\text{HgTe}) \end{aligned} \tag{3.108}$$

The electronic energy of HgCdTe is a linear interpolation of the potentials energies of HgTe and CdTe. While the potential varies linearly with x, this does not mean the energy gap will also do so. Further analysis in a tight binding method, shows VCA produced a bowing term (Zhong 1982). If in addition, the effect of disorder induced random potentials in the crystal are added there appears another contribution to the bowing. Hill (1974) and Wu (1983) adopted a crystal lattice potential with non-linear characteristics, and obtained a result indicating that the energy gap of ternary semiconductors should include a bowing term proportional to x^2. Chen and Sher (1978) using CPA derived quantitative results, with no adjustable parameters, for the gap bowing parameter.

Magneto-optical experimental data taken at 4.2°K shows the $E_g - x$ curve does not follow perfectly a linear plus x^2 fitted curve, but shows some decline in the bowing (Elliott and Spain 1970; Dornhaus and Nimtz 1983). This effect can be taken into account by the addition of a x^4 term into the empirical formula. Chen and Sher (1995) have shown that a full CPA calculation based on the HPT band structure method produces a concentration dependence of the gaps of alloys that do not perfectly fit an x^2 bowing behavior.

Figure 3.20 compares some calculated results to experiment data at 4.2 K. Of the calculations tested, the result calculated from the CXT formula provides the best fit (Seiler et al. 1990). This comparison was done in 1990, and there are more recent results (Chen and Sher 1995).

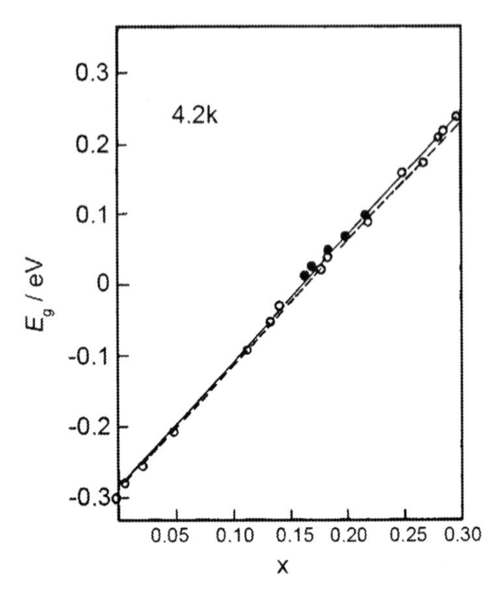

Fig. 3.22. Comparison of E_g obtained from magneto-optical experimental data and calculated results at temperature 4.2 K. The *solid* and *dashed curves* are calculated from (3.107) and (3.102), respectively

Table 3.8. Comparison of energy gap expressions with magneto-optical data for $0 \leq x \leq 0.30$ and $4.2\,\mathrm{K} \leq T \leq 24\,\mathrm{K}$

	Average deviation/eV	Standard error of estimate/eV
Equation (3.107)	0.0014	0.0082
Equation (3.102)	–0.0080	0.0120

Figure 3.21 compares energy gap values obtained from some optical absorption experimental data with values calculated from different $E_g(x, T)$ formulas. In the figure, the solid curve is calculated from (3.107) obtained by Chu et al. and the dashed curve is calculated from $E_g(x, T)$ formula (3.102) of Hansen et al. From the figure it is clear that the Chu result lies a little higher than that of Hansen et al. and fits this data set better.

We can also check the fit of a curve deduced from (3.107) with magneto-optical experimental data taken at low temperatures. These experimental energy gaps were measured on a series of samples by the magneto-optical transmission method at 4.2 K by Guldner (1979), and Dornhaus and Nimtz (1977). The data is shown in Fig. 3.22 by dots and square. In the figure, the solid line is obtained by substituting T=4.2°K into (3.107), and the dashed line is from (3.102).

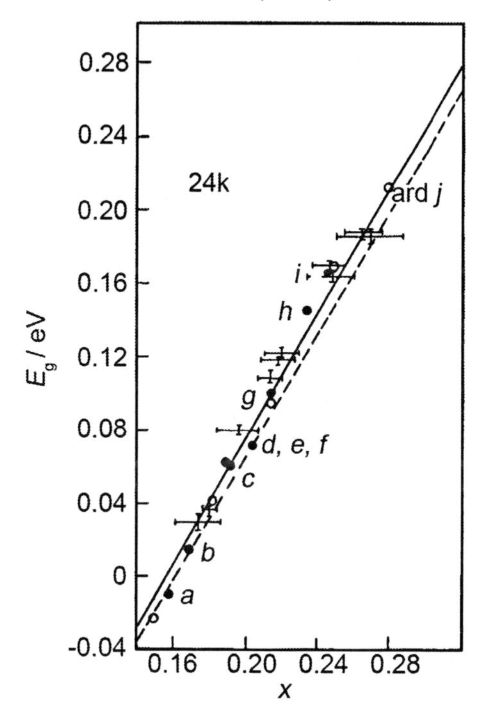

Fig. 3.23. Comparison of E_g obtained from magneto-optical experiment data and calculated results at 24 K. The *solid* and *dashed curves* are calculated from (3.107) and (3.102), respectively

Figure 3.23 has values of the energy gap E_g obtained from (3.107) and from magneto-optical experiments at 24°K. The experimental dots are taken from experiments of many authors (Weiler 1981; Antcliffe 1970; Groves et al. 1971; Strauss et al. 1962; McCombe et al. 1970; Kinch and Buss 1971; Kahlert and Bauer 1973; Weber et al. 1975; Poehler and Apel 1970; Swierkowski et al. 1978). The dashed curve is from the HSC expression (3.101) and the solid curve is from the CXT expression (3.107). The CXT curve is slightly higher than that of the HSC curve.

To compare the calculated results from (3.107) and (3.102) with magneto-optical experimental data in the composition range $0 \le x \le 0.30$, and temperature range $4.2°\text{K} \le \text{T} \le 24°\text{K}$, the average deviation and standard error are listed in Table 3.8 (Chu et al. 1983).

From the above comparison, the CXT expression (3.107) is seen to be the more accurate fit to the data.

Based on (3.107), at 4.2°K, $E_g = 0$ at composition $x_0 = 0.161$. This agrees well with the experimental value $x_0 = 0.161 \pm 0.003$ found by Groves et al. (1971) and the experimental value, $x_0 = 0.165 \pm 0.005$, of Guldner et al. (1977; Guldner 1979). Thus (3.107) is applicable for $0.19 \le x \le 0.443$, and $4.2°\text{K} \le \text{T} \le 300°\text{K}$. Another analysis indicates that the CXT expression is also applicable to the additional region $0.165 \le x \le 0.194$ and $77°\text{K} \le T \le 300°\text{K}$. It stems from the two-photon magnetic absorption experiments of Seiler et al. (1989, 1990; Seiler 1986).

The energy gap is the most important physical parameter of semiconductors. This is especially true for the series of narrow gap $Hg_{1-x}Cd_xTe$ semiconductor alloys, because it is closely related to the spectral response of infrared devices and to their optimum design (Xu 1996; Gong et al. 1996; Yang et al. 1996; Tang 1985; Maxey et al. 1989; Sharma et al. 1994; Herman and Pessa 1985).

3.4.2 The Electron Effective Mass of the Conduction Band

The electron effective mass of the conduction band of narrow gap semiconductors is an important physical quantity. Particularly for the conduction band, the shape is parabolic only very near the band edge. Equation (3.73) has an expression for the conduction band effective mass for small k. In the narrow gap approximation where $E_g \ll \Delta$, it becomes:

$$\frac{1}{m_c^*} \cong \frac{2P^2}{\hbar^2 E_g} \tag{3.109}$$

The 1/m term in (3.73) has been dropped as small compared to the one that has been retained.

Measuring electron effective mass of the conduction band and the band gap is an approach to obtain the energy band parameter P. Alternatively, if we know the energy band parameter, we can obtain the electron effective mass of the conduction band.

If the spin–orbit splitting parameter $\Delta \gg E_g, k\boldsymbol{P}$, an approximate conduction band solution of Kane's cubic equation energy expression (3.70) is given in (3.73) (Kane 1957):

$$E_c = \frac{\hbar^2 k^2}{2m} + \frac{1}{2}\left[E_g + \sqrt{E_g^2 + \frac{8}{3}k^2 P^2} \right] \tag{3.110}$$

This equation holds for a small band gap but finite k, rather than the earlier one we got, valid for small k but arbitrary E_g, and Δ. There are two interesting cases, (a) $E_g{}^2 \gg \frac{8}{3}P^2k^2$, and (b) $E_g{}^2 \ll \frac{8}{3}P^2k^2$. Case (a) reduces to (3.109) and corresponds to $E_c \sim k^2$. Case (b) is different, it yields:

$$E_c \cong \frac{E_g}{2} + \frac{\hbar^2 k^2}{2m} + \sqrt{\frac{2}{3}}Pk \tag{3.111}$$

This equation tells us that in the narrow band gap limit, for an intermediate range of k the conduction band varies nearly linearly with k, since the third term is larger than the second one. In this theory the effective mass for large k becomes the free electron mass:

$$\frac{1}{m_c^*(k)} \cong \frac{1}{m} \tag{3.112}$$

In general (3.110) is the equation of a hyperbola. If one uses the most general relation between the k variation of a band and its effective mass tensor:

$$\frac{1}{\tilde{m}^*} = \frac{1}{\hbar^2}\nabla^2 E(\mathbf{k}) \tag{3.113}$$

then from (3.110), and since $E_g(k)$ depends only on the magnitude of k the tensor reduces to a scalar, and we find:

$$\frac{1}{m_c^*(k)} \cong \frac{1}{m} + \frac{P^2}{\hbar^2\left(E_g{}^2 + \frac{8}{3}P^2k^2\right)^{3/2}}\left(E_g{}^2 - \frac{40}{9}P^2k^2\right) \tag{3.114}$$

This equation reduces to (3.109) in the small k limit, and to (3.111) in the large k limit. Equation (3.72) is a more accurate expression for the small k effective mass, since it holds independent of the relative size of E_g and Δ.

Krishnamurthy and Sher (1994) have calculated the band structures of $Hg_{0.78}Cd_{0.22}Te$ with $E_g = 0.1$ eV at 77 K using the combined HPT and CPA approximations. The results are compared with a Kane prediction in Fig. 3.24 based on (3.110) with E_g, and P selected to fit data taken on a sample with $E_g = 0.1$ eV. The Krishnamurthy calculation includes all the bands not just the four near the gap.

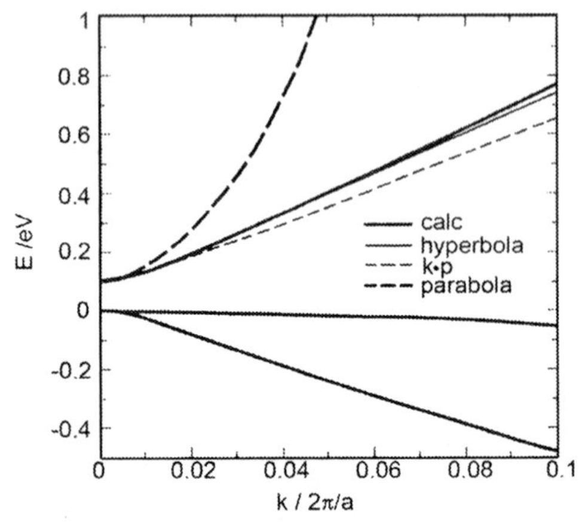

Fig. 3.24. The conduction and valence bands in the (100) direction calculated from HPT/CPA theory, compared to the conduction band based on Kane theory and a parabola. All have an effective mass that agrees with experiment at small k

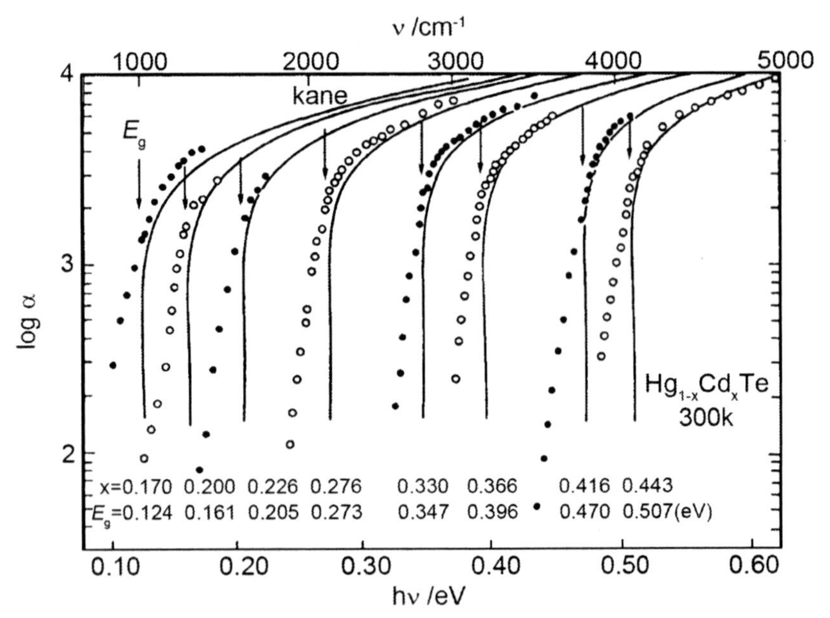

Fig. 3.25a. Experimental absorption spectra and calculated absorption curves of $Hg_{1-x}Cd_xTe$ samples with $x = 0.170$–0.443 at 300 K

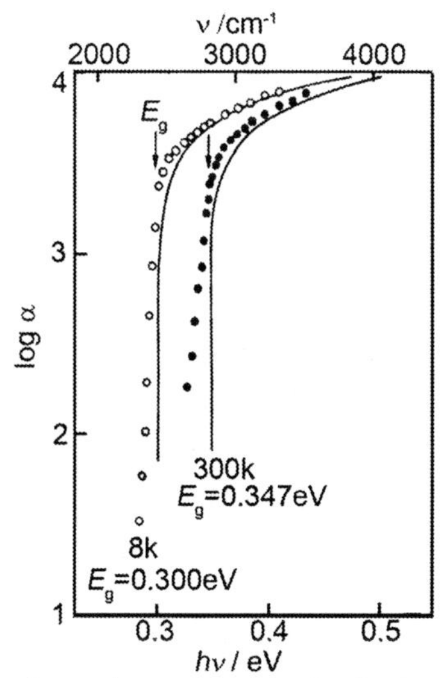

Fig. 3.25b. Experimental absorption spectra and calculated absorption curves for an $x = 0.330$ sample

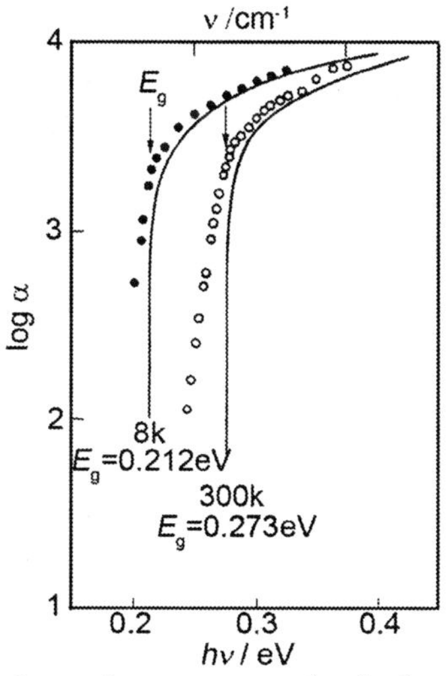

Fig. 3.25c. Experimental absorption spectra and calculated absorption curves for an $x = 0.276$ sample

Chu et al. (1983, 1991, 1992) continued measuring the intrinsic absorption spectrum of a series of $Hg_{1-x}Cd_xTe$ thin samples with different compositions. Because the measured samples are very thin, the most complete intrinsic absorption spectra to date have been obtained. They display a clear steep absorption edge and a flat intrinsic absorption band. This data offers an experimental basis for fitting band-to-band transition calculations, and to extract energy band parameters.

Related experiments were described in earlier references (Chu 1984, 1985) and in Sect. 4.2.2. In Figs. 3.25a–c, squares and dots indicate absorption spectra obtained at room temperature, and solid lines are theoretical absorption curves at 300 K.

The theoretical absorption curves are calculated based on an optical transition theory, and Kane's energy band model of narrow gap semiconductors (Kane 1957; Bassani and Pastori 1975):

$$\alpha = \frac{4\pi^2 e^2 \hbar}{m_0^2 ncE} \sum_j |Mj|^2 \int \frac{2d^3k}{(2\pi)^3} \delta\left(E_c - E_j - \hbar\omega\right) \tag{3.115}$$

in which $|Mj|^2$ is the square of an optical matrix element, and the sum is over the heavy-hole and the light-hole bands. The integral generates the joint initial and final density of states. The value of the energy gap used in this calculation is determined from the experimental absorption spectrum. The photon energy in the region, where the steep absorption edge ends and it begins to bend, is E_g. Actually the values taken are marked on the figures. Details of the calculational procedure are presented in Sect. 4.2. This treatment is for the composition range, x = 0.170–0.443, and temperature range, T = 8–300 K. The calculation takes the momentum matrix element to be $P = 8 \times 10^{-8}$ eV cm, the heavy-hole effective mass to be $m_{hh}^* = 0.55m$, and the spin–orbit splitting to be $\Delta = 1$ eV. The calculated results coincide well with the experiments in the Kane region where the photon energy is larger than energy gap. However the agreement is bad near the absorption edge, where scattering induced band tails exist. A first principles theory of the absorption coefficient also exists that predicts the observed behavior of several semiconductors including a few concentrations of $Hg_{1-x}Cd_xTe$ (Krishnamurthy et al. 1996).

By substituting the values for P and Δ obtained by fitting absorption spectra to (3.115), we can obtain a simplified expression for the electron effective mass at the bottom of the conduction band:

$$\frac{m_c^*}{m} = 0.05966 \frac{E_g\left(E_g + 1\right)}{E_g + 0.667} \tag{3.116}$$

in which E_g can be taken to be given by (3.107), that is

$$E_g(\text{eV}) = -0.295 + 1.87x - 0.28x^2 + (6 - 14x + 3x^2)(10^{-4})T + 0.35x^4$$

By using (3.116), we can calculate the value of (m_c^*/m) for HgCdTe samples with different compositions and temperatures. Again using (3.113) we can obtain the electron effective mass of the conduction band at wave-vector k or at energy E_k.

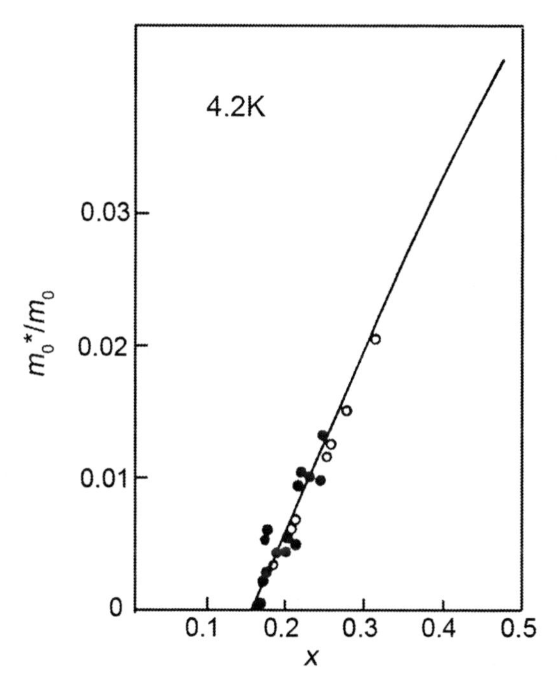

Fig. 3.26. The electron effective mass of the conduction band versus composition x at 4.2 K (The curve is from (3.115), and the data are taken from references.)

Figure 3.26 illustrates the relation between the electron effective mass of the conduction band bottom with composition x at 4.2°K. In the figure, the curve is calculated from (3.116), dots are taken from references (Antcliffe 1970; Ellis and Moss 1971; Suizu and Narita 1972, 1973; Strauss et al. 1962; Kinch and Buss 1971; Kahlert and Bauer 1973; Weber et al. 1975; Poehler and Apel 1970; Verie and Decamps 1965; Long 1968; Narita et al. 1971).

Tables 3.9–3.11 are comparisons of experimental and calculated results for the conduction band effective masses in degenerate cases. Antcliffe (1970) measured the SdH effect for samples with the same composition but with different electron concentrations at 4.2 K, and then obtained the electron concentration n and electron effective mass $m_c^*(n)$ at wave vector $k=(3\pi^2 n)^{1/3}$. At 77 K, the photoconductive response cut-off wavelength of the samples used in these experiments is $\lambda_{c0} = 13.7$ μm with corresponding

Table 3.9. Electron effective mass of $x = 0.1975$ samples at 4.2 K with different electron concentrations ($E_g = 0.063\ 5$ eV, $m_0^*/m_0 = 0.005\ 55$)

$n(cm^{-3})$	$(m_c^*/m)_{cal}$	$(m_c^*/m)_{exp}$
2.32×10^{15}	0.00726	0.00702
3.10×10^{15}	0.00756	0.00755
5.90×10^{15}	0.00847	0.00830
6.60×10^{15}	0.00867	0.00910
8.00×10^{15}	0.00898	0.00920
9.66×10^{15}	0.00938	0.00966

Table 3.10. Experimental and calculated values of the electron effective mass of an $Hg_{1-x}Cd_xTe$ sample at different temperatures

T (K)	x	E_g (eV)	n (cm^{-3})	m_{c0}^*/m	$(m_c^*/m)_{cal}$	$(m_c^*/m)_{exp}$
77	0.135	–0.0154	1.7×10^{16}	0.00136	0.0093	0.010±0.0015
	0.144	–0.0002	8.0×10^{15}	0.0000178	0.0071	0.0068±0.0015
	0.149	0.0083	6.7×10^{15}	0.000739	0.0068	0.0069±0.0004
	0.150	0.0099	2.3×10^{16}	0.000882	0.0102	0.010±0.002
	0.188	0.074	1.4×10^{15}	0.00637	0.0075	0.0076±0.0005
	0.193	0.082	3.5×10^{15}	0.00706	0.0088	0.0085±0.0015
	0.203	0.099	8.9×10^{14}	0.00847	0.0091	0.010±0.003
1.3	0.144	–0.0315	2×10^{15}	0.00277	0.0052	0.0055±0.0015
	0.149	–0.0221	2×10^{15}	0.00195	0.00489	0.0048±0.0007
87	0.149	0.0121	8.3×10^{15}	0.0011	0.00742	0.0073±0.0006
100	0.149	0.0174	1.0×10^{16}	0.0015	0.00755	0.0070±0.0008
125	0.149	0.0273	1.5×10^{16}	0.00241	0.00900	0.0082±0.0010

energy $E_{c0} = 0.0903$ eV. Adopting the relation between the cut-off energy E_{c0} and the composition x obtained by Schmit and Stelzer (1969), the composition of these samples were determined to be $x = 0.1975$. The experimentally measured energy gap of the sample was 0.0635 eV at 4.2 K, which is consistent with the calculated result from (3.107) of $E_g = 0.064$. Using (3.116) the calculated electron effective mass at the bottom of the conduction band is $(m_c^*/m) = 0.00555$, compared to the experimental value of $(m_c^*/m)_{exp}$ = 0056±0.00025. This calculated result agrees well with the experimental result.

At wave vector $k = (3\pi^2 n)^{1/3}$, the measured values of the electron effective mass $(m_c^*(n)/m)_{exp}$ and the calculated values $(m_c^*(n)/m)_{cal}$ are listed in Table 3.9, it can be seen the calculated results is coincides rather well with the experiment results.

Table 3.11. Electron effective mass of samples with high electron concentrations

T (K)	x	E_g (eV)	n (cm^{-3})	$(m_{c0}^*/m)_{cal}$	$(m_c^*/m)_{cal}$	$(m_c^*/m)_{exp}$
4	0	–0.3	4.7×10^{17}	0.024	0.035	0.035
	0	–0.3	1.18×10^{18}	0.024	0.0417	0.043
	0	–0.3	2.82×10^{18}	0.024	0.0516	0.052
	0.15	–0.0192	4.0×10^{17}	0.0017	0.026	0.024
	0.15	–0.0192	1.6×10^{18}	0.0017	0.0418	0.038
	0.15	–0.0192	2.4×10^{18}	0.0017	0.0479	0.047
300	0	–0.115	5.1×10^{17}	0.00978	0.0292	0.029
	0	–0.115	1.2×10^{18}	0.00978	0.0380	0.041
	0	–0.115	5.8×10^{18}	0.00978	0.0625	0.062
	0.1	0.0281	6.2×10^{16}	0.00247	0.0142	0.010
	0.1	0.0281	2.7×10^{17}	0.00247	0.023	0.022
	0.1	0.0281	8.5×10^{17}	0.00247	0.0337	0.033
	0.1	0.0281	2.7×10^{18}	0.00247	0.0495	0.047
	0.19	0.154	1.1×10^{18}	0.0129	0.037	0.034
	0.20	0.168	3.35×10^{18}	0.0140	0.052	0.047
	0.29	0.292	4.8×10^{17}	0.0235	0.0346	0.036
	0.31	0.319	1.0×10^{18}	0.0254	0.040	0.037

Wiley and Dexter (1969) determined the electron effective mass of HgCdTe samples with different compositions and different electron concentrations at wave vector $k=(3\pi^2 n)^{1/3}$, using a helicon wave transmition measurement at 77 and 1.3 K. They also measured an $x = 0.149$ sample at different temperatures, whose results and calculated results from an approximation to (3.114) and (3.116) are listed in Table 3.10. From the table, we can see the calculated results differ from one another and those calculated from (3.116) are in closer agreement with experiment.

Sosnowski and Galazka (1967) also measured samples with high electron concentrations. They measured the relation of thermal electrodynamic potential in a transverse magnetic field, and then deduced electron effective mass at the Fermi energy level. Table 3.11 lists the measured results and their calculated results. The experiment error lies within about ±10%. Theoretical calculated results (except one sample) are all within the same range of experimental errors.

3.4.3 The Momentum Matrix Element and the Heavy-Hole Effective Mass m_{hh}

The momentum matrix element, P, the heavy-hole effective mass, m_{hh}, and the spin–orbit splitting, Δ, are also important energy band parameters. Their values can be obtained by fitting the intrinsic absorption spectra. In Sect. 3.4.2 it has already mentioned that in prescribed composition and temperature regions, the momentum matrix element is $P \cong 8\times 10^{-8}$ eV cm, the heavy-hole effective mass is $m_{hh} \cong 0.55m_0$, and the spin–orbit splitting is $\Delta \cong 1\,\text{eV}$. There have been several direct calculations of P; Katsuki et al. calculated a value $P = 5.9\times 10^{-8}\,\text{eV cm}$ (Katsuki and Kunimune 1971), and Overhof calculated $P = 6.4\times 10^{-8}\,\text{eV cm}$ (Overhof 1971).

There are also some other experimental methods to determine P. For example, the value of P can be determined by measuring the relation between the electron effective mass and the carrier concentration. In the high degeneracy case, (Dornhaus and Nimtz 1983; Hansen and Schmit 1983) it was shown that:

$$\left(\frac{\frac{m_c^*}{m}}{1-\frac{m_c^*}{m}}\right)^2 = \left(\frac{m_{c0}^*}{m}\right)^2 + \frac{1}{2}\left(\frac{3}{\pi}\right)^{2/3}\left(\frac{m_{c0}^*}{m}\right)\frac{\hbar^2}{E_g}n^{2/3} \tag{3.117}$$

where m_{c0}^* is the effective mass at the conduction band edge.

Introducing electron effective mass at the bottom of the conduction band $m_{c0}^* = \frac{3}{4}\frac{\hbar^2 E_g}{P^2}$, and $E_p \equiv \left(\frac{2m}{\hbar^2}\right)P^2$ then the (3.117) changes to:

$$\left(\frac{m_c^*}{m-m_c^*}\right)^2 = \left(\frac{m_{c0}^*}{m}\right)^2 + \frac{3h^2(3\pi^2 n)^{2/3}}{E_p m} \tag{3.118}$$

Therefore, from a plot of $\left(\frac{m_c^*}{m-m_c^*}\right)^2$ versus $(n)^{2/3}$, the slope determines E_p,(and P), and the intercept then yields E_g. Experimental on determination of E_p can be found in the reference (Dornhaus and Nimtz 1983).

Through magneto-optical experiments, one can also determine the value of P. The procedure is to take P, Δ and other parameters as adjustable, to obtain the best theoretical fits to experimental data. Guldner determined $P = 8.5\times10^{-8}$ eV cm through a magneto-optical experiment (Guldner et al. 1977; Guldner 1979).

The heavy-hole effective mass has been assigned a variety of values by different authors. For $x = 0$ semimetal samples, different authors assigned values of m_{hh}^{*}/m to be 0.53, 0.6, and 0.7. Values have been reported for $x = 0.148$ samples of 0.53, 0.71; for $x = 0.161$ samples of 0.28, 0.75; and for $x = 1$ samples of 0.41, 0.5, 0.59, 0.37, 0.46, 0.33. In addition, some authors took 0.4, 0.55, or 0.9 for all compositions. Some of this variation is due to the phenomena being measured. Because the valence band is anisotropic there are different effective masses in different directions. Thus, the effective mass deduced from transport measurements, (conductivity and Hall Effect), and from carrier concentration or Fermi level measurements, involve different combinations of the anisotropic effective masses.

The spin–orbit splitting is generally taken to be $\Delta = 1$ or 0.96 eV.

The same parameters that determine the band structure of narrow gap semiconductors, are also important to material and device design. They enter into many phenomena, such as the radiative recombination mechanism of HgCdTe (Schacham and Finkman 1985), nonlinear optical properties (Seiler et al. 1986; Wang 1986, 1987), the anomalous Hall effect (Pan et al. 1988), HgTe–CdTe superlattice band structures (Harris et al. 1986; Hetzler et al. 1985; Malloy 1989; McGill et al. 1986; Han et al. 1999), the theoretical analysis of HgCdTe luminescence spectrum (Werner and Tomm 1988; Werner et al. 1991; Bouchut et al. 1991; Fuchs et al. 1993; Gille et al. 1990), and the theoretical analysis of HgCdTe carrier transport phenomenon (Schenk 1990; Hoffman et al. 1987; Yadava 1994).

References

Antcliffe GA (1970) Effective mass and spin splitting in $Hg_{1-x}Cd_xTe$. Phys Rev B 2:345–351

Bajaj J, Shin SH, Bostrup G, Cheung DT (1982) Electron mobility in LPE$Hg_{1-x}Cd_xTe$/CdTe layers near zero band-gap crossing. J Vac Sci Technol 21:244–246

Bassani F, Pastori PG (1975) Electronic States and Optical Transitions in Solid. Pergamon Press, Oxford, pp. 149–167

Blair J, Newnham R (1961) In: Metallurgy of Elemental and Compound Semiconductors Vol. 12. Inter Science, New York, pp. 393

Blue MD (1964) Optical Absorption in HgTe and HgCdTe. Phys Rev 134: A226–A234

Bouchut P, Destefanis G, Chamonal JP, Million A, Pelliciari B, Piaguet J (1991) High-efficiency infrared light emitting diodes made in liquid phase epitaxy and molecular beam epitaxy HgCdTe layers. In: AIP Conference Proceedings: Physics and chemistry of mercury cadmium telluride and novel IR detector materials Vol. 235. San Francisco, California pp. 1794–1798

Cadorna M (1963) Phys Rev 129:69

Chadi DJ, Cohen ML (1973) Electronic structure of $Hg_{1-x}Cd_xTe$ alloys and charge-density calculations using representative *k* points. Phys Rev B 7:692–699

Chen AB, Sher A (1978) Gap variation in semiconductor alloys and the coherent-potential approximation. Phys Rev Lett 40:900–903

Chen AB, Sher A (1995) Semiconductor Alloys: Physics and Materials Engineering. Springer, New York

Chu JH (1984) PhD Thesis, Shanghai Institute of Technical Physics

Chu JH (1985) Infrared Res 4A:255

Chu JH, Xu SC, Tang DY (1982) Chinese Sci Bull 27:403

Chu JH, Xu SC, Tang DY (1983) Energy gap versus alloy composition and temperature in $Hg_{1-x}Cd_xTe$. Appl Phys Lett 43:1064–1066

Chu JH, Mi ZY, Tang DY (1991) Infrared Phys 32:195–211

Chu JH, Mi ZY, Tang DY (1992) Band-to-band optical absorption in narrow-gap $Hg_{1-x}Cd_xTe$ semiconductors. J Appl Phys 71:3955–3961

Dornhaus R, Nimtz G (1977) Transverse magnetoresistance of Hg1-xCdxTe in the extreme quantum limit. Solid State Commun 22:41–45

Dornhaus R, Nimtz G (1983) The properties and applications of the HgCdTe alloy system, in narrow gap semiconductors. In: Springer Tracts in Modern Physics, Vol. 98. Springer, New York, pp. 119

Dresselhaus G (1955) Spin-orbit coupling effects in zinc blende structures. Phys Rev 100:580–586

Elliott RJ (1954) Spin-orbit coupling in band theory-character tables for some "double" space groups. Phys Rev 96:280–287

Elliott CT, Spain, IL (1970) Electrical transport properties of semiconducting $Cd_xHg_{1-x}Te$ alloys. Solid State Commun 8:24

Elliott CT, Melngailis I, Harman TC, Foyt AG (1972) Carrier freeze-out and acceptor energies in nonstoichiometric p-type HgCdTe. J Phys Chem Sol 33:1527–1531

Ellis B, Moss TS (1971) In: Pell EM (ed.) Proc. III Int. Conf. On Photoconductivity, Stanford, California 1969. Pergamon, New York, pp. 211

Finkman E (1983) Determination of band-gap parameters of $Hg_{1-x}Cd_xTe$ based on high-temperature carrier concentration. J Appl Phys 54:1883–1886

Finkman E, Nemirovsky Y (1979) Infrared optical absorption of $Hg_{1-x}Cd_xTe$. J Appl Phys 50:4356–4361

Fuchs F, Kheng K, Schwarz K, Koidl P (1993) Fourier transform photoluminescence excitation spectroscopy of medium-bandgap $Hg_{1-x}Cd_xTe$ and InSb. Semi Sci Technol 8:S75–S80

Galazka RR, Sosnowski L (1967) Conduction band structure of Cd sub 0.1 Hg sub 0.9 Te (Conduction band structure and scattering processes of cadmium mercury telluride mixed crystal determined from thermoelectric power, effective mass and electron mobility). Physica Status Solidi 20:113–120

Gille P, Herrmann KH, Puhlmann N, Schenk M, Tomm JW, Werner L (1990) E_g versus x relation from photoluminescence and electron microprobe investigations in p-type $Hg_{1-x}Cd_xTe$ ($0.35 \leq x \leq 0.7$). J Crystal Growth 86:593–598

Gong HM, Hu XN, Li YJ, Shen J, et al. (1996) In: Basic applications studies of HgCdTe materials and devices, 95 collection of the theses, Shanghai Institute of Technical Physics, pp. 276

Groves SH, Harman TC, Pidgeon CR (1971) Interband magnetoreflection of $Hg_{1-x}Cd_xTe$. Solid State Commun 9:451–455

Guldner Y (1979) Science Doctor Thesis, al 'universite Pierre et Marie Curie' (Paris VI) Paris

Guldner, Y, Rigaux C, Mycielski A, Couder Y (1977) Magneto-optical investigation of $Hg_{1-x}Cd_xTe$ mixed crystals. Pt. 1. Semiconducting configuration and semimetal plus semiconductor transition; semimetallic configuration. Phys Status Solidi B 82:149–158; 81:615–627

Han MS, Kang TW, Kim TW (1999) The dependence of the structural properties on the period numbers and the CdTe thicknesses in HgTe/CdTe superlattices. Appl Surf Sci 153:35–39

Hansen GL, Schmit JL (1983) Calculation of intrinsic carrier concentration in $Hg_{1-x}Cd_xTe$. J Appl Phys 54:1639–1640

Hansen GL, Schmit JL, Casselman TN (1982) Energy gap versus alloy composition and temperature in $Hg_{1-x}Cd_xTe$. J Appl Phys 53:7099–7101

Harman TC, Strauss AJ, Dickey DH, Dresselhaus MS, Wright GB, Mavroides JG (1961) Low electron effective masses and energy gap in $Cd_xHg_{1-x}Te$. Phys Rev Lett 7:403–405

Harris KA, Hwang S, Blanks DK, Cook JW Jr, Schetzina JF, Otsuka N, Baukus JP, Hunter AT (1986) Characterization study of a HgTe–CdTe superlattice by means of transmission electron microscopy and infrared photoluminescence. Appl Phys Letts 48:396–398

Harrison WA (1970) Electronic Structure. W.H. Freeman and Co, San Francisco

Herman MA, Pessa M (1985) $Hg_{1-x}Cd_xTe$-$Hg_{1-y}Cd_yTe$ ($0 \leq x, y \leq 1$) heterostructures: Properties, epitaxy, and applications. J Appl Phys 57:2671–2694

Herman F, Kuglin CD, Cuff KF, Kortum RL (1963). Relativistic Corrections to the Band Structure of Tetrahedrally Bonded Semiconductors. Phys Rev Lett 11:541–545

Herring C (1940) A new method for calculating wave functions in crystals. Phys Rev 57:1169–1177

Hetzler SR, Baukus JP, Hunter AT, Faurie JP, Chow PP, McGill TC (1985) Infrared photoluminescence spectra from HgTe–CdTe superlattices. Appl Phys Letts 47:260–262

Hill R (1974) Energy-gap variations in semiconductor alloys. J Phys C: Solid State Phys 7:521–526

Hoffman CA, Bartoli FJ, Meyer JR (1987) Photo-hall determination of acceptor densities in n-type HgCdTe. J Appl Phys 61:1047–1054

Kane EO (1957) Band structure of indium antimonide. J Phys Chem Solids 1:249–261

Kane EO (1966) In: Semiconductors and Semimetals Vol. 1. Academic Press, London, pp. 75

Kahlert H, Bauer G (1973) Magnetophonon effect in n-Type $Hg_{1-x}Cd_xTe$ (x=0.212). Phys Rev Lett 30:1211–1214

Katsuki S, Kunimune M (1971) The band structures of alloy system $Hg_{1-x}Cd_xTe$ calculated by the psedopotential method. J Phys Soc Jpn 31:415–421

Kinch MA, Buss DD (1971) J Phys Chem Solids 32(Suppl 1):461

Kohn W, Rostocker N (1954) Solution of the Schrödinger equation in periodic lattices with an application to metallic lithium. Phys Rev 94:1111–1120

Korringa J (1947) On the calculation of the energy of a Bloch wave in a metal. Physica 13:392–400

Krishnamurthy S, Sher A (1994) Electron mobility in $Hg_{0.78}Cd_{0.22}Te$ alloy. J Appl Phys 75:7904–7909

Krishnamurthy S, Chen AB, Sher A, van Schilfgaarde (1995) Temperature dependence of band gaps in HgCdTe and other semiconductors. J Electron Mater 24:1121–1125

Krishnamurthy S, Chen AB, Sher A (1996) Near band edge absorption spectra of narrow-gap III–V semiconductor alloys. J Appl Phys 80:4045–4048

Krishnamurthy S, Berding MA, Yu ZG (2006) Private communication

Long D (1968) Calculation of ionized-impurity scattering mobility of electrons in $Hg_{1-x}Cd_xTe$. Phys Rev 176:923–927

Long D, Schmit JL (1973) Infrared detectors. Defense Industry Press, Beijing

Luttinger JM (1956) Quantum theory of cyclotron resonance in semiconductors: general theory. Phys Rev 102:1030–1041

Mallon CE, Naber JA, Colwell JF, Green BA (1973). Effects of electron radiation on the electrical and optical properties of HgCdTe. IEEE Trans Nucl Sci 20:214–219

Mavroides JG (1972) In: Abeles F (ed.) Optical Properties of Solids. North-Holland Publishing Company-Amsterdam, London p. 355

Maxey CD, Capper P, Easton BC, Whiffin PAC (1989) Band edge absorption in $Cd_xHg_{1-x}Te$ grown by metal-organic vapour phase epitaxy. Infrared Phys 29:961–964

McCombe BD, Wagner RJ, Prinz GA (1970) Far-infrared observation of electric-dipole-excited electron-spin resonance in $Hg_{1-x}Cd_xTe$. Phys Rev Lett 25:87–90; Infra-red pulsed gas-laser studies of combined resonance and cyclotron-phonon resonance in nonstoichiometric HgCdTe. Solid Stat Commun 8:1687–1691

McGill TC, Wu GY, Hetzler SR (1986) Superlattices: progress and prospects. J Vac Sci Technol A 4:2091–2095

Muller MW, Sher A (1999) Mesoscopic composition fluctuations in semiconductor alloys: effect on infrared devices. Appl Phys Lett 74:2343–2345

Narita S, Kim RS, Ohtsuki O, Ueda R (1971) Far-infrared cyclotron mass of $Cd_xHg_{1-x}Te$ near the $Lambda_6$ – $Lambda_8$ crossover. Phys Lett 35:203–204

Nordheim L (1931) The electron theory of metals. Annalen der Physik 9:607–641
Overhof H (1971) Model calculation for the energy bands in the nonstoichiometric HgCdTe mixed-crystal system. Phys Stat Sol B 45:315–322
Pan DS, Lu Y, Chu M (1988) Constant-current-density model for the anomalous Hall effects in $Hg_{0.8}Cd_{0.2}Te$. Appl Phys Lett 53:307–309
Parmenter RH (1955) Symmetry properties of the energy bands of the blende structure. Phys Rev 100:573–579
Pawlikowski JM, Popko E (1977) Temperature shift on the $Cd_xHe_{1-x}Te$ energy gap. Solid Stat Commun 22:231–233
Pawlikowski JM, Becla P, Dudziak E (1976) Opt Appl 6:3
Phillips JC (1973) Bonds and Bands in Semiconductors. Academic Press, London, p. 212
Pidgenon CR, Groves SH (1967) In: Thomas DG (ed.) Int Conf II–VI Semioonducting Compounds. Benjamin, New York, p. 1080
Poehler TO, Apel JR (1970) Far infrared cyclotron resonance in $Hg_{1-x}Cd_xTe$. Phys Lett A 32:268–269
Schacham SE, Finkman E (1985) Recombination mechanisms in p-type HgCdTe – Freezeout and background flux effects. J Appl Phys 57:2001–2009
Schenk A (1990) Spatially variable drift mobility model for $Hg_{1-x}Cd_xTe$ diodes I. Analytical base and fit to hall data. Phys Stat Sol (A) 122:413–425
Schmit JL, Stelzer EL (1969) Temperature and alloy compositional dependences of the energy gap of $Hg_{1-x}Cd_xTe$. J Appl Phys 40:4865–4869
Scott MW (1969) Energy gap in $Hg_{1-x}Cd_xTe$ by optical absorption. J Appl Phys 40:4077–4081
Seiler DG (1989) In: Willardson RK, Beer AC (eds.) Semiconductors and Semimetals Vol. 36. Academic, New York
Seiler DG, McClure SW, Justice RJ, Loloee MR (1986) Nonlinear optical determination of the energy gap of $Hg_{1-x}Cd_xTe$ using two-photon absorption techniques. Appl Phys Letts 48:1159–1161
Seiler DG, Lowney JR, Littler CL, Loloee MR (1990) Temperature and composition dependence of the energy gap of $Hg_{1-x}Cd_xTe$ by two-photon magnetoabsorption techniques. J Vac Technol A 8:1237–1244
Sharma RK, Verma D, Sharma BB (1994) Observation of below band gap photoconductivity in mercury cadmium telluride. Infrared Phys Technol 35:673–680
Slater JC (1937) Wave functions in a periodic potential. Phys Rev 51:846–851
Sosnowski L, Galazka RR (1967) Proc Inter Conf Phys of II–VI Semiconductors, Providence
Spicer WE, Silberman JA, Morgen J, Lindau I, Wilson JA, Chen AB, Sher A (1982) Dominance of atomic states in a solid: selective breakdown of the virtual crystal approximation in a semiconductor alloy, $Hg_{1-x}Cd_xTe$. Phys Rev Lett 49:948–951
Strauss AJ, Harman TC, Mavroides JG, Dickey DH, Dresselhaus MS (1962) In: Strickland AC (ed.) Proceedings of the International Conference on the Physics of Semiconductors, Exeter. The Institute of Physics and the Physical Society, London, p. 703

Suizu K, Narita S (1972) Shubnikov-de haas oscillation in nonstoichiometric CdHgTe. Solid State Commun 10:627–631

Suizu K, Narita S (1973) Rep of the departm of mat phys. In: The Faculty of Eng Science of Osaka University Toyonaka. Osaka, Japan

Swierkowski L, Zawadski W, Guldner Y, Rigaux C (1978) Two-mode resonant polaron in the $Hg_{0.72}Cd_{0.28}Te$ semiconductor. Solid State Commun 27:1245–1247

Tang DY (1985) Research on IR physics in China. Infrared Phys 25:3–12

Tang TY, Kao KY (1958) Sonderdruck aus Festkorperphysik und Physik der Leuchtstoffe. Akademic-Verlag, Berlin

Tsay YF, Mitra SS, Vetelino JF (1973) Temperature dependence of energy gaps in some II–VI compounds. J Phys Chem Solids 34:2167–2175

Vasilff HD (1957) Electron self-energy and temperature-dependent effective masses in semiconductors: n-Type Ge and Si. Phys Rev 105:441–446

Verie C, Decamps E (1965) Effective electron mass in mercury telluride. Phys Stat Sol 9:797–803

Wang WL (1986) Infrared Res 5:241

Wang W L (1987) Chinese Phys 7:524

Weber BA, Sattler JP, Nemarich J (1975) Magnetically tunable laser emission from [Hg,Cd]Te. Appl Phys Lett 27:93–95

Weiler MH (1981) Magnetooptical properties. In: Willardson RK, Beer AC (eds.) Semiconductors and Semimetals Vol. 16. Academic, New York, p. 119

Werner L, Tomm JW (1988) Photoluminescence in p-$Hg_{0.42}Cd_{0.58}Te$. Phys Stat Solid A 106:K83–K87

Werner L, Tomm JW, Herrmann KH (1991) Identification of the nature of the optical transitions in $Hg_{0.42}Cd_{0.58}Te$. Infrared Phys 31:49–58

Wiley JD, Dexter RN (1969) Helicons and nonresonant cyclotron absorption in semiconductors II $Hg_{1-x}Cd_xTe$. Phys Rev 181:1181–1190

Woolley JC, Ray B (1960) Solid solution in $A^{II}B^{VI}$ tellurides. J Phys Chem Solids 13:151–153

Wu S (1983) On the energy bandgap bowing in $Hg_{1-x}Cd_xTe$ Solid. State Commun 48:747–749

Wu CC, Chu DY, Sun CY, Yang TR (1995a) Infrared spectroscopy of $Hg_{1-x}Zn_xTe$ alloys. Jpn J Appl Phys 34:4687–4693

Wu CC, Chu DY, Sun CY, Yang TR (1995b) Optical characterization of $Hg_{1-x}Zn_xTe$ crystals grown by the travelling heater method. Mater Chem Phys 40:7–12

Yadava RDS (1994) Expression for intrinsic carrier concentration in $Hg_{1-x}Cd_xTe$. Solid State Commun 92:357–360

Yang JR, Wang SL, Guo SP, He L (1996) J Infrared Milli Waves 15:328

Zhong XF (1982) Chinese J Semicond 3:453

Appendix 3A: Crystallography and the Bloch Theorem

Elementary Crystallography

Start from an arbitrary point in a crystal. It is easy to identify a large number of other points P', P'', etc. that are equivalent to P. The crystal looks just the same (except for surface effects) when viewed from any of these equivalent points. A vector ***PP'*** between any two of these points is called a "lattice translation vector." The set of lattice translation vectors is independent of the choice of P.

The set of all equivalent lattice points form a pattern called a Bravais lattice. At each Bravais lattice point there exists a basis of atoms that are identical in number and orientation to ones at every other Bravais lattice point. Once the Bravais lattice and basis are identified the crystal structure is completely specified. There are only 14 distinct Bravais lattice types, space groups; and an unlimited number of basis types, point groups. The focus in this book is on zincblende structures, where the Bravais lattice is the fcc structure, and the basis has two atoms.

Primitive Lattice Vectors

For any Bravais lattice it is possible to find three vectors $\vec{\ell}_1, \vec{\ell}_2$ and $\vec{\ell}_3$ in terms of which any lattice translation vector $\vec{\ell}$ can be expressed as:

$$\vec{\ell} = n_1\vec{\ell}_1 + n_2\vec{\ell}_2 + n_3\vec{\ell}_3 \tag{3A.1}$$

with n_1, n_2, and n_3 being integers. The construction of $\vec{\ell}$ proceed as follows:

- Choose any $\vec{\ell}_l$ that is the shortest lattice vector in its direction. Successive application of $\vec{\ell}_l$ generates a line of equivalent points. This line cannot contain any lattice points not generated by $\vec{\ell}_l$ without iolating the criteria by which it was chosen. Moreover, all Bravais lattice points lie on lines parallel to this one with identical spacing.
- Find a second line, not parallel to the first. Chose $\vec{\ell}_2$ to be the shortest vector connection two Bravais lattice point along this second line. The vectors $\vec{\ell}_1$, and $\vec{\ell}_2$ define a plane.

- Find a second plane parallel to the first with no lattice points separating them. Chose a lattice vector $\vec{\ell}_3$ to be the shortest vector connecting a lattice point in the first plane to one in the second plane.

These three vectors $\vec{\ell}_1$, $\vec{\ell}_2$ and $\vec{\ell}_3$ are called primitive lattice translation vectors. They are not unique. From (3A.1) and the definitions of $\vec{\ell}_1$, $\vec{\ell}_2$ and $\vec{\ell}_3$ it follows that all Bravais lattices have inversion symmetry though lattice points.

The parallelepiped defined by $\vec{\ell}_1$, $\vec{\ell}_2$ and $\vec{\ell}_3$ whose volume is:

$$\Omega = \vec{\ell} \cdot \left(\vec{\ell}_2 \times \vec{\ell}_3 \right) \tag{3A.2}$$

contains no points except its eight corners. From this one can deduce that all primitive lattices contain only one basis independent of how the primitive vectors are chosen. As an example examine the fcc lattice. A primitive vector set often chosen is:

$$\begin{aligned} \vec{\ell}_1 &= \frac{a}{2}(1,1,0) \\ \vec{\ell}_2 &= \frac{a}{2}(1,0,1) \\ \vec{\ell}_3 &= \frac{a}{2}(0,1,1) \end{aligned} \tag{3A.3}$$

where “a” is the length of a cube edge, and the vectors are in Cartesian coordinates. Note that this in not an orthogonal set, nor is it normalized. For example, the point $\vec{\ell}_1 + \vec{\ell}_2 = a(2,1,1)/2$ is a face centered lattice point, and the point a(1,0,0) is reached by the combination $\vec{\ell}_1 + \vec{\ell}_2 - \vec{\ell}_3$. The volume of the primitive cell is $a^3/8$.

Wigner–Seitz Primitive Cell Construction

To construct the Wigner–Seitz cell draw lines from one lattice point to all other Bravais lattice points. Then construct planes that intersect these lines at their mid-points and are perpendicular to the lines. The surface generated by the intersections of these planes is the Wigner–Seitz primitive cell. This cell can be translated on the Bravais lattice to fill all space. Uses for this cell will become evident in later sections.

Reciprocal Lattice Vectors

Primitive reciprocal lattice vectors are defined by the relations:

$$\mathbf{g}_j \bullet \vec{\ell}_j \equiv 2\pi : j = 1,2,3 \tag{3A.4}$$

and they can be written as:

$$\begin{aligned} \mathbf{g}_1 &= 2\pi \frac{\vec{\ell}_2 \times \vec{\ell}_3}{\Omega} \\ \mathbf{g}_2 &= 2\pi \frac{\vec{\ell}_3 \times \vec{\ell}_1}{\Omega} \\ \mathbf{g}_3 &= 2\pi \frac{\vec{\ell}_1 \times \vec{\ell}_2}{\Omega} \end{aligned} \tag{3A.5}$$

where Ω is given by (3A.2). The volume of the primitive reciprocal lattice cell $\Omega^{\#}$ is:

$$\Omega^{\#} = \mathbf{g}_1 \bullet (\mathbf{g}_2 \times \mathbf{g}_3) = \frac{(2\pi)^3}{\Omega} \tag{3A.6}$$

A general reciprocal lattice vector $\boldsymbol{G}$ is defined as:

$$\mathbf{G} \equiv m_1 \mathbf{g}_1 + m_2 \mathbf{g}_2 + m_3 \mathbf{g}_3 \tag{3A.7}$$

where the m_1, m_2, m_3 are all $\pm$ integers or zero.

A Wigner–Seitz primitive cell can also be constructed in the reciprocal lattice; it is referred to as a Brillouin zone. The import of this zone will be discussed below. Because the reciprocal lattice of a face-centered lattice is a body-centered lattice the shape of this reciprocal lattice cell is that shown in Fig. 3.2.

Electron Motion in Crystals

When approximations to the electron–electron interaction are made that allow the potential seen by each electron to be treated as periodic, then it can be expressed as:

$$V(\mathbf{r}) = V\left(\mathbf{r} + \vec{\ell}\right) \tag{3A.8}$$

Because $V(\mathbf{r})$ is invariant under a lattice translation it can be expanded in a Fourier series in terms of reciprocal lattice vectors:

$$V(\mathbf{r}) = \sum_{\mathbf{G}} v_{\mathbf{G}} e^{i\mathbf{G}\cdot\mathbf{r}} \tag{3A.9}$$

The sum over all **G** is the same as a sum over all values of m_1, m_2, and m_3.

Now the one electron Hamiltonian can be written as:

$$H(\mathbf{r}) = KE + V(\mathbf{r}) = H(\mathbf{r} + \vec{\ell}) \tag{3A.10}$$

There is no restriction on the magnitude of V in this discussion. Define a lattice translation operator T as:

$$\mathrm{T} \equiv e^{\vec{\ell}\cdot\nabla} \tag{3A.11}$$

Remember that exponential operators are defined by their Taylor expansions:

$$e^{\mho} \equiv 1 + \mho + \frac{1}{2}\mho^2 + \ldots = \sum_j \frac{1}{j!}\mho^j$$

Let's examine properties of this operator. Since $\left[\ell_x \frac{\partial}{\partial x} . \ell_y \frac{\partial}{\partial y}\right] = 0$, they commute, as do the other components, the translation operator takes the form:

$$\mathrm{T} = e^{\ell_x \frac{\partial}{\partial x}} e^{\ell_y \frac{\partial}{\partial y}} e^{\ell_z \frac{\partial}{\partial z}} \tag{3A.12}$$

Operating with $e^{\ell_x \frac{\partial}{\partial x}}$ on a function $\Phi(x, y, z)$ yields:

$$e^{\ell_x \frac{\partial}{\partial x}} \Phi(x, y, z) = \left(1 + \ell_x \frac{\partial}{\partial x} + \frac{\ell_x^{\ 2}}{2!}\frac{\partial^2}{\partial x^2} + \frac{\ell_x^{\ 3}}{3!}\frac{\partial^3}{\partial x^3} + \ldots\right)\Phi(x, y.z)$$
$$= \Phi(x + \ell_x, y, z).$$

Thus for the whole translation operator we find:

$$\mathrm{T}\Phi(r) = \Phi(\mathbf{r} + \vec{\ell}) \tag{3A.13}$$

as one would expect for something called a lattice translation operator. The lattice translation operator and the Hamiltonian, $H = \frac{\hbar^2}{2m}\nabla^2 + V(\mathbf{r})$ commute, $[\mathrm{T}, H] = 0$, so they share a common set of eigen functions.

Next we start from the Schrödinger equation

$$H(\mathbf{r})\Psi_{\mathbf{k}}(\mathbf{r}) = E_{\mathbf{k}}\Psi_{\mathbf{k}}(\mathbf{r}) \tag{3A.14}$$

for a cubic crystal of side lengths $L_x = N_x a_x, L_y = N_y a_y$, and $L_z = N_z a_z$, where the a_j's are the edge lengths of cubic unit cells, and the N_j's are the number of cubic cells in the j direction. A cubic cell in a zincblende lattice contains eight atoms, four cations and four anions. The boundary conditions will be taken as periodic.

$$\Psi_{\mathbf{k}}(\mathbf{r}) = \Psi_{\mathbf{k}}(\mathbf{r} + \mathbf{L}) \tag{3A.15}$$

These boundary conditions simplify the arguments but are not realistic. However it has been demonstrated that they produce the same results for large crystals containing many atoms as are found with proper conditions.

Next solve for the eigenvalues $C_{\mathbf{k}}$ of the translation operator

$$\mathrm{T}\Psi_{\mathbf{k}}(\mathbf{r}) = C_{\mathbf{k}}\Psi_{\mathbf{k}}(\mathbf{r}) \tag{3A.16}$$

Start with the case $\vec{\ell} = a_x\hat{x}$, where $\hat{x}$ is a unit vector in the x-direction

$$e^{\vec{a}_x \bullet \nabla}\Psi_{\mathbf{k}}(\mathbf{r}) = \Psi_{\mathbf{k}}(\mathbf{r} + \vec{a}_x) = C_{\mathbf{k}}(\vec{a}_x)\Psi_{\mathbf{k}}(\mathbf{r}) \tag{3A.17}$$

and

$$(e^{\vec{a}_x \bullet \nabla})^{N_x}\Psi_{\mathbf{k}}(\mathbf{r}) = \Psi_{\mathbf{k}}(\mathbf{r} + L_x\hat{x}) = [C_{\mathbf{k}}(\vec{a}_x)]^{N_x}\Psi_{\mathbf{k}}(\mathbf{r}) \tag{3A.18}$$

The boundary condition in the x direction requires $\Psi_{\mathbf{k}}(L_x\hat{x}) = \Psi_{\mathbf{k}}(0)$, so $\left[C_{\mathbf{k}}(\vec{a}_x)\right]^{N_x} = 1$, and the $C_{\mathbf{k}}(\vec{a}_x)$ values are the N_x roots of unity.

$$C_{\mathbf{k}}(\vec{a}_x) = e^{2\pi i n_x a_x} = e^{ik_x a_x} \tag{3A.19}$$

where

$$k_x \equiv 2\pi n_x / a_x N_x, \quad n_x = 0, \pm 1, \pm 2, \ldots \pm N_x/2 \tag{3A.20}$$

Next we examine the more general case:

$$\vec{\ell} = m_x a_x\hat{x} + m_y a_y\hat{y} + m_z a_z\hat{z}, \ m_x, m_y, m_z = 0, \pm 1, \pm 2, \ldots \tag{3A.21}$$

to find:

$$e^{\vec{\ell} \bullet \nabla}\Psi_{\mathbf{k}}(\mathbf{r}) = \Psi_{\mathbf{k}}\left(\mathbf{r} + \vec{\ell}\right) = e^{i\mathbf{k} \bullet \vec{\ell}}\Psi_{\mathbf{k}}(\mathbf{r}) \tag{3A.22}$$

The last equality in (3A.22) is one statement of the Bloch theorem.

We now seek solutions to the Schrodinger equation of the form:

$$\Psi_{\mathbf{k}}(\mathbf{r}) = e^{i\mathbf{k}\cdot\vec{\ell}} u_{\mathbf{k}}(\mathbf{r}) \tag{3A.23}$$

and require:

$$\begin{aligned}\Psi_{\mathbf{k}}\left(\mathbf{r}+\vec{\ell}\right) &= e^{i\mathbf{k}\cdot\left(\mathbf{r}+\vec{\ell}\right)} u_{\mathbf{k}}\left(\mathbf{r}+\vec{\ell}\right) \\ &= e^{i\mathbf{k}\cdot\vec{\ell}}\Psi_{\mathbf{k}}(\mathbf{r}) = e^{i\mathbf{k}\cdot\vec{\ell}} e^{i\mathbf{k}\cdot\mathbf{r}} u_{\mathbf{k}}(\mathbf{r})\end{aligned} \tag{3A.24}$$

This equation can be satisfied only if $u_{\mathbf{k}}(\mathbf{r})$ is periodic so $u_{\mathbf{k}}(\mathbf{r}) = u_{\mathbf{k}}\left(\mathbf{r}+\vec{\ell}\right)$. The most often quoted statement of the Bloch theorem is, "The solution of the Schrödinger equation for a periodic Hamiltonian has the form (3A.23) with $u_{\mathbf{k}}(\mathbf{r})$ being a periodic function."

Reduced Zone

Now we are in a position to examine one of the principal virtues of the Bloch theorem. The solution to the Schrödinger equation has states **k** that lie inside the first Brillouin zone and also outside this zone. Examine a **k** = **k'** + $\mathbf{G}_\beta$, with the reciprocal lattice vector $\mathbf{G}_\beta$ chosen to make **k'** lie in the first zone. Then we have:

$$\begin{aligned}\Psi_{\mathbf{k}}(\mathbf{r}) &= \Psi_{\mathbf{k}'+\mathbf{G}_\beta}(\mathbf{r}) = u_{\mathbf{k}'+\mathbf{G}_\beta}(\mathbf{r}) e^{i(\mathbf{k}'+G_\beta)\cdot\mathbf{r}} \\ &\equiv u_{\mathbf{k}'\beta}(\mathbf{r}) e^{i\mathbf{k}'\cdot\mathbf{r}} = \Psi_{\mathbf{k}'\beta}(\mathbf{r})\end{aligned} \tag{3A.25}$$

where the $u_{\mathbf{k}'\beta}(\mathbf{r})$ defined above is:

$$u_{\mathbf{k}'\beta}(\mathbf{r}) \equiv u_{\mathbf{k}'+\mathbf{G}_\beta}(\mathbf{r}) e^{i\mathbf{G}_\beta\cdot\mathbf{r}} = u_{\mathbf{k}'\beta}\left(\mathbf{r}+\vec{\ell}\right) \tag{3A.26}$$

Thus $\Psi_{\mathbf{k}'\beta}(\mathbf{r})$ is also a Bloch wave function, with β designating the zone from which the function was shifted. In the extended zone method only the wave vector **k** (along with the spin orientation) is needed to specify the electron state. In the reduced zone method all wave vectors are shifted into the first zone, and then states are designated by a wave vector in the first zone, an index that tells from which zone they originated, and a spin index.

Number of Electrons per Spin in a Band

To be more precise in (3A.20), k_x has unique values for n_x between $-(N_x+1)/2$ and $N_x/2$ ranging over a total of N_x values. Because there are $N_xN_yN_z$ primitive cells in the Brillouin zone, this says that each band in the zone can accommodate one electron. This assumes that bands in the reduced zone are designated by a **k** vector, a band index β, and a spin orientation index. If the spin orientation is not designated then each band in the zone accommodates two electrons. While this result was derived for a cubic Bravais crystal lattice it is valid for any Bravais lattice. In particular for a II–VI zincblende compound with two atoms per primitive cell and a total of eight valence electrons, there are 16 bands in the valence plus the lowest conduction bands of the Brillouin zone. At 0 K, there are eight filled valence bands, and eight empty conduction bands.

Appendix 3B: Overview of the *k·p* Method

In the "narrow" band gap region, the Γ_8, Γ_7 and Γ_6 bands are close to each other at the Γ point of the Brillouin zone. Their interactions can be treated using $\mathbf{k}\cdot\mathbf{p}$ perturbation method enabling the solutions of the dispersion relation at Γ. In a study of the band structure of InSb the method proposed by Kane is to solve the Schrödinger equation in an eight Bloch function basis for the Γ_6, Γ_7 and Γ_8 bands. Later the interactions from the remote bands are also included. The Schrödinger equation with the spin–orbit coupling included in the Hamiltonian is:

$$\left[\frac{p^2}{2m}+V(\boldsymbol{r})+\frac{\hbar}{4m^2c^2}(\nabla V(\mathbf{r})\times \boldsymbol{p})\cdot\sigma\right]e^{i\boldsymbol{k}\cdot\boldsymbol{r}}u_k(\boldsymbol{r})=E_k e^{i\boldsymbol{k}\cdot\boldsymbol{r}}u_k(\boldsymbol{r}) \tag{3B.1}$$

where the $\mathbf{p}$ is the momentum operator $\boldsymbol{p}=-\hbar\nabla$, $V(\mathbf{r})$ is the periodic lattice potential, and $\boldsymbol{\sigma}$ is the electron spinor vector. In the $\boldsymbol{k}\cdot\boldsymbol{p}$ representation, the Schrödinger equation for the periodic part of the wave function $u_k(\mathbf{r})$ in the unit cell is:

$$\left[\frac{p^2}{2m}+V(\vec{r})+\frac{\hbar}{m}\boldsymbol{k}\cdot\boldsymbol{p}+\frac{\hbar}{4m^2c^2}(\nabla V\times\boldsymbol{p})\cdot\boldsymbol{\sigma}+\frac{\hbar}{4m^2c^2}(\nabla V\times\boldsymbol{k})\cdot\boldsymbol{\sigma}\right]u_k(\vec{r}) \\ =E_k^{'}u_k(\vec{r}) \tag{3B.2}$$

where

$$E_k^{'}=E_k-\frac{\hbar^2k^2}{2m} \tag{3B.3}$$

In (3B.2) the third term comes from the $\boldsymbol{k}\cdot\boldsymbol{p}$ interaction, the fourth and the fifth terms come from the spin–orbit interaction. Further, the fifth term depends on $\boldsymbol{k}$, and it can be omitted at first. Considering the $\boldsymbol{k}\cdot\boldsymbol{p}$ and spin–orbit terms as perturbations, employing first order degenerate perturbation theory, letting the $\boldsymbol{k}$ vector lie along z axis, and defining:

$$P\equiv -i\left(\frac{\hbar}{m}\right)\langle s|p_z|z\rangle \tag{3B.4}$$

$$\Delta\equiv\frac{3\hbar i}{4m^2c^2}\left\langle x\left|\frac{\partial V}{\partial x}p_y-\frac{\partial V}{\partial y}p_x\right|y\right\rangle \tag{3B.5}$$

then in the proper representation the 8×8 interaction matrix becomes block diagonal and can be written as $\begin{pmatrix} H & 0 \\ 0 & H \end{pmatrix}$, where

$$H = \begin{bmatrix} E_s & 0 & k_F & 0 \\ 0 & E_p - \Delta/3 & \sqrt{2}\Delta/3 & 0 \\ k_F & \sqrt{2}\Delta/3 & E_p & 0 \\ 0 & 0 & 0 & E_p + \Delta/3 \end{bmatrix} \tag{3B.6}$$

In this approach E_s and E_p are the eigenvalues of the unperturbed Hamiltonian at the Γ point in the Brillouin zone. E_s is the energy at the bottom of the unperturbed conduction band and E_p is the energy at the top of the unperturbed valence band.

When $\Delta >> kp$ and E_g, the $E-k$ dispersion relation to this level of approximation is:

$$\begin{cases} E_c = \dfrac{\hbar^2 k^2}{2m} + \dfrac{1}{2}\left(E_g + \sqrt{E_g^2 + \dfrac{8}{3}P^2 k^2} \right) \\ E_{v1} = \dfrac{h^2 k^2}{2m} \\ E_{v2} = \dfrac{\hbar^2 k^2}{2m} + \dfrac{1}{2}\left(E_g - \sqrt{E_g^2 + \dfrac{8}{3}P^2 k^2} \right) \\ E_{v3} = -\Delta + \dfrac{\hbar^2 k^2}{2m} - \dfrac{P^2 k^2}{3(E_g + \Delta)} \end{cases} \tag{3B.7}$$

The six-fold degeneracy of the valence band has been lifted partly. It is separated into three two-fold degenerate bands, a heavy hole band E_{v1}, a light hole band E_{v2}, and a spin–orbit split-off band E_{v3}. Here E_c and E_{v2} are non-parabolic. To this level of approximation E_{v1} has the wrong curvature. The E_{v3} splitting is due in part to the symmetry between E_c and E_{v2}.

If the second order perturbation from the $\boldsymbol{k} \cdot \boldsymbol{p}$ interaction is added, all of the two-fold degeneracies are lifted. Then the approximate energies for the conduction band, the v_2 band and the v_3 band are:

$$\begin{aligned}E_i^{\pm} &= E_i^{'} + \frac{\hbar^2 k^2}{2m} + a_i^2 A' k^2 \\ &+ b_i^2 [Mk^2 + (L - M - N)\cdot(k_x^2 k_y^2 + k_y^2 k_z^2 + k_z^2 k_x^2)/k^2] \\ &+ c_i^2 [L' k^2 - 2(L - M - N)\cdot(k_x^2 k_y^2 + k_y^2 k_z^2 + k_z^2 k_x^2)/k^2] \\ &\pm \sqrt{2} a_i b_i B [k^2 (k_x^2 k_y^2 + k_y^2 k_z^2 + k_z^2 k_x^2) - 9 k_x^2 k_y^2 k_z^2]^{\frac{1}{2}} / k\end{aligned} \tag{3B.8}$$

Where A', B, L, M, N and L' are constants related to the interaction matrixes. i represents c, v_2 and v_3. The parameters a_i, b_i and c_i are the coefficients in the expression for the wave functions (see Sect. 3.2). For zincblende structures, because $B \neq 0$, the two-fold degeneracy in c, v_2 and v_3 are lifted. For more precise calculations, the dispersion relations in bands c, v_2 and v_3, have to be corrected. For the heavy hole band, the relation is:

$$E_{v1} = \frac{\hbar^2 k^2}{2m} + Mk^2 + (L - M - N)\frac{k_x^2 k_y^2 + k_y^2 k_z^2 + k_z^2 k_x^2}{k^2} \tag{3B.9}$$

This formula is also often written as

$$E_{v1} = -\frac{\hbar^2 k^2}{2m_{hh}} \tag{3B.10}$$

where m_{hh} is the effective mass of the heavy hole band. Equations (3B.7) and (3B.10) are often used expressions for the energy bands in Kane's model. In (3B.2), if the first order perturbation correction from the fifth term $\frac{\hbar^2}{2\sqrt{3}m^2c^2}(\nabla V \times \boldsymbol{k})\cdot\boldsymbol{\sigma}$ is included, a term linear in $\boldsymbol{k}$, has to be added to the E_{v1} energy expression. This term is proportional to C_a

$$C_a = \frac{\hbar^2}{2\sqrt{3}m^2c^2}\left(x \left| \frac{\partial V}{\partial y} \right| z \right) \tag{3B.11}$$

and is:

$$\Delta E_{v1} = \pm C_a [k^2 \pm \sqrt{3}\left(k_x^2 k_y^2 + k_y^2 k_z^2 + k_z^2 k\right)^{1/2}]^{1/2} \tag{3B.12}$$

Second order perturbation terms from the $\boldsymbol{k}\cdot\boldsymbol{p}$ and the spin–orbit interactions also make contributions to the linear energy term. Due to this linear term, Kane theory predicts the peak energy of the heavy hole band is not located at Γ, but is displaced about $0.003\times\frac{2\pi}{a}$ along the (111) direction. The value of the energy trough is 10^{-5} eV. This prediction has never been seen in any narrow gap semiconductor, and is not predicted by more exact theories.

Kane's model predicts most features of InSb and HgCdTe semi-quantatively. In this model, the band gap E_g, the spin–orbit split-off energy Δ, the momentum matrix element P as well as the effective mass of the conduction m_c and light hole m_{lh} bands can be calculated. However, most often they are set by experiments when the band structure parameters are used in device designs. These procedures are treated in detail in sections of this chapter.

4 Optical Properties

4.1 Optical Constants and the Dielectric Function

4.1.1 Fundamentals

The complex dielectric function $\tilde{\varepsilon}$ of materials is a bridge connecting microscopic and macroscopic quantities, and is central to the investigation of the optical properties of semiconductors.

When a polarized plane electromagnetic wave at angular frequency ω propagates along the z-axis of a material, the electric field vector is:

$$E_x = E_0 \exp(-\omega\kappa z / c) \exp\left[i\omega(t - nz / c)\right] \tag{4.1}$$

where n is the refractive index and κ is the extinction coefficient. The quantities n and κ are the real and imaginary parts of the complex refractive $\tilde{n}$. In general the relations in cgs units are:

$$\begin{cases} \tilde{n} = n - i\kappa \\ \tilde{\varepsilon} = \varepsilon_1 - i\varepsilon_2 \\ \varepsilon_1 = n^2 - \kappa^2 \\ \varepsilon_2 = 2n\kappa = \dfrac{4\pi\sigma}{\omega} \end{cases} \tag{4.2}$$

where ε_1, ε_2 are the real and imaginary parts of the dielectric function, respectively, and σ is the electric conductivity of the material. In cgs units ε, n, and κ are dimensionless, and σ has dimensions of 1/time.

The optical constants can be obtained from the dielectric function:

$$\begin{cases} n = \sqrt{\dfrac{\sqrt{\varepsilon_1^2 + \varepsilon_2^2} + \varepsilon_1}{2}} \\ \kappa = \sqrt{\dfrac{\sqrt{\varepsilon_1^2 + \varepsilon_2^2} - \varepsilon_1}{2}} \end{cases} \tag{4.3}$$

Optical constants are related directly to the transmission T and the reflection R coefficients, which can be measured experimentally. The light intensity at z is proportional to the square of the amplitude of the electric vector, given in (4.1):

$$I = I_0 \exp(-2\omega\kappa\sigma\omega / c) = I_0 \exp(-\alpha z) \tag{4.4}$$

where I_0 is the light intensity at $z = 0$, and α is the absorption coefficient, with relations among α, T and R, and κ, n, wave length λ = c/ω, and the thickness of the material d given by:

$$\begin{cases} \alpha = \dfrac{2\omega k}{c} = \dfrac{4\pi\kappa}{\lambda} \\ R = \dfrac{(n-1)^2 + \kappa^2}{(n+1)^2 + \kappa^2} \\ T = \dfrac{(1+R)^2 e^{-\alpha d}}{1 - R^2 \exp(-2\alpha d)} \end{cases} \tag{4.5}$$

These equations for R and T are only valid for an incident wave in vacuum impinging on the dielectric material at normal incidence. The relations between the complex dielectric function and macroscopic parameters are captured in (4.2)–(4.5).

In terms of Maxwell equations, the energy loss of electromagnetic waves in media is $(\sigma E_0)/2$, and according to quantum statistical mechanics, the loss is $w \cdot \hbar\omega$, where w is a transition probability per unit time between all levels coupled by the wave at frequency ω, so:

$$\frac{1}{2}\sigma E_0 = w \cdot \hbar\omega \tag{4.6}$$

The frequency dependent conductivity σ is related to the imaginary part of dielectric function ε_2, or the product of refractive index and extinction coefficient $2nk$. Thus it is related to the material's microstructure, and to T and R.

The real part ε_1, and the imaginary part ε_2, of the complex dielectric function are connected through the Kramers–Kronig (KK) relation (Landau and Lifshitz 1969). The KK relation is a consequence of causality; a response to a perturbation introduced at time to must occur at a later time $t > t_0$. The complex dielectric function $\tilde{\varepsilon} = \varepsilon_1 + i\varepsilon_2$ describes the response of matter to an electromagnetic field $E(r)$. In general, $\tilde{\varepsilon}$ is a tensor, and is a function of frequency and wave vector, but when the field variation over atomic dimensions is very small, the wave vector dependence can be neglected, so $\tilde{\varepsilon} = \tilde{\varepsilon}(\omega)$. $\tilde{\varepsilon}(\omega)$ is a scalar for crystals with cubic symmetry.

In polar dielectric crystals, low frequency incident radiation interacts with optical phonons and free carriers. When the photon energy increases into the intrinsic region, an electromagnetic wave interacts with the valence band electrons causing them to be excited into the conduction band. At still higher frequencies, in the UV or X-ray energy range, an electromagnetic wave excites transitions between core levels and the conduction band. All these interactions contribute to the dielectric function. Therefore, the complex dielectric function can be written as:

$$\tilde{\boldsymbol{\varepsilon}}(\boldsymbol{\omega}) = \tilde{\boldsymbol{\varepsilon}}_{\infty} + \Delta\tilde{\boldsymbol{\varepsilon}}_{\text{inter}} + \Delta\tilde{\boldsymbol{\varepsilon}}_{\text{intra}} + \Delta\tilde{\boldsymbol{\varepsilon}}_{\text{phonon}} \tag{4.7}$$

where $\tilde{\varepsilon}_{\infty}$ is the high frequency dielectric constant (the total contribution above those from the intrinsic transitions); $\Delta\varepsilon_{\text{inter}}$ is the inter-band contribution at photon energies near those separating the valence and conduction bands; $\Delta\varepsilon_{\text{intra}}$ is the free carrier inter-band contribution; and $\Delta\varepsilon_{\text{phonon}}$ is the phonon contribution.

ε_{∞} is a constant, for which it is difficult to obtain a detailed microscopic process based expression. The contribution from the inter-band transitions to the imaginary part of the dielectric function is:

$$\Delta\boldsymbol{\varepsilon}_{2inter} = \frac{4\hbar^2 e^2}{\boldsymbol{\pi} m^2 \boldsymbol{\omega}^2} \int dk \left| e \cdot M_{cv} \right|^2 \boldsymbol{\delta}(E_c - E_v - \hbar\boldsymbol{\omega}) \tag{4.8}$$

$\left| e \cdot M_{cv} \right|^2$, defined in Sect. 4.2, is proportional to the square of the momentum matrix element. In the inter-band transition region; this contribution dominates the intrinsic absorption of the samples. For $Hg_{1-x}Cd_xTe$ samples with small Cd compositions, it also affects the far-infrared spectrum. Using (4.8), the inter-band transition contribution to the real part of the dielectric function can be calculated from the KK relation.

The contributions of intra-band transition induced lattice absorption to the dielectric function are:

$$\Delta\varepsilon_{\text{intra}} = -\frac{ne^2}{\pi m^* c^2} \frac{1}{\omega^2 - i\Gamma_p \omega} \tag{4.9}$$

and

$$\Delta\varepsilon_{\text{phonon}} = \sum_j \frac{S_j \omega_{TO}^2}{\omega_{TO,j}^2 - \omega^2 - i\omega\Gamma_j} \tag{4.10}$$

where Γ_p is the damping constant of plasma oscillation, S_j, $\omega_{TO,j}$ and Γ_j are the intensity, frequency and damping constant of the j th lattice oscillation. Equations (4.6)–(4.10) display the relation between the dielectric function and microscopic properties of matter.

Therefore, the dielectric function is a bridge connecting measurable macroscopic variables and microscopic parameters. Measurement of the dielectric function is an important tool in the study of some important properties of matter.

4.1.2 Kramers–Kronig (KK) Relation and Optical Constants

The KK relations provide a useful tool to deal with the optical constants. If $Z(\omega) = Z'(\omega) + iZ''(\omega)$ is analytic, converges at infinity, all poles of $Z(\omega)$ lie below the real axis, and $Z'(\omega)$ is an even function and $Z''(\omega)$ is an odd function of real ω, then the KK relations are:

$$\begin{cases} Z'(a) = \dfrac{2}{\pi} \wp \displaystyle\int_0^\infty \frac{\omega Z''(\omega)}{\omega^2 - a^2} d\omega \\ Z''(a) = -\dfrac{2a}{\pi} \wp \displaystyle\int_0^\infty \frac{Z'(\omega)}{\omega^2 - a^2} d\omega \end{cases} \tag{4.11}$$

Where $\wp$ designates a principal part integral. It can be demonstrated that all the properties ascribed to $Z(\omega)$, stem directly from the causality principle (Landau and Lifshitz 1969) in linear response fluctuation-dissipation theory, and are not separate approximations. The principal part designation avoids the divergence in the integrand when $\omega \to a$. It can be removed when one term with integral value of zero is added to make the integrand remain finite at $\omega = a$:

$$\begin{cases} Z'(a) = \dfrac{2}{\pi}\displaystyle\int_0^{\infty} \dfrac{\omega Z''(\omega) - aZ''(a)}{\omega^2 - a^2} d\omega \\ Z''(a) = -\dfrac{2a}{\pi}\displaystyle\int_0^{\infty} \dfrac{Z'(\omega) - Z'(a)}{\omega^2 - a^2} d\omega \end{cases} \tag{4.12}$$

Hence, if the imaginary part of the response function is known over the whole frequency range, then the real part of the response function can be calculated, or if the real part is known, then the imaginary part can be calculated. Thus the real part ε_1 and imaginary part ε_2 of the complex dielectric function $\tilde{\varepsilon}$ satisfy the expressions:

$$\begin{cases} \varepsilon_1(a) - 1 = \dfrac{2}{\pi}\wp\displaystyle\int_0^{\infty} \dfrac{\varepsilon_2 \cdot \omega}{\omega^2 - a^2} d\omega \\ \varepsilon_2(a) = -\dfrac{2a}{\pi}\wp\displaystyle\int_0^{\infty} \dfrac{\varepsilon_1}{\omega^2 - a^2} d\omega \end{cases} \tag{4.13}$$

Therefore, when ε_1 is known, ε_2 can be calculated, or when ε_2 is known, ε_1 can be calculated.

The reflection coefficient, defined as:

$$\tilde{\gamma}(\omega) \equiv E_{re} / E_{in} = \gamma(\omega) e^{i\theta(\omega)}$$

has real and imaginary parts, $\ln\tilde{\gamma} = \ln\sqrt{R} + i\theta$, that satisfy the KK relations (Moss et al. 1973), so

$$\theta(a) = -\frac{2a}{\pi}\wp\int_0^{\infty} \frac{\ln\sqrt{R}}{\omega^2 - a^2} d\omega \tag{4.14}$$

or

$$\theta(a) = -\frac{1}{\pi}\wp\int_0^{\infty} \ln\left|\frac{\omega + a}{\omega - a}\right| \frac{d\ln\sqrt{R(\omega)}}{d\omega} d\omega \tag{4.15}$$

From (4.15), only a limited spectral range makes the principal contribution to phase angle θ. When ω is near a, or a rapidly varying part of $R(\omega)$, there will be a major contribution to the integral. If the frequency interval (ω_1, ω_2) makes the major contribution, other regions $(0, \omega_1)$ and (ω_2, ∞) can be treated as small perturbations. The integral can be written as:

$$\int_0^\infty = \int_0^{\omega_1} + \int_{\omega_1}^{\omega_2} + \int_{\omega_2}^\infty$$

or

$$\boldsymbol{\theta}(a) \cong A \ln\left|\frac{\boldsymbol{\omega}_1 + a}{\boldsymbol{\omega}_2 - a}\right| + \boldsymbol{\phi}(a) + B \ln\left|\frac{\boldsymbol{\omega}_2 + a}{\boldsymbol{\omega}_2 - a}\right| \tag{4.16}$$

with

$$\phi(a) = \frac{1}{\pi}\int_{\omega_1}^{\omega_2} \ln\sqrt{R(\omega)}\frac{d}{d\omega}\ln\left|\frac{\omega + a}{\omega - a}\right| d\omega \tag{4.17}$$

In (4.16) A and B are the median values of the integrated functions, respectively, in the region $0 \le \boldsymbol{\omega}_1 \le \boldsymbol{\omega} \le \boldsymbol{\omega}_2 \le \infty$, which can be determined by the method of undetermined coefficients. They can also be determined by the Philipp method (Philipp 1972).

After the phase angle θ has been calculated, the refractive index n and extinction coefficient κ can be obtained:

$$\begin{cases} n = \dfrac{1 - R}{1 + R - 2\cos\boldsymbol{\theta}\sqrt{R}} \\ \boldsymbol{\kappa} = \dfrac{2\sin\boldsymbol{\theta}\sqrt{R}}{1 + R - 2\cos\boldsymbol{\theta}\sqrt{R}} \end{cases} \tag{4.18}$$

The KK relation can also deal with the complex function (Moss et al. 1973):

$$\tilde{n} - 1 = (n - 1) + i\boldsymbol{\kappa} \tag{4.19}$$

Therefore, we have:

$$n_a - 1 = \frac{2}{\boldsymbol{\pi}}\wp\int_0^\infty \frac{k\boldsymbol{\omega}}{\boldsymbol{\omega}^2 - a^2} d\boldsymbol{\omega} \tag{4.20}$$

If the absorption coefficient over the whole frequency interval is measured, the extinction coefficient κ can be calculated, and then the refractive index n is obtained from (4.20). Equation 4.20 can also be written as:

$$n_a - 1 = \frac{1}{2\boldsymbol{\pi}^2}\wp\int_0^\infty \frac{\boldsymbol{\alpha}(\boldsymbol{\lambda})}{1 - (\boldsymbol{\lambda}/\boldsymbol{\lambda}_a)^2} d\boldsymbol{\lambda} \tag{4.21}$$

Thus n_a can be calculated from the measured absorption spectrum $\alpha(\lambda)$. When $\lambda_a \to \infty$, (4.21) reduces to:

$$n_0 - 1 = \frac{1}{2\pi^2}\int_0^{\infty} \alpha(\lambda) d\lambda \tag{4.22}$$

and n_0 is the refraction index at zero frequency, which can be calculated from (4.22). The main contributions to the integral are those from the intrinsic absorption and reststrahlen bands, which is:

$$\int_0^{\infty} \alpha(\lambda) d\lambda = \int_{\text{intrin}} \alpha(\lambda) d\lambda + \int_{\text{phonon}} \alpha(\lambda) d\lambda \tag{4.23}$$

n_0 can be calculated from the above equation.

For the HgCdTe alloy, the reststrahlen band lies in the range 60–100 μm. Depending on the Cd concentration, intrinsic absorption lies in the range of several to more than 10 μm. Therefore, in the wavelength range λ_a = 20–50 μm, that is in the intrinsic absorption and reststrahlen band wavelength range, (4.10) can be simplified to:

$$\int_0^{\infty} \frac{\alpha(\lambda)}{1-\left(\frac{\lambda}{\lambda_a}\right)^2} = \int_{\text{intrin}} \frac{\alpha(\lambda)}{1-\left(\frac{\lambda}{\lambda_a}\right)^2} d\lambda + \int_{\text{phonon}} \frac{\alpha(\lambda)}{1-\left(\frac{\lambda}{\lambda_a}\right)^2} d\lambda \tag{4.24}$$

In the first term on the right, where $\lambda << \lambda_a$, the denominator of the integrand is approximately unity. In the second term, where $\lambda >> \lambda_a$, the denominator of the integrand is much greater than 1, which makes its value only about 1/10 that of the first term, and so it can be neglected to lowest order.

Then (4.11) simplifies to:

$$n_0 - 1 = \frac{1}{2\pi^2}\int_{\text{intrin}} \alpha(\lambda) d\lambda \tag{4.25}$$

where n_0 is the refractive index in the long wavelength range. It is the refractive index beyond the long wavelength threshold for intrinsic absorption, $\lambda << \lambda_\omega$, i.e., it is the refractive index between the wavelengths 20~50 μm, which is different from n_0 in (4.22).

Chu (1983a) calculated the refractive indices from the absorption spectra $\alpha(\lambda)$ at different wavelength, and found that n_0 decreases slowly when the band gap E_g increases, it satisfies the relation:

Table 4.1. Experimental and calculated values of refractive index n_0 vs. temperature for x = 0.205

T(K)	80	100	150	200	250	300
Experimental n_λ=20 μm	3.68	3.66	3.61	3.55	3.51	3.48
Calculation n_0	3.69	3.65	3.52	3.41	3.32	3.24

$$n_0^4 E_g = C \tag{4.26}$$

where C is a constant. Moss (1952, 1959) suggested that $n_0^4 E_g$ should be constant and found that the product of n_0^4 and E_g lies between 49 (InAs) and 210 (PbTe) in the long wavelength region for most compound semiconductors. For HgCdTe, C is related to x:

$$n_0^4 E_g = 55.5x + 7.8 \tag{4.27}$$

Substituting the $E_g(x,T)$ expression, one can calculate the refractive index at long wavelength of HgCdTe at different compositions and different temperatures.

Finkman and Nemirovsky (1979) measured the transmission of a sample with x = 0.205 and deduced the refractive index at different wavelengths from the interference fringes.

The results at λ = 20 μm are shown in Table 4.1. The values calculated from (4.40) are also listed in Table 4.1. At low temperature, the calculations agree well with experiment. The calculated values become progressively lower than the experimental values as the temperature increases. The reason for the discrepancy is that the carrier concentration increases as the temperature increases, but the contribution of free carriers is neglected in the calculation.

4.1.3 Dispersion of the Refractive Index

The refractive index is an important parameter for HgCdTe, both in helping to understand its fundamental physical properties and in device applications. It is a function of composition, temperature, and wavelength. There are several experimental ways to deduce the refractive index in the extrinsic region (photon energies less than E_g). These include an interference fringe method, reflectance spectroscopy, and absorption spectroscopy plus the KK relation. There have been many reports of refractive index measurements in the extrinsic region (Baars and Sorger 1972; Finkman and

Nemirovsky 1979; Jensen and Torabi 1983; Chu 1983c; Finkman and Schacham 1984). Finkman deduced the refractive index of HgCdTe at different composition by using the interference fringes observed in the transmission data. He also obtained the refractive index dispersion caused by free carrier absorption at frequencies below the onset of band gap absorption. Liu (1994) derived the refractive index near the band gap from calculated transmission coefficients and the KK relation. He then derived an empirical formula effective over a more useful range.

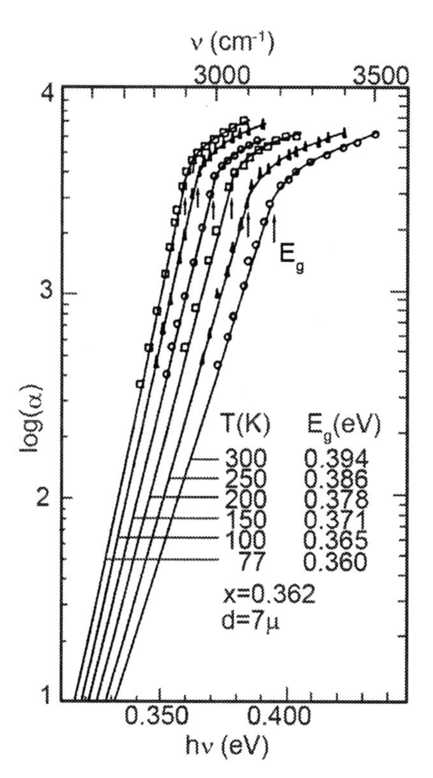

Fig. 4.1. Absorption spectra of HgCdTe at different temperatures with x = 0.362, d = 7 μm

To derive the refractive index from the KK relation, it is necessary to have complete absorption spectra, including the absorption spectra in the intrinsic absorption region, the exponential absorption region, the free carrier absorption region, and the phonon absorption region, especially the absorption in the intrinsic absorption region. So the thickness of the sample must be thin enough to obtain the intrinsic absorption coefficient. Chu (1983b), Chu and Mi (1987), and Chu et al. (1982) measured the transmission spectra of $Hg_{1-x}Cd_xTe$ with different compositions x = 0.170–0.443, and different thicknesses $d = 2.5 \sim 8\,\mu\text{m}$. They obtained the absorption spectra with the absorption coefficient reaching values up to 8,000 cm^{-1}. Figure 4.1 shows the typical absorption spectra of a HgCdTe x = 0.362, $d = 7\,\mu\text{m}$ sample at different temperatures. It shows that the absorption spectra include two parts: a steep exponential absorption edge and a flater intrinsic absorption region.

In terms of an exponential absorption rule, the absorption spectra could be also drawn as $\log(\alpha) \sim \nu$, and can be extrapolated to the short wavelength (large ν) region. According to the KK relation (4.21), the refractive index at wavelength λ can be obtained from the absorption spectra. Because the absorption coefficient of free carriers is only 1–10 cm^{-1} and the energy range of phonons is far from the investigated energy range; the contributions of free carriers and phonons to the refractive index can be neglected. The refractive index can be obtained accurately in terms of the absorption coefficient at λ~0–20 μm, with the absorption spectra at short wavelength extrapolated exponentially.

The question of how to evaluate the refractive index from the singular integral in (4.21) remains. It can be solved through an analytic extension to the complex plane, or as a principal part integral by setting it equal to the average of the values on the left and right sides of the singularity. Equation 4.21 also can be expressed as:

$$n(\nu) = 1 + \frac{c}{\pi} \int_0^{\infty} \frac{d\alpha(\nu')}{d\nu'} \log\left(\frac{\nu' + \nu}{\nu' - \nu}\right) d\nu' \qquad (4.28)$$

where ν' and ν are frequencies.

From (4.21) or (4.28) and measured absorption spectra, the refractive index dispersion of HgCdTe can be deduced for different compositions and temperatures. Figures 4.2–4.4 are the refractive index dispersion curves of HgCdTe with compositions 0.330, 0.362, and 0.416 at different temperatures. From the figures, we can see clearly that the refractive index curves have peaks whose position is at the wavelength corresponding to the band-gap energy for each of these HgCdTe samples. This can be understood from (4.28). In the wavelength region near the absorption edge, the value of $d\alpha(\nu')/d\nu'$ is large and the refractive index increases as the photon energy increases. However, when the photon energy reaches the gap energy, the absorption curve changes its slope and becomes flatter, so the value of $d\alpha(\nu')/d\nu'$ decreases above the gap energy. Thus, a peak appears in the refractive index curve at the gap energy.

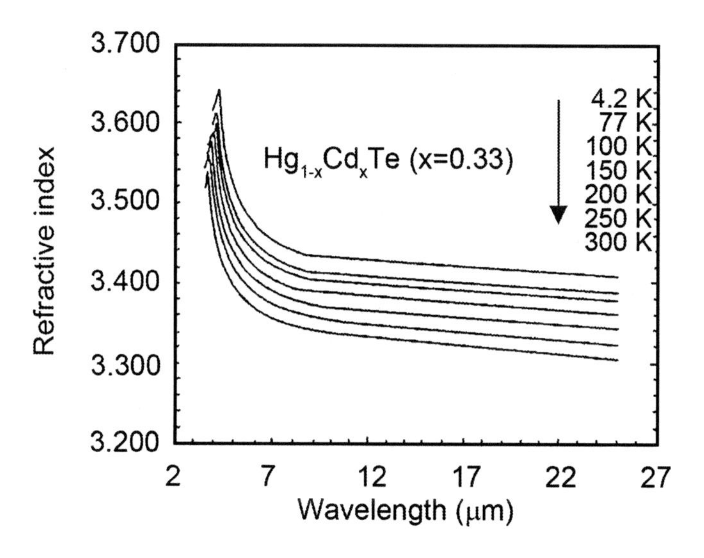

Fig. 4.2. Refractive index dispersion for $Hg_{1-x}Cd_xTe$ at x = 0.330

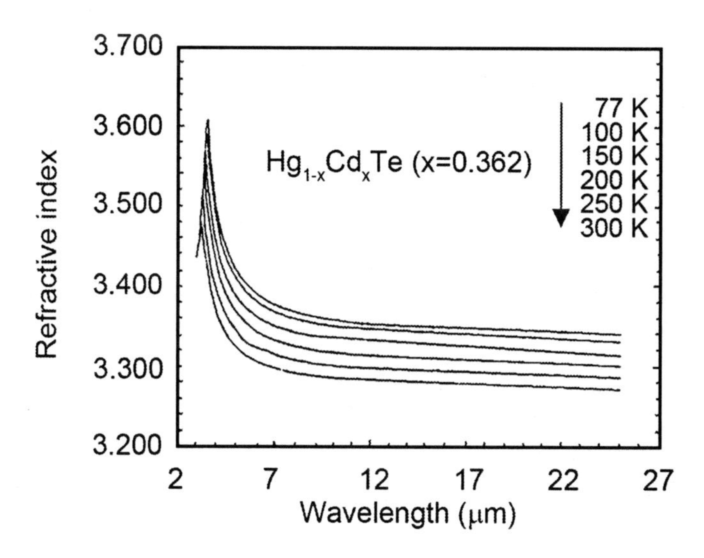

Fig. 4.3. Refractive index dispersion for $Hg_{1-x}Cd_xTe$ at x = 0.362

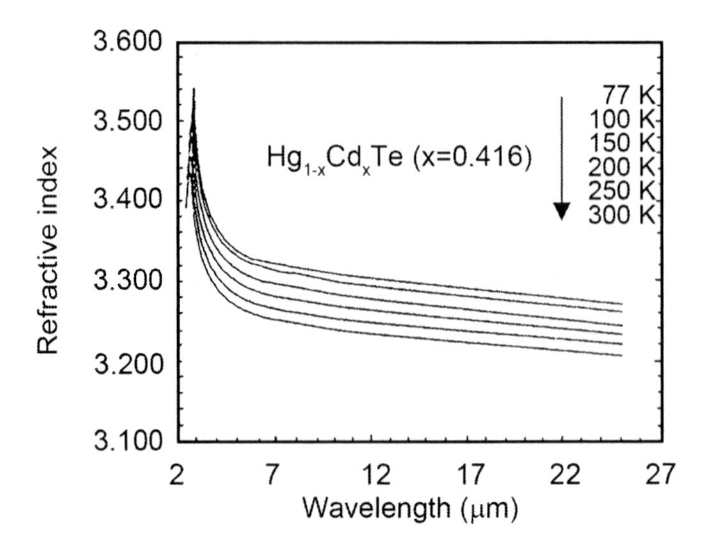

Fig. 4.4. Refractive index dispersion for $Hg_{1-x}Cd_xTe$ at x = 0.416

From these figures, the refractive index has a negative temperature coefficient, and this can be understood by examining (4.28). Due to the positive temperature coefficient of the energy gap for HgCdTe in this composition range, the absorption edge moves to shorter-wavelengths as the temperature increases. Therefore the integral of (4.28) becomes smaller.

The dispersion of the refractive index can be described by an empirical formula:

$$n(\lambda,T)^2 = A + B/[1-(C/\lambda)^2] + D\lambda^2 \tag{4.29}$$

where A, B, C, and D are fitting parameters, which vary with composition x and temperature T, as follows:

$$\begin{aligned}
A &= 13.173 - 9.852x + 2.909x^2 + 10^{-3}(300-T),\\
B &= 0.83 - 0.246x - 0.0961x^2 + 8\times10^{-4}(300-T),\\
C &= 6.706 - 14.437x + 8.531x^2 + 7\times10^{-4}(300-T),\\
D &= 1.953\times10^{-4} - 0.00128x + 1.853\times10^{-4}x^2.
\end{aligned} \tag{4.30}$$

From the formulas in (4.29) and (4.30), the refractive index calculated for HgCdTe with composition at different temperatures, is shown as solid lines in Figs. 4.5 and 4.6. The dots are experimental data (Baars and Sorger 1972; Finkman and Sachacham 1984). The dashed lines are the calculated results of Jensen Torabi (JT) (Jensen and Torabi 1983). The empirical formulas,

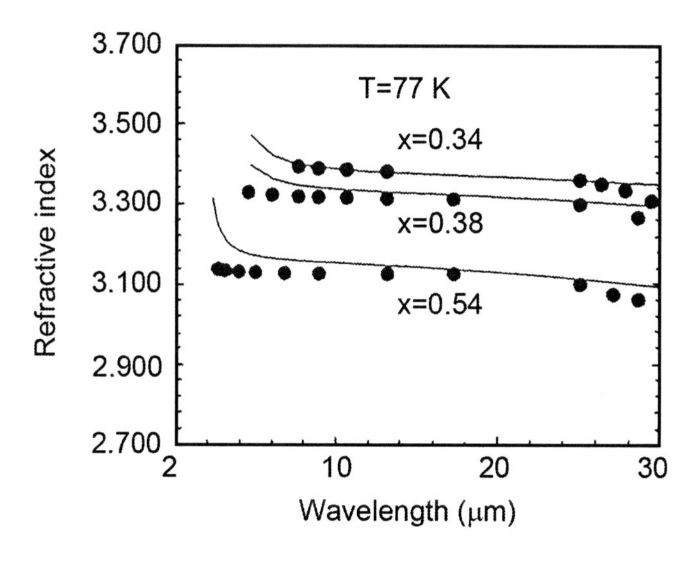

Fig. 4.5. Refractive index for HgCdTe with x = 0.33, 0.38, and 0.54 at 77 K

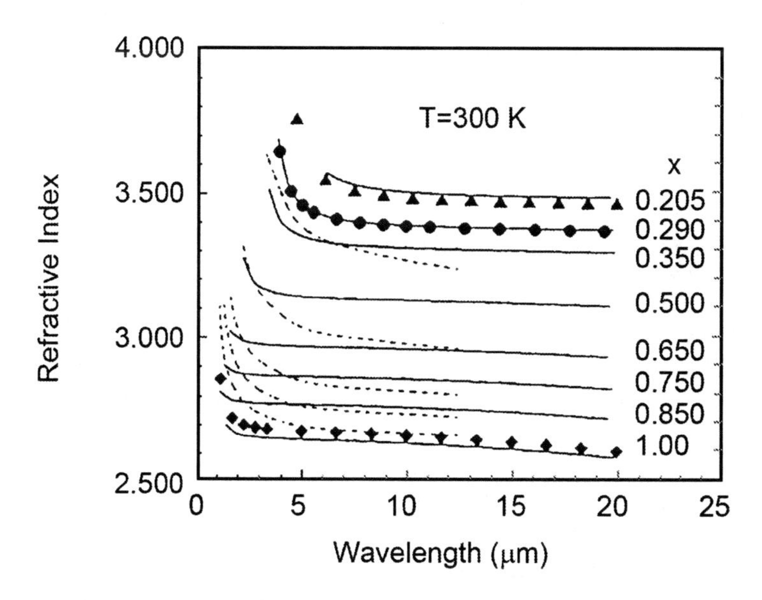

Fig. 4.6. Refractive index for HgCdTe with x = 0.33, 0.38, and 0.54 at 300 K

(4.29) and (4.30), fit the refractive index for HgCdTe very well with the compositions of 0.34, 0.38, and 0.54 at 77 K, and for the compositions 0.205, 0.390, and 1.00 at room temperature, while the data have a different behavior from that of the JT model. Equations (4.29) and (4.30) reflect a behavior of the refractive index that decreases quickly in a wavelength region near the band gap, and then decreases slowly as the wavelength increases, while the JT model predicts a quick decrease of the refractive index in a relatively wide-wavelength region, not only near the band gap but also at longer-wavelengths. The main reason for the difference between them may be that Jensen used quite different parameters. In his calculation, Jensen used a constant electron concentration for HgCdTe with different Cd mole fractions.

As emphasized in Chap. 3, it is not easy to determine the band gap of HgCdTe accurately, but there are several empirical formulas. Among them, the CXT formula, (3.106), and the HSC formula, (3.101), are the most accurate. Because the energy at the peak of the refractive index corresponds to the band gap energy, it too can be used to determine the band gap accurately. If the refractive index can be measured directly and its peak observed, then the band gap can be obtained directly from such experiments.

Liu (1994), Liu et al. (1994a) obtained the refractive index near the absorption edge. But using the KK relation it is difficult to calculate the refractive index from their absorption spectra. The best way to obtain the refractive index above the band gap is with an infrared spectroscopic ellipsometer, which can measure the complete optical constants of materials directly. Huang and Chu (2000, 2001) obtained a refractive index above the band gap of several HgCdTe compositions at room temperature. The results of these measurements will be presented in following chapters.

4.1.4 Effect of Electric and Magnetic Fields on Optical Constants

The effect of external conditions on the optical response and dielectric constants of semiconductors can be used to deduce aspects of their physical properties. We will discuss the impact of electric and magnetic fields.

Optical absorption in the presence of an electric filed was first investigated by Franz (1958) and Keldysh (1958a), and their finding is called the Franz–Keldysh effect. When an electric field is applied to semiconductor materials, the electric field modifies the electrons and hole spatial distributions. Translational symmetry of the crystal is broken by the direction of the electric field $\mathbf{F}$. The one electron Hamiltonian has a potential energy term eFx added, if $\mathbf{F}$ is in the x direction. The perturbed wave functions are linear combinations of those unperturbed Bloch functions with wave vectors parallel to the x direction, and the bands increase in the x direction as illustrated in Fig. 4.7. An electron can absorb a

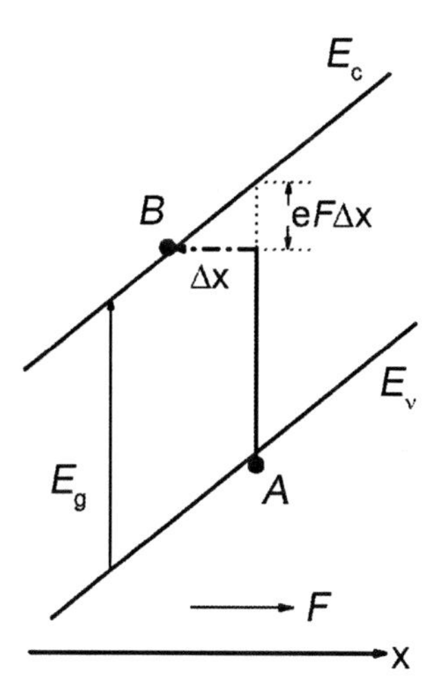

Fig. 4.7. Diagram of optical absorption of a semiconductor in an electric field

photon at an energy less than E_g to be excited into a virtual state, from which it either re-emits the photon or tunnels into the conduction band as shown in Fig. 4.7. If the tunneling distance Δx is small enough so the tunneling rate is faster than, or at least comparable to, the re-combination rate of the virtual state, then there is significant radiation absorption at photon energies less than E_g. In the electric field, an electron in the valence band at point A transitions to point B in the conduction band.

We have:

$$eF\Delta x = E_g - \hbar\omega \tag{4.31}$$

The absorption coefficient for a transition from A in the valence band to B in the conduction band due to radiation at $\hbar\omega < E_g$, is (Franz 1958):

$$\alpha = \frac{A\hbar eF}{8(2m_r)^{1/2}(E_g - \hbar\omega)}\exp\left[-\frac{4(2m_r)^{1/2}(E_g - \hbar\omega)^{3/2}}{3\hbar eF}\right] \tag{4.32}$$

where $A \propto P^2/\hbar\omega$, P is a momentum matrix element, and m_r is the reduced effective mass. A detailed derivation of (4.33) can be found in the paper of Franz, or semiconductor textbooks (Seeger 1989).

In the presence of an electric field, the dielectric function will be changed. Aspens derived the expression for $\Delta\varepsilon$ as functions of energy E, under a weak field F condition:

$$\Delta\varepsilon = \frac{e^2F^2\hbar^2}{24\mu_{//}E^2}\frac{\partial^3}{\partial E^3}[F^2\varepsilon(E)] \tag{4.33}$$

and $\mu_{//}$ is the reduced effective mass parallel to the electric field defined as:

$$\frac{1}{\mu_{//}} \equiv \left(\frac{F_x^2}{\mu_x} + \frac{F_y^2}{\mu_y} + \frac{F_z^2}{\mu_z}\right)\frac{1}{F^2} \tag{4.34}$$

Equation (4.34) can also be written as:

$$\begin{cases} \Delta\varepsilon = \sum_j C_j \left|eP_j\right|^2 (E - E_{g,j} + i\Gamma)^{-n} \\ C_j = \dfrac{F^2}{(\hbar^2\mu_j)} \end{cases} \tag{4.35}$$

Subscript j denotes different critical points.

The change of the dielectric function will induce changes in n, κ, and the absorption:

$$\begin{aligned} &\Delta\alpha_a \propto \Delta\varepsilon_2 \\ &\Delta\kappa_a = \lambda_a \frac{\Delta\alpha_a}{4\pi} \\ &\Delta n_a = \frac{1}{4\pi^2}\int_0^\infty \frac{\Delta\alpha_a}{1 - \frac{\lambda^2}{\lambda_a^2}} d\lambda \end{aligned} \tag{4.36}$$

It will also change the reflection constant:

$$\begin{aligned} \Delta R/R &= a\Delta\varepsilon_1 + b\Delta\varepsilon_2 \\ &= \frac{4(n^2 - \kappa^2 - 1)\Delta n + \Delta\kappa \cdot 8n\kappa}{[(n+1)^2 + \kappa^2][(n-1)^2 + \kappa^2]} \end{aligned} \tag{4.37}$$

The curve of $\Delta R/R$ vs. $\hbar\omega$ is called the electro-reflectance spectrum.

When a magnetic field is applied to a semiconductor material, the energy bands of the semiconductor are also changed. The field will affect transition matrix elements, the density of states (DOS), and the dielectric function. For a non-parabolic band semiconductor, a magnetic field has the following effect:

$$E_n(1+\frac{E_n}{E_g})=\frac{\hbar^2k^2}{2m_0^*}+(n+\frac{1}{2})\hbar\omega_c\pm\frac{1}{2}\beta g_0^*H \tag{4.38}$$

where $\beta = e\hbar/2m_c^*$ is the Bohr magneton, ω_c is the cyclotron frequency, and g_0^* is an effective cyclotron factor at the conduction band bottom. Therefore, due to cyclotron resonance effects, the energy band is split into a series of magnetic sub-bands, corresponding to $n = 0,1,2\ldots$

The double degeneracy of each sub-band due to spin is eliminated, and the magnetic field ***H*** splits them into spin up and spin down bands with a Zeeman energy splitting βgH. According to selection rules, inter-band transitions will take place only between valence sub-bands and conduction bands with the same index *n*, $\Delta n = 0$. A transition matrix element (actually a tensor in coordinate space) is:

$$eM_{cv}(\mathbf{H})=\int\Psi_c(\mathbf{r},\mathbf{H})e(\mathbf{p}+\frac{e}{c}\mathbf{A})\Psi_v(\mathbf{r},\mathbf{H})d\mathbf{r} \tag{4.39}$$

The transition joint DOS, when ***H*** is perpendicular to the surface of the sample, is:

$$J_{cv}(E,H)=\frac{2eH}{h^2c}(2\mu)^{1/2}\sum_n[E-Eg-\hbar\omega_c\cdot(n+\frac{1}{2})]^{-\frac{1}{2}} \tag{4.40}$$

$\omega_c = \frac{eH}{\mu c}$ is the cyclotron frequency, and μ is the reduced effective mass of electron and hole. Then the dielectric function near the Γ point is:

$$\varepsilon_2(\omega,H)$$
$$=\frac{4\pi^2e^2}{m^2\omega^2}\frac{2eH}{h^2c}(2\mu)^{1/2}|e\cdot M_{cv}(H)|^2\sum_n[\hbar\omega-E_g-\hbar\omega_c\cdot(n+\frac{1}{2})]^{-\frac{1}{2}}. \tag{4.41}$$

The above two expressions produce sharp peaks in the transition energy spectrum due to the selection rules. The reduced effective mass can be calculated from these singularities, and E_g can be obtained by projecting the experimental curves to $H = 0$.

The presence of a magnetic field complicates the dielectric function tensor, because the dielectric properties depend on the configuration of magnetic field, light propagation direction and polarization, and crystal orientation. When light is parallel to the magnetic field direction, the

Faraday configuration, the polarization plane rotates. The rotation angle θ is given by the expression:

$$\frac{n\theta}{Bd} = B\lambda^2 + \frac{A}{\lambda^2} \tag{4.42}$$

B and A are the Faraday rotation coefficients of free carriers and inter-band transition, respectively. n, B and d are the refractive index, the magnetic field, and sample thickness. When the wave vector is perpendicular to the magnetic field, and the angle between the polarization direction and the magnetic field is 45°, this is the Voigt configuration. As it passes through the sample, the plane-polarized light is changed to elliptic-polarized light, and the phase shift is φ. By measuring φ and θ, the effective mass m^* can be deduced from $\phi / \theta = \omega_c / \omega$.

4.2 Theory and Experiment of Interband Optical Transitions

4.2.1 The Theory of Direct Interband Optical Transitions

Based on the general theory of optical transitions in semiconductors, and choosing proper Kane model energy band coefficients, E_g, P, m_{hh} and Δ, the absorption spectrum can be deduced for comparison with experimental results. In the calculation, E_g can be determined from the intrinsic absorption spectrum. Therefore, the energy band coefficients, P, m_{hh} and Δ, can be obtained from the best conformity of theoretical and experimental curves. In what follow, we discuss the calculation of the absorption coefficients for direct interband optical transitions.

The absorption coefficient is defined as the ratio of the absorbed photon energy per unit volume and unit time to the energy–flux density. If the probability of photon induced electron transitions per unit time are known, then summing over all $\boldsymbol{k}$ and spin states in a unit volume in the conduction and the valence bands, we can obtain the number of transitions induced by photons with frequency ω in unit volume and unit time, which multiplied by $\hbar\omega$ is the power absorbed per unit volume. Then division by the energy–flux density determines the absorption coefficient (Bassani and Parravicini 1975).

The Schrodinger equation of an electron in a crystal subject to the field of an electromagnetic wave is:

$$\begin{cases} (H_0 + H')\psi = i\hbar \dfrac{\partial \psi}{\partial t} \\ H_0 = -\dfrac{\hbar^2}{2m}\nabla^2 + V(\boldsymbol{r}) \\ H' = \dfrac{ie\hbar}{mc}\boldsymbol{A}\cdot\nabla \end{cases} \tag{4.43}$$

The vector potential contributing to the absorption is:

$$\begin{aligned} \boldsymbol{A}(r,t) &= \boldsymbol{A}_0 \exp[i(\boldsymbol{q}\cdot\boldsymbol{r} - \omega t)] \\ \boldsymbol{A}_0 &= A_0 \hat{\mathbf{A}} \end{aligned} \tag{4.44}$$

$\hat{\mathbf{A}}$ is an unit vector in the wave's polarization direction. The action of the perturbation H' induces a transition probability per unit time of an electron from the initial state *i* to the final state *f*:

$$\begin{aligned} P_{i\to f} &= \frac{2\pi}{\hbar}\left|\langle f|H'|i\rangle\right|^2 \delta(E_f - E_i - \hbar\omega) \\ &= \frac{2\pi}{\hbar}\left(\frac{eA_0}{mc}\right)^2 \left|\langle \psi_{ck_f}|e^{i\boldsymbol{q}\cdot\boldsymbol{r}}\hat{\mathbf{A}}\cdot\mathbf{p}|\psi_{vk_i}\rangle\right|^2 \delta(E_f - E_i - \hbar\omega) \end{aligned} \tag{4.45}$$

In this formula ψ_{ck_f} belongs to the irreducible representation of the k_f vector translation group, ψ_{vk_i} and its differential quotient $\boldsymbol{a}\cdot P\psi_{vk_i}$ belong to the irreducible representation of the k_i vector translation group, and $e^{i\boldsymbol{q}\cdot\boldsymbol{r}}$ belongs to irreducible representation of vector, $\boldsymbol{q}$. The direct product of the irreducible representation of k_f and the irreducible representation of $\boldsymbol{q}$ is the irreducible representation of $k_i + q$, thus the condition for which the matrix element is nonzero is just the conservation of crystal momentum:

$$\mathrm{k}_f = \mathrm{k}_i + \mathrm{q} + l \tag{4.46}$$

l is any a reciprocal lattice vector. When considering the case of the first Brillouin zone, l vanishes. For a photon with an energy of order ~1 eV, so $\lambda \approx 10^4$ A, and the ranges of k_f and k_i are of order $\frac{2\pi}{a}$, where the lattice constant a is approximately several Å, then $k \approx \frac{2\pi}{a} >> |q|$. Therefore the wave functions of the connected bands are slowly varying functions of the relevant Δk, and $|q|$ can be ignored:

$$k_i \cong k_f \tag{4.47}$$

The photon-connected transitions are vertical in the reduced zone, and are referred to as direct transitions. In (4.45), the initial state energy is the electron energy in the valence band plus the photon energy, and the final state energy is the electron energy in the conduction band. The δ function guarantees conservation of energy:

$$E_C - E_V = \hbar\omega \tag{4.48}$$

and then the transition probability per unit time $P_{i\to f}$ can be written as

$$P_{V\mathbf{k}_i \to C\mathbf{k}_f} = \frac{2\pi}{\hbar}\left(\frac{eA_0}{mc}\right)^2 \left|\hat{\mathbf{A}}\cdot\mathbf{M}_{CV}(\mathbf{k})\right|^2 \delta[E_c(\mathbf{k}) - E_V(\mathbf{k}) - \hbar\omega] \tag{4.49}$$

in this formula $\left|\hat{\mathbf{A}}\cdot\mathbf{M}_{CV}(\mathbf{k})\right|^2 = \left|\left\langle \psi_{C\mathbf{k}_f}\right|\hat{\mathbf{A}}\cdot\mathbf{p}\left|\psi_{V\mathbf{k}_i}\right\rangle\right|^2$.

To obtain the umber of transitions per unit time $W(\omega)$ excited in a unit volume by photons with frequency ω, there still needs to be a summation over the transition probability per unit time over all occupied initial and connected empty final states in a unit volume. At low temperature where the valence band is full and the conduction band is empty, summing $P_{V\mathbf{k}\to C\mathbf{k}}$ over all $\boldsymbol{k}$ and spin states in the conduction and the valence bands yields:

$$W(\omega) = \frac{2\pi}{\hbar}\left(\frac{eA_0}{mc}\right)^2 \sum_{C,V}\int \frac{2dk^3}{(2\pi)^3}\left|\hat{\mathbf{A}}\cdot\mathbf{M}_{CV}\right|^2 \delta[E_C(\mathbf{k}) - E_V(\mathbf{k}) - \hbar\omega] \tag{4.50}$$

Then we have

$$\alpha = \frac{\text{the absorbed photon energy per unit time and volume}}{\text{the energy - flux density}} = \frac{\hbar\omega\cdot W(\omega)}{U\cdot\left(\frac{c}{n}\right)} \tag{4.51}$$

In this formula: the energy density is:

$$U = \frac{n^2 A_0^2 \omega^2}{2\pi c^2} \tag{4.52}$$

Combining (4.50), (4.51) and (4.53) yields:

$$\alpha = \frac{4\pi^2 e^2}{ncm^2 \omega} \sum_{C,V} \int \frac{2d^3k}{(2\pi)^3} \left| \boldsymbol{a} \cdot M_{CV}(k) \right|^2 \delta[E_c(k) - E_V(k) - \hbar\omega] \tag{4.53}$$

in which, $\left| \hat{\mathbf{A}} \cdot \mathbf{M}_{CV}(k) \right|^2$ is a slowly varying function of $\boldsymbol{k}$ and can be brought out of the integration sign. Then $\left| \hat{\mathbf{A}} \cdot \mathbf{M}_{CV}(k) \right|^2$ can be written as $\left| M_j \right|^2$, which is square of the optical matrix element averaged over all directions. In (4.53), the rest of the integral is the joint DOS (Moss et al. 1973):

$$\rho_{CV}(k) \equiv \int \frac{2d^3k}{(2\pi)^3} \delta[E_c(k) - E_V(k) - \hbar\omega] \tag{4.54}$$

defined as the density of a pair of states formed by an empty conduction band state and a filled valence band state whose energy difference is $\hbar\omega$. Finally we have:

$$\alpha = \frac{4\pi^2 e^2 \hbar}{ncm^2 E} \sum_j \left| M_j \right|^2 \rho_{CV}(k) \tag{4.55}$$

The dielectric function of a material is $\varepsilon_2 = 2nk = 2n \cdot \frac{\lambda}{4\pi} \alpha = \frac{nc}{\omega} a$. Therefore we have:

$$\varepsilon_2 = \frac{4\pi^2 e^2}{cm^2} \sum_j \left| M_j \right|^2 \rho_{CV}(k) \tag{4.56}$$

In this way, we have obtained a theoretical expression for the absorption coefficient α and in turn the imaginary part of complex dielectric function ε_2, thereby devising a bridge connecting macroscopic measurable quantities to microcosmic variables.

In what follows, we discuss the joint DOS $\rho_{CV}(\mathrm{k})$ and the optical matrix element M_j.

Based on the character of the δ function, $\rho_{CV}(k)$ (4.54) can be written as:

$$\rho_{CV}(\mathrm{k}) = \int \frac{2d^3k}{(2\pi)^3}\delta(E_C - E_V - \hbar\omega) = \frac{2}{(2\pi)^3}\int_{E_C - E_V = \hbar\omega} \frac{ds}{\nabla_k(E_C(\mathrm{k}) - E_V(\mathrm{k}))} \tag{4.57}$$

ds is the surface element in k-space of the surface on which $E_C(k) - E_V(k) = \hbar\omega$. Then in an isotropic band approximation we find:

$$\rho_{CV}(k) = \frac{k^2}{\pi^2}\left(\frac{\partial E_C}{\partial k} - \frac{\partial E_j}{\partial k}\right)^{-1} \tag{4.58}$$

where E_c, E_j are the energies of the conduction band and the three top valence bands $j = 1,2,3$, respectively.

For direct transitions with a spherical constant-energy surface approximation, parabolic bands, and $\hbar\omega \geq E_g$:

$$\begin{aligned} &E_C - E_V = E_g + \frac{\hbar^2 k^2}{2m_\gamma} \\ &\frac{1}{m_\gamma} = \frac{1}{m_e} + \frac{1}{m_h} \\ &\rho_{CV}(k) = 4\pi(2m_\gamma)^{3/2}(\hbar\omega - E_g)^{1/2}/\hbar^3 \end{aligned} \tag{4.59}$$

For narrow gap semiconductors, and based on the Kane model in the approximation $\Delta >> E_g, kP$, then we find:

$$\begin{cases} E_c = \dfrac{\hbar^2 k^2}{2m} + \dfrac{1}{2}[E_g + (E_g^2 + \dfrac{8}{3}k^2P^2)^{\frac{1}{2}}] \\ E_{hh} = -\dfrac{\hbar^2 k^2}{2m_{hh}} \\ E_{lh} = \dfrac{\hbar^2 k^2}{2m} + \dfrac{1}{2}[E_g - (E_g^2 + \dfrac{8}{3}k^2P^2)^{\frac{1}{2}}] \\ E_{SO} = -\Delta + \dfrac{\hbar^2 k^2}{2m} - \dfrac{P^2k^2}{3(E_g + \Delta)} \end{cases} \tag{4.60}$$

In this formula, each symbol has its customary meaning defined in Chap. 3, m is the free electron mass, m_{hh} is the heavy-hole effective mass, P is the momentum matrix element, and E_g is the bandgap energy. The energy of the heavy-hole band is independent of temperature. The energy of the light-hole band and the conduction band depends on temperature through E_g. Substituting (4.60) into (4.59), allows us to write:

$$\rho_{CV1}(E)=\frac{k_1}{\pi^2}\left[\frac{\hbar^2}{m}\left(1+\frac{m}{m_{hh}}\right)+\frac{4}{3}\frac{P^2}{\sqrt{E_g^2+\frac{8}{3}P^2k_1^2}}\right]^{-1} \tag{4.61}$$

$$\rho_{CV2}(E)=\frac{3}{8}\frac{k_2\sqrt{E_g^2+\frac{8}{3}P^2k_2^2}}{\pi^2P^2} \tag{4.62}$$

where k_1 and k_2 are calculated from:

$$\hbar\omega=E_c(k_1)-E_{hh}(k_1) \tag{4.63}$$

$$\hbar\omega=E_c(k_2)-E_{lh}(k_2) \tag{4.64}$$

For HgCdTe with x=0.330, at 8 K and 300 K, $E-k$ curves deduced from (4.60) are shown in Fig. 4.8. Figure 4.9 shows the relation between the joint DOS and *k*, for ρ_{CV1}, labeled by ρ_1, and ρ_{CV2}, labeled by ρ_2, at temperatures of 8 K and 300 K.

ρ_1 is the joint DOS of the heavy-hole band and the conduction band, while ρ_2 is the joint DOS of the light-hole band and the conduction band. They rarely vary much with temperature. From the figure we can see that when wave vector k changes from 0 to 1×10^5 cm^{-1}, the joint DOS increase rapidly from 0 to 10^{17} eV^{-1} cm^{-3}. When k continues to increase, the joint DOS still increases but rather slowly. Hence when the photon energy first reaches and then exceeds E_g, the absorption increases rapidly from zero to rather high values, then increases more slowly.

Next we examine the behavior of the optical matrix element M_j:

$$\begin{aligned}M_j&=\hat{\mathbf{A}}\cdot\mathbf{M}_{CV}(\mathbf{k})\\ \mathbf{M}_{CV}(\mathbf{k})&=<\psi_{C\mathbf{k}}|\mathbf{p}|\psi_{V\mathbf{k}}>\end{aligned} \tag{4.65}$$

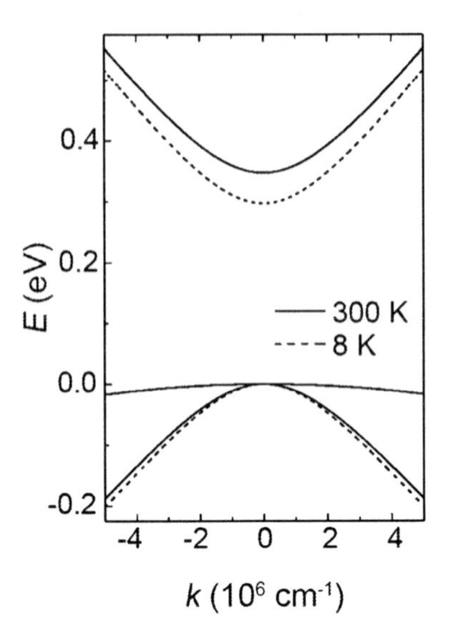

Fig. 4.8. The relation of the band energies to wave vectors for HgCdTe with x = 0.330 at two temperatures 8 K and 300 K (Chu 1983b,d)

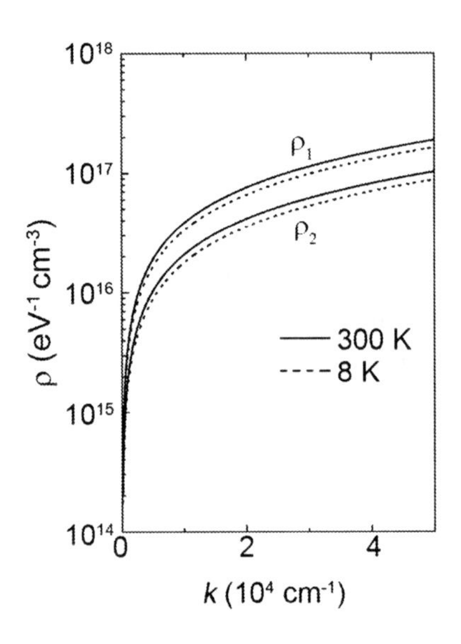

Fig. 4.9. The relation of the joint DOS to wave vectors for HgCdTe with x = 0.330 at two temperatures 8 and 300 K (Chu 1983b,d)

Kane's momentum matrix element P is defined as (Kane 1957) $P = -\frac{i\hbar}{m} < s|p_z|z >$, thus (Johnson 1967):

$$\begin{aligned} |M_j(0)|^2 &= \frac{2m^2P^2}{3\hbar^2} \\ |M_j(k)|^2 &= \frac{2m^2P^2}{3\hbar^2}[(a_c c_j + c_c a_j)^2 + (a_c b_j - b_c a_j)^2] \end{aligned} \tag{4.66}$$

In (4.66), the factor of 3 in the denominator comes from a three dimensional directional average, and the factor of 2 in the numerator comes from the sum over spin orientations. But in the analysis of joint DOS spin was already included. In an entire calculation process spin can be considered only once. So when this result is inserted into (4.56) for ε_2, an adjustment must be made by dropping one factor of 2. In (4.66), j runs over several valence bands, a, b, c are the coefficients in the wave function expressions arising from the k•p and spin–orbit perturbations. Their values are given by the following formulas:

$$\begin{cases} a_i = kp(E_i + \frac{2}{3}\Delta)/N_i \\ b_i = \frac{\sqrt{2}}{3}\Delta(E_i - E_g)/N_i \\ c_i = (E_i - E_g)(E_i + \frac{2}{3}\Delta)/N_i \end{cases} \tag{4.67}$$

in which $i = 1,3$ indicate the conduction band and the light-hole band. For the heavy-hole band the results are:

$$a_2 = 0, \quad b_2 = 1, \quad c_2 = 0 \tag{4.68}$$

N_i are normalization factors, determined from the condition:

$$a_i^2 + b_i^2 + c_i^2 = 1 \tag{4.69}$$

The values of the coefficients obtained from (4.67) are shown in Fig. 4.9, and vary little with temperature. If from (4.66), it is supposed that T_{cj} is defined as:

$$T_{cj}^2 \equiv [(a_c c_j + c_c a_j)^2 + (a_c b_j - b_c a_j)^2] \tag{4.70}$$

then when $k = 0$, $T_{cj}^2 = 1$, and when k increases, T_{cj}^2 gradually decreases. This is another of the factors making the intrinsic absorption band flat at higher frequencies.

The theoretical prediction from (4.55) of the intrinsic absorption spectrum can compared with the experimental spectrum. E_g is determined from the location of the onset of the intrinsic absorption spectrum. The parameters P, m_{hh} and Δ can be obtained from a best fit of the theoretical curve to the experimental curve.

It is not easy to reduce the expression for the intrinsic absorption coefficient deduced in the above process to a simple but accurate analytic expression. But for one approximate condition, we can make a simple analysis. The expression for the joint DOS includes the wave vectors where the optical transition takes place. In the general case, the expression for the wave vector of narrow gap semiconductors is complex. For a transition from the heavy-hole band to the conduction band, a term enters that is approximated by:

$$\sqrt{E_g^2 + \frac{8}{3}P^2k_1^2} \geq \hbar^2 k_1^2 \left(\frac{1}{m} + \frac{1}{m_{hh}} \right) \tag{4.71}$$

Then from the conservation of energy, expression (4.63), we find:

$$(2\hbar\omega - E_g) = \sqrt{E_g^2 + \frac{8}{3}P^2k_1^2} \tag{4.72}$$

Thus, k_1 reduces to the simple expression:

$$k_1 = \frac{1}{P}\sqrt{\frac{3}{2}\hbar\omega(\hbar\omega - E_g)} \tag{4.73}$$

In addition, again using the conservation of energy expression (4.64), we find:

$$\begin{aligned} \hbar\omega &= \sqrt{E_g^2 + \frac{8}{3}P^2k_1^2} \\ k_2 &= \frac{1}{P}\sqrt{\frac{3}{8}(\hbar^2\omega^2 - E_g^2)} \end{aligned} \tag{4.74}$$

The joint densities of states can be written as simple expressions in which the photon energy is a variable.

$$\rho_{C-hh}(E)$$
$$=\frac{1}{\pi^2}\frac{1}{P}\sqrt{\frac{3}{2}\hbar\omega(\hbar\omega-E_g)}\left[\frac{\hbar^2}{m}\left(1+\frac{m}{m_{hh}}\right)+\frac{4}{3}\frac{P^2}{(2\hbar\omega-E_g)}\right]^{-1} \tag{4.75}$$

$$\rho_{C-lh}(E)=\frac{3\sqrt{3}}{16\sqrt{2}}\cdot\frac{\hbar\omega}{\pi^2P^3}\sqrt{\hbar^2\omega^2-E_g^2} \tag{4.76}$$

For one approximate condition, T_{cj}^2 can also be written as analytic expressions. For the transition from the heavy-hole band to the conduction band, and from the light-hole band to the conduction band they become:

$$T_{c-hh}^2=\frac{1}{2}\left[\left(1+\frac{8P^2k^2}{3E_g^2}\right)^{-\frac{1}{2}}+1\right]=\frac{\hbar\omega}{2\hbar\omega-E_g} \tag{4.77}$$

$$T_{c-lh}^2=\frac{1}{3}+\frac{2}{3}\left(\frac{E_g}{\hbar\omega}\right)^2 \tag{4.78}$$

Then from (4.55), the absorption coefficient can be obtained. The net contribution to the absorption coefficient is a sum of those from the heavy-hole band and the light-hole band.

$$\alpha=\alpha_{hh}+\alpha_{lh} \tag{4.79}$$

After simplification, they are:

$$\alpha_{hh}=\frac{\sqrt{3/2}}{137n}\cdot\frac{1}{P}\frac{\sqrt{\hbar\omega}\sqrt{\hbar\omega-E_g}}{1+\frac{m_c^*}{m}\left(1+\frac{m_c^*}{m}\right)\left(\frac{2\hbar\omega}{E_g}-1\right)}$$
$$\alpha_{lh}=\frac{1}{137\sqrt{6}n}\cdot\frac{1}{4P}\left[1+2\left(\frac{E_g}{\hbar\omega}\right)^2\right]\sqrt{\hbar^2\omega^2-E_g^2} \tag{4.80}$$

in which m_{hh} is the heavy-hole effective mass, P is the momentum matrix element, m_c^* is the effective mass of an electron in the conduction band that depends on the band gap energy and the momentum matrix element.

In general, for an allowed direct transition, the absorption coefficient can be written as (Moss et al. 1973):

$$\alpha = Af(\hbar\omega)\left(\hbar\omega - E_g\right)^{1/2} \tag{4.81}$$

where A is a coefficient, and $f(\hbar\omega)$ is a increasing function on $\hbar\omega$.

The absorption spectrum of an interband transition can be calculated from above theory. However, some energy band parameters are undetermined. Therefore, the intrinsic transition absorption spectrum measurements are needed to extract them. Because the number of undetermined parameters band parameters is large, nearly perfect fits to the data can be obtained. Of course this leaves some ambiguity in the set of fitted parameters.

4.2.2 Experimental Investigations of Interband Optical Transitions

Kane compared his theoretical calculation of the InSb intrinsic absorption band to the experimental results (Fig. 4.10). For ternary semiconductors, the samples used in such comparisons should have homogeneous compositions and electrical characteristics. Thus for HgCdTe narrow gap semiconductors bulk materials were often adopted for these measurements. The HgCdTe samples used in these experiments have been crystals grown by solid recrystallization, tellurium solvent, and semi-molten methods. The samples must be cut, ground, polished and etched. For intrinsic absorption spectrum measurements, thin samples are used. The samples are attached to a substrate that is transparent in the measured wavelength region, and its coefficient of linear expansion must be close to that of HgCdTe to avoid stress or dilapidation at low temperature. Substrates used include diamond, ZnSe, Si, and KRS5. The glue used has a transparent window in the 2~6 μm wavelength region. For measurements above 6 μm, ZnSe substrates are often used. The glue is spread only around the perimeter of the sample, or a glue is chosen with less absorption in the long wavelength region. It is also possible to adopt Si or KRS5 substrates. Using this fabrication method, samples can be thinned to ~2.5–20 μm. After measuring the transmittance ratio, the absorption coefficient is deduced from a proper analytic method presented in Chap. 3.

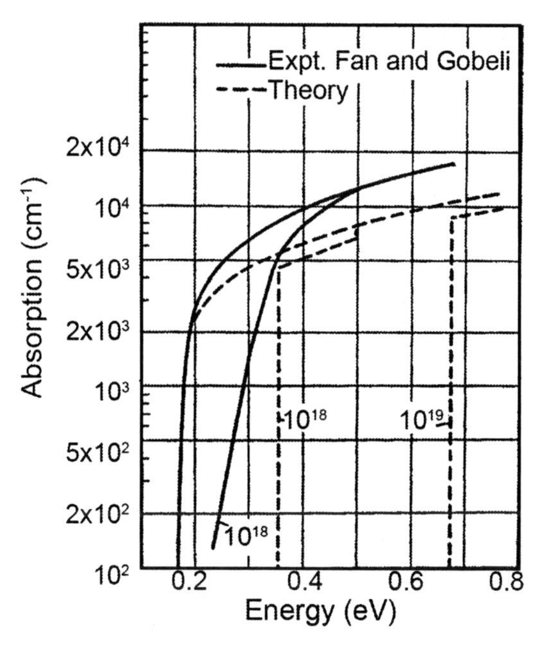

Fig. 4.10. Comparison of the theoretical calculation and experimental results for the intrinsic absorption band of InSb (Kane 1957)

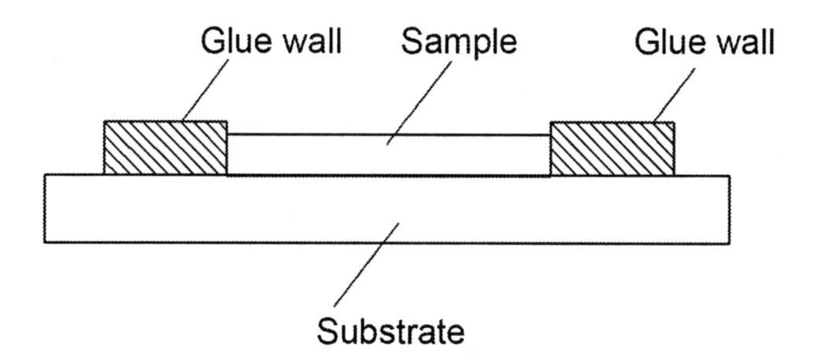

Fig. 4.11. A sketch of the loop glue method

It is rather difficult to prepare 10~15 μm thin samples of the rather "soft" HgCdTe material. It tends to crack, or its rim begins to shrink causing its area to become smaller. An improved method is to prepare samples in a kind of "loop glue method" adopted by the author. It protects thin samples in the process of reducing their thickness. This method uses epoxy or "cryogenic glue" to build a glue wall around the perimeter of the sample with a same thickness as that of sample to adhere it to the silicon substrate, as shown in Fig. 4.11. Then in the process of polishing, this glue wall's thickness will also be reduced as the sample is thinned. The glue serves to protect the semiconductor sample. After that, an etch method can be used to reduce the sample's thickness still more. In this way thin HgCdTe samples can be prepared with the desired thickness.

For thicker samples a micrometer screw gauge is often used to determine the thickness of samples. For very thin samples the thickness is determined by use of an interference microscope. When using an interference microscope, the light dot should be at the step edge of the sample on the substrate. Generally it is chosen at a defect in the sample where the surface of the substrate shows through, or at the rim of the sample. Then we can observe two groups of stripes, one from the surface of the sample, and the other comes from the surface of the substrate. Knowing the wavelength of the illumination, the difference in the periods of these two groups determines the thickness.

The composition of samples is determined using a densitometer. The compositional uniformity is determined using a scanning electron microscope. The relation between the density and the composition of HgCdTe was shown in Fig. 2.4. The analytic expression is (Blair and Newnham 1961):

$$\mathrm{x} = 3.628 - 0.449249d \tag{4.82}$$

It is usual to choose samples with a compositional uniformity of about ±0.003 for further measurements. A typical example is shown in Fig. 4.52 later in the text in Sect. 4.6.2. Samples used to measure the intrinsic absorption spectrum, their composition, thickness and carrier concentration at 77 K are shown in Tables 4.2a–c.

The intrinsic absorption spectrum is measured using an infrared spectrophotometer. Samples are placed in a temperature variable Dewar. The temperature is controlled and determined using a digital thermometer.

For a sample without a substrate, or a sample with a substrate but the glue is around the rim of sample so the effect of the substrate can be subtracted, the formula for the transmittance:

$$T = \frac{(1-R)^2 e^{-\alpha d}}{1-R^2 e^{-2\alpha d}} \tag{4.83}$$

can be used to calculate the absorption coefficient α:

$$\alpha = \frac{1}{d} \ln \frac{\sqrt{(1-R)^4 + 4T^2R^2} + (1-R)^2}{2T} \tag{4.84}$$

Table 4.2a. Partial coefficients of different samples

Sample#	No.1	No.2	No.3	No.4	No.5	No.6	No.6	No.7
x	0.165	0.170	0.194	0.200	0.218	0.226	0.258	0.264
d(μm)	10	9	24	25	24	15	70	370
$n_{77\,K}$ (cm^{-3})	4×10^{16}	5.85×10^{15}	2.2×10^{16}	1.4×10^{15}	2.8×10^{15}	2.2×10^{15}	1.3×10^{15}	1×10^{16}

Table 4.2b. Partial of coefficient of different samples

Sample#	No.9	No.10	No.11	No.12	No.13	No.14	No.15
x	0.276	0.330	0.344	0.362	0.366	0.416	0.443
d(μm)	6	8	6.5	7	9	8	2.5
$n_{77\,K}$(cm^{-3})	7.2×10^{14}	6.2×10^{14}	5×10^{14}	2.4×10^{14}	4.4×10^{15}	(P)1.1×10^{16}	(P)2×10^{16}

Table 4.2c. Partial coefficients of different samples

Sample#	No.16	No.17	No.18	No.19	No.20	No.21
X	0.19	0.21	0.25	0.28	0.34	0.37
d(μm)	8	10	7	7.5	8	6
$n_{77\,K}$ (cm^{-3})	1.64×10^{15}	8.5×10^{15}	1.6×10^{15}	7.2×10^{14}	5×10^{14}	2×10^{16}

The effect of multiple reflections is included in this formula, where:

$$R = \frac{(n-1)^2 + k^2}{(n+1)^2 + k^2} \tag{4.85}$$

For $\alpha \simeq 10^4\ \text{cm}^{-1}$, $\lambda \simeq 10^{-4}$ cm, $k \simeq 10^{-1}$, (4.85) can be written as:

$$R = \left(\frac{n-1}{n+1}\right)^2 \tag{4.86}$$

n is the index of refraction, and $n = \sqrt{\varepsilon_\infty}$, where ε_∞ is the high-frequency dielectric constant. The ε_∞ value of HgCdTe at different compositions is given in Baars and Sorger (1972) experimental results. It can also be calculated from the approximate imperial expression:

$$\varepsilon_\infty = 15.2 - 15.5x + 13.76x^2 - 6.32x^3 \tag{4.87}$$

For thin samples with the sample-glue-substrate structure, the total transmittance of the structure can be found approximately using the formula (Packard 1969):

$$T=\frac{(1-R_{as})(1-R_{sc})e^{-\alpha_s d_s}}{1-R_{as}R_{sc}e^{-2\alpha_s d_s}}\times\frac{(1-R_{sc})(1-R_{cb})e^{-\alpha_c d_c}}{1-R_{sc}R_{cb}e^{-2\alpha_c d_c}}\times\frac{(1-R_{cb})(1-R_{ba})e^{-\alpha_b d_b}}{1-R_{cb}R_{ba}e^{-2\alpha_b d_b}} \tag{4.88}$$

Subscript a, s, c and b denotes air, semiconductor sample, glue and substrate, respectively.

The corresponding expression for the transmittance of a glue-substrate system is:

$$T'=\frac{(1-R_{ac})(1-R_{cb})e^{-\alpha_c d_c}}{1-R_{ac}R_{cb}e^{-2\alpha_c d_c}}\times\frac{(1-R_{cb})(1-R_{ba})e^{-\alpha_b d_b}}{1-R_{cb}R_{ba}e^{-2\alpha_b d_b}} \tag{4.89}$$

The ratio of these transmittance expressions is:

$$\frac{T}{T'}=\frac{(1-R_{as})(1-R_{sc})^2 e^{-\alpha_s d_s}(1-R_{ac}R_{cb}e^{-2\alpha_c d_c})}{(1-R_{as}R_{sc}e^{-2\alpha_s d_s})(1-R_{sc}R_{cb}e^{-2\alpha_c d_c})(1-R_{ac})} \tag{4.90}$$

For a sapphire substrate, n_b=1.89, n_c=1.65, $R_{cb}\approx 0$, the above formula simplifies to:

$$\frac{T}{T'}=\frac{Ae^{-\alpha d}}{1-Be^{-2\alpha d}} \tag{4.91}$$

where

$$A=\frac{(1-R_{as})(1-R_{sc})^2}{1-R_{ac}},\quad B=R_{as}R_{sc}. \tag{4.92}$$

For a Si substrate, in the spectral region where α_c ~0 and $e^{-2\alpha_c d_c}\to 1$, the expression for A is:

$$A=\frac{(1-R_{as})(1-R_{sc})^2}{1-R_{ac}}\times\frac{1-R_{ac}R_{cb}}{1-R_{sc}R_{cb}} \tag{4.93}$$

Here A and B are functions of the indicated reflectance, all of which can be calculated based on the materials' indices of refraction. T/T' is obtained by measuring T and T', and dividing. Then we can calculate the absorption spectrum from (4.91). (Chu 1983c; Chu et al. 1991)

There is another practical measurement method that can be adopted, the "$\alpha_c d_c$" method (Chu 1983b). It involves measuring the transmittance of a sample consisting of a ZnS+glue+ZnS structure. Then we have:

$$T'' = \left(\frac{(1-R_{ab})(1-R_{bc})e^{-\alpha_b d_b}}{1-R_{ab}R_{bc}e^{-2\alpha_b d_b}} \right)^2 \cdot \frac{(1-R_{bc})^2 e^{-\alpha_c d_c}}{1-R_{bc}^2 e^{-2\alpha_c d_c}} \tag{4.94}$$

In the region where ZnS is transparent, $\alpha_b \to 0$, the above formula can be written as:

$$T'' = \frac{A' e^{-\alpha_c d_c}}{1-B' e^{-2\alpha_c d_c}} \tag{4.95}$$

Here A', and B' are just functions of reflectance, which can be obtained knowing the materials' refractive indices. From (4.95), we can use T'' to deduce $\alpha_c d_c$, and substitute it into (4.89). Then we can obtain the absorption spectrum $\alpha(v)$ of the sample. This procedure supposes the glue thickness d_c is the same for the two samples, but actually there is always a difference. In the spectral region where $\alpha_c = 0$, (that is the transparent region of the glue) the impact of this difference is rather small. But when $\alpha_c \neq 0$, the differences between the d_c values will introduce a distortion in the part of spectral curve where the transmission is high. In this case special care is needed in the data analysis.

There is still another method to determine the absorption coefficient. Because the reflectivity of the glue-substrate interface $R_{cb} \approx 0$, to analyze the transmittance of a sample-glue-substrate layered structure, we can reduce the formula as follows (Hougen 1989):

$$T = \frac{(1-R_1)(1-R_2)(1-R_s)\exp[-(\alpha d + \alpha' d')]}{(1-R_1 R_2 \exp(-2\alpha d))(1-R_1 R_2 \exp(-2\alpha' d')) - (1-R_2)^2 R_1 R_3 \exp[-2(\alpha d + \alpha' d')]} \tag{4.96}$$

Here $R_1 = R_{as}$, $R_2 = R_{sc}$, $R_3 = R_{ba}$. α, d are, respectively, the absorption coefficient and thickness of semiconductor layer, and $\alpha' d' = \alpha_c d_c + \alpha_b d_b$, $\alpha_b \cong 0$. To eliminate the effect of the glue, we need to determine the transmittance T' of an air-glue-substrate-air combination:

$$T' = \frac{(1-R_c)(1-R_3)\exp(-\alpha' d')}{(1-R_c R_s \exp(-2\alpha' d))} \tag{4.97}$$

in which $R_c = R_{ac}$. From (4.97), $\exp(-\alpha' d')$ can be obtained and substituted into (4.96), from which the absorption coefficient of semiconductor can be deduced to be:

$$\alpha = \frac{1}{d}\ln\left(\frac{2B}{\sqrt{\left(\frac{T'}{t}A\right)^2 + 4B} - \left(\frac{T'}{T}\right)A}\right) \tag{4.98}$$

where

$$\begin{aligned} A &= \frac{(1-R_1)(1-R_2)}{1-R_c}; \\ B &= R_1R_2 + \frac{R_1R_3(1-R_2)^2T'^2}{(1-R_c)(1-R_3)} \end{aligned} \tag{4.99}$$

The samples used in these experiments have already been described, and details can be found in the references (Chu 1984; Chu et al. 1985). Samples were put in a low temperature Dewar whose temperature can be varied from liquid helium to room temperature. When we extract the absorption spectrum from the measured transmittance spectrum, the substrate correction of (Chu 1984; Chu et al.1985) must be taken into account. The results of this data reduction are shown in Figs. 4.12–4.14. In the figures, the square and the dot symbols denote the absorption spectrum deduced from the measurements, and the solid lines are theoretical absorption curves.

Figure 4.12 displays the intrinsic absorption spectra of HgCdTe with x = 0.330, at 8 K and 300 K. The band-gap energy is determined from the experimental absorption spectrum as explained in detail in Sect. 3.4. In Fig. 4.12, at 8 K, E_g =0.300 eV and at 300 K, E_g =0.347 eV. In the calculation the best fit of the calculated curve to the experimental curve is obtained, if the momentum matrix element P is taken as 8×10^{-8} eV cm, the heavy-hole effective mass is $m_{hh} = 0.55m_0$, and the spin–orbit splitting Δ is 1 eV.

Figure 4.13 shows a similar comparison for x = 0.276, taking the same values of P, m_{hh}, and Δ as used for x = 0.276. Figure 4.14 shows the experimental and the calculated absorption spectra of HgCdTe samples with x = 0.170–0.443 at 300 K. Again the same values of P, m_{hh} and Δ were used.

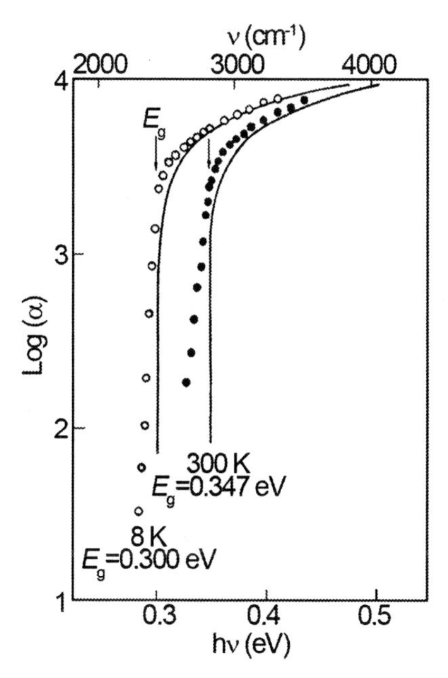

Fig. 4.12. The experimental and the calculated absorption spectra of an x = 0.330 sample at 8 K, and 300 K

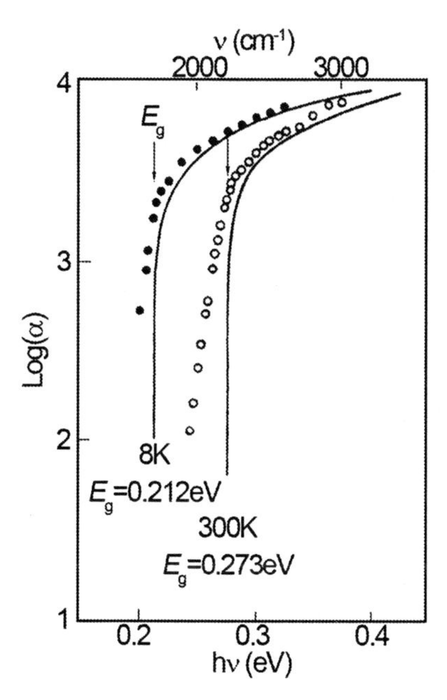

Fig. 4.13. The experimental and the calculated absorption spectra of an x = 0.276 sample at 8, and 300 K

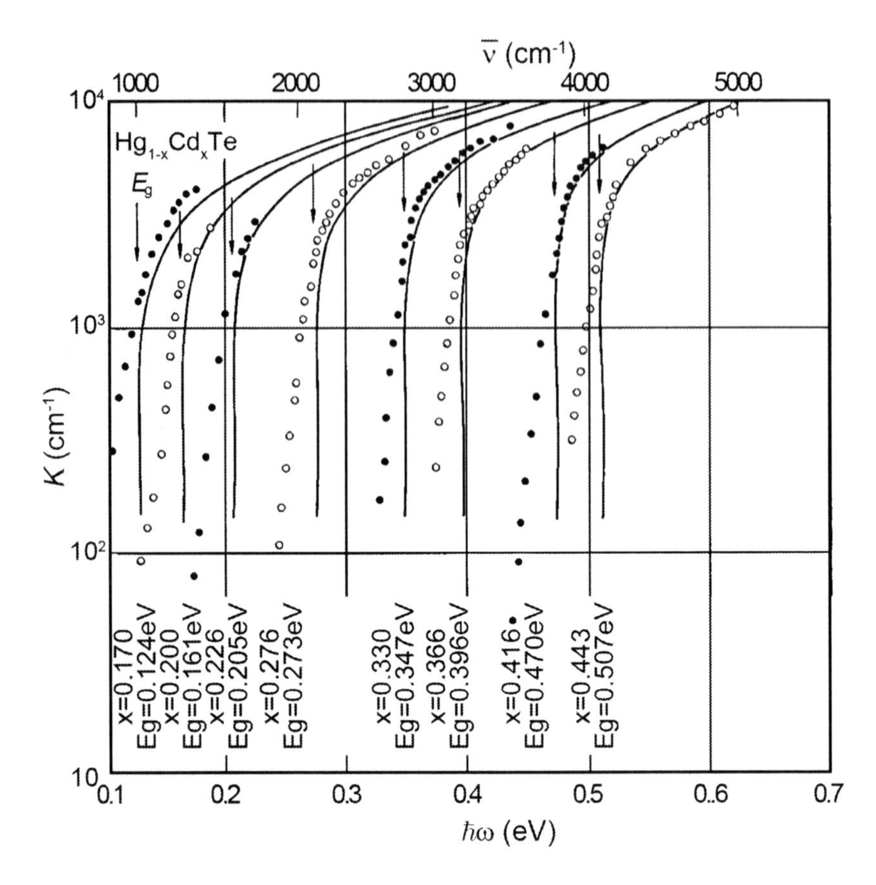

Fig. 4.14. The experimental and the calculated absorption spectra of x = 0.170–0.443 samples at 300 K

4.2.3 Indirect Interband Transitions

HgCdTe alloys and InSb all belong to the class of direct band gap semiconductors. For narrow gap semiconductors with indirect band gaps indirect phonon assisted transition theory is required. Bardeen (1957), devised such a theory to explain the absorption tail of Ge.

For some semiconductor materials, the top of the valence band is at $k = 0$ and the bottom of the conduction band is not at the same k value. These materials are called indirect band gap semiconductors, and transitions at the absorption edge do not correspond to vertical electronic transitions in the reduced zone. Since the wave vectors of photons with energies corresponding to the band gap are small and can be ignored, the electronic transition requires the assistance of a phonon to ensure conservation of momentum. During an electronic indirect transition where an electron is excited from the valence band to the conduction band, there are two independent perturbation processes. One is the absorption of a photon, which conserves the momentum of the electron and excites it into a virtual state. The other is the absorption or emission of a phonon, which provides the momentum to excite the electron from the virtual state into a real conduction band state. The net interaction Hamiltonian is:

$$H' = H_{eP} + H_{eL} \tag{4.100}$$

$H_{aP}(= H')$ was introduced, (4.43), when the direct transition was treated. Phonon-assisted absorption is present in direct gap semiconductors, but is usually small and ignored compared to the contribution from the direct transitions.

In the case of a crystal with several atoms per primitive cell, the perturbation Hamilton for the deformation potential interaction of an electron with the excited crystal lattice is:

$$H_{eL} = \sum_{\substack{\{\mathbf{R}+\tau_{\mathbf{R},j,\alpha_j}+\eta_{\mathbf{R},j,\alpha_j}\}, \\ \{\mathbf{R}'+\tau_{\mathbf{R}',j',\alpha_{j'}}\}}} V(\{\mathbf{r}-\mathbf{R}+\tau_{\mathbf{R},j,\alpha_j}+\eta_{\mathbf{R},j,\alpha_j}\}) - V(\{\mathbf{r}-\mathbf{R}'+\tau_{\mathbf{R}',j',\alpha_{j'}}\}) \tag{4.101}$$

Where:

- $\boldsymbol{R}$ is the equilibrium displacement of a unit cell
- $\tau_{\mathbf{R},j,\alpha_j}$ is the equilibrium displacement of the jth type of atom in the unit cell relative to $\boldsymbol{R}$
- α_j is the electronic state of the atom located at $\boldsymbol{R}+\tau_{\mathbf{R},j,\alpha_{j'}}$
- $\eta_{\mathbf{R},j,\alpha_j}$ is the vibrational displacement of the atom located at $\boldsymbol{R}+\tau_{\mathbf{R},j,\alpha_j}$
- V is the potential between an electron located at $\boldsymbol{r}$ in the set of any of the ions' localized states, with the ions located in their distorted (or equilibrium) positions
- The sums run over the set all equilibrium atom sites
- H_{eL} is the difference between the electron's interaction with the distorted lattice and the equilibrium lattice. There is an additional contribution to the electron–phonon interaction in polar solids that will be introduced presently.

For a lattice with only one atom per unit cell the τ's vanish and j has only one value. For a zinc blende polar lattice j takes on two values, one for the anion and the other for the cation, and the τ's are finite. The α states can in principle designate any of the atomic states of the atoms, but in most calculations only range over those states that contribute to the valence and conduction bands, e.g. the outer ***s*** and ***p*** atomic states of the anion and cation. In more complex solids ***j*** can take on many values.

For small displacements of the lattice ions from their equilibrium sites, and in an orthogonalized local orbital basis $\left|\mathbf{R}, j, \alpha_j\right\rangle$ a matrix element of H_{eL} become:

$$\left\langle \mathbf{R}', j', \alpha_{j'} \middle| H_{eL} \middle| \mathbf{R}, j, \alpha_j \right\rangle = \frac{\partial V}{\partial d} \hat{\mathbf{d}} \cdot \left(\eta_{\mathbf{R},j,\alpha_j} - \eta_{\mathbf{R}',j',\alpha_{j'}} \right) \tag{4.102}$$

where

$$\mathbf{d} \equiv \mathbf{R} + \tau_j + \eta_{\mathbf{R},j,\alpha_j} - \mathbf{R}' - \tau_{j'} - \eta_{\mathbf{R}',j',\alpha_{j'}} \tag{4.103}$$

Note that the numerous subscripts on ***d*** have been suppressed. If one uses atomic orbitals, which overlap for most zinc blende semiconductors, and are therefore not orthogonal, the calculations are complicated by having to keep track of the overlap integrals. Chen and Sher (1982) devised a way of transforming the atomic states into an orthonormal local orbital (OLO) basis set. These orbitals form a natural basis set for this calculation.

For small displacements $\eta_{\mathbf{R},j,\alpha_j}$, the orbitals can be expanded in terms of phonon unit eigen vectors $\hat{e}_{\lambda,\mathbf{q}}$:

$$\eta_{\mathbf{R},j.\alpha_j} = \sum_{\lambda,\mathbf{q}} \left(\frac{\hbar}{2NM_j\omega_{\lambda,\mathbf{q}}} \right)^{1/2} \left[\hat{\mathbf{e}}_{\lambda\mathbf{q}} e^{i\mathbf{q}\cdot(\mathbf{R}+\tau_j)} A(\lambda,\mathbf{q}) + \hat{\mathbf{e}}_{\lambda\mathbf{q}}^{\ *} e^{-i\mathbf{q}\cdot(\mathbf{R}+\tau_j)} A^{\dagger}(\lambda,\mathbf{q}) \right] \tag{4.104}$$

where λ is the polarization index (longitudinal, or one of the two transverse polarizations); **q** is the wave number of the phonon; $A^{\dagger}$, and A are the annihilation and creation operators respectively with matrix elements $\left\langle n_{\lambda\mathbf{q}} - 1 \middle| A(\lambda,\mathbf{q}) \middle| n_{\lambda\mathbf{q}} \right\rangle = \sqrt{n_{\lambda\mathbf{q}}}$ and $\left\langle n_{\mathbf{q},\eta_\mathbf{q}} + 1 \middle| A^{*}(\mathbf{q},\eta_\mathbf{q}) \middle| n_{\mathbf{q},\eta_\mathbf{q}} \right\rangle = \sqrt{n_{\mathbf{q},\eta_\mathbf{q}} + 1}$, where $\left|n_{\lambda,\mathbf{q}}\right\rangle$ is a phonon state $\lambda,\mathbf{q}$ with occupation number $n_{\lambda,\mathbf{q}} = (e^{\frac{\hbar\omega_{\lambda,\mathbf{q}}}{kT}} - 1)^{-1}$ (a Bose–Einstein factor).

Expanding free electron basis sets in terms of the OLO basis sets and including the phonon state, (4.102), taking (4.104) into account, become:

$$\begin{aligned}&\left\langle n_{\lambda\mathbf{q}}-1,\mathbf{k}',j',\alpha_{j'}\right|H_{eL}\left|n_{\lambda\mathbf{q}},\mathbf{k},j,\alpha_j\right\rangle\\&=\lambda_{\mathbf{k}',\mathbf{k}+\mathbf{q}}V_{j'j}^{\alpha_{j'}\alpha_j}\left(\mathbf{k}+\mathbf{q},\mathbf{k},\lambda\right)\sqrt{n_{\lambda\mathbf{q}}}\end{aligned}\tag{4.105}$$

where

$$V_{j'j}^{\alpha_{j'}\alpha_j}\left(\mathbf{k}+\mathbf{q},\mathbf{k},\lambda\right)\equiv\sum_{\mathbf{d}}\frac{\partial V}{\partial d}\hat{\mathbf{d}}\bullet\hat{\mathbf{e}}_{\lambda\mathbf{q}}\left(\frac{e^{-i\mathbf{k}\bullet\mathbf{d}}}{\sqrt{M_{j'}}}-\frac{e^{-i\mathbf{k}\bullet\mathbf{d}}}{\sqrt{M_j}}\right)$$

for phonon annihilation, and:

$$\begin{aligned}&\left\langle n_{\lambda\mathbf{q}}+1,\mathbf{k}',j',\alpha_{j'}\right|H_{eL}\left|n_{\lambda\mathbf{q}},\mathbf{k},j,\alpha_j\right\rangle\\&=\delta_{\mathbf{k}',\mathbf{k}-\mathbf{q}}V_{j'j}^{\alpha_{j'}\alpha_j}\left(\mathbf{k}-\mathbf{q},\mathbf{k},\lambda\right)\sqrt{n_{\lambda\mathbf{q}}+1}\end{aligned}\tag{4.106}$$

for phonon creation.

This is not quite what is needed for insertion into the Golden Rule. The matrix elements required are those between the energy eigenvalues of the unperturbed Hamiltonian, e.g. $\left\langle n_{\lambda\mathbf{q}}-1,m',\mathbf{k}'\right|H_{eL}\left|n_{\lambda\mathbf{q}},m,\mathbf{k}\right\rangle$, where m is a band index. These matrix elements can be obtained from (4.105) and (4.106) by replacing $V_{j'j}^{\alpha_{j'}\alpha_j}\left(\mathbf{k}+\mathbf{q},\mathbf{k},\lambda\right)$ by:

$$V_{m',m}\left(\mathbf{k}+\mathbf{q},\mathbf{k},\lambda\right)=\sum_{j',\alpha_{j'}}\sum_{j,\alpha_j}C_{j',\alpha_{j'},m'}C_{j,\alpha_j,m}V_{j'j}^{\alpha_{j'}\alpha_j}\left(\mathbf{k}+\mathbf{q},\mathbf{k},\lambda\right)\tag{4.107}$$

The $C_{j,\alpha_j,m}$'s are expansion coefficients.

The transition probability per unit time for an electron initially in the valence band with momentum ***k*** and phonon occupation numbers $n_{\lambda q}$, accompanied by a phonon annihilation, to the conduction band into a state with momentum ***k+q*** is:

$$\begin{aligned}&P_{val,\mathbf{k}\to con,\mathbf{k}+\mathbf{q}/}^{annih}\left(\omega\right)=\\&\frac{2\pi}{\hbar}\left(\frac{eA_0}{mc}\right)^2\sum_{b,c,v,\lambda}\left[\frac{\left\langle n_{\lambda\mathbf{q}}+1,c,\mathbf{k}+\mathbf{q}\right|H_{eL}\left|n_{\lambda\mathbf{q}},b,\mathbf{k}\right\rangle\left\langle n_{\lambda\mathbf{q}},b,\mathbf{k}\right|\hat{\mathbf{A}}\bullet\mathbf{p}\left|n_{\lambda\mathbf{q}},v,\mathbf{k}\right\rangle}{E\left(n_{\lambda\mathbf{q}},m,\mathbf{k}\right)-E\left(n_{\lambda\mathbf{q}},v,\mathbf{k}\right)}\right]^2\\&\bullet\delta\left(E\left(n_{\lambda\mathbf{q}}+1,c,\mathbf{k}+\mathbf{q}\right)-E\left(n_{\lambda\mathbf{q}},v,\mathbf{k}\right)-\hbar\omega_{\lambda\mathbf{q}}\right)\end{aligned}\tag{4.108}$$

If the transition is accompanied by phonon creation the transition probability per unit time becomes:

$$P^{creation}_{val,\mathbf{k}\to con,\mathbf{k}-\mathbf{q}}(\omega)=$$

$$\frac{2\pi}{\hbar}\left(\frac{eA_0}{mc}\right)^2\sum_{b,c,v,\lambda}\left[\frac{\left\langle n_{\lambda\mathbf{q}}-1,c,\mathbf{k}-\mathbf{q}\middle|H_{eL}\middle|n_{\lambda\mathbf{q}},b,\mathbf{k}\right\rangle\left\langle n_{\lambda\mathbf{q}},b,\mathbf{k}\middle|\hat{\mathbf{A}}\bullet\mathbf{p}\middle|n_{\lambda\mathbf{q}},v,\mathbf{k}\right\rangle}{E\left(n_{\lambda\mathbf{q}},m,\mathbf{k}\right)-E\left(n_{\lambda\mathbf{q}},v,\mathbf{k}\right)}\right]^2 \bullet\delta\left(E\left(n_{\lambda\mathbf{q}}-1,c,\mathbf{k}\right)-E\left(n_{\lambda\mathbf{q}},v,\mathbf{k}\right)+\hbar\omega_{\lambda\mathbf{q}}\right) \quad (4.109)$$

If only absorption and the band edge is being considered, in the sum the c values run over the spin degenerate conduction band minima, which if the band edge is at X or L points have 6×2 or 8×2 values respectively. The sum over v includes the heavy and light hole bands. More generally it also includes the split-off band and the even the lower bands. The sum over b includes all bands as possible virtual states.

Note that $\hat{\mathbf{A}}\bullet\mathbf{p}$ only couples electron states with the same $\boldsymbol{k}$, and H_{eL} only couples phonon states with the same polarization index.

While the deformation potential couples the electron to all phonon states, there is a stronger coupling mechanism between the longitudinal optical phonons and electrons in polar zinc blende semiconductors. This mechanism results from the fact that in these polar solids longitudinal optical modes have the anion and cation sub-lattices oscillating against one another. This adds an extra field that induces an extra restoring force to the ion lattice, increasing the longitudinal optical phonon frequencies relative to those of the transverse optical phonons with the same wave number. The extra fields also interact with the electrons, generating a far stronger interaction coupling constant than that of the deformation potential. This interaction is known as the Frohlich Hamiltonian (Frohlich 1937) given by (Callen 1949; Ehrenreich 1957):

$$H^{LO}_{eL}=i\left(4\pi^2e^2\frac{\hbar\omega_{LO}}{2\mathcal{V}}\right)^{1/2}\sum_{\mathbf{q}}\frac{1}{q}\left(b_{LO,\mathbf{q}}e^{i\mathbf{q}\bullet\mathbf{r}}+b^{\dagger}_{LO,\mathbf{q}}e^{-i\mathbf{q}\bullet\mathbf{r}}\right) \quad (4.110)$$

In (4.110), $\mathcal{V}$ is the volume of the crystal, ω_{LO} is the longitudinal optical phonon frequency assumed to be constant over the whole Brillouin zone.

Equations (4.108) and (4.109) must be modified for polar semiconductors to include the Frohlich Hamiltonian. So H_{eL} has two contributions $H_{eL}=H^{D}_{eL}+H^{LO}_{eL}$, one from the deformation potential and the other from the longitudinal optical polar interaction. Because the polar interaction coupling constant is much larger than that of the deformation potential, one expects the absorption coefficients of indirect gap polar semiconductors like GaP, to be greater than those of non-polar ones like Si. Care must be taken in making this kind of assessment because selection rules depending on the location in the Brillouin zone of the conduction band edge can influence the results.

Then the expression for the phonon assisted absorption coefficient at frequency ω is:

$$\alpha_{ph,abs}(\omega) = \frac{2\pi c\hbar \nu}{n\omega A_0^{\,2}} \int_{BZ}\int_{BZ} \frac{d^3k d^3q}{(2\pi)^3(2\pi)^3} \left[P^{anniih}_{val\mathbf{k}\to con\mathbf{k}+\mathbf{q}} + P^{create}_{val\mathbf{k}\to con\mathbf{k}-\mathbf{q}} \right] \quad (4.111)$$

where ν is the volume of the crystal.

The interpretation of the absorption measurements of indirect gap semiconductors to extract the band gap is much more complex than that for direct gap materials. Phonons must assist the transition. A phonon must provide at least a momentum of magnitude q_{min} to enable the photon absorption. The energy of an acoustic phonon with this momentum is approximately $\hbar v_{phonon} q_{min}$ where v_{phonon} is the velocity of sound in the solid. If the phonon is annihilated in the process, the minimum photon energy required to excite an electron from the top of the valance band to the bottom of one of the conduction band minima is $E_g - \hbar v_{phonon} q_{min}$. Depending on phonon occupations and coupling constants, the acoustic phonon energy may be replaced by one of the larger optical phonon energies. At low temperatures where there are no excited photons to absorb, the assistance must come from the creation of a phonon with momentum q_{min} and the minimum photon energy required for the transition is $E_g + \hbar v_{phonon} q_{min}$. This is an addition effect contributing to the apparent temperature variation of band gaps that must be included in the interpretation of absorption data to extract the actual temperature variation of the band gap. For polar indirect gap semiconductors the larger longitudinal optical phonon energy replaces that of the acoustic phonon, so the effective minimum measured gap is reduced. In either case the real band gap cannot be deduced from absorption measurements unless the phonon energies are known.

4.3 Intrinsic Absorption Spectra Expressions

4.3.1 The Absorption Edge

The absorption spectrum has a steep low energy edge feature. As discussed earlier, above this steep edge there is a slowly ascending intrinsic absorption band. This section first discusses the absorption edge.

Urbach in 1953 published a half page article (Urbach 1953) pointing out that the absorption coefficient of AgBr follows an exponential frequency dependence:

$$\frac{d\log\alpha}{d\nu} = -\frac{1}{kT} \tag{4.112}$$

and that the absorption of AgCl, Ge, CdS and TiO2 also follow this rule. In 1957, Martienssen (1957) through research on the alkali metal halide KBr, and based on the above formula, wrote:

$$\alpha = \alpha_0 \exp\left[\frac{\sigma(E - E_0)}{k_B T}\right] \tag{4.113}$$

Equation (4.113) shows, that when $E = E_0$, then $\alpha = \alpha_0$, independent of temperature. The logarithmic graph $(E_0, \log\alpha_0)$, is the low energy focal point of absorption edges measured at different temperatures. Marple (1966) afterward found that the II–VI compounds, CdTe etc., also follow this rule.

Experiments also found that the absorption edge of HgCdTe follows the Urbach exponential rule (Finkman and Sachacham 1984; Chu 1983a). Figure 4.15 shows the absorption spectrum of HgCdTe with composition x = 0.443 from liquid nitrogen to room temperature. We can see that projections to low energies, of absorption edges at different temperatures intersect at one point, E_0 = 0.441 eV, $\alpha_0 = 10\,\text{cm}^{-1}$. This phenomenon is also found in HgCdTe samples with different compositions. The values of E_0 and α_0 for the absorption edge focal points for samples with different compositions are listed in Table 4.3 (Chu et al. 1991).

In addition a sample with x = 0.19, has E_0=0. 0148 eV, $\alpha_0 = 10^{-4}\,\text{cm}^{-1}$. The fitted relation E_0 vs. x is found to be:

$$E_0(\text{eV}) = -0.355 + 1.77x \tag{4.114}$$

Similarly, it is found that there is a linear relation between $\ln\alpha_0$ and x:

$$\ln\alpha_0 = -18.5 + 45.68x \tag{4.115}$$

In (4.113), when $E = E_g$, then $\alpha = \alpha_g$, that is:

$$\alpha_g = \alpha_0 \exp[\sigma(E_g - E_0)/k_B T] \tag{4.116}$$

Table 4.3. The values of E_0 and α_0 of absorption edge focal points for HgCdTe alloys at different compositions

x	0.200	0.264	0.276	0.330	0.344	0.362	0.366	0.416	0.443
E_0(eV)	–0.0124	0.089	0.1178	0.255	0.214	0.285	0.304	0.397	0.441
α_0(cm^{-1})	1.5×10^{-4}	10^{-3}	2.4×10^{-3}	4×10^{-2}	10^{-4}	1.4×10^{-1}	3×10^{-1}	1.1	10

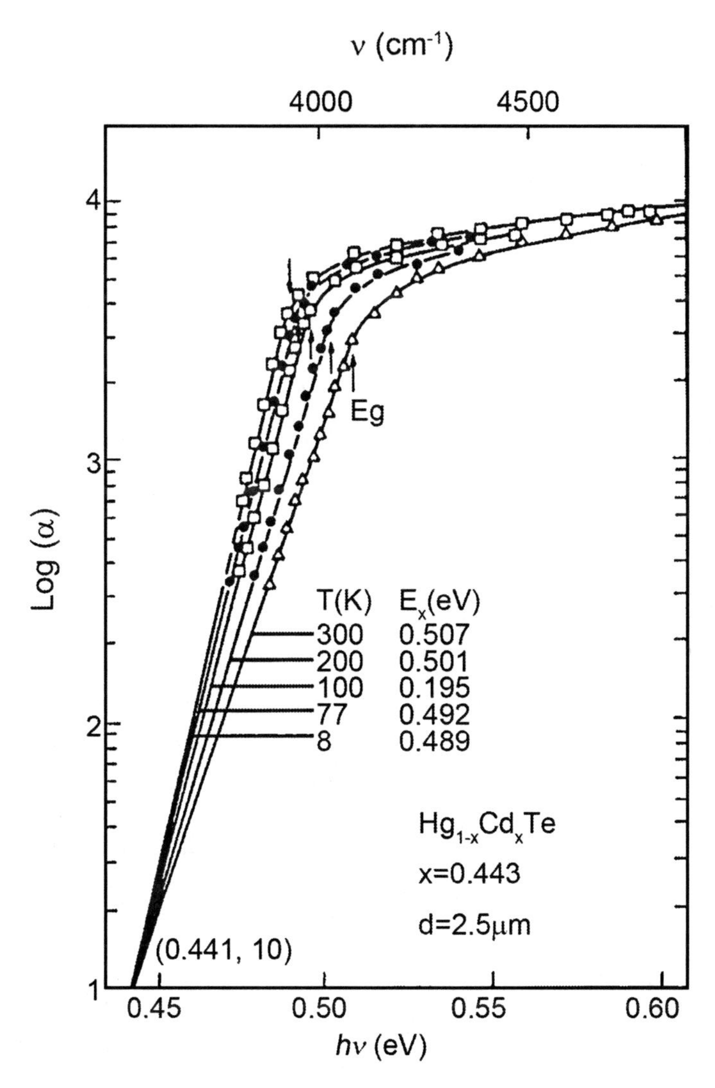

Fig. 4.15. The focal point of the absorption edges for a HgCdTe sample with composition $x = 0.443$

Table 4.4. E_g and $\alpha(E_g)$ for $Hg_{1-x}Cd_xTe$ samples with different composition at room temperature

x	0.170	0.200	0.226	0.276	0.330	0.366	0.416	0.443
E_g/eV	0.124	0.161	0.205	0.273	0.347	0.396	0.470	0.507
$\alpha(E_g)$ /cm^{-1}	1400	1600	1750	2150	2450	2650	2700	2850

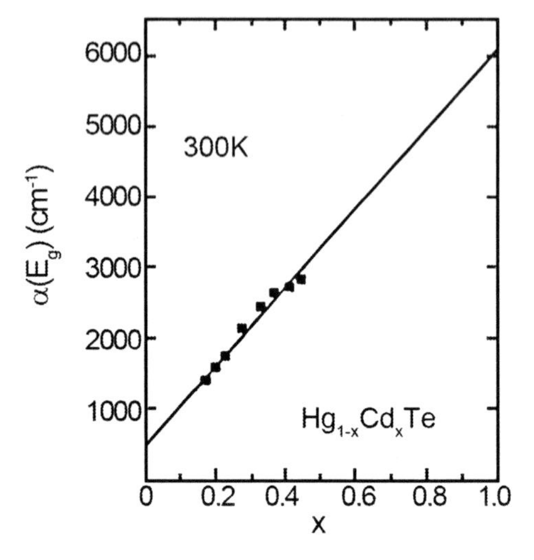

Fig. 4.16. The relation between $\alpha(E_g)$ and x at 300 K

From above formula, we can deduce the slope σ / kT to be:

$$\frac{\sigma}{k_B T} = \frac{\ln \alpha_g - \ln \alpha_0}{E_g - E_0} \tag{4.117}$$

in which E_g is given by (3.106), E_0 and $\ln \alpha_0$ are given by (4.114) and (4.115), respectively. α_g is the absorption coefficient when the photon energy is equal to the band gap energy. Then we can calculate the absorption edge slope at different temperatures for samples with different compositions using (4.117).

At room temperature, the experimental intrinsic absorption spectrum and the calculated absorption curve for HgCdTe samples with x = 0.170–0.443 are shown in Fig. 4.14. From Fig. 4.14, we can see that the absorption edges begin to change their slopes between 1,600~3,500 cm^{-1}. For these eight samples, the composition x, the band gap E_g at 300 K and the absorption coefficient $\alpha(E_g)$ at their band gap energies are listed in Table 4.4.

Based on Table 4.4, the relation between $\alpha(E_g)$ and composition x displayed in Fig. 4.16 can be fitted by the expression:

$$\alpha_g = 500 + 5{,}600x \tag{4.118}$$

This formula is applicable to samples at 300 K in the composition range, x = 0.170–0.443. If this formula is extrapolated to x = 1, then $\alpha_g = 6{,}100\,\text{cm}^{-1}$, which is consistent with data for the CdTe intrinsic absorption spectrum (Kireev 1978). At other temperatures, the relation between α_g and x is approximately:

$$\alpha_g = -65 + 1.88T + (8694 - 10.31T)x \tag{4.119}$$

Now collecting results, the fitted formula for the HgCdTe absorption edge when $E \le E_g$, can be written as:

$$\alpha = \alpha_0 \exp[\sigma(E - E_0)/k_B T] \tag{4.120}$$

in which:

$$\begin{aligned}
&\ln\alpha_0 = -18.5 + 45.68x \\
&E_0 = -0.355 + 1.77x \\
&\sigma / k_B T = (\ln\alpha_g - \ln\alpha_0)/(E_g - E_0) \\
&\alpha_g = -65 + 1.88T + (8694 - 10.31T)x \\
&E_g(x,T) = -0.295 + 1.87x - 0.28x^2 \\
&\qquad\qquad + (6 - 14x + 3x^3)(10^{-4})T + 0.35x^4
\end{aligned} \tag{4.121}$$

There are several different exponential expressions describing the absorption edge, and only the parameters are somewhat different.

The mechanisms responsible for the shape of the absorption edge are worth investigating. Many authors have proposed mechanisms responsible for band tails attached to the bottom of the semiconductor conduction band and the top of the valence band. The formation of band tails may be induced by the exciton–phonon interaction (Toyazawa 1959), or the effect of charged or neutral impunity scattering (Redfield 1963). Despite the fact that alloy scattering modifies the band gap it does not in lowest order contribute to band tails (Chen and Sher 1978). In narrow gap semiconductors, there are shallow impurity energy levels that are nearly degenerate with the conduction band edge. The impurity wave functions can couple to those of the conduction band to form an impurity band. The net effect is to broaden the conduction band and form a band tail extending into the band gap. Kane (1960) speculated that the DOS of this band tail is a Gaussian, but others found that it is a simple exponential distribution (Moss et al. 1973) of the form:

$$\rho \propto \exp[a(E-v)^n] \tag{4.122}$$

where a is a constant, v is the mean potential energy, and n is between $1/2$ and 2. Based on the observed HgCdTe exponential absorption edge, we can verify that the band tail of the bottom of the conduction band is an exponential distribution. Because the absorption coefficient is directly proportional to the conduction and valence bands joint DOS, and assuming the square of the optical absorption matrix element is approximately constant, what is known as the "Urbach rule" can be written as:

$$\rho = \rho_0 \exp[\sigma(E-E_0)/k_B T] \tag{4.123}$$

When $E = E_g$, then

$$_g = \rho_0 \exp[\sigma(E_g - E_0)/k_B T] \tag{4.124}$$

From this formula, we can solve for ρ_0 and substitute it into (4.123)

$$\rho = \rho_g \exp[-\sigma(E_g - E)/k_B T] \tag{4.125}$$

ρ_g is related to the real joint DOS at energy $E = E_g$, which can be estimated from the value of the absorption coefficient when the absorption coefficient begins to be dominated by fundamental processes.

Concentration fluctuations are another cause of band tails. These fluctuations are a natural consequence of statistics. There are small regions of space (clusters) with cation concentrations different from the average concentration x. These fluctuations are most prevalent in lattice-matched alloys like HgCdTe where there is no strain penalty associated with the existence of a region with a concentration differing from that of its surroundings (Muller and Sher 1999). In regions of $Hg_{1-x}Cd_xTe$ where the local x is smaller than the average the gap is narrowed, and the opposite is true in regions where the concentration exceeds the average. If the size of some of these clusters (or quantum dots) is smaller than that allowed to assign them a specific band gap, then the situation requires special averaging differing from the traditional local "molecular CPA" (MCPA) (Chen and Sher 1995). Treatment of these fluctuations, coupled to the effects of native point defects and impurities, present a challenging unresolved problem.

4.3.2 An Analytic Expression for the Intrinsic Absorption Band

Above the steep absorption edge at the low energy terminal of the intrinsic absorption spectrum, there is a slowly ascending intrinsic absorption band, known as the Kane region. This section focuses on the expression for the absorption coefficient in the Kane region.

When the photon energy is larger than E_g, the relation between the absorption coefficient and the frequency of the incident light can be calculated from the theory of optical transitions and the Kane model (Kane 1957; Chu 1983b; Chu et al. 1991, 1992a). Kane (1957) first devised this theory for the intrinsic absorption band, and applied it to InSb. He used the $k \cdot p$ method and a theory of optical transitions, and compared the results with experimental results. Blue (1964) extended the method to HgCdTe. He also measured the intrinsic absorption spectrum of HgCdTe, but because his samples had inhomogeneous compositions, the measured absorption edges were distorted. The absorption coefficient of HgCdTe measured by Scott (1969) ranged up to 2×10^3 cm^{-1}, which is still in the absorption edge region. Finkman and Nemirovsky (1979) also measured the absorption edge. These experimental results still did not constitute a comprehensive intrinsic absorption band data set of HgCdTe to which theory could be compared. After 1980, Chu (1983b) and Chu et al. (1991, 1992a) embarked on an extensive study obtaining experimental results about the intrinsic absorption band of $Hg_{1-x}Cd_xTe$ on well prepared samples with different composition x, over a wide range of temperatures. They also extended the theory of band-to-band transitions, and compared the two results.

The absorption coefficients for band-to-band transitions were presented in Sect. 4.2. Within the confines of the kane model, the calculation is rigorous, but it is not easy to reduce it into an analytic expression, without a significant extension. Anderson, based on the Kane model, deduced an approximate relation between α and $\hbar\omega$ in the intrinsic absorption region (Anderson 1980). The optical absorption above the absorption edge includes two contributions, one from the heavy hole band and another from the light hole band. Assuming the band gap is far smaller than the spin–orbit splitting ($E_{g«}$ «$\Delta\approx1$eV), and $2-\frac{E_g}{\hbar\omega}>>\frac{3}{4}\frac{\hbar^2}{m_0}\frac{\hbar\omega-E_g}{P^2}\left(1+\frac{m_0}{m_{hh}}\right)$, relations can be obtained between the absorption coefficient and photon energy:

$$\alpha_{lh} = \frac{1+2(E_g/\hbar\omega)^2}{137\sqrt{6}\sqrt{\varepsilon_\infty}} \cdot \frac{\sqrt{\hbar^2\omega^2 - E_g^2}}{4P} \cdot BM_{lh}$$

$$\alpha_{hh} = \frac{1}{137\sqrt{\varepsilon_\infty}} \cdot \frac{\sqrt{3/2}}{P} \sqrt{\hbar\omega(\hbar\omega - E_g)} \cdot BM_{hh} \left[1 + \frac{3}{4}\frac{\hbar^2 E_g}{m_0 P^2}\left(1+\frac{m_0}{m_{hh}}\right)\left(\frac{2\hbar\omega}{E_g} - 1\right)\right] \tag{4.126}$$

in which BM_{lh} and BM_{hh} are Burstein–Moss factors for the light hole band and heavy hole band, respectively, which depend on the Fermi level:

$$BM_{lh} = \frac{1-\exp\left(\frac{\hbar\omega}{k_B T}\right)}{\left[1+\exp\left(-\frac{\hbar\omega + E_g - 2E_F}{2k_B T}\right)\right]\left[1+\exp\left(-\frac{\hbar\omega - E_g + 2E_F}{2k_B T}\right)\right]}$$

$$BM_{hh} = \frac{1-\exp\left(\frac{\hbar\omega}{k_B T}\right)}{\left[1+\exp\left(-\frac{E_F + \left(\hbar^2 k_\omega^2/2m_{hh}\right)}{k_B T}\right)\right]\left[1+\exp\left(-\frac{\hbar\omega - E_F - \left(\hbar^2 k_\omega^2/2m_{hh}\right)}{k_B T}\right)\right]} \tag{4.127}$$

$$k_\omega^2 = \frac{\frac{4P^2}{3} + \frac{\hbar^2 E_g}{m_0}\left(1+\frac{m_0}{m_{hh}}\right)\left(\frac{2\hbar\omega}{E_g}-1\right)}{\frac{\hbar^4}{m_0^2}\left(1+\frac{m_0}{m_{hh}}\right)^2}\left[1 - \sqrt{1 - \frac{\frac{4\hbar^4}{m_0^2}\left(1+\frac{m_0}{m_{hh}}\right)^2 \hbar\omega(\hbar\omega - E_g)}{\left[\frac{4P^2}{3} + \frac{\hbar^2 E_g}{m_0}\left(1+\frac{m_0}{m_{hh}}\right)\left(\frac{2\hbar\omega}{E_g}-1\right)\right]^2}}\right]$$

In the above expressions m_0 is being used to represent the free electron mass rather than m which was used before. The total absorption coefficient is:

$$\alpha = \alpha_{lh} + \alpha_{hh} \tag{4.128}$$

It had been hoped there would be a simpler expression that could be used for practical applications. For direct gap semiconductor materials with parabolic energy bands, the intrinsic absorption coefficient was found to be proportional to the square root of the photon energy (Moss et al. 1973):

$$\alpha = A(E - E_g)^{1/2} \tag{4.129}$$

So it was first presumed this square root rule could also be applied to $Hg_{1-x}Cd_xTe$ material (Schacham and Finkman 1985; Sharma et al. 1994). Schacham and Finkman considered a relation between the absorption coefficient α and energy in the Kane region, it was (Schacham and Finkman 1985):

$$\alpha = \beta(E - E_g)^{1/2} \tag{4.130}$$

in which $\beta = 2.109 \times 10^5[(1+x)/(81.9+T)]^{1/2}$.

However, their experimental data were taken only near the absorption edge, and do not include the data higher in the intrinsic absorption band. Actually, the absorption coefficient calculated from this formula lies far above experimental values in the literature (Chu et al.1994). Moreover the calculated absorption coefficient is zero when $E = E_g$, which differs from the projection of the observed absorption edge to the point $E = E_g$.

Chu et al. measured the absorption spectrum of a series of thin $Hg_{1-x}Cd_xTe$ samples with different compositions, which not only includes the absorption edge, but also the intrinsic absorption band (Chu et al. 1991, 1992a). Sharma, based on the Chu's experimental data (Chu et al. 1994), proposed a modified square root rule:

$$\alpha = \beta(E - E_g^{'})^{1/2} \tag{4.131}$$

In this formula $E_g^{'} = E_g - (E_g - E_0)/2/\ln(\alpha_g/\alpha_0)$, β is a fitting parameter, and $E_g^{'}$ is an effective band gap. This modification does not require α to vanish when $E = E_g$.

Chu pointed out (Chu et al. 1994), that an exponential version of the square root rule:

$$\alpha = \alpha_g \exp[\beta(E - E_g)]^{1/2} \tag{4.132}$$

fits the experimental date well if the parameter β depends on composition, x, and temperature, T, as follows:

$$\begin{aligned} T &= 300K \qquad \beta = 24 - 18x \\ T &= 77K \qquad \beta = 5.4 + 11x \end{aligned} \tag{4.133}$$

Supposing the relation of β to x, and T is linear, then using an interpolation method, the follow formula is obtained:

$$\beta(T,x) = -1 + 0.083T + (21 - 0.13T)x \tag{4.134}$$

The rational for (4.132) is:

- When $E = E_g$, then $\alpha = \alpha_g$, so the intrinsic absorption section and band tail section link up at E_g.
- When $E > E_g$, and the exponent is still small the exponential can be expanded with only the first correction term retained. This term is proportional to the square root of energy $(E - E_g)^{1/2}$. In this energy range (4.132) has the same functional dependence as (4.131) but with a constant offset.
- At slightly higher energies the functional dependence of α also has a term proportional to $(E - E_g)^1$, mocking the region where the conduction band becomes linear in k.
- At still higher energies the exponential must break down, since it has no justification in physical mechanisms.

Figures 4.17a, b are the intrinsic absorption spectra of the $Hg_{1-x}Cd_xTe$ samples with compositions in the range $x = 0.170$–0.443, at 300 K and 77 K, respectively. In the figures, the dots are the measured data of Chu et al. (1991, 1992a); the absorption edges are calculated from (4.120) and (4.121); the solid lines are calculated from the exponential square root rule (4.132), and conforms best with the experimental results; and the dash lines are calculated from Sharma's square root rule (4.131), and obviously lie higher than the experimental values. The lower dashed dot line was calculated from Anderson's expression (4.125). The parameters chosen were: the momentum matrix element $P = 8 \times 10^{-8}$ eV cm, the heavy hole effective mass $m_{hh} = 0.55 m_0$; the high frequency dielectric constant $\varepsilon_\infty = 15.2 - 15.5x + 13.76x^2 - 6.32x^3$; and the Burstein–Moss factors BM_{lh} and BM_{hh} were both supposed to be 1 (the non-degenerate case).

As the figure shows, the Kane plateau calculated from the square root rule, (4.131), and the exponential square root rule, (4.132), link up well with the Urbach band tail calculated from (4.120). The absorption curve calculated from (4.126) descends rapidly near E_g, because the band tail effect in Anderson's model is ignored. In addition, at room temperature, the intrinsic absorption coefficient calculated from (4.126), (4.131), and (4.132) conform well with the experimental results. At 77 K, the intrinsic absorption coefficient calculated from the Anderson model and the exponential square root rule of Chu et al. still fits the experimental values, but the result of the Sharma square root rule has large deviations from the measured results. Because the square root relation between the absorption coefficient and the energy are only expected to model the behavior for parabolic band semiconductors, and not that for non-parabolic band semiconductors such as $Hg_{1-x}Cd_xTe$, it is not surprising that it fails at low concentrations and temperatures.

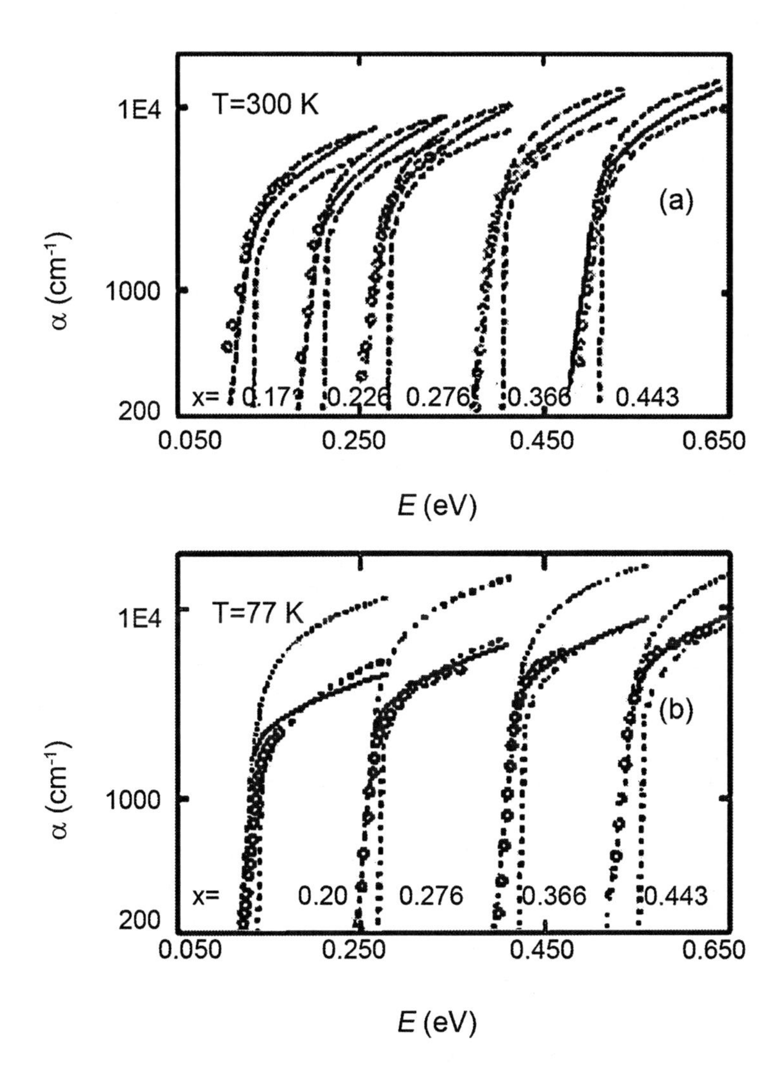

Fig. 4.17. The intrinsic absorption spectra of $Hg_{1-x}Cd_xTe$ samples with compositions ranging from x = 0.17 to 0.443 at 300K (Fig. 4.17a) and 77K (Fig. 4.17b). In the figure, the *dots* are the measured data of Chu et al. (1991, 1992a); the *dot-dashed curves*, the *dashed curves*, and the *solid line curves* are the intrinsic absorption coefficients calculated from (4.126), (4.131), (4.132), respectively. The dashed line, located at energies smaller than E_g, are Urbach absorption tails calculated from (4.120).

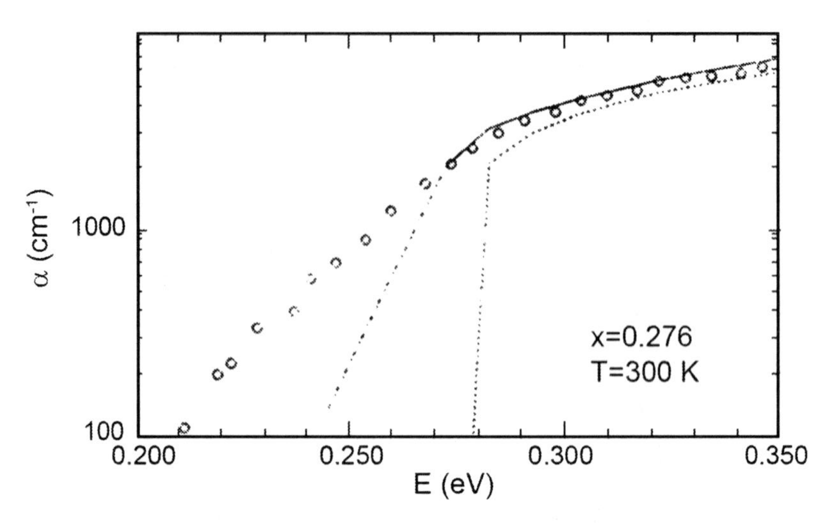

Fig. 4.18. The absorption curve of a $Hg_{0.724}Cd_{0.276}Te$ liquid phase epitaxial grown film sample at room temperature. The thickness of the sample is 5 μm . The experimental data, dots, are those of Mollmann and Kissel (1991); the *dashed line* below E_g is the Urbach absorption edge calculated from (4.120); and the *dashed line* and *solid line* above E_g are the intrinsic absorption spectra calculated from (4.126) and (4.131), respectively

It is well-known that thin film material can be conveniently fabricated using MBE, MOCVD, and LPE etc. epitaxial methods. However, these $Hg_{1-x}Cd_xTe$ epitaxial films are generally compositionally inhomogeneous through their depth. This causes the absorption edge to broaden. Therefore, we cannot obtain an ideal intrinsic absorption spectrum from these epitaxial films.

An example of an absorption curve taken at room temperature for an LPE grown $Hg_{1-x}Cd_xTe$ ($x = 0.276$) film is shown in Fig. 4.18. In the figure, the experimental dots are the measured data of Mollmann and Kissel (1991), the dashed line below E_g is the Urbach absorption band tail calculated from (4.120), and the dashed line and the solid line above E_g are the intrinsic absorption spectra calculated from (4.126), and (4.132), respectively. From the figure, we can see that in the Urbach absorption region, the calculated curve does not coincide with the experimental dots. The measured absorption edge is broadened due to the compositional inhomogeneity through the depth of the sample. In the Kane region, the Anderson model and the empirical formula of Chu et al. conform well to the experimental date. This result indicates that the exponential rule between the intrinsic absorption band and energy, (4.132), can still be used to extract fitted parameters for LPE films in the Kane region.

4.3.3 Other Expressions for the Intrinsic Absorption Coefficient

Most first principles energy band structures of semiconductors are calculated at 0 K. Extensions to finite temperature are complex. Consequently it has proved to be useful to augment these theories with empirical data. One of the better experimental techniques for gathering helpful data is measurements of absorption coefficients. To extract the information from the measurements, models are employed. Here one such method not treated previously is introduced.

Nathan (1998) devised another expression for the intrinsic absorption coefficient. It is based on Anderson's method and the experimental results of Chu. For direct band gap semiconductors, the absorption coefficient of electromagnetic radiation with angular frequency ω can be expressed as:

$$\alpha(\omega)=\frac{\sqrt{\varepsilon_\infty}}{c}\int w(k)\left[\frac{1}{1+\exp[(E_v(k)-E_F)/kT]}-\frac{1}{1+\exp[(E_c(k)-E_F)/k_BT]}\right]\frac{2\mathcal{V}d^3k}{(2\pi)^3} \tag{4.135}$$

where $\mathcal{V}$ is the volume of the solid, and $w(\mathbf{k})$ is the transition probability per unit time:

$$w(\mathbf{k})=\frac{2\pi}{\hbar}\left|H_{vc}\right|^2\delta[E_c(\mathbf{k})-E_v(\mathbf{k})-\hbar\omega] \tag{4.136}$$

c is the velocity of light in vacuum, $\hbar$ is Planck's constant divided by 2π, ε_∞ is the dielectric constant at high frequency, k_B is Boltzmann's constant, E_F is the Fermi energy, k is the wave vector, and T is the temperature. H_{vc} is the electron photon interaction Hamiltonian given by:

$$H=\frac{e}{mc}\mathbf{A}\cdot\mathbf{p} \tag{4.137}$$

where $\mathbf{A}$ is the vector potential of the incident radiation, $\mathbf{p}$ is the electron momentum operator, m is the free electron mass, and e is the electric charge. To calculate the integral in (4.136), we need to know the functional forms of $E_c(\mathbf{k})$ and $E_v(\mathbf{k})$. To simplify the analytic result, Nathan adopted the expressions of Keldysh (1958b):

$$E_v(\mathbf{k}) = -\frac{E_g}{2}\left(1 + \frac{\hbar^2 k^2}{\mu^* E_g}\right)^{0.5}$$
$$E_c(\mathbf{k}) = \frac{E_g}{2}\left(1 + \frac{\hbar^2 k^2}{\mu^* E_g}\right)^{0.5} \tag{4.138}$$

where the zero of energy is taken at the center of band gap, and μ^* is the reduced effective mass of the conduction band and the valence band. Then by simplifying, we obtain Nathan's expression:

$$\alpha(\omega) = \frac{2e^2}{c\hbar^2}\sqrt{\frac{\mu^* E_g}{\varepsilon_\infty}}\sqrt{\left(\frac{\hbar\omega}{E_g}\right)^2 - 1} \cdot BM \tag{4.139}$$

where BM is Burstein–Moss offset, which is given by the formula:

$$BM = \left\{\left[1 + \exp\left[\frac{-\hbar\omega - 2E_F}{2k_B T}\right]\right]^{-1} - \left[1 + \exp\left[\frac{\hbar\omega - 2E_F}{2k_B T}\right]\right]^{-1}\right\}^{-1} \tag{4.140}$$

in which the Fermi energy E_F is:

$$E_F = \frac{E_c + E_v}{2} + \frac{3k_B T}{4}\ln\left(\frac{m_v^*}{m_c^*}\right) \tag{4.141}$$

Equations (4.139) and (4.140) are the theoretical analytic expressions for calculating the intrinsic absorption coefficient of direct band gap semiconductors. This expression is simpler than the theoretical formula of Anderson.

To test this calculated result, Nathan compared its predictions with the experimental results of Chu et al. Figure 4.19a–c for the $Hg_{1-x}Cd_xTe$ intrinsic absorption spectra with compositions $x = 0.265$, 0.33 and 0.433 at 300 and 4.2 K. The result of this comparison shows Nathan's expression can be applied.

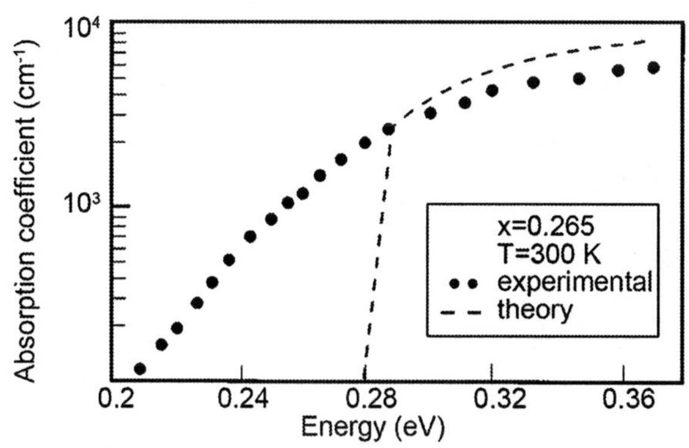

Fig. 4.19a. The experimental and calculated absorption spectrum of a sample with $x = 0.265$ at 300 K

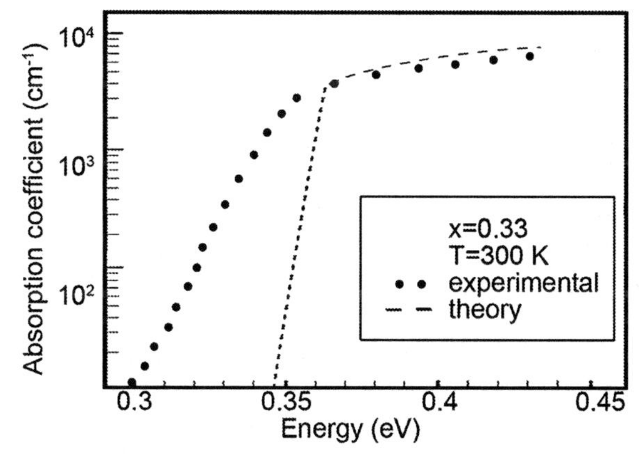

Fig. 4.19b. The experimental and calculated absorption spectrum of a sample with $x = 0.33$ at 300 K

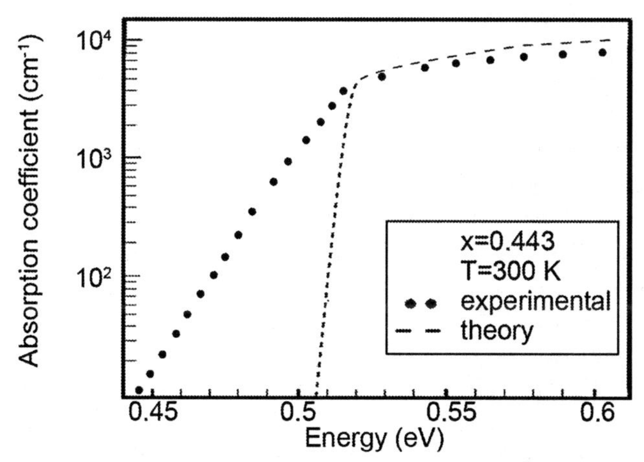

Fig. 4.19c. The experimental and calculated absorption spectrum of a sample with $x = 0.443$ at 300 K

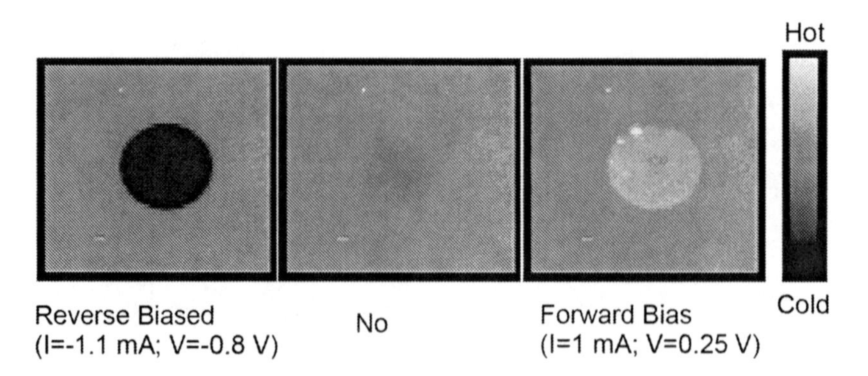

Fig. 4.20. Thermal images of a 1 mm array under (**a**) saturated reverse bias; (**b**) zero bias; and (**c**) forward bias conditions using a 64×64 InSb camera with a 3–5 mm bandpass cold filter. The *grey bar* on right indicates the apparent temperature. (From Lindle et al. 2004)

Recently, the infrared negative luminescence phenomenon has been investigated for a HgCdTe p–n junction. This phenomenon occurs when an object and its surroundings are thermally isolated from the rest of the universe. Then if the object absorbs net radiation from its surroundings the surroundings will cool. To treat this phenomenon quantitatively the intrinsic absorption expression is used. For a photo-diode in radiative thermal contact with its surroundings, in equilibrium where the diode is unbiased, the radiation it absorbs from its surroundings is equal to its emitted radiation. If the diode is forward biased, the electron and hole recombination rate increases, and the emitted radiation will increases. The diode is an LED, and the surroundings, when it absorbs this radiation, are heated. If reverse bias is applied, the diode will absorb net radiative energy from the surroundings to cool the surroundings. Lindle et al. (2003a,b, 2004) and Bewley et al. (2003) investigated this phenomenon for shortwave- and midwave-infrared HgCdTe photo-diodes. Figure 4.20 shows thermal image pictures for a diode at forward bias and reverse bias. It is clear that the diode is a cooling source at reverse bias, and a heating source at forward bias. This phenomenon can be applied to the cold shielding of focal plane arrays, sources for IR spectroscopy, and IR scene projection.

Lindle et al. measured and calculated the negative luminescence radiation for a diode. Figure 4.21 shows the measured negative luminescence spectrum (solid line) and the calculated spectrum (dashed line) at 296 K under a saturation current of 3.3 mA. In the calculation the black body Planck radiation law and emissivity formula were used. The negative luminescence radiation is just the product of the black body radiation $M(\lambda,T)$ and the emissivity $\varepsilon(\lambda,T)$:

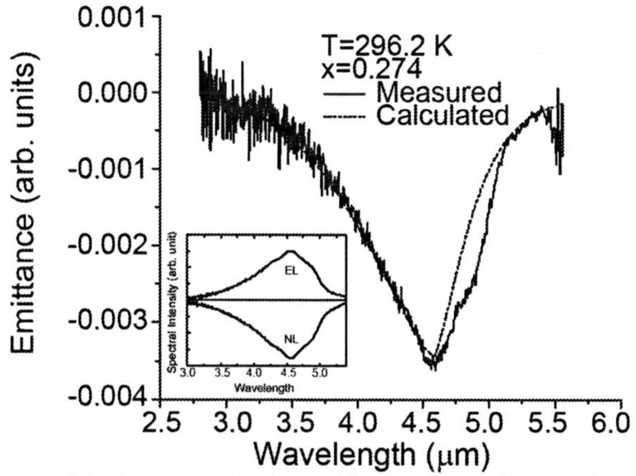

Fig. 4.21. Measured (*solid*) and calculated (*dashed*) negative luminescence spectra at 296 K and 3.3 mA of current (saturated). The inset illustrates the forward-bias (EL) and reverse-bias (NL) emission spectra at the same current (0.06 mA) below saturation (Lindle et al. 2004)

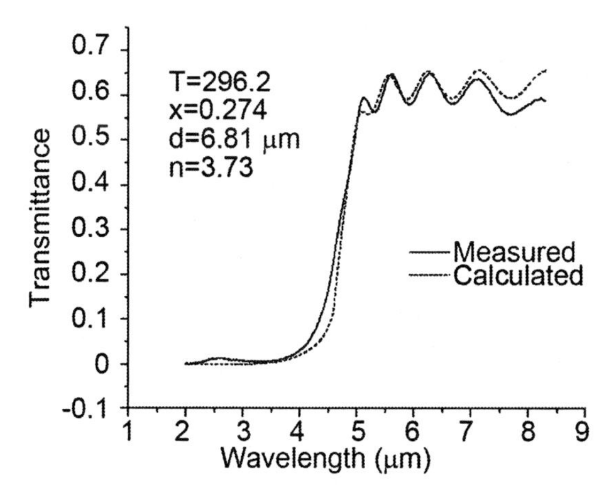

Fig. 4.22. Measured (*solid*) and calculated (*dashed*) spectral dependences of the device's transmission at 296 K (Lindle et al. 2004)

$$\varepsilon(\lambda,T) = (1 - R(\lambda,T))(1 - e^{-\alpha(\lambda,x,T)d}) \tag{4.142}$$

$$M(\lambda,T) = \frac{2\pi hc^2}{\lambda^5}\frac{1}{e^{hc/\lambda kT} - 1} \tag{4.143}$$

In (4.142) and (4.143), R is the reflectance, d is the sample thickness, λ is the wavelength, T is the temperature, h is Planck's constant, and k is Boltzmann's constant. There is one additional parameter that enters

$\alpha(\lambda,x,T)$, the intrinsic absorption coefficient given by Chu's exponential square root rule expression (4.132). Figure 4.21 shows that the calculated results agree with the measurements. Figure 4.22 shows the agreement between the measured and calculated transmittance curves using (4.132).

Much research has been done on the optical constant, the absorption coefficient, the absorption band tail, the BM offset, and the influence of the absorption spectrum on technology, e.g. ion inter-diffusion at the interface following ion implantation (Viña et al. 1984; Gering et al. 1987; Herrmann et al. 1993a; Mao et al. 1998; Huang et al. 2000; Djurisic and Li 1999; Djurisic et al. 2002).

4.4 Direct Measurements of Optical Constants

4.4.1 Introduction

The optical constants of materials can be deduced from transmission and reflection measurements, but they are indirect measurement methods. The optical constants can be measured directly by spectroscopic ellipsometry, an important technique for the measurement of narrow gap semiconductors and other functional materials.

Spectroscopic ellipsometry became popular at the end of the last century. It is nondestructive, sensitive to surface layers, and can be operated under all kinds of environments, such as air, vacuum and high pressure. A traditional ellipsometer was operated in an inconvenient null configuration. From the 1960's, photoelectric technology became better developed, and some of the new detectors were incorporated into ellipsometers. Their operation is now much easier and their sensitivity has improved greatly. Different kinds of configurations have been devised, tailored for specific applications. In addition, the incorporation of computers has increased their efficiency. Now, ellipsometry is applied widely, to fundamental physics, chemistry, and material science studies, to photographic science, to biology, and to fields related to optics, electrics, metallurgy, and biomedicine.

Usually, an ellipsometer operates under ex situ conditions, where the surface of the samples is held steady. For example, Alterovitz et al. (1983) investigated Si_3N_4 thin films deposited by magnetic sputtering on GaAs substrates, and GaAs–AlGaAs multilayers using a variable angle spectroscopic ellipsometer. Merkel et al. (1989) investigated GaAs/AlGaAs superlattices. Aspnes and Studna (1983) investigated the optical constants of HgCdTe with oxide surface over layers prepared using different processing methods. Orioff et al. (1994) investigated hydrogenized HgCdTe surfaces. Using infrared spectroscopic ellipsometry, the optical constants of HgCdTe alloys above the band gap energy can be obtained.

Recently, ellipsometers to monitor in situ processing have been developed. They can monitor surface corrosion processes, growth procedures, and so on. For example, in recent years ellipsometers have become standard attachments to epitaxial growth systems. The related research is mainly focused on the MBE growth of group III–V and group II–VI compounds and alloys.

Maracas et al. (1992) monitored the growth of GaAs using a spectroscopic ellipsometer with a 250–1,000 nm wavelength range. The ellipsometric parameters ψ and Δ reliably follow the changes of all processes. From the measured optical constants and substrate temperature, the actual surface growth temperature was determined. By fitting the interference fringes, the growth rate was obtained. The difference in the growth rate determined this way and by RHEED oscillations was 10%. An advantage of ellipsometry is that it can measure growth rates under both layered and island growth conditions, but RHEED oscillations can be observed only under island growth condition. In addition, ellipsometry is also useful when the substrate is rotating. Furthermore, it also enables determination of interface changes of quantum wells.

Several teams have investigated HgCdTe MBE growth processes using an ellipsometer. Demay et al. (1987) in France initiated the study. They introduced polarized light from one window of RIBER 2300P cavity with an incident angle of 70° to the sample surface, the reflected light from the sample then reflects from a mirror to exit through the same window in the cavity. They studied the inter-diffusion of CdTe/HgTe, and determined that: at 260 K, when a CdTe layer is growing over an underlying HgTe layer of a (111) growth surface, the inter-diffusion rate bears no relation to the Hg partial pressure. In the other extreme, when a HgTe layer is covering a CdTe layer, the concentration of Hg holes in the growing island, which depends on the Hg partial pressure, affects the inter-diffusion. In 1992, Hartley et al. (1992) in Australia monitored the growth of HgCdTe on a ZnCdTe substrate, and a CdTe/HgTe superlattice using a monochromic polarization-modulated ellipsometer. The key element is the polarization modulation. It is accomplished with a piezoelectric quartz crystal energy converter that is bonded to quartz and driven by an alternating voltage. A mono-axial sinusoidal strain standing wave envelope is built in the rectangular quartz glass. The oscillating strain induces an oscillating birefringence, which causes the quartz to act as a linear retarder. Its delay varies with time, the direction of the strain determines the direction of the fast and slow axis, the strain is proportional to the amplitude of the sinusoidal envelope, and so to the voltage on the energy converter.

The ellipsometer parameters the Hartley group observed under their growth conditions indicated a variation of the surface topography as well as the composition. Their measured composition deviation was about ±0.003. In 1996, Bevan et al. (1996) at Texas Instruments Company measured a composition deviation of ±0.0015 using multi-wavelength 400–850 nm modulated spectroscopic ellipsometer. Benson et al. (1996) at the US Night Vision Electronic Sensor Laboratory investigated the composition variation, growth rate, and surface cleanliness of HgCdTe alloys using a rotating analyzer ellipsometer with multi-wavelength capabilities.

4.4.2 The Principles of Spectroscopic Ellipsometry

In the theory of electromagnetic fields, an electromagnetic wave is described using four basic field vectors: the electric-field intensity E, the electric displacement vector D, the magnetic intensity H and the magnetic flux density B. The direction of the electric field vector E defines the polarization direction of the light. According to the Maxwell equations, if we know the direction of E and the diamagnetic characteristics of the media in which the light is propagating, the directions of the other three vectors can be calculated. For a normal light source, where electrons transition from an excited to a lower atomic level by spontaneous emission, the polarization, and the emitted directions are random.

Any polarization can be decomposed to two mutually perpendicular polarizations, or into two opposite circularly rotating polarizations. To describe a monochromatic propagating light plane wave, its frequency, direction, amplitude, phase, and polarization must be known. This information is captured in the ellipsometric parameters ψ and Δ.

$$\rho = \frac{r_p}{r_s} = \frac{|r_p|}{|r_s|}\exp(i\delta_p - i\delta_s) = \tan\psi \exp(i\Delta) \tag{4.144}$$

where p and s denote respectively the directions parallel and perpendicular to the incident plane, Δ is the phase difference between the p and the s components, and $\tan\psi$ is the ratio of the amplitudes r_p and r_s.

We can use the Jones matrix to describe the polarization of the transmitted and reflected light of a sample. For example, the Jones matrix for the transmission through a phase retarder is:

$$T = \begin{bmatrix} e^{-j\delta_1} & 0 \\ 0 & e^{-j\delta_2} \end{bmatrix} \tag{4.145}$$

with

$$\delta_1 = \frac{2\pi n_e d}{\lambda}, \delta_2 = \frac{2\pi n_0 d}{\lambda}.$$

When the light passes through a series of optical components, the Jones matrix connecting the incident and exiting polarization states is the product of Jones matrices of all the optical components.

Early ellipsometry operated in a null configuration. The principle is to find the azimuth angles of a polarizer, a compensator, and an analyzer (PCA) to minimize the photocurrent of the detector. From the azimuth angles of the polarizer and analyzer, ellipsometric parameters ψ and Δ can be deduced. A null ellipsometer is simple in principle, and was popular before the 1970s. But the measurement procedure is complicated, not easy to automate, and the null condition reached is only minimal, not strictly zero. Consequently the precision is not high. The advent of photometric ellipsometers has greatly improved the sensitivity (Cahan and Spanier 1969; Hauge and Dill 1973; Aspnes 1974). These instruments can scan wavelength, and determine complete materials optical constants (real and imaginary parts). They can be automated easily by rotating the optical components, such as the polarizer, and the analyzers.

Consider an RAE (rotating analyzer ellipsometer) system, consisting of a monochromator, a polarizer, a sample, an analyzer, and a detector. Assume the s and p axis are perpendicular and parallel to the incident plane, respectively, and with P being the azimuth angle of the polarizer, and A being the azimuth angle of analyzer, both relative to the s axis (the light propagation direction). Assume all components are ideal, and the optical system is in perfect alignment. The Jones matrix of the light is:

$$[\mathbf{E}_D] = [\tilde{A}][\tilde{R}(A)][\tilde{S}][\tilde{R}(P)][\mathbf{E}_i] \tag{4.146}$$

$\mathbf{E}_D$ is the electric field at the detector, $\tilde{R}$ is the angle transfer matrix, $\tilde{S}$ is the sample matrix, $\tilde{A}$ and $\tilde{P}$ are the matrices of the polarizer and the analyzer.

$$E_D = \begin{pmatrix} 1 & 0 \\ 0 & 0 \end{pmatrix} \begin{pmatrix} \cos A & \sin A \\ -\sin A & \cos A \end{pmatrix} \begin{pmatrix} r_s & 0 \\ 0 & r_p \end{pmatrix} \begin{pmatrix} \cos P & -\sin P \\ -\sin P & \cos P \end{pmatrix} \begin{pmatrix} E_i \\ 0 \end{pmatrix} \tag{4.147}$$

The intensity at the detector is:

$$I = [E_D][E_D]^* = \frac{I_0}{2}\{|r_p|^2 \cos^2 P \cos^2 A + |r_s|^2 \sin^2 P \sin^2 A \\ +(r_p r_s^* + r_p^* r_s)\sin P \cos P \sin A \cos A\} \tag{4.148}$$

$$I(t) = \frac{I_0}{2}\left(|r_s|^2 \cos^2 P + |r_p|^2 \sin^2 P\right)\left[1 + \alpha\cos(2A) + \beta\sin(2A)\right], \tag{4.149}$$

here

$$A = A(t) = \omega_0 t + \theta \tag{4.150}$$

where I_0 is the incident light flux, r_p and r_s are the electric field components parallel and perpendicular to the incident plane, respectively. α and β are normalized Fourier coefficients, ω_0 is the optical circular frequency (twice that of the light frequency), and θ is an arbitrary phase factor.

From the Jones matrix analyses of the RAE ellipsometer system (Azzam and Bashara 1977), the following expressions are obtained:

$$\alpha = \frac{1 - \tan^2\psi \tan^2 P}{1 + \tan^2\psi \tan^2 P}, \\ \beta = \frac{2\tan\psi \cos\Delta \tan P}{1 + \tan^2\psi \tan^2 P}. \tag{4.151}$$

where ψ and Δ are ellipsometric parameters. α and β are obtained by a Fourier transform of the light intensity:

$$\alpha = \frac{2}{N}\sum_{t=1}^{N} I(t)\cos(2A(t)), \\ \beta = \frac{2}{N}\sum_{t=1}^{N} I(t)\sin(2A(t)). \tag{4.152}$$

Then ψ and Δ can be calculated as follows:

$$\tan\psi = \sqrt{\frac{1-\alpha}{1+\alpha}}\frac{1}{|\tan P|}, \\ \cos\Delta = \frac{\beta}{\sqrt{1-\alpha^2}}\frac{\tan P}{|\tan P|}. \tag{4.153}$$

The above ellipsometric parameters are measured. On the other hand, the ellipsometric data can be calculated from sample structures from Fresnel formulas. In the following, we will give an example of an epitaxial film on a thick substrate, to illustrate the relation between ψ and Δ optical constants, and film thickness of sample structures.

When light is incident on a multi-layered structured sample, part of the light is reflected and part is transmitted at each interface. The reflected electric field components of the light waves after different multiple passes can be denoted as follows:

$$\begin{aligned}\tilde{E}_1^r &= \tilde{r}_{01}\tilde{E}_0\\ \tilde{E}_2^r &= \tilde{t}_{10}\tilde{t}_{01}\tilde{r}_{12}e^{-j2\beta}\tilde{E}_0\\ \tilde{E}_3^r &= \tilde{t}_{10}\tilde{t}_{01}\tilde{r}_{10}\left(\tilde{r}_{12}\right)^2 e^{-j4\beta}\tilde{E}_0\\ &\cdots\cdots\\ \tilde{E}_n^r &= \tilde{t}_{10}\tilde{t}_{01}\left(\tilde{r}_{10}\right)^{n-2}\left(\tilde{r}_{12}\right)^{n-1} e^{-j(2n-2)\beta}\tilde{E}_0\end{aligned} \tag{4.154}$$

β is the phase thickness of the film,

$$\beta = 2\pi\tilde{n}_1\frac{d}{\lambda}\cos\tilde{\phi}_1 = 2\pi\frac{d}{\lambda}\sqrt{\tilde{n}_1^2 - \tilde{n}_0^2\sin^2\phi_0} \tag{4.155}$$

and ϕ_1 is the refracted angle of the light in the film n_1. The total reflected field is the sum of the above components:

$$\tilde{E}_R^r = \left[\tilde{r}_{01} + \tilde{t}_{10}\tilde{t}_{01}\exp(j2\beta)\sum_{n=2}^{\infty}\left(\tilde{r}_{10}\right)^{n-2}\left(\tilde{r}_{12}\right)^{n-2}\exp(-j2n\beta)\right]\tilde{E}_0 \tag{4.156}$$

where r_{01} ... and t_{01} ... are the complex reflection coefficient and transmission coefficient amplitudes at the different interfaces. Referenced to interface normal direction they have the properties:

$$\begin{aligned}\tilde{r}_{01} &= -\tilde{r}_{10}\\ \tilde{t}_{10}\tilde{t}_{01} &= 1 - \tilde{r}_{01}^2\end{aligned} \tag{4.157}$$

After a simple calculation one finds:

$$\tilde{E}_R^r = \left(\frac{\tilde{r}_{01} + \tilde{r}_{12}e^{-j2\beta}}{1 + \tilde{r}_{01}\tilde{r}_{12}e^{-j2\beta}}\right)\tilde{E}_0 \tag{4.158}$$

So the reflection coefficient amplitude is:

$$r = \frac{\tilde{E}_R^r}{\tilde{E}_0} = \frac{\tilde{r}_{01} + \tilde{r}_{12} e^{-j2\beta}}{1 + \tilde{r}_{01}\tilde{r}_{12} e^{-j2\beta}} \tag{4.159}$$

A version of the above formula is useful for both the p and the s components once it is modified to take account of effects at the front surface. For example, for the interface between the film and air, the reflection and transmission coefficient amplitudes of the p and s components are as follows:

$$\begin{aligned}
\tilde{r}_{01p} &= \frac{\tilde{n}_1 \cos\phi_0 - \tilde{n}_0 \cos\tilde{\phi}_1}{\tilde{n}_1 \cos\phi_0 + \tilde{n}_0 \cos\tilde{\phi}_1} \\
\tilde{r}_{01s} &= \frac{\tilde{n}_0 \cos\phi_0 - \tilde{n}_1 \cos\tilde{\phi}_1}{\tilde{n}_0 \cos\phi_0 + \tilde{n}_1 \cos\tilde{\phi}_1} \\
\tilde{t}_{01p} &= \frac{2\tilde{n}_0 \cos\phi_0}{\tilde{n}_1 \cos\phi_0 + \tilde{n}_0 \cos\tilde{\phi}_1} \\
\tilde{t}_{01s} &= \frac{2\tilde{n}_0 \cos\phi_0}{\tilde{n}_0 \cos\phi_0 + \tilde{n}_1 \cos\tilde{\phi}_1}
\end{aligned} \tag{4.160}$$

First calculate $\tilde{r}_p$ then use the following relation:

$$\rho = \frac{\tilde{r}_p}{\tilde{r}_s} = \frac{|\tilde{r}_p|}{|\tilde{r}_s|} \exp(i\delta_p - i\delta_s) = \tan\psi \exp(i\Delta) \tag{4.161}$$

from which predicted ψ and Δ values are deduced. In the analysis of experimental ellipsometric data, we build a sample structure model, then using the above methods attempt to find parameters that fit its predictions to the experimental ψ and Δ data. In this way refractive indices and film thicknesses are determined.

4.4.3 Operational Configuration

Different ellipsometer operational configurations are used to obtain the ellipsometric parameters. For example, Chen et al. (1994) developed an improved ellipsometry technique with a rotating analyzer and polarizer (RAP) with a rotation rate ratio of 2:1. It has the advantage that the data are self-consistent, the DC component is not involved, and the calibration is simple. In application, data precision is important because it is often differentiated one or several times to deduce details of the structures.

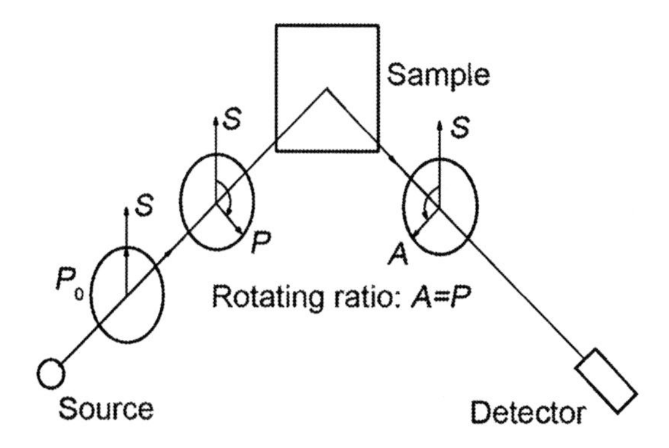

Fig. 4.23. Schematic of an FPRPRA for infrared ellipsometry

In order to increase the data precision, the sampling repetitions (1,000 times at each wavelength) (Aspnes and Studna 1975), or the number of data points per repetition (up to 10,000 points per circle) (Chen et al. 1994) are increased. But in the infrared region, where the detector detectivities and light source intensities are lower than those in the visible and near UV region, it is essential to improve the S/N ratio of the operational configuration, Huang et al. (2000) developed an ellipsometry configuration by rotating polarizer and analyzer with 1:1 rotation ratio (FPRPRA). It has a higher S/N ratio leading to a higher precision at a given measuring rate, or faster measuring rate for a given precision.

Figure 4.23 is a schematic of the (Fixed Polarizer, Rotating Polarizer, and Rotating Analyzer) FPRPRA system. It starts with a polarizer having a fixed P_0 azimuth angle perpendicular to the incident plane to produce linearly polarized light.

Assuming all components are ideal, then the electric field at the detector is:

$$\begin{aligned} E_f &= \begin{bmatrix} 1 & 0 \\ 0 & 0 \end{bmatrix} \begin{bmatrix} \cos A & \sin A \\ -\sin A & \cos A \end{bmatrix} \begin{bmatrix} \tilde{r}_s & 0 \\ 0 & \tilde{r}_p \end{bmatrix} \begin{bmatrix} \cos P & -\sin P \\ \sin P & \cos P \end{bmatrix} \begin{bmatrix} 1 & 0 \\ 0 & 0 \end{bmatrix} \begin{bmatrix} \cos P & \sin P \\ -\sin P & \cos P \end{bmatrix} \begin{bmatrix} 1 \\ 0 \end{bmatrix} E_0 \\ &= \begin{bmatrix} \tilde{r}_s \cos A \cos^2 P + \tilde{r}_p \sin A \sin P \cos P \\ 0 \end{bmatrix} E_0 \end{aligned} \quad (4.162)$$

so the light intensity at the detector is:

$$\begin{aligned} I = \left|E_f\right|^2 = I_0 \left|r_s\right|^2 (\cos^2 A \cos^4 P + \frac{1}{4}\rho_0^2 \sin^2 A \sin^2 2P \\ + \frac{1}{2}\rho_0 \cos \Delta \sin 2A \sin 2P \cos^2 P) \end{aligned} \quad (4.163)$$

In (4.163) ρ_0 and Δ are defined in terms of the reflection coefficient amplitudes:

$$\rho = \frac{\tilde{r}_p}{\tilde{r}_s} = \frac{r_p}{r_s}\exp(i\Delta) = \rho_0 \exp(i\Delta) = \tan\psi \exp(i\Delta) \tag{4.164}$$

Consider a FPRPRA with $A = P = \omega_0 t$, and a rotating polarizer and a rotating analyzer with 1:1 speeds. Using three ac components ($2\,\omega_0$, $4\,\omega_0$, and $6\,\omega_0$), the ellipsometric parameters ψ and Δ can be determined uniquely (Huang and Chu 2000)

Then

$$\begin{aligned} I &= I_d + I_2\cos 2A + I_4\cos 4A + I_6\cos 6A \\ &= I_d + I_2\cos 2\omega_0 t + I_4\cos 4\omega_0 t + I_6\cos 6\omega_0 t \end{aligned} \tag{4.165}$$

Here the quantities entering are defined as:

$$\begin{aligned} I_d &= \eta\left(5 + \rho_0^2 + 2\rho_0\cos\Delta\right), \\ I_2 &= \frac{\eta}{2}\left(15 - \rho_0^2 + 2\rho_0\cos\Delta\right), \\ I_4 &= \eta\left(3 - \rho_0^2 - 2\rho_0\cos\Delta\right), \\ I_6 &= \frac{\eta}{2}\left(1 + \rho_0^2 - 2\rho_0\cos\Delta\right). \end{aligned} \tag{4.166}$$

where η is a coefficient that accounts for the light energy loss in the optical system.

Therefore we get:

$$\begin{aligned} \rho_0 &= \sqrt{\frac{I_2 - 4I_4 + 9I_6}{I_2 + I_6}}, \\ \cos\Delta &= \frac{I_2 - 4I_4 + 9I_6}{\sqrt{(I_2 - 4I_4 + 9I_6)(I_2 + I_6)}}. \end{aligned} \tag{4.167}$$

The intensity components I_j can also be calculated from a Fourier transform of $I(t)$:

$$I_j = \frac{2}{N}\sum_{t=1}^{N} I(t)\cos\left(jA(t)\right), \qquad j = 2,4,6\ldots \tag{4.168}$$

From the above equation, using the light intensity $I(t)$ and different azimuth angles $A(t)$, except for the DC component I_d, the three ac components, whose frequencies are $2\omega_0$, $4\omega_0$, and $6\omega_0$, respectively are obtained. Then using (4.136), the ellipsometric parameters ψ and Δ are found.

Next the complex dielectric function ε is deduced from a two-phase model:

$$\varepsilon = \varepsilon_a \left\{ \sin^2\phi + \sin^2\phi \tan^2\phi \left[\frac{1-\rho}{1+\rho} \right]^2 \right\}, \tag{4.169}$$

where ε, and ε_a are the dielectric functions of the substrate and the transparent medium, respectively. The real and imaginary parts of complex refractive index can be obtained as follows:

$$\begin{aligned} n &= \frac{1}{\sqrt{2}}\sqrt{\sqrt{\varepsilon_1^2+\varepsilon_2^2}+\varepsilon_1}, \\ k &= \frac{1}{\sqrt{2}}\sqrt{\sqrt{\varepsilon_1^2+\varepsilon_2^2}-\varepsilon_1}. \end{aligned} \tag{4.170}$$

where ε_1, and ε_2 are the real and imaginary parts of the dielectric function ε.

Therefore, the ellipsometric parameters can be obtained uniquely in terms of three ac components of the rotating polarizer and analyzer with a 1:1 speed ratio. This technique has a higher S/N ratio than RAE or RAP for weak reflection samples. It not only applies to visible and near UV ellipsometers, but also to infrared ellipsometers. However this method is sensitive to the polarization of the detector. To overcome this disadvantage, one can use the FPRPFA configuration, which has lower sensitivity, but is not sensitive to the polarization of the detector.

4.4.4 Investigation of the Optical Constants of $Hg_{1-x}Cd_xTe$ by Infrared Spectroscopic Ellipsometry

The optical constants of $Hg_{0.691}Cd_{0.309}Te$ bulk material were measured above, near and below the band gap by infrared spectroscopic ellipsometry (Huang and Chu 2001). A refractive index peak was observed whose energy corresponds approximately to the band gap. Near the absorption edge, the refractive index first drops quickly as the wavelength increases, then decreases more slowly as the wavelength increases further. The experimental results compared favorably with other methods, to validate the method. $Hg_{1-x}Cd_xTe$ bulk materials with different composition were studied and the peak of the refractive index as a function of composition followed the expected behavior.

The refractive index of MCT has been obtained by the interference fringe method (Kucera 1987; Finkman and Nemirovsky 1979; Finkman and Schacham 1984) and through the Kramers–Kronig relation (Liu 1994). Ellipsometry is an established accurate measurement method, and has been widely used to determine the optical constants of materials in the near UV, visible and near infrared region (6–1.5 eV) (Arwin and Aspnes 1984; Viňa et al. 1984). Until recently there was no report on the refractive index of MCT in the mid- and far-infrared regions (2.5–12.5 μm) using a spectroscopic ellipsometer. Next a study of this sort will be introduced on the refractive index of $Hg_{0.691}Cd_{0.309}Te$ bulk material above, near and below the band gap.

A $Hg_{1-x}Cd_xTe$ (x = 0.309) sample was polished mechanically, then polished using a Br-methanol solution (Viňa et al.1984). The back of the sample was roughened by emery paper to reduce the specular reflection contribution from the back-surface. The thickness of the sample was about 0.4 mm. Transmission spectra were measured using a PE983 infrared spectrometer. The composition was obtained by fitting the transmission spectra using the energy band gap and absorption spectra empirical formulas (Chu et al. 1992a,b, 1994). Infrared spectroscopic ellipsometric measurements were carried out using a home-made infrared spectroscopic ellipsometer with variable incident angle. Its operating modes were described in detailed in Sects. 4.4.2 and 4.4.3.

The extinction coefficient k is very small for photon energies below the band gap. The sensitivity of the instrument must be high to measure it. In the free carrier absorption region, one can assume $k \approx 0$. Thin over-layers on the surface, such as oxides, or micro roughness will have an effect on the dielectric function ε. Fortunately at low energy it affects mainly ε_2 (Aspnes and Studna 1983), so there is little correction from the over-layer to the reported dielectric function.

Figure 4.24 shows the refractive index of $Hg_{1-x}Cd_xTe$ (x = 0.309) bulk material above, near and below band gap (2.5–12.5 μm) and the extinction coefficient k above the band gap. A peak is observed on the refractive index curve at a critical wavelength, whose energy corresponds approximately to band gap energy. As the wavelength increases above the critical wavelength, the refractive index first drops quickly, then decreases more slowly. For photon energies below the band gap where the material is

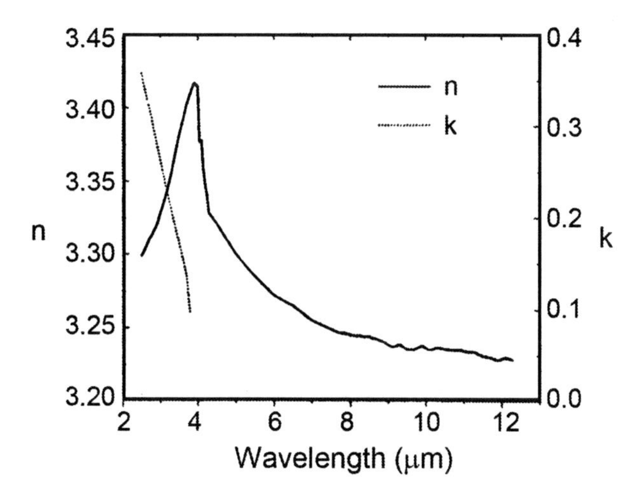

Fig. 4.24. Refractive index n near the band gap of a $Hg_{1-x}Cd_xTe$ ($x = 0.309$) sample taken at room temperature by an infrared spectroscopic ellipsometer, and the extinction coefficient above the band gap

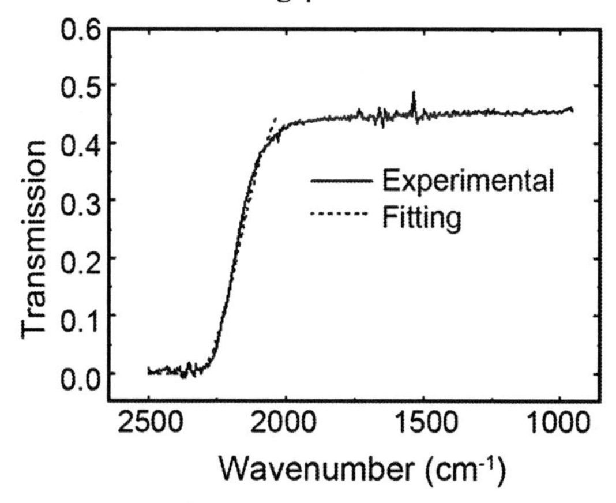

Fig. 4.25. Transmission spectrum and the composition fitting

transparent, the back-surface effect on ρ is very small due to its roughness. The back-surface contribution is estimated to be less than 1%. So no correction is added to the refractive index below the band gap from the roughened back-surface.

In the intrinsic absorption region, the extinction coefficient k decreases as energy decreases which is consistent with the intrinsic absorption spectra (Chu et al. 1994).

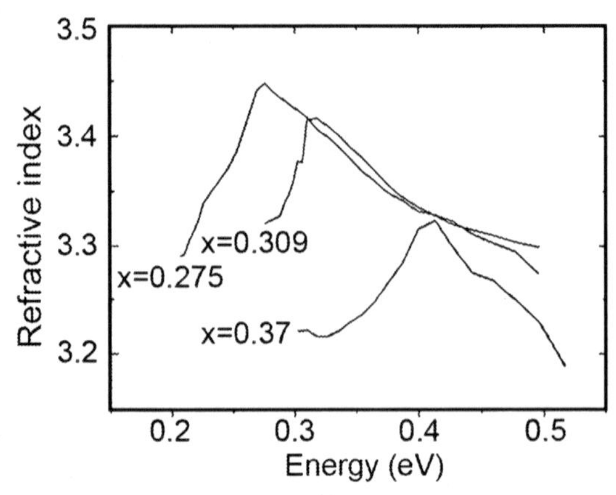

Fig. 4.26. Refractive index spectra near E_g of $Hg_{1-x}Cd_xTe$ at room temperature for three compositions

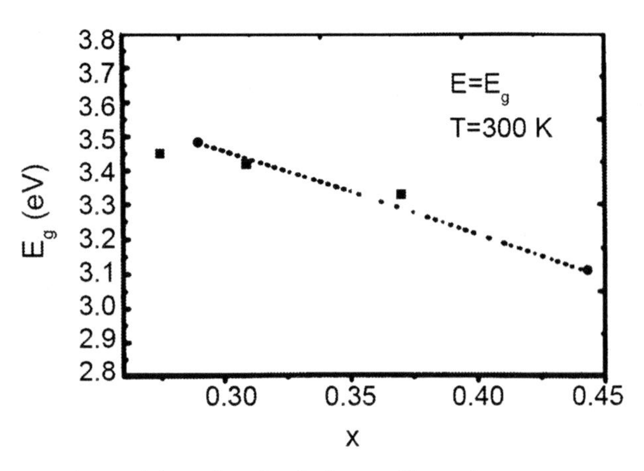

Fig. 4.27. Comparison of the refractive index at E_g and room temperature with the reference (*Solid circle*: Edward, 1991; *solid squares*: experimental data; *dashed line*: point-to-point connections)

The composition dependence of the data was fitted by a least-square fitting method. The fitting was only carried out at the absorption edge, and the effect of free carrier absorption was neglected. Details of the fitting procedure will be presented in the following chapter. Figure 4.25 shows the fitted and measured curves using a composition of $x = 0.309$. From the empirical formula in the reference (Chu et al. 1983a), the band gap is 0.318 eV. According to the refractive index peak position in Fig. 4.26, the energy gap is 0.317 eV. The agreement is quite good.

Further experiments discovered a refractive index enhancement near the band gap in $Hg_{1-x}Cd_xTe$. Recently, the refractive index enhancement near E_g was also observed in the narrow gap semiconductors (Pb,Eu)Se, (Pb,Sr)Se (Herrmann et al. 1993b; Heinz 1991), (Pb,Eu)Te (Yuan et al. 1994) and two dimensional $PbTe/Pb_{1-x}Eu_xTe$ multi quantum well structures (Yuan et al. 1993) grown by MBE. However for the narrow gap semiconductor $Hg_{1-x}Cd_xTe$, because the absorption coefficient is large above the band gap, it is difficult to fix $n(E)$. Herrmann et al. tried to observe the refractive index peaks several times (Herrmann et al. 1993b; Herrmann and Melzer 1996), but failed. Liu et al. (1994a) calculated the refractive index using the KK relation starting from observed intrinsic absorption spectra of thin HgCdTe bulk samples over the composition range x = 0.276–0.443. They found refractive index peaks, with wavelength positions interpreted as corresponding to band gap energies. However, due to the limited data above the band gap, the shape of the deduced refractive index is questionable.

By contrast the infrared spectroscopic ellipsometer measured refractive index curves near E_g for different compositions displayed obvious refractive index peaks. Figure 4.26 shows the data for three compositions. As expected, the observed refractive index peak positions, which correspond to the band gap energy, decrease as the composition decreases.

Finally, Fig. 4.27 compares the refractive index at band gap energy with that taken from the Handbook of Optical Constants (Edward 1991). The solid circle was from the reference (Edward 1991), and the solid squares are experimental data from extensions of the data in Fig. 4.26. In the composition range (x = 0.290–0.443), the refractive index varies nearly linearly with composition, as shown in Fig. 4.27.

4.4.5 In Situ Monitoring of the Composition During $Hg_{1-x}Cd_xTe$ Growth

The energy position of the critical point at the peak of an index spectrum, E_1 of $Hg_{1-x}Cd_xTe$ can be measured by a visible spectroscopic ellipsometer. If the energy position of E_1 vs. composition is known, it can be used to determine the composition. In the MBE growth process, the temperature of a growing sample is about 180°C. Therefore, it is possible to determine the composition in situ using a spectroscopic ellipsometer. For the composition $x = 0.20$, the optical constants different greatly from those at room and high temperature (Wang 1997, see Fig. 4.28).

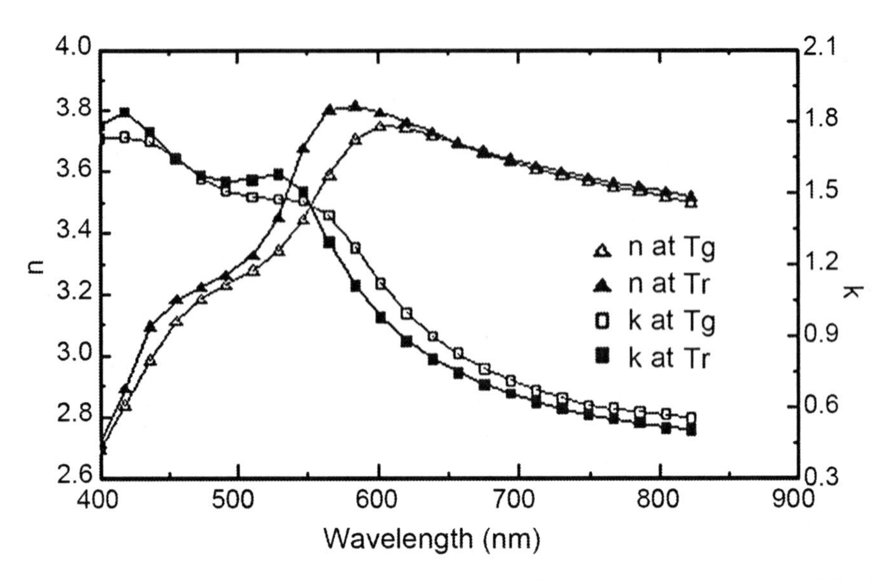

Fig. 4.28. Optical constants of $Hg_{1-x}Cd_xTe$ with x = 0.20 at room and high temperature

Table 4.5. Critical point energy E_1 vs. composition for $Hg_{1-x}Cd_xTe$

Composition	E_1(eV)
0.204	2.207
0.224	2.213
0.246	2.231
0.286	2.253
0.303	2.306
0.585	2.561

To determine the composition, Aspnes et al. (1986) investigated the relation between the optical constants and composition of $Al_xGa_{1-x}As$ in the 1.5–6.0 eV band gap range. They selected nine samples with discrete compositions in the range $0<x<0.8$, and determined the energy positions of the critical points E_1. From this data they obtained an empirical formula for E_1 vs. composition. Snyder et al. (1990) extended the spectral region to 0–6 eV, and from the measured critical points E_1, obtained a similar fit. For $Hg_{1-x}Cd_xTe$ bulk material, Aspnes et al. (1986) investigated surface oxidization effects and critical point energies E_1, and then deduced the composition x.

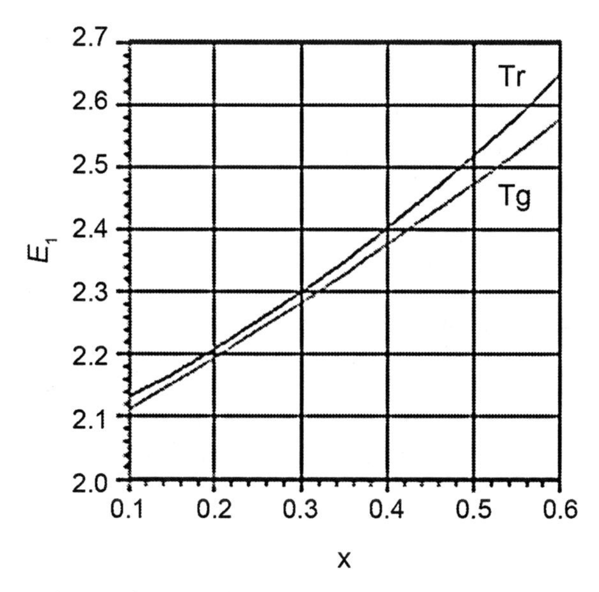

Fig. 4.29. Comparison of E_1 at the growth temperature $T_g \cong 180\,\mathrm{K}$ and room temperature $T_r \cong 300\,\mathrm{K}$

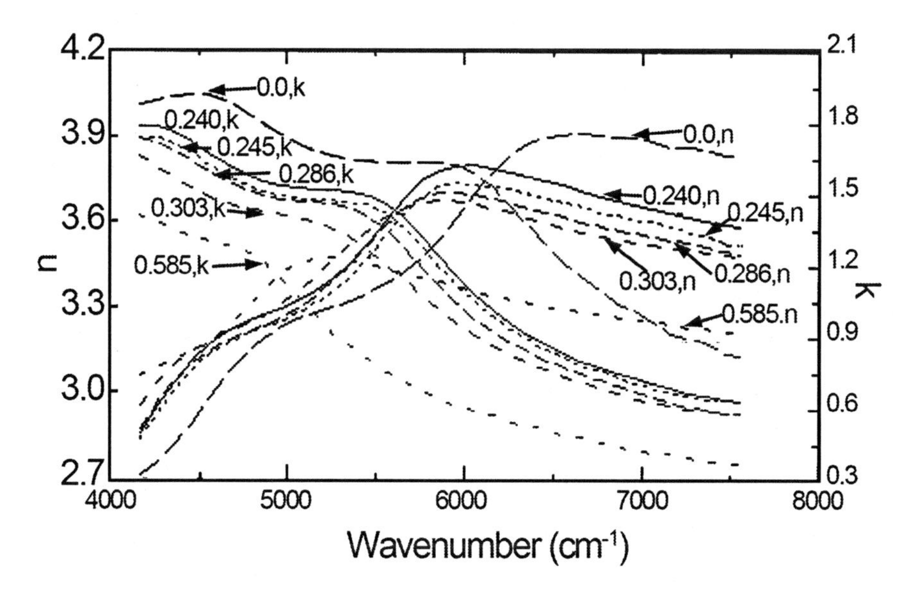

Fig. 4.30. Optical constants *n* and *k* of $Hg_{1-x}Cd_xTe$ for several compositions x

To adopt Snyder's fitting method to determine the composition of $Hg_{1-x}Cd_xTe$ samples from spectroscopic ellipsometry, it is necessary to determine the critical point energy and optical constants at the growth temperature (180 K). Wang (1997) He et al. (1993) measured six samples in situ with compositions in the range $0<x<0.59$. The compositions were determined later, ex situ, from infrared transmission spectra. To avoid composition non-uniformity distortions, the energy gap deduced from the transmission spectral measurement and spectroscopic ellipsometric measurement must be the same.

From the data in Table 4.5, an empirical formula for the critical point $E_1(\mathrm{eV}) = 2.0347 + 0.7853\mathrm{x} + 0.242\mathrm{x}^2$ was obtained for the MBE growth temperature ~180 K. Compared with the E_1 position at room temperature (Aspnes et al. 1986), the position of E_1 at high temperature is red-shifted to lower energies. Figure 4.29 shows a comparison of E_1 at room temperature with that at the growth temperature for the same composition.

Figure 4.30 shows the optical constants n and k for selected compositions x. The composition x is determined by infrared transmission spectroscopy (Wang 1997).

Using the optical constant database for $Hg_{1-x}Cd_xTe$ at high temperature, the in situ spectroscopic ellipsometric data can be fit. Figure 4.31 is the in situ results of ellipsometric data for a sample designated gamct032.

The acquisition of ellipsometric data for the growth began after 4 min, so there are no obvious changes of ψ and Δ from CdTe to $Hg_{1-x}Cd_xTe$ in this short period. In the long wavelength range, oscillating peaks exist, because the absorption coefficient is small at this wavelength, interference fringes between the CdTe substrate and the $Hg_{1-x}Cd_xTe$ interface form. As the thickness of the $Hg_{1-x}Cd_xTe$ increases, only the alloy contributes, so the fringes disappear. From the interference fringes, the film thickness and the growth rate can be determined. As shown in Fig. 4.31, the fitted growth rate is 8.63 ± 0.14 Å s^{-1}.

Using an ellipsometer to monitor HgCdTe growth in situ is a non-destructive method to study the longitudinal distribution of the composition. To monitor the longitudinal distribution at one position on the growing surface, sample rotation is prohibited. Due to the high vacuum process, there is no oxidization of the surface, and the substrate can be viewed as infinite. Fitting the data at any different times, e.g. 5.50 min. and 30.02 min, yielded the compositions 0.2995 ± 0.0018 and 0.3024 ± 0.0017, respectively. By analyzing the composition at different times in the growth process, the longitudinal composition variation can be obtained (See Fig. 4.33). From Fig. 4.32, the compositional variation is less than ±0.002 at different times.

Ellipsometry can also be used to investigate the surfaces of HgCdTe bulk materials, as-grown MBE epitaxial films, and annealed MBE thin films.

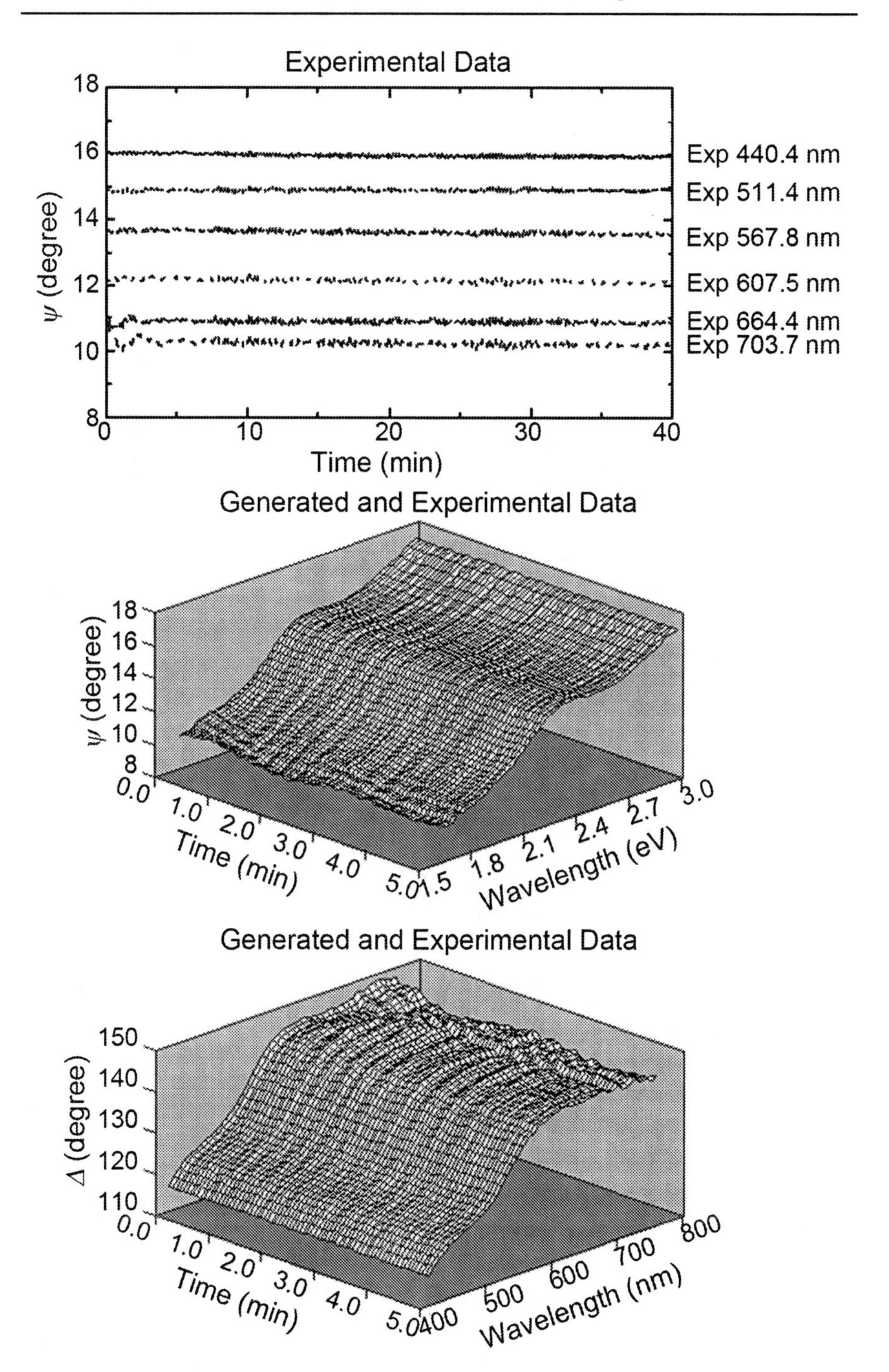

Fig. 4.31. (**a**)–(**c**) In situ ellipsometric data for sample gamct032

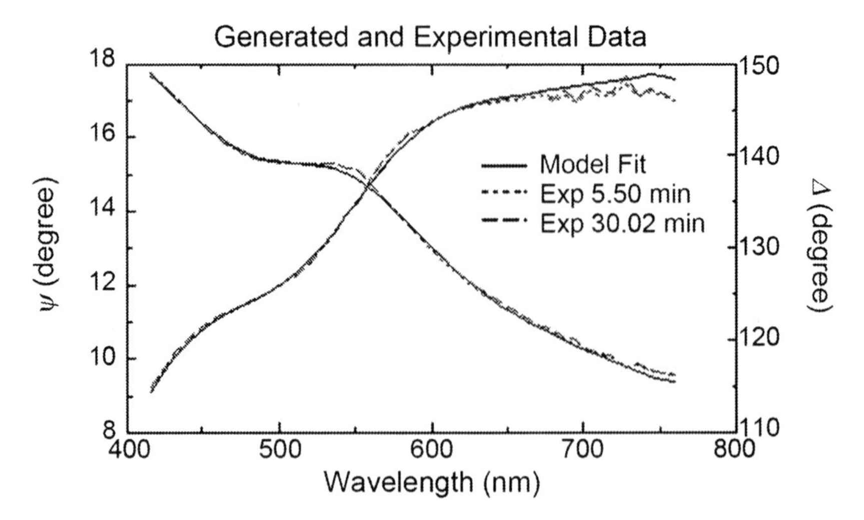

Fig. 4.32. Experimental and theoretical results for ψ and Δ at two different times

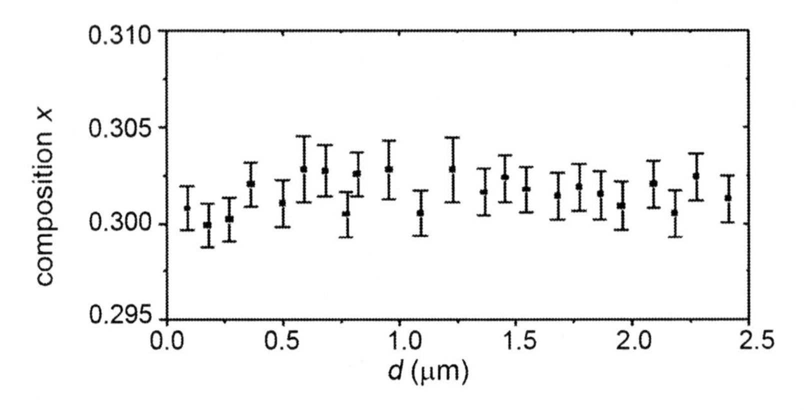

Fig. 4.33. Example of the determination by ellipsometry of the longitudinal composition of a HgCdTe alloy during growth

Surfaces of narrow gap $Hg_{1-x}Cd_xTe$ alloys can be studied using visible ellipsometry because their great absorption in the visible leads to low transmission depths of only several hundred angstroms, making the method surface sensitive. If other materials are deposited on the surface, or there are surface imperfections, they have a great influence on the polarization of the reflected light. Figure 4.34 shows the results of optical constants determined before and after etching using 2% Br–Methanol (BrM) solution. Before etching, the pseudo extinction coefficient on the surface layer is great, the light is absorbed, and the properties of the material cannot be deduced. When an MCT sample is placed in air for a long time, the thickness of a

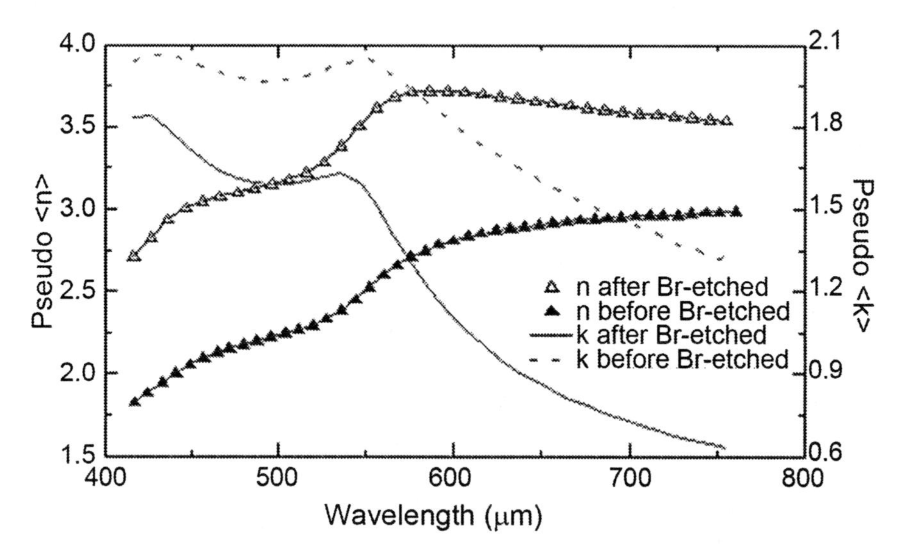

Fig. 4.34. Optical constant changes before and after BrM etching

surface layer can reach several hundred angstroms. This makes it difficult to test an idealized theory by comparison to experiments. Hence, the surface layer must be cleaned at first.

Frequently a CdTe cap layer is grown on HgCdTe surfaces to prevent their surface from oxidizing. The cap also serves as a passivation layer for HgCdTe detectors. During the cap layer growth, some Hg atoms will diffuse into the cap layer. This converts it into a CdTe rich alloy. This adds some difficulty to composition measurements of the narrow gap HgCdTe layer using ex situ methods.

In order to analyze the composition deduced from the observed optical constants of HgCdTe with a surface layer, we can use EMA (effective medium approximation) model (Aspnes and Studna 1983):

$$\sum_{i=1}^{n} f_i \frac{\varepsilon_i - \varepsilon}{\varepsilon_i - 2\varepsilon} = 0$$
$$\sum_{i=1}^{n} f_i = 1 \tag{4.180}$$

where f_i is the ratio of the i th component, ε_i is its dielectric function. Use the word "void" to designate the level of surface roughness. Table 4.6 presents the fitting results using the EMA model with a fixed void of 50%.

Table 4.6. Model fitting results

Sample	Cap Layer			HgCdTe	
	d0(EMA)Å	%(Void)	x_1	d1(Å)	x_2
g009	24	50	0.979	307.8	0.234
g034	69	50	0.993	137	0.314
g035	64	50	0.993	118	0.276
g036	59	50	0.996	93	0.228
g037	40	50	0.976	110	0.236

Table 4.7. Compositions deduced from the in situ ellipsometer and the infrared spectrometer

Sample	Infrared spectra	Ellipsometerin situ
032	0.303	0.301
033	0.305	0.316
034	0.278	0.281
035	0.285	0.253
036	0.231	0.226
037	0.240	0.222
038	0.219	0.212
039	0.240	0.225

Analyzing the surface of an annealed CdTe/HgCdTe sample, using the same model as that for an as-grown sample, results in poor fits to the data. A possible reason is that when the samples were annealed at high temperature, due to the Hg diffusion, the interface between the cap layer and the HgCdTe becomes diffuse. It exists as a broadened transition layer. If we loosen the void parameter, the fitting is much improved. Table 4.7 is a comparison of the fitted results to the infrared and ellipsometer spectra.

4.5 Optical Effects Induced by Free Carriers

The main optical and magneto-optical effects caused by free carriers are absorption edge shifts (the Moss–Burstein [MB] shift), free carrier absorption, plasma reflection, and the free carrier magneto-optical effect.

4.5.1 Moss–Burstein Effect

The MB shift is caused when the free electron concentration is increased, by band-to-band optical absorption or thermal excitation, to a point where the quasi-Fermi energy level in the case of optical excitation or Fermi level for thermal excitation moves into the conduction band.

If the electron quasi-Fermi energy level of a semiconductor moves into the conduction band, the intrinsic optical absorption edge will shift to a shorter wavelength. This is called the Moss–Burstein shift (MB effect), (Burstein 1954; Moss 1954) and has been studied widely. In principle a similar shift will occur if the hole quasi-Fermi level moves into the valence band, but because in most semiconductors the electron mass is far smaller than the hole mass, the effect is dominated by the conduction band. But because low x $Hg_{1-x}Cd_xTe$ semiconductor alloys have very small conduction band effective masses, these alloys need accurately calculated Fermi energy levels to study the MB phenomenon. Initially the expression for the intrinsic carrier concentration will be used to calculate the Fermi level. A detailed derivation will be reported in a later section; here a brief introduction is presented.

Generally, in a situation where there is strong degeneracy, and the approximation of a free electron Fermi gas can be adopted, the conduction band electrons are distributed in a Fermi sphere with wave vector k_f,

$$k_f = (3\pi^2 n)^{\frac{1}{3}} \tag{4.181}$$

where n is electron concentration in the conduction band. According to Kane's model the energy on the Fermi surface is:

$$E_f = \frac{\hbar^2}{2m_0}(3\pi^2 n)^{2/3} + \frac{1}{2}\left[E_g + \sqrt{E_g^2 + \frac{8}{3}P^2(3\pi^2 n)^{2/3}}\right] \tag{4.182}$$

This expression takes the zero of energy to lie at the top of valance band. If the term $\frac{\hbar^2 k^2}{2m_0}$ is neglected, and the zero of energy is taken to lie at the bottom of the conduction band, then the expression for the Fermi energy level reduces to that deduced by Zawadzki and Szymanska (1971) ζ for InSb in case of strong degeneracy:

$$\zeta = E_F - E_g = \frac{1}{2}E_g(\sqrt{\Delta} - 1) \tag{4.183}$$

where in their notation:

$$\Delta = 1 + \frac{8}{3}(3\pi^2 n)^{\frac{2}{3}} \cdot \frac{P^2}{E_g^2} \tag{4.184}$$

Δ can also be written as an often used expression of Zawadzki and Szymanska:

$$\Delta = 1 + 2\pi^2 (\frac{3}{\pi})^{\frac{2}{3}} (\frac{\hbar^2}{m_0^* E_g}) n^{\frac{2}{3}} \tag{4.185}$$

The formulas, (4.182) and (4.183), only apply in strong degeneracy. For non-degenerate or moderate degenerate situations the Fermi energy level is set through its relation to the electron and hole concentrations and effective masses. Focusing on the intrinsic case, the carrier concentration formula for HgCdTe semiconductors (deduced in Sect. 5.1) generates the Fermi energy. The intrinsic carrier concentration for $Hg_{1-x}Cd_xTe$ semiconductors can be calculated from the following expression (Chu 1983b):

$$n_i = \frac{A \times 9.56(10^{14}) E_g^{3/2} T^{3/2}}{1 + \sqrt{1 + 3.6 A E_g^{3/2} \exp(E_g / k_b T)}} \tag{4.186}$$

In (4.186) the unit of the energy gap E_g is in eV, and the intrinsic carrier concentration n_i is in cm^{-3}. The relation between the carrier concentration and the Fermi energy level is:

$$n = A N_C F_{1/2}(\eta) = A \cdot 1.29(10^{14})(E_g T)^{3/2} F_{1/2}(\eta) \tag{4.187}$$

In the intrinsic case,

$$n_i = A \times 1.29(10^{14}) E_g^{3/2} T^{3/2} F_{1/2}(\eta) \tag{4.188}$$

where $\eta = (E_F - E_C)/kT$ is the reduced Fermi energy, and "A" is a modifying factor for a non-parabolic band,

$$A = 1 + \frac{15}{4}\left(\frac{k_B T}{E_g}\right)\frac{F_{3/2}(\eta)}{F_{1/2}(\eta)} + \frac{105}{32}\left(\frac{k_B T}{E_g}\right)^2 \frac{F_{5/2}(\eta)}{F_{1/2}(\eta)} \tag{4.189}$$

and $F_j(\eta)$ is the Fermi–Dirac integral. When $\eta < 1.3$, $F_{1/2}(\eta)$ can be approximated as (Blakemore 1985):

$$F_{1/2}(\eta) = \left[0.27 + \exp(-\eta)\right]^{-1} \tag{4.190}$$

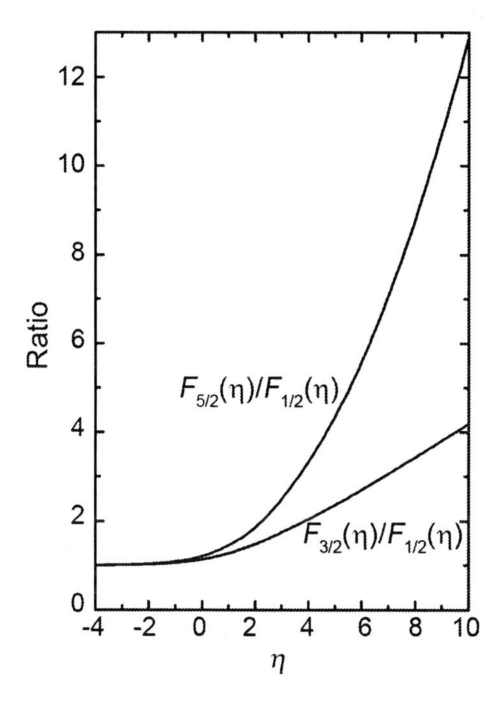

Fig. 4.35. Values of $F_{3/2}(\eta)/F_{1/2}(\eta)$ and $F_{5/2}(\eta)/F_{1/2}(\eta)$

In (4.189), $F_{3/2}(\eta)/F_{1/2}(\eta)$ and $F_{5/2}(\eta)/F_{1/2}(\eta)$ are both functions of η, their values can be deduced from calculated Fermi–Dirac integral tables, and lead to the curves in Fig. 4.35.

From (4.186) to (4.188), we find:

$$F_{1/2}(\eta)\cdot\left[1+\sqrt{1+3.6AE_g^{3/2}\exp\left(E_g/k_BT\right)}\right]=7.41 \tag{4.191}$$

This is a transcendental equation for η. η and A can be calculated by adopting a self-consistent calculational method. In one calculation the CXT expression for the energy band gap:

$$E_g(\text{eV})=-0.295+1.87\text{x}-0.28\text{x}^2 \tag{4.192}$$
$$+(6-14\text{x}+3\text{x}^2)(10^{-4})\text{T}+0.35\text{x}^4$$

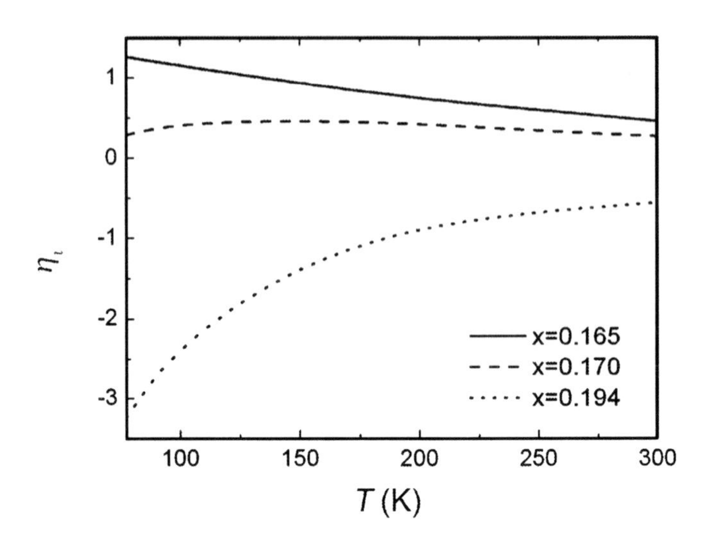

Fig. 4.36. The intrinsic case reduced Fermi levels referenced to the conduction band edge, of $Hg_{1-x}Cd_xTe$ samples with x = 0.165, 0.170, and 0.194 as functions of temperature

was used, (Chu et al. 1983b). It has been verified that this expression is applicable in the range x = 0.19–0.443, T = 4.2–300 K, and in the range of x = 0–0.19, T = 4.2–77 K. Here we assume that the expression is also applicable in the range of x = 0.165–0.19 and T = 77–300 K.

For samples with x = 0.165, x = 0.170, and x = 0.194 that were used in experiments, the reduced Fermi energy η_i, the non-parabolic band modification factor A, the intrinsic carrier concentration n_i, and the band gap E_g in the intrinsic case can be calculated and their values are listed in Tables 4.8–4.10. It can be seen from the tables that for a sample with x = 0.165 the reduced Fermi level η_i =1.26 at 77 K, corresponds to strong degeneracy, so (4.190) and (4.186) are applicable. For samples with x = 0.165, 0.170, and 0.194, the intrinsic Fermi levels are given by the three curves in Fig. 4.36.

In non-intrinsic situations, for small Cd concentrations x of $Hg_{1-x}Cd_xTe$ alloys, the effective donor concentration N^*_D at low temperatures is often far greater than the intrinsic electron concentration. This leads to Fermi levels differing greatly from the intrinsic level. Since the energy levels found in narrow gap HgCdTe alloys often are very shallow, electrons are not bound even at low temperatures, resulting in fully ionized donor states. The relation between n obtained from Hall coefficient measurements of a sample at 77 K and the effective donor concentration N^*_D is (Sze 1981):

$$n_i^2 = n(n - N_D^*) \tag{4.193}$$

Given N^*_D the carrier concentration n at any temperature can be calculated. For the three samples studied here, the carrier concentration n calculated from this equation is consistent with the measured Hall coefficient results at different temperatures. Therefore based on (4.187):

$$n = A \cdot 1.29(10^{14})(E_g T)^{3/2} F_{1/2}(\eta) \tag{4.194}$$

as well as (Blakemore 1985):

$$F_{1/2}(\eta) = \begin{cases} \dfrac{4}{3\sqrt{\pi}}(\eta^2 + 1.7)^{\frac{3}{4}} & (\eta > 1.3) \\ \left[0.27 + \exp(-\eta)\right]^{-1} & (\eta \le 1.3), \end{cases} \tag{4.195}$$

the reduced Fermi level η can be calculated. The results for samples of x = 0.165, 0.170, and 0.194 at 77 K with carrier concentration 4.0×10^{16}, 5.85×10^{15}, 2.2×10^{16} cm^{-3}, respectively, are shown in the tables below.

As shown in the Tables, for degenerate HgCdTe semiconductors, in the temperature range from 77 to 300 K, the real Fermi level (here referenced to the bottom of the conduction band), decreases with increasing temperature, and then gradually approaches the intrinsic Fermi level near room temperature. This behavior is illustrated in Fig. 4.37.

Table 4.8. The Fermi level of $Hg_{1-x}Cd_xTe$ with $x = 0.165$

T (K)	E_g (ev)	Intrinsic		Non-intrinsic (N_D^* = 4.0×10^{16} cm^{-3})				
		ηi	A	ni(cm^{-3})	n(cm^{-3})	A	η	E_F-E_C
77	0.0349	1.26	2.122	2.18×10^{15}	4.0×10^{16}	4.31	7.6	0.050
100	0.0436	1.15	2.155	4.32×10^{15}	4.05×10^{16}	3.45	5.4	0.046
150	0.0624	0.93	2.180	1.21×10^{16}	4.34×10^{16}	2.67	2.97	0.038
200	0.0813	0.74	2.192	2.47×10^{16}	5.17×10^{16}	2.39	1.84	0.032
250	0.100	0.60	2.182	4.28×10^{16}	6.72×10^{16}	2.29	1.24	0.027
300	0.119	0.46	2.186	6.67×10^{16}	8.96×10^{16}	2.25	0.87	0.022

Table 4.9. The Fermi level of $Hg_{1-x}Cd_xTe$ with $x = 0.170$

T (K)	E_g (ev)	Intrinsic			Non-intrinsic ($N_D^* = 5.85\times10^{15}$ cm^{-3})			
		η_i	A	n_i(cm^{-3})	n(cm^{-3})	A	η	E_F-E_C
77	0.044	0.29	1.765	1.39×10^{15}	5.85×10^{15}	2.07	2.5	0.0166
100	0.052	0.44	1.866	3.13×10^{15}	7.20×10^{15}	2.02	1.67	0.0144
150	0.0705	0.47	1.985	$9{,}83\times10^{15}$	1.32×10^{16}	2.02	0.9	0.0116
200	0.089	0.43	2.038	2.14×10^{16}	2.45×10^{16}	2.04	0.62	0.0106
250	0.1075	0.34	2.070	3.77×10^{16}	4.08×10^{16}	2.07	0.45	0.0097
300	0.126	0.27	2.09	6.07×10^{16}	6.37×10^{16}	2.07	0.33	0.0087

Table 4.10. The Fermi level of $Hg_{1-x}Cd_xTe$ with $x = 0.194$

T (K)	E_g (ev)	Intrinsic			Non-intrinsic ($N_D^* = 2.2\times10^{16}$ cm^{-3})			
		η_i	A	n_i(cm^{-3})	n(cm^{-3})	A	η	E_F-E_C
77	0.084	−3.24	1.316	1.08×10^{14}	2.2×10^{16}	1.656	3.88	0.0257
100	0.091	−2.27	1.384	4.92×10^{15}	2.2×10^{16}	1.634	2.63	0.0226
150	0.108	−1.31	1.500	3.17×10^{15}	2.24×10^{16}	1.652	1.05	0.0135
200	0.1245	−0.86	1.594	9.69×10^{15}	2.45×10^{16}	1.680	0.25	0.0043
250	0.142	−0.67	1.703	2.08×10^{16}	4.08×10^{16}	1.730	−0.09	−0.002
300	0.160	−0.56	1.755	3.71×10^{16}	6.37×10^{16}	1.755	−0.22	−0.006

As shown in the figure, the larger the effective donor concentration, the higher the reduced Fermi level; a feature especially prominent at low temperature. To see the Fermi level variation with the effective donor concentration more clearly, take the x = 0.194 sample as an example. Assuming N_D^*=1×10^{16} and 2.26×10^{15} cm^{-3}, the calculated reduced Fermi level for N_D^* =1×10^{16} cm^{-3} has its onset of degeneracy occur at a temperature lower than 150 K, while for the sample with N_D^* = 2.26×10^{15} cm^{-3}, degeneracy does not occur until the temperature falls below 77 K. The temperature for the onset of degeneracy needs to be considered in the preparation of very long wavelength HgCdTe devices.

Knowledge of the Fermi level for degenerate semiconductors is essential to a quantitative analysis of the MB shift of HgCdTe and to its optical absorption edge. For high electron concentrations where the Fermi level lies in the conduction band, the electron states in the conduction band for energies below the Fermi level are fully occupied, so no transitions to these states are possible. Then the MB effect occurs and the absorption edge shifts to a shorter wavelength. For a HgCdTe semiconductor with compositions $x \leq 0.194$, the Fermi level rises into the conduction band when $N_D^* \sim 10^{16}$ cm^{-3}. Thus the MB effect must be considered for HgCdTe samples with low alloy concentrations.

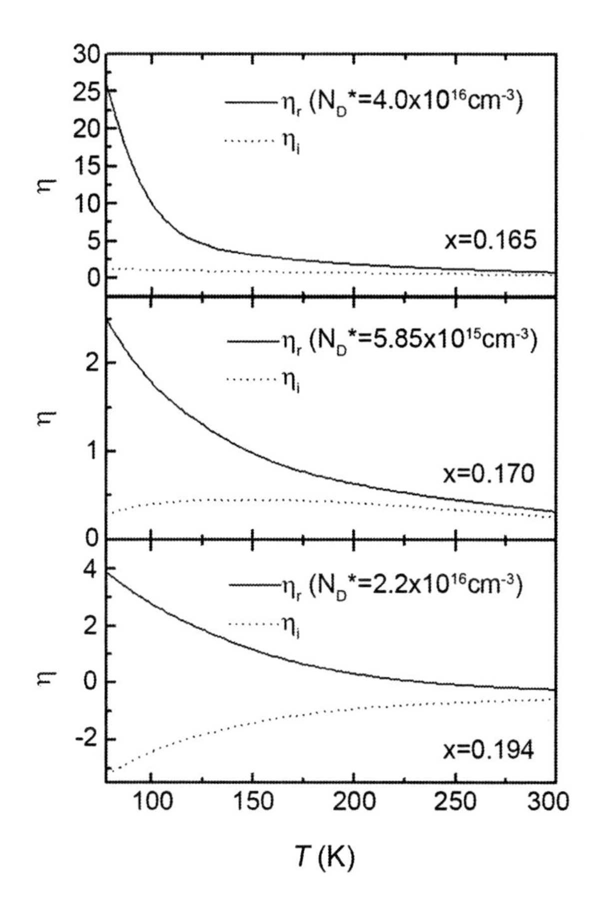

Fig. 4.37. The relation between the reduced Fermi level and temperature for intrinsic and non-intrinsic $Hg_{1-x}Cd_xTe$ samples with $x = 0.165$, 0.170, and 0.194

HgCdTe samples with $x = 0.165$, 0.170, and 0.194 and effective donor concentrations $N_D^* = 4\times10^{16}$, 5.85×10^{16}, and 2.2×10^{16} cm^{-3}, respectively, were measured. The samples were pasted on transparent Si, KRS-5, and ZnSe substrates respectively. Then the thickness of the samples was reduced to 10, 9, and 24 μm, respectively by grinding, polishing, and etching. The sample thicknesses were measured by an interference microscope and a spiral micrometer. A PE983 infrared spectrophotometer was used to measure their transmittance in the wave number range 2,000–300 cm^{-1} and over the temperature range 77–300 K. The absorption coefficient was calculated from (4.136). Figures 4.38 and 4.39 display the measured absorption spectra for the samples with x = 0.165 and 0.194 at different temperatures.

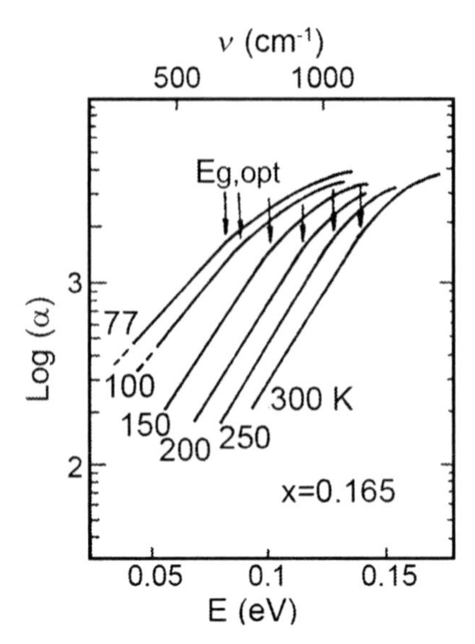

Fig. 4.38. Intrinsic absorption spectra of a $Hg_{1-x}Cd_xTe$ sample with x = 0.165 (d = 10 μm) at several temperatures between 77 and 300 K

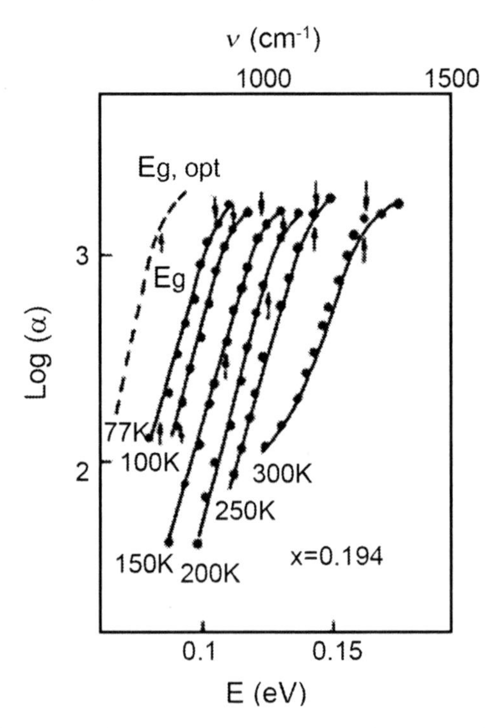

Fig. 4.39. Intrinsic absorption spectra of a $Hg_{1-x}Cd_xTe$ sample with x = 0.194 (d = 24 μm) at several temperatures between 77 and 300 K

Table 4.11. Measured E_{opt} and η compared with calculated results

x	T(K)	E_g(eV) cal	E_{opt}(eV) exp	η_{exp}	η_{cal}	E_{opt}-E_g (eV)	E_F-E_C(eV) cal
0.165	77	0.0349	0.083	7.23	7.6	0.048	0.050
	100	0.0436	0.0893	5.29	5.4	0.0457	0.046
	150	0.0624	0.1016	3.04	2.97	0.039	0.038
	200	0.0813	0.1153	1.97	1.84	0.034	0.032
	250	0.100	0.1277	1.29	1.24	0.0277	0.027
	300	0.119	0.140	0.81	0.87	0.021	0.022
0.194	77	0.084	0.1054	3.22	3.88	0.0214	0.0257
	100	0.091	0.1116	2.39	2.63	0.0206	0.0226
	150	0.108	0.1227	1.06	1.05	0.0147	0.0135
	200	0.1245	0.1302	0.33	0.25	0.0057	0.0043
	250	0.142	0.1426	0.027	–0.09	0.0006	–0.002
	300	0.160	0.1612	0.046	–0.22	0.0012	–0.006

As shown in the figures, around α = 1,200–1,700 cm^{-1}, each absorption curve shows a photon energy where the slope of the absorption curve becomes small. Arrows in the figures point to these transition energies. In degenerate situations the energy at which the slope changes identifies the optical gap energy E_{opt}. This energy is:

$$E_{opt} = E_g + (1+\frac{m_C}{m_{hh}})E_f \approx E_g + E_f \tag{4.196}$$

In Figs. 4.36 and 4.37, the gap energies E_g obtained can be identified with the CXT gap expression at different temperatures, (4.192). Then from:

$$\eta = \frac{E_{opt} - E_g}{k_B T} \tag{4.197}$$

the reduced Fermi level can be obtained through a measurement that actually finds E_{opt}. Table 4.11 lists the calculated E_g values determined this way, and the measured values E_{opt}, $E_{opt} - E_g$ and $\eta_{\exp}$. It also lists the previously calculated results, $E_F - E_C$ and η_{cal} for the two samples.

It can be seen from the table and figures that for the degenerate $Hg_{1-x}Cd_xTe$ semiconductors with x≈0.165–0.194, the optical gap obtained is consistent with the calculated Fermi levels. Also the CXT equation for E_g leads to agreement between the experimental and the calculated results, indicating that the CXT equation is applicable over the extended range of x = 0.165–0.194, and T = 77–330 K. To sum up, the CXT equation for $E_g(x,T)$is applicable over the following composition and temperature ranges:

$$\begin{cases} 0.165 \le x \le 0.443 \\ 4.2K \le T \le 300K \end{cases} \tag{4.198}$$

and

$$\begin{cases} 0 \le x \le 0.165 \\ 4.2K \le T < 77K \end{cases} \tag{4.199}$$

Taking account of the MB shift, amends the predicted absorption coefficient. The photo-electronic induced transition from the valance band to the conduction band leads to an absorption coefficient (Kane 1960):

$$\alpha = \frac{4\pi^2 e^2 \hbar}{m^2 cnE} \sum_j M_j^2 \int \frac{2d^3k}{(2\pi)^3} \delta\left[E_C(\mathbf{k}) - E_j(\mathbf{k}) - \hbar\omega\right] \tag{4.200}$$

Kane's energy band expression enables a calculation of the absorption coefficient. When the MB effect is taken into account, the absorption coefficient calculation is modified. According to the Fermi–Dirac distribution function, the vacant electron state probability in the conduction band is:

$$1 - f_c = \frac{1}{1 + \exp\left(\dfrac{E_F - E}{k_B T}\right)} = \frac{1}{1 + \exp(\eta - \varepsilon)} \tag{4.201}$$

where $\varepsilon = \dfrac{E - E_C}{k_b T}$ is a reduced energy. The MB effect is accounted for by the modification:

$$\alpha = \frac{\alpha_0}{1 + \exp(\eta - \varepsilon)} \tag{4.202}$$

where α_0 is the absorption coefficient without the MB effect. The complete expression is $\alpha_0 = \alpha_{hh} + \alpha_{lh}$, and after inclusion of the MB effect, it becomes:

$$\alpha = \frac{\alpha_{hh}}{1 + e^{\eta - \varepsilon_{hh}}} + \frac{\alpha_{lh}}{1 + e^{\eta - \varepsilon_{lh}}} \tag{4.203}$$

where ε_{hh} and ε_{lh} are the reduced energies in the conduction band due to transitions from the heavy and the light hole bands, respectively. In

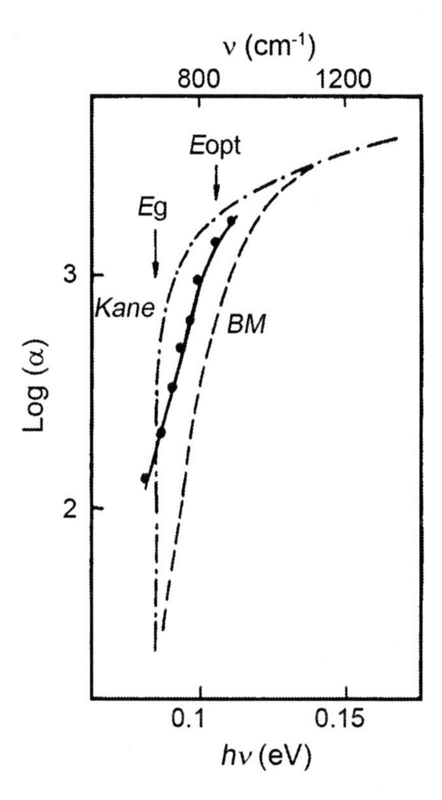

Fig. 4.40. The *dashed dot curve* denotes the direct absorption results predicted by the Kane model, the *dashed curve* denotes the results predicted with the MB correction included, and the dots are the measured results

calculations $\varepsilon_{hh} = \dfrac{E_{hh \to condoctionband} - E_C}{k_b T}$ and $\varepsilon_{lh} = \dfrac{E_{lh \to condoctionband} - E_C}{k_b T}$ were used to arrive at the curves in Fig. 4.40. In Fig. 4.40, for $x = 0.194$ and $T = 77$ K, the calculated absorption curve without the MB correction is denoted by a dash-dotted curve, and with the correction, (4.203), it is denoted by a dashed curve. The parameters used were $E_g = 0.084$ eV, $P = 8 \times 10^{-8}$ eV cm, and m_{hh} = 0.55m$_0$. The measured result is denoted by dots. The measurements are bracketed by the theory curves with and without the MB correction. However both of the theoretical curves have sharper slopes in the low photon energy range than the experimental values. Because the curve for the MB corrected theory, which should be included for a degenerate semiconductor, lies below the experimental points and its slope at low energies is too high, it is likely that there is another mechanism acting on the electrons that needs to be added.

Because the optical gap has a strong dependence on the Fermi level and therefore on the conduction band electron concentration, the gap's cutoff can be used to characterize the quality of samples by measuring it at different temperatures. If the variation of absorption edge with temperature is the same as the variation of energy gap with temperature, then the sample is non-degenerate. If the absorption edge moves in the direction of shorter wavelengths, it indicates that the Fermi level is moving into the conduction band in this temperature range. This occurrence depends on the composition, the temperature and the impurity types and densities. Adopting the above method enables a quantitative analysis.

4.5.2 General Theory of Free Carrier Absorption

Free carrier absorption can be dealt with by a semi-classical method, and the same results are obtained from quantum theory. For narrow gap semiconductors free carrier optically assisted transitions chiefly take place in the far-infrared spectral region. Under a parabolic band approximation, and assuming there is a velocity relaxation time τ (Chen and Sher 1995) related to properties of the medium, the free carrier equation of motion in an optical field is:

$$m^* \frac{dv}{dt} + m^* \frac{v}{\tau} = eE \tag{4.204}$$

where the electric field varies as $E = E_0 \exp(i\omega t)$, and the carrier speed is $v = v_0 \exp(i\omega t)$. Then for a system with N carriers per unit volume, the current density is given by $J = Nev = \sigma E$, and thus the complex electrical conductivity is:

$$\sigma = \frac{Ne^2\tau}{m^*}\left(\frac{1}{1+i\omega\tau}\right) = \frac{\sigma_0}{1+i\omega\tau} \tag{4.205}$$

The quality σ_0 is the electrical conductivity for a DC current. In general, the relaxation time τ deduced from the Boltzmann transport equation depends on energy. A commonly used approximation is to replace τ in (4.205) by its mean value:

$$\sigma_0 = \frac{Ne^2\langle\tau\rangle}{m^*} \tag{4.206}$$

$$\langle\tau\rangle = -\frac{2}{3}\frac{\int_0^\infty \tau E^{3/2}\left(\partial f_0/\partial E\right)dE}{\int_0^\infty f_0 E^{1/2} dE} \tag{4.207}$$

For high frequencies, $\langle\tau\rangle$ becomes $\left\langle\frac{\tau}{1+i\omega\tau}\right\rangle$, so from (4.204) we have:

$$\sigma = \sigma_1 + i\sigma_2 = \frac{Ne^2}{m^*}\left[\left\langle\frac{\tau}{1+\omega^2 c^2}\right\rangle - i\omega\left\langle\frac{\tau^2}{1+\omega^2 c^2}\right\rangle\right] \tag{4.208}$$

Since $\varepsilon = n^2 - \kappa^2 - i\frac{4\pi}{\omega}\sigma$, we find:

$$\varepsilon_1 = n^2 - \kappa^2 = \varepsilon_\infty - \frac{4\pi}{\omega}\sigma_2 = \varepsilon_\infty - \frac{4\pi Ne^2}{\omega m^*}\left\langle\frac{\omega\tau^2}{1+\omega^2 c^2}\right\rangle \tag{4.209}$$

$$\varepsilon_2 = 2n\kappa = \frac{4\pi}{\omega}\sigma_1 = \frac{4\pi Ne^2}{\omega m^*}\left\langle\frac{\tau}{1+\omega^2 c^2}\right\rangle \tag{4.210}$$

Then the free carrier absorption coefficient is:

$$\alpha = \frac{2\omega\kappa}{c} = \frac{4\pi Ne^2}{ncm^*}\left\langle\frac{\tau}{1+\omega^2 c^2}\right\rangle \tag{4.211}$$

Assuming the relaxation time has no energy dependence, and $\omega\tau >> 1$, we obtain a simple classical expression:

$$\alpha = \frac{Ne^2\lambda^2}{\pi nc^3 m^*}\left(\frac{1}{\tau}\right) \tag{4.212}$$

The free carrier absorption is directly proportional to the carrier concentration and the square of the wavelength. The quantity $(1/\tau)$ depends on the carrier scattering mechanism. To complete the analysis one needs to evaluate the contributions from different scattering mechanisms. The main scattering mechanisms have different functional dependences leading to different material and temperature behaviors.

Free carriers also influence the optical reflection. In general, the free carrier absorption coefficient is small, so $n^2 >> \kappa^2$. The reflectance of normally incident light is given by:

$$R = \frac{(n-1)^2}{(n+1)^2} \tag{4.213}$$

From (4.209) we get:

$$n^2 = \varepsilon_\infty - \frac{4\pi}{\omega}\sigma_2 = \varepsilon_\infty - \frac{4\pi Ne^2}{\omega^2 m^*} = \varepsilon_\infty \left(1 - \frac{\omega_p^2}{\omega^2}\right) \tag{4.214}$$

where the definition of the plasma frequency ω_p is:

$$\omega_p = \left(\frac{4\pi Ne^2}{m^* \varepsilon_\infty}\right)^{1/2} \tag{4.215}$$

When the frequency lies below the plasma frequency n is pure imaginary and there is both reflection and absorption. It is obvious that $R = 1$ when $\omega = \omega_p$, so $n = 0$. This indicates that the reflectance will increase to 1 rapidly as the frequency decreases to the plasma frequency. While in the region where the frequency is slightly higher than ω_p, there is also a possibility that $n = 1$, in which case $R = 0$. Thus a minimum in the reflectivity will occur at a frequency $\omega_{\min}$ defined as:

$$\omega_{\min} = \omega_p \left(1 - \frac{1}{\varepsilon_\infty}\right)^{-1/2} \tag{4.216}$$

So knowing ε_∞, ω_p can be determined from the frequency $\omega_{\min}$ at which there is a minimum reflectance. Then the effective mass of the free carriers can be determined from (4.215). (Remember that this analysis was predicated on the effective mass approximation, so it must be modified in narrow gap materials where near the bottom of the conduction band it deviates from a parabolic k behavior.) The range from $\omega_{\min}$ to ω_p where the reflectance increases rapidly is called the plasma oscillation reflection edge or more simply, the plasma reflection edge.

Free carrier absorption is a kind of indirect transition process. Because photons are mass-less particles and have little momentum, it is imposable to conserve momentum in an event in which a photon is destroyed to deliver its energy to a free electron. To conserve momentum (or wave vector), a second quasi-particle's participation is required. Typical scattering mechanisms are shown in Fig. 4.41.

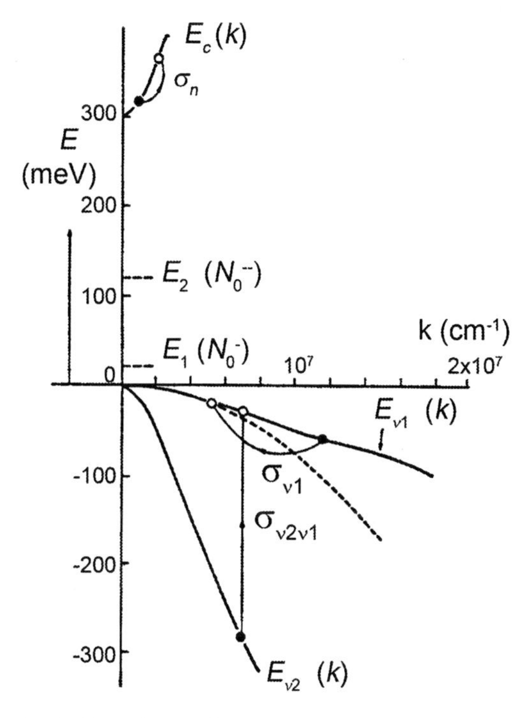

Fig. 4.41. $Hg_{1-x}Cd_xTe$ free carrier absorption processes

For n-type $Hg_{1-x}Cd_xTe$, Baranskii et al. (1990) considered various mechanisms contributing to free carrier absorption: ionized impurity scattering, optical and acoustic phonon scattering coupled through polar and deformation potentials. They lead to the following absorption coefficient expressions:

- Ionized impurity

$$\alpha_I = 3B(\pi e^2\hbar^2\varepsilon^{-1}m^{*-1})^2 N_I(\hbar\omega)^{-2} \tag{4.217}$$

- Longitudinal optical phonons, polar coupled

$$\alpha_{OP} = 3Be^2\hbar^2\varepsilon^{-1}m^{*-1}(\varepsilon_0\varepsilon_\infty^{-1}-1)\omega_{LO}\omega^{-1}\chi_{OP} \tag{4.218}$$

- All optical phonons, deformation potential coupled

$$\alpha_{DP} = BD^2\hbar(\rho\omega_{TO})^{-1}\chi \tag{4.219}$$

- All acoustic phonons, deformation potential coupled

$$\alpha_{AP} = B\Xi_u^2 k_B T(\rho v^2)^{-1} \tag{4.220}$$

Table 4.12. Parameters for $Hg_{1-x}Cd_xTe$ free carrier absorption calculations

Momentum matrix P	8.0×10^{-8} eV cm (Chu et al. 1992a)
Spin–orbit splitting Δ	1 eV (Chu et al. 1992a)
Deformation potential for optical phononsD	14.8 eV (Baranskii et al. 1990)
Deformation potential for acoustic phonons Ξ_u	14 eV (Baranskii et al. 1990)
Longitudinal sound speed v	3.01×10^5 cm s^{-1}
Longitudinal optical phonon frequency of HgTe ω_{LO1}	137.2 cm^{-1}
Transverse optical phonon frequency of HgTe ω_{TO1}	120.4 cm^{-1}
Longitudinal optical phonon frequency of CdTe ω_{LO2}	151.2 cm^{-1}
Transverse optical phonon frequency of CdTe ω_{TO2}	147.3 cm^{-1}
Low frequency dielectric constant ε_0	$20.8-16.8x+10.6x^2-9.4x^3+5.3x^4$ (Brice and Capper 1987)
Optical frequency dielectric constant ε_∞	$15.1-10.3x-2.6x^2+10.2x^3-5.2x^4$ (Brice and Capper 1987)
Ground state degeneracy factor	4 (Kim et al. 1994)
Acceptor ionized energy E_A	11 (Bartoli et al. 1986)
Intrinsic carrier concentration n_i	$9.56\times10^{14}\dfrac{\left(1+3.25k_BT/E_g\right)E_g^{3/2}T^{3/2}}{1+1.9E_g^{3/4}\exp\left(E_g/2k_BT\right)}$ (Chu et al. 1992a)
Energy band gap E_g	$-0.295+1.87x-0.28x^2+(6-24x+3x^2)(10^{-4})T+0.35x^4$ (Chu et al. 1992a)

The physical meanings of all the parameters can be found in the literature (Baranskii et al. 1990), and in Table 4.12.

For n-type $Hg_{1-x}Cd_xTe$ free carrier absorption is due chiefly to electron transitions in the conduction band. The electron absorption cross-section σ_n is obtained from (4.217) to (4.220). For p-type $Hg_{1-x}Cd_xTe$, the influence of the light-heavy hole inter-band, and the heavy hole intra-band carrier transitions on the absorption cannot be ignored.

The light-heavy hole inter-band absorption coefficient can be calculated from the following equation:

$$\alpha = \frac{1}{137}\frac{1}{\sqrt{\varepsilon}}\sum_{E_h - E_l = \hbar\omega} d\delta \frac{M_{hl}^2}{\left|\nabla_k (E_h - E_l)\right|}\left[f(E_h) - f(E_l)\right] \tag{4.221}$$

It can also be used to calculate the absorption cross section. For $Hg_{1-x}Cd_xTe$ with compositions x<0.4, according to above equation, the light-heavy hole inter-band transition absorption cross-section σ_{v2v1} is (Mroczkowski and Nelson 1983):

$$\sigma_{v2v1} = \frac{4\alpha_k K_p^2}{3nE}\left(\frac{dE}{dK}\right)^{-1} a_s^2 \exp\left[\frac{E_{v1}(K_p)/k_B T}{N_v}\right] \tag{4.222}$$

where $N_v = 2(2\pi m_{hh}^* k_B T / h^2)^{3/2}$ is the valence band terminal state density; for the meaning of other physical quantities refer to the literature (Mroczkowski and Nelson 1983).

For $Hg_{1-x}Cd_xTe$ with compositions $x > 0.4$, the contributions from terms higher order in wave vector k must be added to the calculation. Then the absorption cross-section expression becomes:

$$\sigma_{v1v2} = K\eta^{1/2}(1+\eta^{1/2})\exp\left[-\gamma\eta(1+\eta)\right]\left[1-\exp(-\beta\eta)\right] \tag{4.223}$$

where

$$K = \frac{1}{137\sqrt{\varepsilon}}\frac{1}{PN_v}\sqrt{\frac{3}{2}E_g},$$

$$\gamma = \frac{3}{2}\hbar^2 E_g^2 m_v^* E_p k_B T,$$

$$\beta = E_g / k_B T, \quad \eta = \hbar\omega / E_g, \text{ and}$$

$$Ep = 2m_0 P^2 / \eta^2 .$$

The heavy hole inter-band absorption cross-section σ_{v1} is given by the equation:

$$\sigma_{v1} = \frac{e^3}{4\pi c n \varepsilon (m_{hh}/m)^2 \mu_p} \tag{4.224}$$

Then the net absorption cross-sections of electron and hole free carrier absorption coefficients:

- α_n of an n-type $Hg_{1-x}Cd_xTe$ semiconductor is:

$$\alpha_n = (\sigma_{v2v1} + \sigma_{v1})p_n + \sigma_n n_n \tag{4.225}$$

- α_p of a p-type $Hg_{1-x}Cd_xTe$ semiconductor is:

$$\alpha_p = (\sigma_{v2v1} + \sigma_{v1})p_p + \sigma_n n_p \tag{4.226}$$

where n_n and p_n are the electron and the hole concentrations, respectively, of an n-type $Hg_{1-x}Cd_xTe$, and n_p and p_p are the electron and the hole concentrations, respectively, of a p-type $Hg_{1-x}Cd_xTe$. According to the electrical neutrality condition, for an n-type material:

$$\begin{aligned} n_n &= n_i^2 / n_n + N_D - N_A \\ p_n &= n_i^2 / n_n \end{aligned} \tag{4.227}$$

and for a p-type material:

$$\begin{aligned} p_p &= n_i^2 / p_p + \frac{N_A}{1 + gp_p \exp(E_A / k_B T) / N_A} - N_D \\ n_p &= n_i^2 / p_p \end{aligned} \tag{4.228}$$

where g is the ground state degeneracy factor, E_A is the acceptor ionization energy, N_A is the acceptor concentration of a p-region, N_D is the donor concentration of an n-region, and n_i is the intrinsic carrier concentration. Using the parameters listed in Table 4.12, the free carrier absorption coefficients of $Hg_{1-x}Cd_xTe$ can be calculated from (4.217) to (4.228).

4.5.3 Free Carrier Absorption of $Hg_{1-x}Cd_xTe$ Epitaxial Films

The characteristics of optical absorption due to free carrier intra-band transitions can be used for judging scattering mechanisms and related physical properties of $Hg_{1-x}Cd_xTe$. The wavelength range ($10-40\,\mu m$) of $Hg_{1-x}Cd_xTe$ free carrier absorption lies between the quantum limit ($\hbar\omega >> k_BT, \hbar\omega >> E_F$) and the classical limit ($\hbar\omega << k_BT, \hbar\omega << E_F$), which makes it difficult to deduce the scattering mechanism. But the mechanisms can be studied through quantitative comparisons between experimental results and theoretical calculations. Also because of the many competing phenomena, intrinsic defects, composition fluctuations, tellurium inclusions, etc. in $Hg_{1-x}Cd_xTe$, sorting out which dominates the spectrum in a given crystal is difficult. There have been many studies of the free carrier absorption optical spectra of $Hg_{1-x}Cd_xTe$ bulk crystals (Mroczkowski and Nelson1983; Huga and Kimura 1963; Gurauskas et al. 1983; Baranskii et al. 1990; Belyaev et al. 1991; Brossat and Reymond 1985; Qian 1986; Tian et al. 1991), however, there have been few reports about the free carrier absorption of $Hg_{1-x}Cd_xTe$ epitaxial films. The impurity defect concentrations of epitaxial materials are generally smaller than those of bulk crystals; therefore, their absorption spectra can be more reflective of the properties of $Hg_{1-x}Cd_xTe$ material itself. However, there are often longitudinal composition inhomogeneities in epitaxial films, so the analysis method used must be different from that of bulk crystals. In the following paragraphs both theoretical and experimental aspects the free carrier absorption properties of $Hg_{1-x}Cd_xTe$ epitaxial films are examined.

For $Hg_{1-x}Cd_xTe$ epitaxial films, longitudinal compositional inhomogeneities are always present, and must be taken into account when calculating the free carrier absorption. Hougen has deduced the longitudinal compositional distribution by modifying a room temperature infrared transmitted spectrum (Hougen 1989). When a light beam irradiates the surface of a HgCdTe two-layer structure, an epitaxial layer/CdTe substrate, its transmittance $T_{1,3}$ is:

$$T_{1,3} = \frac{(1-R_1)(1-H)T_{2,3}a_1}{1-R_1(1-H)R_{2,3}a_1^2} \tag{4.229}$$

R_1, R_2, R_3 are the reflectances at the three interfaces, which can be found from the refractive index equation (Hougen 1989; Li et al. 1995a,b). The parameter H, accounts for the loss of incident light at the surface of the epitaxial layer, and is an adjustable parameter taken to fix the maximum transmittance of the calculated curve to coincide with that of the measured curve. The longitudinal compositional distribution of the $Hg_{1-x}Cd_xTe$ epitaxial film can be expressed as:

$$x(z) = \frac{1-(x_s+sd)}{1+4(z/\Delta z)^2} + (x_s+sd) - sz \tag{4.230}$$

where z is the normal distance from the substrate, d is the thickness of the epitaxial layer, and $x_s, s,$ and Δz are fitting parameters that have meaning only in the context of shape dictated by (4.230). The composition x_s represents approximately the composition at the surface of the epitaxial layer, Δz represents the width of transition region, s approximates the compositional slope of the longitudinal variation of the epitaxial layer following the transition region. The three parameters can be deduced by fitting absorption edge and intrinsic absorption region transmittance data taken at room temperature (Li et al. 1995b). Equations (4.210) and (4.211) are applicable for $Hg_{1-x}Cd_xTe$ epitaxial samples grown by MBE, LPE and MOCVD.

For an epitaxial layer sample with a longitudinal compositional inhomogeneity, if at point z the composition is $x(z)$, the local free carrier absorption coefficient is $\alpha[x(z)]$, and the total absorption coefficient of the epitaxial layer is:

$$\alpha d = \int_0^d \alpha[x(z)]dz \tag{4.231}$$

In the model represented by (4.229), the parameter H is used to denote the optical loss at the sample's surface. If the surface of a sample has been exposed to a polishing treatment, H is generally less than 0.02. If the sample is well polished, the H value is 0 for p-type or n-type $Hg_{1-x}Cd_xTe$ samples at 77 K and room temperature. For well polished samples where $H=0$, the measured free carrier absorption coefficients α_m agree with the calculated result α_c from (4.217) to (4.228) at 300 and 77 K. But for some less well polished samples, the calculated transmission curves do not agree well with the experimental data at different temperatures unless a finite H value (H >0.02) is used. In this case, the measured absorption coefficient, α_m, is often greater than the calculated values, α_c. However, even then the difference between α_m and α_c does not vary with temperature. It is likely that there are other optical mechanisms at play for samples with H >0.02.

If tellurium inclusions (or deposits) exist inside the film, they impact the optical absorption. From the HgCdTe phase diagram, it is easy to create tellurium inclusion during HgCdTe crystal growth from tellurium-rich solutions (Schaake and Tregilgas 1983). Tellurium is an insulator with a wide energy gap compared to those of the HgCdTe alloys of interest, so inclusions scatter much more radiation than they absorb. The contribution to the extinction cross-section from Te inclusions has almost no temperature dependence. Therefore, the net effect of the Te deposits is to increase the path length of the radiation through the absorbing medium, thereby increasing the net absorption. In this model the effect is captured by enhancing H, which accounts for $H > 0.02$ samples. If the Te deposits scatter uniformly in a spherical form, and the distance between them is greater than the wavelength of the incident light, then according to an aerosol model and Mie scattering theory, the relative extinction cross-section C_{ext} (Kerker 1969) is:

$$\begin{aligned} C_{ext} &= \frac{\lambda_m^2}{2\pi}\sum_{n=1}^{\infty}(2n+1)\mathrm{Re}(a_n+b_n) \\ &= \frac{3\pi V\varepsilon_m^{1/2}}{\lambda\alpha^3}\sum_{n=1}^{\infty}(2n+1)\mathrm{Re}(a_n+b_n) \end{aligned} \tag{4.232}$$

where $\lambda_m = \lambda / n_2$ is the wavelength of the incident light in the HgCdTe medium, $V = (4/3)\pi a^3$ is the deposit's bulk volume, a is the deposit's radius, ε_m and ε are the dielectric constants of the HgCdTe medium and the deposit, respectively, $\varepsilon_m^{1/2} = n_1 + in_2$; a_n, b_n include Bessel functions $\psi_n(\alpha), \psi_n(\beta)$ and a Hankel function $\xi_n(\alpha)$ and their derivatives, $\alpha = 2\pi a\varepsilon^{1/2}/\lambda$, $\beta = 2\pi m_1 a/\lambda$. If $\alpha < 1$, then the first three terms of the above equation are enough to describe the light scattering (Yadava et al. 1994), that is:

$$C_{ext} = C_{1a} + C_{2a} + C_{1b} \tag{4.233}$$

where C_{1a}, C_{2a}, C_{1b} represent the extinction cross-sections of the electric dipole oscillations, the electric quadruple oscillations, and the magnetic dipole oscillations, respectively. They are:

$$\begin{aligned} C_{1a} &= V\frac{18\pi\varepsilon_m^{3/2}}{\lambda}\frac{\varepsilon_2}{(\varepsilon_1+2\varepsilon_m)^2\varepsilon_2^2} \\ C_{2a} &= V\frac{1.25\pi\varepsilon_m^{3/2}}{\lambda}\frac{\varepsilon_2\alpha^2}{(\varepsilon_1+1.5\varepsilon_m)^2\varepsilon_2^2} \\ C_{1b} &= -V\frac{\pi\varepsilon_m^{-1/2}\varepsilon_2}{\lambda}\alpha^2 \end{aligned} \tag{4.234}$$

The radii of tellurium precipitates are distributed in a range from a_1 to a_2. The radial distribution of these micro-precipitates can be represented by a log-normal distribution *P(a)*, and then the total apparent absorption coefficient can be written as:

$$\alpha_{Mie} = N_{pre} \int C_{ext}(a) P(a) da \tag{4.235}$$

where

$$P(a) = \frac{1}{\sqrt{2\pi}\sigma} \frac{1}{a} \exp\left[-\frac{(\ln a - \mu)^2}{2\sigma^2} \right] \tag{4.236}$$

The argument of the exponential must be dimensionless. Defining the quantity $\mu \equiv \ln a^*$ fixes this because $\ln a - \ln a^* = \ln(a/a^*)$ is dimensionless. The quantity σ is also dimensionless. σ and a^* are related to the mean radius and its mean square variance of the Te particles. The mean particle radius, $\bar{a} = \exp(\mu + \sigma^2/2) = a^* \exp\left(\sigma^2/2\right) \cong a^*$, where the last approximate equality holds for small σ. The square variance, $\Delta^2 = \exp(2\mu + \sigma^2)\left[\exp(\sigma^2) - 1\right] = \bar{a}^2\left[\exp\left(\sigma^2\right) - 1\right] \cong \bar{a}^2\sigma^2$, and once again the last approximate equality holds for small σ. Then by using the tellurium optical constants (Caldwell and Fan 1959) the tellurium deposit's absorption coefficient α_{Mie} can be calculated from (4.233) to (4.236).

Sometimes an abnormal optical absorption phenomenon is found in $Hg_{1-x}Cd_xTe$ epitaxial layers. From (4.217) to (4.228), we find that for a pure n-type $Hg_{1-x}Cd_xTe$ sample, the absorption coefficient is predicted to decrease with decreasing temperature; while for a pure p-type sample, the coefficient is predicted to increase with decreasing temperature (Mroczkowski and Nelson 1983). But in practice, sometimes an abnormal temperature behavior occurs and with decreasing temperature, for an n-type sample the absorption coefficient first decreases then increases, and for a p-type sample the absorption coefficient first increases then decreases. This behavior is possibly caused by imperfect growth and annealing, leaving p-type defects in n-type samples, or n-type defects in p-type samples (Tian et al. 1991; Huang 1990).

Generally inclusions have arbitrary shapes, and their carrier concentrations are not uniform. But it can be assumed that inclusion areas are uniform in the plane illuminated by the light beam spot. As long as the lighted spot is small enough, the inclusion areas can be treated as bulk areas with uniform carrier distributions, as show in Fig. 4.42. Some important properties of samples with inclusions can be gleaned from this simple model.

For samples with inclusions, assuming that the volume ratio for p-type regions is D_P, the total absorption coefficient α_T can be expressed as:

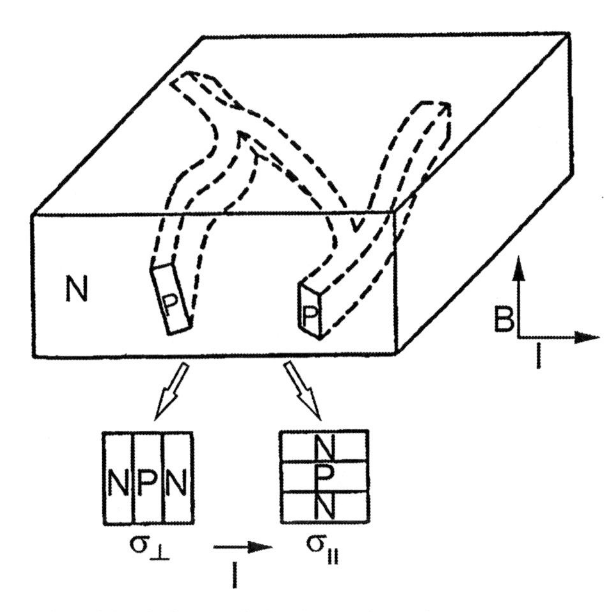

Fig. 4.42. Schematic of the infrared light absorption of a $Hg_{1-x}Cd_xTe$ sample with inclusions

$$\alpha_T = D_p\alpha_p + (1 - D_p)\alpha_n \tag{4.237}$$

where α_n, α_p are given by (4.225) and (4.226), respectively. According to this model the total apparent absorption coefficient α_T of a $Hg_{1-x}Cd_xTe$ sample will differ from that of a pure p-type or a pure n-type sample, consequently it leads to an abnormal absorption effect. Actually p–n junctions will form, so depending on the carrier concentrations and the sizes of the inclusions the depletion layers of the junctions could significantly modify the interpretation of the parameters deduced from this first step model.

Some experimental results follow. The samples used in these experiments were $Hg_{1-x}Cd_xTe$ epitaxial films grown by LPE, MOCVD and MBE. By using a far infrared Fourier transform spectrometer (FTIR), and a matched variable temperature Dewar, transmission spectra of samples were measured over the temperature range 77—300 K. Note that sometimes absolute transmittance values of samples cannot be accurately measured by an FTIR spectrometer, but the data can be corrected by comparisons to room temperature infrared transmission spectra measured by a grating infrared spectrometer. The transmission spectrum of a sample can be transformed into its absorption spectrum by employing (4.229), providing the light loss

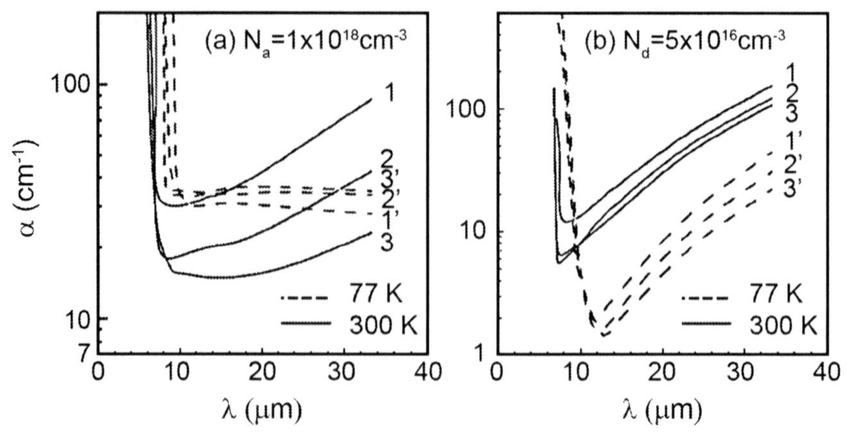

Fig. 4.43. The free carrier absorption coefficient spectra at 300 and 77 K for (**a**) p-type epitaxial film, and p-type bulk samples with acceptor concentrations of $N_A = 1\times10^{16}$ cm^{-3} for all the samples; (**b**) n-type epitaxial film and n-type bulk samples with donor concentration of $N_D = 5\times10^{16}$ cm^{-3} for all the samples (In both (**a**) and (**b**) figures the *solid lines* labeled 1 stand for bulk samples with x = 0.236, the *solid lines* labeled 2 and 3 stand for epitaxial film samples. The slope of the sample labeled with *solid line* 2 is s = 0.003 μm^{-1}, while that for the sample labeled with solid line 3, is s = 0.01 μm^{-1}. Other parameters are: the composition of the surface of the samples is $x_S = 0.236$, the width of the compositional transition regions is Δz = 2 μm, and the thickness of the samples is d = 25 μm)

at the sample's surface need not be considered, i.e. if $H = 0$ in (4.229) (Mollmann and Kissel 1991). This approximation is valid for well polished surfaces.

It is assumed that the $Hg_{1-x}Cd_xTe$ bulk samples have a uniform compositional distribution in both transverse and longitudinal directions, while the epitaxial film samples have a uniform compositional distribution in the transverse direction, but a non-uniform distribution in the longitudinal direction. Figures 4.43a, b show the absorption spectra for a p-type and an n-type epitaxial film samples, as well as for a p-type and an n-type bulk $Hg_{1-x}Cd_xTe$ samples at 300 and 77 K. In the figures two solid lines labeled 1 display the absorption curves for bulk $Hg_{1-x}Cd_xTe$ ($x = 0.236$) samples, and solid lines 2 and 3 label the optical absorption curves for the two epitaxial samples with the same parameters except for different slopes of their longitudinal compositional distributions. For the sample corresponding to solid line 2, the slope is s = 0.003 μm^{-1}, while the sample corresponding to solid line 3 has slope s = 0.01 μm^{-1}. The thickness of the samples is 25 μm^{-1}. The acceptor concentration of the p-type material is $N_A = 1\times10^{16}$ cm^{-3}, and the donor concentration of the n-type material is $N_D = 5\times10^{16}$ cm^{-3}.

As shown in the figures, the free carrier absorption coefficients of $Hg_{1-x}Cd_xTe$ follow a power law relation with the incident light wavelength, λ, namely, $\alpha \sim \lambda^r$, where the power r depends on the scattering mechanism (Gurauskas et al. 1983). For p-type $Hg_{1-x}Cd_xTe$ samples, $r \approx 2$ at room temperature, and $r \approx 0$ (independent of wavelength) at 77 K. For n-type samples, $r \approx 2.7$ at room temperature, and the r value basically does not change when the temperature changes. Holes are the chief cause of free carrier absorption of the p-type $Hg_{1-x}Cd_xTe$. At room temperature the intra-band heavy hole contribution to the absorption coefficient, $\sigma_{\nu 1}$, is of the same order of magnitude as the inter-band coefficient, $\sigma_{\nu 1 \nu 2}$. The coefficient, $\sigma_{\nu 1}$, varies as the square of the wavelength λ. Since narrow gap $Hg_{1-x}Cd_xTe$ is intrinsic at room temperature, the contribution from the electron absorption coefficient σ_n to the total at elevated temperatures cannot be ignored.

The total absorption coefficients relation to wavelength and temperature is complex because of the multitude of many contributions. Some of this complexity arises because of the heavy-hole concentration distribution near the top of valence band. This concentration sets the final state density for the transition from the light hole band to the heavy hole band, and therefore to the absorption coefficient $\sigma_{\nu 1 \nu 2}$.

The dependence of $\sigma_{\nu 1 \nu 2}$ on wavelength λ is rather weak, and leads to the slow variation of the p-type sample absorption coefficient with wavelength. The free carrier absorption of n-type $Hg_{1-x}Cd_xTe$ is mainly determined by electron transitions within the conduction band, assisted by polar optical phonon scattering, acoustic phonon scattering, and ionized impurity scattering. From (4.217) to (4.228), it is known that $\alpha_{OP} + \alpha_{AP}$ and α_I have power law dependences on the wavelength with powers of 2.7 and 3.7, respectively. Furthermore, these relationships do not vary appreciably with temperature (Baltz and Escher 1972). Thus to lowest order, the calculated α_n wavelength dependence is $\lambda^{2.7}$. The free carrier absorption behavior for $Hg_{1-x}Cd_xTe$ epitaxial layers is different from those of bulk materials. For a p-type sample, the larger the longitudinal composition slope s is at room temperature (bulk materials with uniform longitudinal composition have s=0), the smaller is the free carriers absorption coefficient α_p. In the relation of α_p to wavelength, the power r also decreases with an increase of s. The absorption coefficient decreases rapidly for a sample with small s when the temperature decreases, and the values of

r are all nearly zero. That leads to a tendency of the free carrier absorption of various p-type samples to be the same.

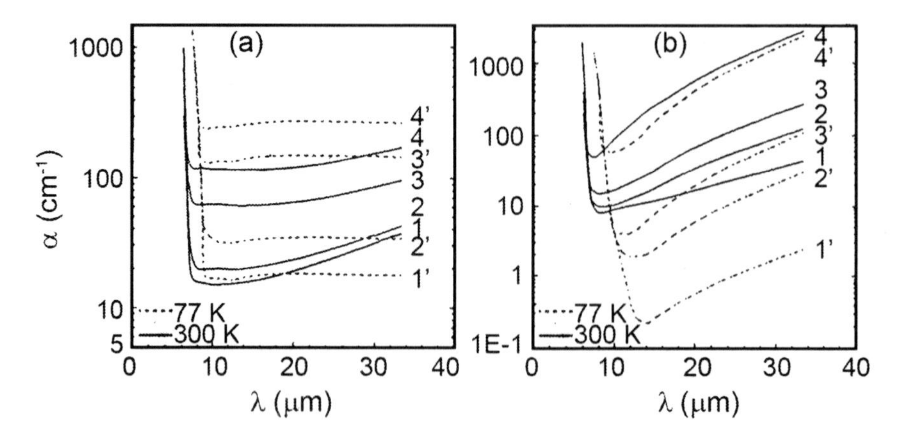

Fig. 4.44. All samples have the same longitudinal compositional distribution parameters which are: surface composition $x_s = 0.236$, composition slope $s = 0.003\,\mu m^{-1}$, transition region width $\Delta z = 1.5\,\mu m$, sample thickness $d = 25\,\mu m$. (**a**): The free carrier absorption curves of p-type $Hg_{1-x}Cd_xTe$ epitaxial samples at 300 and 77 K. The concentrations of the samples for curves 1,1' is $N_A = 5\times10^{15}\ cm^{-3}$, for curves 2,2' is $N_A = 1\times10^{16}\ cm^{-3}$, for curves 3,3' is $N_A = 5\times10^{16}\ cm^{-3}$, and for curves 4,4' is $N_A = 1\times10^{17}\ cm^{-3}$. (**b**): The free carrier absorption curves of n-type $Hg_{1-x}Cd_xTe$ samples at 300 and 77 K. The concentrations of the samples for curves 1,1' is $N_D = 1\times10^{16}\ cm^{-3}$, for curves 2,2' is $N_A = 5\times10^{17}\ cm^{-3}$, for curves 3,3' is $N_D = 1\times10^{17}\ cm^{-3}$, and for curves 4,4' is $N_D = 5\times10^{16}\ cm^{-3}$

For n-type samples, an increase of the longitudinal compositional slope s causes the free carrier absorption to decrease, and tends to be independent of temperature and wavelength. This happens because large s values mean that there are larger volumes of higher x material, and the light-heavy hole inter-band absorption cross-section σ_{v1v2} is insensitive to x, while the electron absorption cross-section σ_n and the intrinsic carrier concentration n_i decrease with an increase of x. Free carrier absorption of n-type $Hg_{1-x}Cd_xTe$ is dominated by optical phonon, acoustic phonon and ionized impurity scattering, as can be seen from its relation to α_{OP}, α_{AD} and α_I in (4.217)–(4.228) to composition x.

For p-type samples, the intrinsic absorption is strong at room temperature; the absorption coefficient is large for samples with small s; as the temperature decreases the free carrier concentration decreases; and the contribution from σ_{v1v2} is enhanced. The net effect is that the difference among the absorption constants of a variety of samples is small. However, the absorption coefficients of samples with small s do exhibit somewhat larger temperature dependences. Also the absorption does become weaker for samples with larger s. The carrier concentration can be estimated from free carrier absorption of $Hg_{1-x}Cd_xTe$ samples. The free carrier absorption spectrum of p-type and n-type $Hg_{1-x}Cd_xTe$ epitaxial samples, with different doping concentrations, at 300 and 77 K, are shown in Fig. 4.44.

The sample's longitudinal compositional distribution parameters used in the data reduction are set as: surface composition $x_S = 0.236$, composition slope $s = 0.003\,\mu\text{m}^{-1}$, transition region width $\Delta z = 1.5\,\mu\text{m}$, and sample thickness $d = 25\,\mu\text{m}$. From the figure it can be seen that for p-type $Hg_{1-x}Cd_xTe$, the absorption coefficient α_P increases as the acceptor concentration N_A increases, and at the lower temperature the absorption constant variation with concentration is larger. For n-type material, as the donor concentration N_D increases, the absorption coefficient α_n increases, and the power r of the α_n relationship to the wavelength decreases. At the lower temperature, the concentration dependence increases.

For p-type samples varying N_A just changes the carrier concentration, and that has little influence on the σ_{v1} and σ_{v1v2} contributions to the absorption coefficient. However, for n-type samples changing N_D not only changes the carrier concentration, but also influences the magnitude of the ionized impurity scattering contribution, and consequently the net absorption coefficient's dependence on the wavelength. At lower temperatures the influence of ionized impurity scattering on free carrier absorption becomes dominant. This result indicates that the ionized impurity concentration can be deduced by fitting the absorption spectrum at 77 K or even at somewhat higher temperatures (Brossat and Reymond 1985). Due to the fact that the donor ionization energy for n-type HgCdTe samples is very small, there is hardly any freezeout at low temperature. So unless there is compensation, the total donor and ionized donor concentration are the same. Because the hole effective mass for p-type HgCdTe is large, the acceptor ionization energies are larger, and carrier freeze out and competing mechanisms makes it difficult to measure acceptor compositions the same way. The measurement of acceptor concentrations by direct optical methods is preferable.

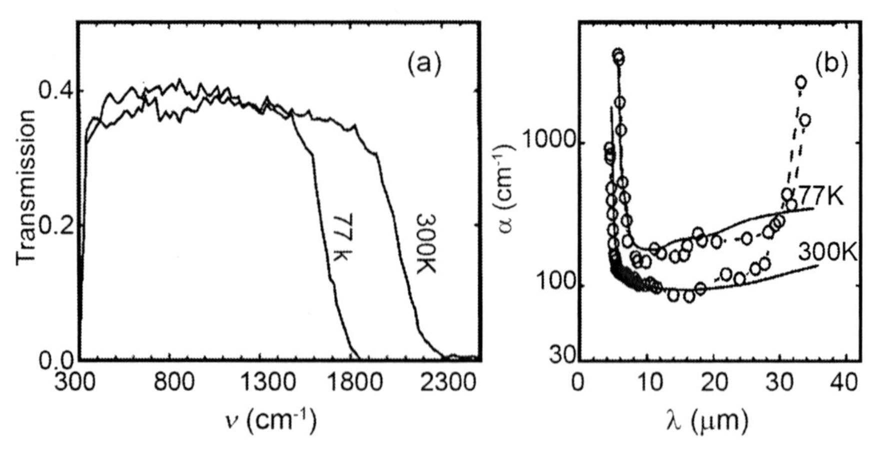

Fig. 4.45. A p-type liquid phase epitaxial grown $Hg_{1-x}Cd_xTe$ sample's infrared (**a**) transmission spectra and (**b**) absorption spectra at 300 and 77 K. The sample's thickness is $d = 15\,\mu m$. The solid lines in Fig. 4.45b are calculated, and the dots are absorption coefficients deduced from the transmittance data in Fig. 4.45a

The infrared transmission spectra of a p-type $Hg_{1-x}Cd_xTe$ liquid phase epitaxially grown sample measured at 300 and 77 K are show in Fig. 4.45a. The sample's thickness is $d = 15\,\mu m$. The sample's longitudinal composition distribution parameters were obtained by fitting the absorption edge and intrinsic absorption region of transmission spectrum at room temperature. The fitted parameters are listed as: a surface alloy composition $x_s = 0.265$, a compositional slope $s = 0.003\,\mu m$, a transition region width $\Delta z = 1.0\,\mu m$, and $H = 0$. Based on the transmission curve, the absorption coefficient can be calculated from (4.229), and are shown in Fig. 4.45b. The dots are experimental values, and the solid lines are absorption coefficients deduced by fitting the parameters in (4.217)–4.228. In (4.217)–(4.228), after the longitudinal composition distribution parameters are found, only the ionized acceptor concentration N_A is unknown, and the calculated curve can be made to agree well with the experimental data by adjusting the N_A value. The effective ionized acceptor concentration deduced by fitting the data for this sample is $N_A - N_D = 8 \times 10^{16}\,cm^{-3}$, which agrees with the acceptor concentration of $8.32 \times 10^{16}\,cm^{-3}$ obtained from a Hall measurement. In view of the epitaxial layer being quite thin ($< 30\,\mu m$) so its transmittance is large, and because it is a sensitive function of temperature, the precision of the impurity concentration deduced by fitting the absorption spectrum is high (<10%).

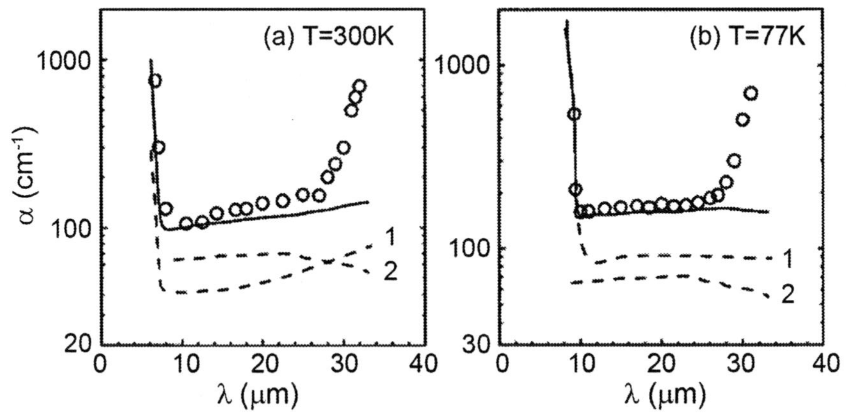

Fig. 4.46. A p-type $Hg_{1-x}Cd_xTe$ epitaxial sample's absorption spectra at (**a**) 300 K and (**b**) 77 K. *Dots* are experimental results, the *dashed curves* labeled 1 and 2 are the free carrier absorption and tellurium precipitate absorption contributions, respectively, and the *solid curve* is the total absorption coefficient

From the contribution to the absorption spectra ascribed to tellurium precipitates, their concentration can also be deduced. The absorption spectra of p-type $Hg_{1-x}Cd_xTe$ epitaxial samples are shown in Fig. 4.46a at 300 K and 4.46b at 77K. The measured ionized acceptor concentration is $N_A = 3\times10^{16}\,\text{cm}^{-3}$, and the thickness is $d = 20\,\mu\text{m}$.

The epitaxial layer's longitudinal compositional distribution parameters obtained by fitting the room temperature transmittance curve are: the surface composition $x_s = 0.22$, the composition slope $s = 0.004\,\mu\text{m}^{-1}$, and the transition region width $\Delta z = 2\,\mu\text{m}$. It was found that to fit the transmittance maximum, a value of $H = 0.12$ was needed in (4.210). Substituting these parameters into (4.217)–(4.228) produces the absorption coefficient α_c shown as the dashed curve labeled 1 in the figure. It is obvious that this α_c is much smaller than the measured value at 300 and 77 K. If account is taken of the absorption from tellurium precipitates, using (4.233)–(4.235) the tellurium's absorption coefficient α_{Te} can be calculated and is shown as the dashed curve labeled 2 in the figure. The total absorption coefficient $\alpha_s = \alpha_c + \alpha_{Te}$, denoted by the solid curve in the figure, is seen to agree well with the experimental results at both 300 and 77 K. In the calculation the tellurium precipitate concentration $N_{pre} = 2\times10^{18}\,\text{cm}^{-3}$ was used, and the radius distributional dimensions in (4.235) were taken as $a_1 = 0.25\,\mu\text{m}$, and $a_2 = 1.25\,\mu\text{m}$. Based on this

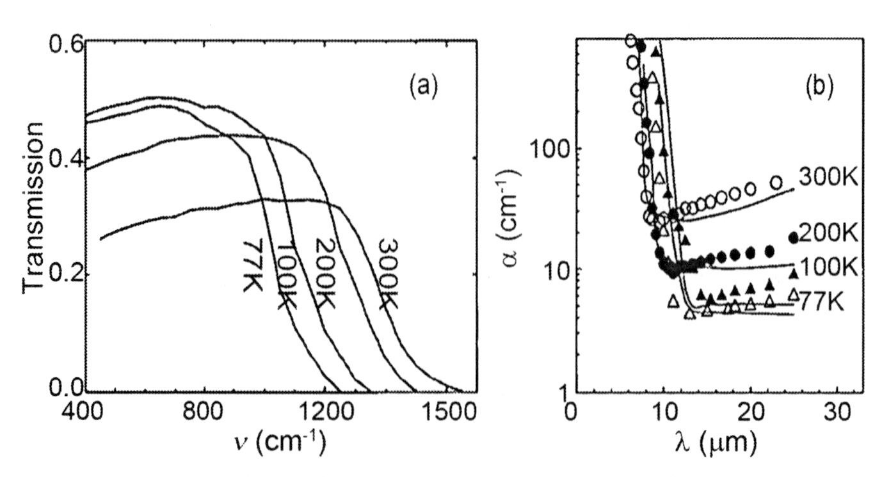

Fig. 4.47. A $Hg_{1-x}Cd_xTe$ epitaxial sample's infrared (**a**) transmission spectra and (**b**) absorption spectra at different temperatures. The sample thickness is $d = 25\,\mu m$. The *solid curves* in (**b**) are calculated, dots are absorption coefficient data deduced by fitting the transmittance curves in (**a**)

calculation the excess tellurium occupying the bulk is 1.8×10^{-3}, or the equivalent concentration is $1.8\times10^{19}\,cm^{-3}$. This value is very close to the value estimated by Anderson et al. (1982). It shows that the absorption spectrum of this epitaxial film sample with H>0.02 is explained satisfactorily by free carrier absorption and micro-precipitate scattering.

For a $Hg_{1-x}Cd_xTe$ epitaxial film with an abnormal optical absorption spectrum, the n and p inclusion situation can be also determined. The infrared transmission spectra of a $Hg_{1-x}Cd_xTe$ epitaxial sample of thickness, $d = 25\,\mu m$, at different temperatures are shown in Fig. 4.47a. The sample's longitudinal compositional distribution parameters, obtained by fitting the room temperature transmission spectrum, are: the surface composition is, $x_s = 0.215$, the composition slope is, $s = 0.003\,\mu m^{-1}$, and the transition region width is, $\Delta z = 2\,\mu m$. As shown in the figure, the sample's transmittance increases when the temperature decreases from 300 to 100 K, which is the absorption property of an n-type material. However, from 100 to 77 K the transmittance decreases, which is a p-type behavior. So the sample is n-type but with p-type inclusions in it. Based on the sample's transmittance characteristic, its absorption coefficient can be fitted by (4.229), as show in Fig. 4.47b. The dots in Fig. 4.47b are the experimental results, and the solid curve is calculated. The parameters adopted in the calculation are: in the n-type region the ionized donor concentration is

$N_D = 5.8 \times 10^{14}$ cm^{-3} (from a Hall measurement), and in the p-type region the inclusion parameter is $D_p = 0.08$, and the ionized acceptor concentration is $N_A = 2 \times 10^{16}$ cm^{-3}. There are two adjustable parameters D_p and N_A, their choices can be confirmed by testing the quality of the fittings to the measured absorption coefficient data at different temperatures.

Using epitaxial technology, as-grown $Hg_{1-x}Cd_xTe$ films have carrier concentrations and mobilities that are generally unsuitable for device preparation. Anneal treatments to improve the quality of the as-grown samples are needed to convert them into weak p-type or n-type samples suitable for device preparation. If a thermal treatment process is not perfect, it is easy to generate p-type inclusions in n-type samples; or if the as-grown p-type crystal is non-uniform, it may have p-type islands after treatment. All these effects will influence the device quality (Yang 1988). It is difficult to measure a sample's inclusions by Hall measurements. However, based on the free carrier absorption properties of p-type and n-type $Hg_{1-x}Cd_xTe$, the effect of inclusions on their absorption spectra can be measured. The optical model and the inclusion model reported in this section can be used to estimate the inclusion properties of epitaxial samples.

4.5.4 Magneto-Optic Effect of Free Carriers

When a magnetic field is applied to a semiconductor, the moving free electrons experience the action of an external applied Lorentz force. Then the response to an optical frequency field is characterized by a complex conductivity parameter. If the magnetic field is parallel to the z-axis, an electron with effective mass m^* has an equation of motion in a magnetic field which is:

$$m^* \frac{dv}{dt} + m^* \frac{v}{\tau} = e\left(E + \frac{v \times \boldsymbol{H}}{c} \right) \tag{4.238}$$

where τ is the electron's relaxation time, assuming that the scattering is isotropic. For a material with cubic symmetry, the conductivity tensor is:

$$\sigma = \begin{pmatrix} \sigma_{xx} & \sigma_{xy} & 0 \\ \sigma_{yx} & \sigma_{yy} & 0 \\ 0 & 0 & \sigma_{zz} \end{pmatrix} \tag{4.239}$$

where the components are:

$$\begin{cases} \sigma_{xx} = \sigma_{yy} = \dfrac{\sigma_0(1+i\omega\tau)}{(1+i\omega\tau)^2 + \omega_c^2\tau^2} \\ \sigma_{yx} = -\sigma_{xy} = \dfrac{\sigma_0\omega_c\tau}{(1+i\omega\tau)^2 + \omega_c^2\tau^2} \\ \sigma_{zz} = \dfrac{\sigma_0}{1+i\omega\tau} \end{cases} \tag{4.240}$$

Here the zero frequency conductivity is $\sigma_0 = \dfrac{Ne^2\tau}{m^*}$, ω is the optical frequency, and $\omega_c = \dfrac{eB}{m^* c}$ is the cyclotron frequency.

The relevant Maxwell equations are:

$$\begin{aligned} \nabla \times \boldsymbol{E} &= -\frac{1}{c}\frac{\partial \boldsymbol{B}}{\partial t} \\ \nabla \times \boldsymbol{H} &= \frac{4\pi}{c}\boldsymbol{J} + \frac{1}{c}\frac{\partial \boldsymbol{D}}{\partial t} \end{aligned} \tag{4.241}$$

Taking the parameters to be:

$$\begin{aligned} \boldsymbol{J} &= \boldsymbol{\sigma} \cdot \boldsymbol{E} \\ \boldsymbol{B} &= \mu \boldsymbol{H} \\ \boldsymbol{D} &= \varepsilon \boldsymbol{E} \end{aligned} \tag{4.242}$$

and substituting them into (4.241), yields the equation for the field of the electromagnetic wave propagating in the medium:

$$\begin{aligned} \nabla \times \nabla \times \boldsymbol{E} &= \nabla(\nabla \cdot \boldsymbol{E}) - \nabla^2 \boldsymbol{E} \\ &= -\left(\frac{4\pi\mu}{c^2}\right)\boldsymbol{\sigma} \cdot \frac{\partial \boldsymbol{E}}{\partial t} - \left(\frac{\mu\varepsilon}{c^2}\right)\boldsymbol{\sigma} \cdot \frac{\partial^2 \boldsymbol{E}}{\partial t^2} \end{aligned} \tag{4.243}$$

Examine a trial plane wave solution:

$$\mathbf{E} = \mathbf{E}_0 \exp\left[i(\omega t - \boldsymbol{k} \cdot \boldsymbol{r})\right] \tag{4.244}$$

Substituting (4.244) into (4.243) one obtains an expression for the dispersion relation between the frequency and the wave number:

$$-\boldsymbol{k}(\boldsymbol{k} \cdot \boldsymbol{E}_0) + k^2 \boldsymbol{E}_0 = \left(\frac{\omega^2}{c^2}\right)\mu\varepsilon\left[\boldsymbol{I} - \frac{4\pi i \boldsymbol{\sigma}}{\omega\varepsilon}\right] \cdot \boldsymbol{E}_0 \tag{4.245}$$

Let

$$\boldsymbol{\varepsilon} \equiv \varepsilon \boldsymbol{I} - \left(\frac{4\pi i}{\omega}\right)\boldsymbol{\sigma} \tag{4.246}$$

$\boldsymbol{I}$ is the unit matrix, $\boldsymbol{\varepsilon}$ and $\boldsymbol{\sigma}$ are, respectively, the dielectric tensor and the conductivity tensor. Then (4.245) becomes:

$$-\boldsymbol{k}(\boldsymbol{k}\cdot \boldsymbol{E}_0) + k^2\boldsymbol{E}_0 = \left(\frac{\omega^2}{c^2}\right)\mu\boldsymbol{\varepsilon}\boldsymbol{E}_0 \tag{4.247}$$

When an electromagnetic wave propagates parallel to the ***B*// z** direction ($\boldsymbol{k}\,//\,\boldsymbol{B}$), substituting (4.241) into (4.247), and writing the result for an electromagnetic circularly-polarized electric field wave in *xy* plane. $E_{0\pm} = E_{0x} \pm iE_{0y}$, yields:

$$k_\pm^2 = \left(\frac{\omega^2}{c^2}\right)\mu\varepsilon\left[1 - \frac{4\pi i}{\omega\varepsilon}\sigma_\pm\right] \tag{4.248}$$

where $\sigma_{xx} = \sigma_{yy}$, $\sigma_{xy} = -\sigma_{yx}$, $\sigma_\pm = \sigma_{xx} \pm i\sigma_{xy}$, and:

$$\begin{aligned} k &= \frac{\omega}{c}n \\ k_\pm^2 &= \frac{\omega^2}{c^2}n_\pm^2 \end{aligned} \tag{4.249}$$

where $n_\pm$ are the complex index of refraction responses to the circularly polarized fields, $n_\pm \equiv n - ik_\pm$. Comparing (4.248) and (4.249), we find:

$$\begin{cases} n_\pm^2 - k_\pm^2 = \mu\varepsilon \\ 2n_\pm k_\pm = \dfrac{4\pi\mu\sigma_\pm}{\omega} \end{cases} \tag{4.250}$$

When an electromagnetic wave propagates perpendicular to the $\boldsymbol{B}$ (or **z**) direction, so $\boldsymbol{k}$ lies in the *x*-*y* plane the result is a different. Assume the wave propagates along the *y* direction, then the electromagnetic wave's electric vector lies in the *z*–*x* plane, and there are electric field component parallel and perpendicular to $\boldsymbol{B}$ which introduce modifications to the dispersion relations:

$$k_{//}^2 = \omega^2 \varepsilon \left[1 - \frac{4\pi i}{\omega \varepsilon} \sigma_{zz} \right]$$
$$k_{\perp}^2 = \omega^2 \varepsilon \left[1 - \frac{4\pi i}{\omega \varepsilon} \left(\sigma_{xx} + \frac{\sigma_{xy}^2}{\sigma_{xy} + i\omega\varepsilon / 4\pi} \right) \right] \tag{4.251}$$

The relation of k to the complex index of refraction $n - i\kappa$ is:

$$k = \left(\frac{i\omega}{c} \right) (n - i\kappa), \text{or}$$
$$(n - i\kappa) = -\frac{ick}{\omega} \tag{4.252}$$

Then we find:

$$\varepsilon_{\pm} = \mu\varepsilon \left[1 - \frac{4\pi i}{\omega \varepsilon} \sigma_{\pm} \right]$$
$$k_{//}^2 = c\varepsilon \left[1 - \frac{4\pi i}{\omega \varepsilon} \sigma_{zz} \right] \tag{4.253}$$
$$k_{\perp}^2 = c\varepsilon \left[1 - \frac{4\pi i}{\omega \varepsilon} \left(\sigma_{xx} + \frac{\sigma_{xy}^2}{\sigma_{xy} + i\omega\varepsilon / 4\pi} \right) \right]$$

The power absorbed by an electron from the electromagnetic wave is:

$$P = \frac{1}{2} \mathrm{Re}(\mathbf{J} \cdot \mathbf{E}) = \frac{1}{2} \mathrm{Re}(\boldsymbol{\sigma} \cdot \mathbf{E} \cdot \mathbf{E}) \tag{4.254}$$

For linearly polarized light, this equation becomes:

$$P = \frac{1}{2} E_{0x}^2 \sigma_{xx}^R = \frac{1}{2} E_{0x}^2 \sigma_0 \, \mathrm{Re} \left[\frac{1 + i\omega\tau}{(1 + i\omega\tau)^2 + \omega_c^2 \tau^2} \right] \tag{4.255}$$

The fraction of the incident power absorbed is:

$$\frac{P}{P_0} = \frac{1 + (\omega^2 + \omega_c^2)\tau^2}{\left[1 + (\omega^2 + \omega_c^2)\tau^2 \right]^2 + 4\omega^2\tau^2} \tag{4.256}$$

where P_0 is the incident power. This formula was first derived by Lax et al. (1954). Equation (4.256) can be rewritten as:

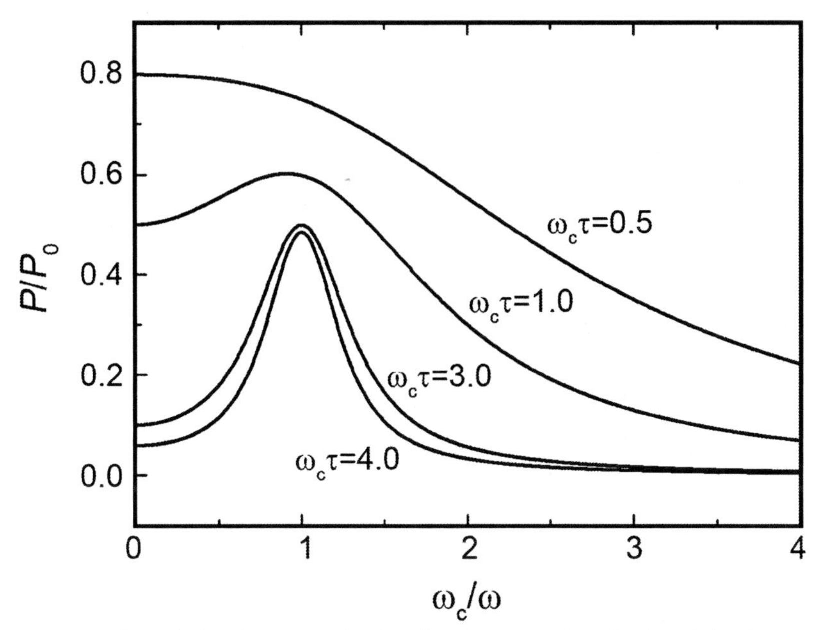

Fig. 4.48. The relation between the fractional power absorbed and the frequency, for various values of $\omega\tau_c$

$$\frac{P}{P_0}=\frac{1+\omega_c^2\tau^2\left(\dfrac{\omega^2}{\omega_c^2}+1\right)}{\left[1+\omega_c^2\tau^2\left(1-\dfrac{\omega^2}{\omega_c^2}\right)\right]^2+4\omega^2\tau^2\dfrac{\omega^2}{\omega_c^2}} \tag{4.257}$$

Figure 4.48 is the curve of the fractional absorption vs. frequency at different $\omega_c\tau$ values. An absorption peak at $\omega/\omega_c=1$ appears when $\omega_c\tau>1$. This is the cyclotron resonance. When the frequency ω_c is determined from experiment and knowing the magnitude of the B field, the effective mass m^* can be deduced.

In the experiments ω/ω_c is scanned by varying B, rather than ω. A fixed frequency laser sets ω, and ω_c is determined by B. A typical result is in Fig. 4.49.

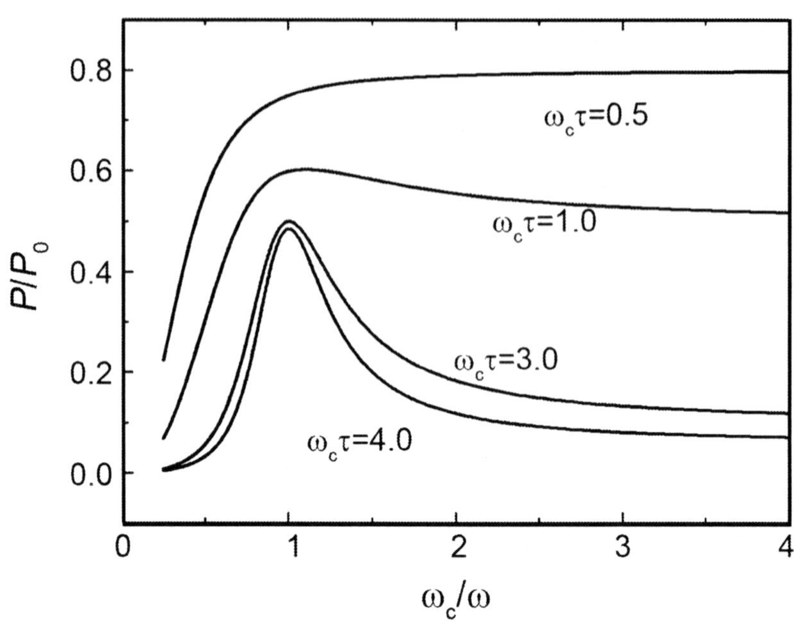

Fig. 4.49. The relation between the fractional absorption and the magnetic field

When the optical wave vector is parallel to the magnetic field ***k***//***B***, the Faraday configuration, the absorption coefficient can be derived by using the conductivity tensor in (4.240). In this case, we have:

$$\left(n_{\pm} - ik_{\pm}\right)^2 = \mu\varepsilon_{\infty}\left[1 - \frac{\omega_p^2}{\omega(\omega \mp \omega_c - i/\tau)}\right] \tag{4.258}$$

where ω_p is plasma oscillation frequency that determines the reflection edge, and is given by (4.215).

In the situation where $n^2 >> k^2$, for non-magnetic material, we have:

$$n_{\pm}^2 = \varepsilon_{\infty}\left[1 - \frac{\omega_p^2}{\omega\left[\left(\omega \mp \omega_c\right)^2 + \left(\frac{1}{\tau}\right)^2\right]}\right] \tag{4.259}$$

When $\omega >> \tau^{-1}, \omega_c$, (4.260) becomes:

$$n_{\pm}^2 = \varepsilon_\infty \left[1 - \frac{\omega_p^2}{\omega} \frac{1}{\omega \mp \omega_c}\right] \tag{4.260}$$

$$(2nk)_{\pm} = \varepsilon_\infty \frac{\omega_p^2}{\omega} \frac{\frac{1}{\tau}}{\frac{1}{\tau^2} + (\omega \mp \omega_c)^2} \tag{4.261}$$

The meaning of these formulas is that a linearly polarized incident light beam can be resolved into two oppositely rotating circularly polarized branches. When the wave vector is parallel to an applied magnetic field, in the medium the two circular polarizations encounter different refractive indices, and consequently have different light speeds in the medium. This causes the polarization of the light exiting the medium to be rotated relative to the incident polarization.

Also the two polarizations have different absorption coefficients in the medium:

$$\alpha_{\pm} = \frac{2\omega}{c} k_{\pm} \cong \frac{4\pi}{nc}\left(\sigma_{xx}^R + \sigma_{xy}^I\right) \tag{4.262}$$

or

$$\alpha_{\pm} = \frac{\varepsilon_\infty}{nc} \frac{\omega_p^2 \frac{1}{\tau}}{\frac{1}{\tau^2} + (\omega \mp \omega_c)^2} \tag{4.263}$$

where σ_{xx}^R and σ_{xy}^I are the real part of σ_{xx} and the imaginary part of σ_{xy}, respectively. σ_{xx} and σ_{xy} are defined by (4.240).

From the above equations, it can be seen that for the α_- branch (right circularly polarized light), a cyclotron resonance peak can occur when $\omega = \omega_c$, while for the α_+ branch there is no cyclotron resonance peak.

At the plasma edge, as modified by the applied magnetic field, where from (4.260) $n_{\pm} = 0$, we have:

$$\omega^2 \mp \omega\omega_c - \omega_p^2 = 0 \tag{4.264}$$

which has the solution:

$$\omega = \frac{1}{2}\left[\pm\omega_c + \left(\omega_c^2 + 4\omega_p^2\right)^{1/2} \right] \tag{4.265}$$

In a cyclotron resonance transmission experiment, one requires $\omega_c >> \omega_p$ in order to allow the light to penetrate the sample. Then, the solution is:

$$\omega \cong \omega_c + \frac{\omega_p^2}{\omega_c^2} \tag{4.266}$$

Given these conditions the effective plasma frequency deviates only slightly from the cyclotron resonance frequency.

In the opposite extreme where, $\omega_c << \omega_p$, we have:

$$\omega_\pm \cong \omega_p \pm \frac{1}{2}\omega_c + \frac{\omega_c^2}{8\omega_p} \tag{4.267}$$

This indicates that the two effective plasma edges for the different circular polarizations are split roughly by ω_c. This phenomenon provides another means to measure the cyclotron frequency ω_c, and therefore the effective mass m^*.

If the optical wave-vector is perpendicular to the magnetic field, $\boldsymbol{k} \perp \boldsymbol{B}$, there are two possibilities: the electric field vector is parallel to ***B***, or perpendicular to ***B***. If ***E//B***, the optical constant derivation in the former section applies, and:

$$\left(n - ik\right)_{//}^2 = \varepsilon_\infty \left[1 - \frac{\omega_p^2}{\left(\omega - i/\tau\right)\omega} \right] \tag{4.268}$$

or when $\omega\tau << 1$:

$$\left(n - ik\right)_{//}^2 = \varepsilon_\infty \left[1 - \omega_p^2 / \omega^2 \right] \tag{4.269}$$

This equation is the same as that in the ordinary plasma situation with no external magnetic field. Because the electron motion driven by the electric field is parallel to the direction of magnetic field, the magnetic force contribution to the Lorentz force, $e\mathbf{V} \times \mathbf{B}$, vanishes.

When the electric field vector is perpendicular to ***B***, $\boldsymbol{E} \perp \boldsymbol{B}$, and for $\omega\tau >> 1$, the situation is different:

$$n_{\perp}^{2} \cong \varepsilon_{\infty}\left[1-\frac{\omega_{p}^{2}}{\omega^{2}}\left(\frac{\omega^{2}-\omega_{p}^{2}}{\omega^{2}-\omega_{p}^{2}-\omega_{c}^{2}}\right)\right] \tag{4.270}$$

At the effective plasma edge, where $n_{\perp}=0$, (4.270) reduces to:

$$\omega^{4}-\left(2\omega_{p}^{2}+\omega_{c}^{2}\right)\omega^{2}+\omega_{p}^{4}=0 \tag{4.271}$$

The solutions are:

$$\omega_{\pm}^{2}=\omega_{p}^{2}+\frac{\omega_{c}^{2}}{2}\pm\frac{1}{2}\omega_{c}\sqrt{\omega_{c}^{2}+4\omega_{p}^{2}} \tag{4.272}$$

The plasma edge now is split into two branches, and by measuring the splitting and knowing ω_{p}, the frequency ω_{c} can be deduced.

The above discussion is an approximate analysis of the frequency dependence of the absorption near the plasma edge in a magnetic field. From the refractive index the reflection spectrum can be also be determined. Doing this enables a determination of the relaxation time τ.

In the very long wavelength range, $\omega << \omega_{c}$, the so called spiral or Helicon wave phenomenon is predicted by (4.260). In this limit we find:

$$n_{\pm}^{2} \cong \varepsilon_{\infty}\left(1\mp\frac{\omega_{P}^{2}}{\omega\omega_{C}}\right) \tag{4.273}$$

The index n_{-}, the one with the plus sign in the bracket on the right of the equation, (n_{+} is the one with the minus sign), has a phase velocity for $\omega_{p}^{2} \gg \omega\omega_{c}$:

$$v_{p}=\frac{\omega}{k}=\frac{c}{n_{-}}\cong\frac{c}{\varepsilon_{\infty}\omega_{p}}\left(\omega\omega_{c}\right)^{1/2} \tag{4.274}$$

Substituting ω_{p} and ω_{c} into the above expression yields:

$$v_{p}=\sqrt{\frac{c}{4\pi\varepsilon_{\infty}eN}}\sqrt{\omega B} \tag{4.275}$$

Obviously, v_{p} is not related to the effective mass m^{*}. If there are two kinds of carriers with same carrier densities, $N_1=N_2=N$, and their effective masses are m_1 and m_2, respectively, then we have:

$$n_{\pm}^{2} \cong \varepsilon_{\infty}\left[1-\frac{\omega_{p_1}^{2}}{\omega\left(\omega \pm \omega_{c_1}\right)}-\frac{\omega_{p_2}^{2}}{\omega\left(\omega \pm \omega_{c_2}\right)}\right] \tag{4.276}$$

and if $\omega << \omega_{c_1}, \omega_{c_2}$, then:

$$n_{\pm}^{2} \cong \varepsilon_{\infty}\left[1+\frac{4\pi N}{\varepsilon B^{2}}\left(m_1+m_2\right)\right] \tag{4.277}$$

In this situation the propagating wave is known as an Alfven wave. The earliest experimental research on Helicon waves and non-resonant cyclotron absorption of InSb and HgCdTe materials are in papers by Wiley (Wiley et al. 1969; Wiley and Dexter 1969).

Faraday (1846) discovered the polarization plane rotation phenomenon when light passed through glass along the direction of a magnetic field early in 1845. Lorentz (1906) provided the physical explanation in 1906, in which he showed that the polarization rotation angle per unit length is:

$$\theta=\frac{\omega}{2c}\left(n_{-}-n_{+}\right) \tag{4.278}$$

Where n_- and n_+ are the material's refractivity for left-handed and right-handed polarizations.

Because we have $n_+^2-n_-^2=\left(n_+ + n_-\right)\left(n_+ - n_-\right) \cong 2n\left(n_+ - n_-\right)$, where n is the mean value of n_- and n_+. Assuming that $\omega\tau >> 1$, $\omega >> \omega_c$, and $n^2 >> k^2$, then it can be shown that:

$$\theta=\frac{2\pi}{nc}\frac{Ne^{2}}{m^{*}}\left(\frac{\omega_c}{\omega^{2}}\right)=\frac{2\pi e^{3}NH}{nc^{2}m^{*2}\omega^{2}} \tag{4.279}$$

If a material's refractivity n and carrier concentration N are given, then the effective mass m^* can be obtained by measuring the polarization rotation angle per unit path length θ of the light with frequency ω propagating in along the direction of the magnetic field H.

4.6 Optical Characterization of Materials

The optical properties of materials (in particular $Hg_{1-x}Cd_xTe$) are useful for characterizing some important parameters. In Sects. 4.4 and 4.5, we have discussed the use of spectroscopic ellipsometry to determine alloy compositions and to study the surfaces. Also introduced were investigations of the absorption spectra due to free carriers that give information about compositional distributions of epilayers, carrier concentrations, and doping. If we find a rule governing a material property, then a study of that property can be used to determine material parameters. It is important for $Hg_{1-x}Cd_xTe$ related technology to determine the alloy composition. In this section, we will discuss a refined technique that enables the determination of the transverse compositional profile through an analysis of infrared-absorption spectra.

4.6.1 Using Infrared-Absorption Spectra to Determine the Alloy Composition of $Hg_{1-x}Cd_xTe$

It is very important for technology developments to determine the alloy composition of narrow gap semiconductor materials like $Hg_{1-x}Cd_xTe$. Many methods have been devised to determine these compositions, such as density measurements, electronic probe, cut-off wavelength, modulated optical reflectance, Hall measurements, etc. (Dornhaus and Nimtz 1976; Schmit and Stelzer 1969; Zhong et al. 1983) In this subsection, we introduce a refinement of a previously discussed (Sect. 4.3) two-step method: the first step is to obtain the band gap E_g through the measurement of infrared optical absorption, and the second step is to get the composition x of the material from the equation relating E_g to x (Chu 1983a). There is still a question of the choice of the way to determine E_g from the absorption spectrum. Different people have used different methods to extract E_g from the spectra.

The complete intrinsic absorption spectra of a $Hg_{1-x}Cd_xTe$ sample includes not only the Urbach exponential absorption edge tail, but also the rather flat intrinsic absorption band. These are the two main regions of the spectra. The first higher energy region (the Kane region) is relatively flat, which reflects the characteristics of band-to-band optical absorption, and the second is the Urbach tail absorption due to band-tail states produced by electron–phonon interactions, impurity effects, disorder effects, etc. So the energy band-gap can be identified with the turning point on the absorption curves, where the exponential absorption edge and the Kane plateau region meet. There the absorption curve begins to change its slope. To observe the turning point, the infrared optical spectra for the samples with a thickness of about 10 μm need to be measured. This is difficult. However, there is an easier way. Because from prior measurements we have already found the relationship $\alpha(E_g)$ at different x and T, the energy gap E_g of a test sample can be determined just by locating the photon energy at which the measured absorption coefficient equals $\alpha(E_g)$. This energy can be obtained by extrapolating the exponential curve to higher energies. Then, the value of x can be extracted from the previously determined E_g (x, T). In Sect. 4.3, the relation $\alpha(E_g)$ for different x at 300 K was shown to be:

$$\alpha(E_g) = 500 + 5{,}600x \tag{4.280}$$

The CXT expression (Chu 1983a) is the relation between the energy gap E_g and composition x at temperature T:

$$\begin{aligned} E_g(eV) = &-0.295 + 1.87x - 0.28x^2 \\ &+ (6 - 14x + 3x^2)(10^{-4})T + 0.35x^4 \end{aligned} \tag{4.281}$$

Or neglecting the term, $0.35x^4$, (4.281) can be solved approximately for x at $T = 300$ K:

$$x = 3.8158 - \sqrt{13.9547 - 5.263E_g(300°K)} \tag{4.282}$$

The procedure for extracting x from the observed absorption tail spectra is:

1. Extrapolate the observed $log\ \alpha(E_g)$ curve taken at T =300 K to higher energies.
2. Pick an initial guess for E_{g0} from the observed spectrum ($E_{g0}<E_g$ by construction).
3. From (4.281) or (4.282) solve for x_0.
4. Plug x_o into (4.280) and solve for α_0.
5. From the extrapolated absorption curve find the $h\nu$ corresponding to α_o and set it equal to E_{g1}.
6. Solve for x_1 from (4.281).
7. Iterate.

This procedure converges and often the first iteration is quite close the final x value. It yields x values such that (4.280) and (4.281) are self consistently satisfied under the constraint of the observed absorption tail.

A sample with a thickness of $d = 0.43$ mm measured. The diameter of the light beam was 2 mm and the transmission spectra of three regions on the sample are shown in Fig. 4.50. Following steps 1 through 7 listed above, the concentrations for the three regions on the sample deduced after three iterations are x_1=0.208, x_2=0.216, and x_3=0.204(Fig. 4.51). An SEM measurement finds the average composition of the sample is $\bar{x} = 0.206$.

The composition and its uniformity are reliably measured by infrared spectra measurements. The approximate quality of the measured samples can also be judged. The gradient of the absorption edge is an indicator of the compositional uniformity in the region illuminated by the light beam. When

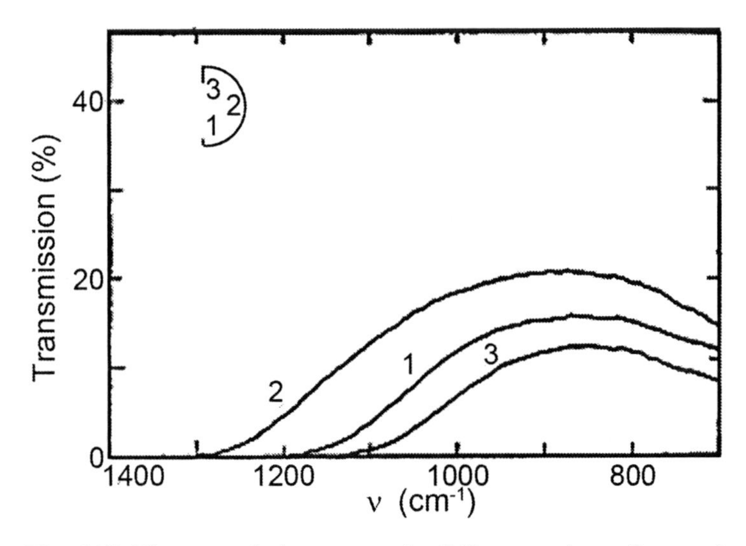

Fig. 4.50. The transmission spectra for different regions of a sample

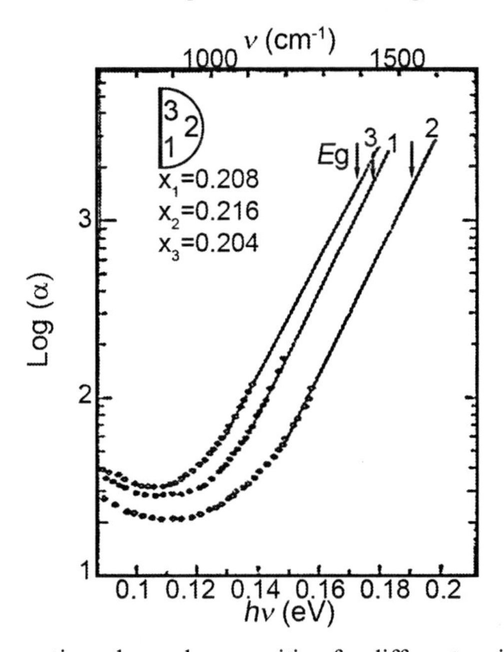

Fig. 4.51. The absorption edge and composition for different region of a sample

for a given carrier concentration and temperature, the absorption edge is sharper, the composition of the sample is more uniform. Well below the absorption edge the absorption is determined by free carriers. If the absorption coefficient at low wave numbers is smaller in a region, it indicates the concentration of free carriers there is smaller. If the absorption coefficient at higher wave numbers is also smaller, it is one indication that the sample in that region is better. However in this case one cannot distinguish the contributions from a non-uniform doping concentration in the region and a non-uniform concentration profile. A more accurate indication of a sample with a uniform concentration is to find a uniform carrier concentration throughout the sample, and with little variation in the absorption coefficients of the band tail spectra in different regions. Then the regions with the lowest absorption spectra would be the most uniform.

This method can be applied only when the Urbach absorption edge can be resolved in the absorption spectra. Usually this requires samples with thickness less than 0.5 mm. Furthermore, for poor quality samples whose free carrier absorption is large, the transmission is very small, and there is no clear absorption edge for the determination of E_g. This method can only be applied to good samples. If a sample is not good enough, it could be thinned to about 0.2 mm, which often allows the absorption edge to be resolved.

Another method to determine the composition is by fitting the transmission spectra as follows. It is known that when a light beam strikes a surface at normal incidence, the relation between the reflectance R and the transmittance T_r is:

$$T_r = \frac{(1-R)^2 e^{-\alpha d}}{1-R^2 e^{-2\alpha d}} \tag{4.283}$$

where d is the thickness of sample, and α is the absorption coefficient. When $E < E_g$ near the absorption edge, α can be expressed as:

$$\alpha = \alpha_0 \exp\left[\delta(E-E_0)/k_B T\right] \tag{4.284}$$

where

$$\begin{aligned} &\ln\alpha_0 = -18.5 + 45.68x \\ &E_0 = -0.355 + 1.77x \\ &\delta/k_B T = \ln(\alpha_g/\alpha_0)/(E_g - E_0) \\ &\alpha_g = -65 + 1.88T + (8694 - 10.31T)x \\ &E_g(x,T) = -0.295 + 1.87x - 0.28x^2 + 10^{-4}(6 - 14x + 3x^2)T + 0.35x^4 \end{aligned} \tag{4.285}$$

For thin samples the data can even be fit in the range, $E > E_g$. In the energy range, $E > E_g$ but close to the edge where the bands are nearly parabolic, it has been found (Chu et al. 1994) that to a good approximation:

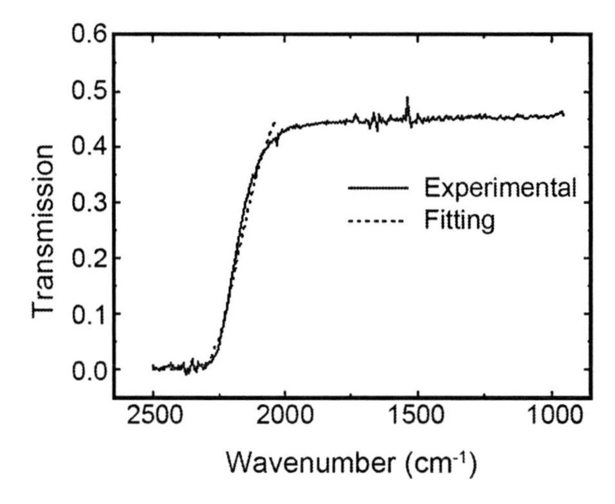

Fig. 4.52. Experimental transmission data and fitted curve

$$\alpha = \alpha_g \exp\left[\beta\left(E - E_g\right)\right]^{1/2} \tag{4.286}$$

where

$$\beta(T,x) = -1 + 0.083T + (21 - 0.13T)x \tag{4.287}$$

The compositions of samples have been obtained by fitting the experimental data using a least square method, as shown in Fig. 4.52.

When the region being fitted is near the absorption edge, free carrier absorption does not need to be considered. The average composition can be obtained by fitting the absorption edge. For example, the composition of the sample data fitted in Fig. 4.52 is $x = 0.309$.

The above procedure assumes that the composition of the sample is homogeneous. If it is inhomogeneous, this must be considered in the fitting procedure. The extended procedure also reveals the magnitude of the inhomogeneity.

4.6.2 Transverse Compositional Uniformity of $Hg_{1-x}Cd_xTe$ Samples

The determination of the transverse compositional uniformity of $Hg_{1-x}Cd_xTe$ samples is critical to research and to infrared detector focal plane array technology. Because the absorption spectra of a beam with dimensions that are small compared to the surface area of the sample can be measured, the composition in these small regions can be determined. Then a

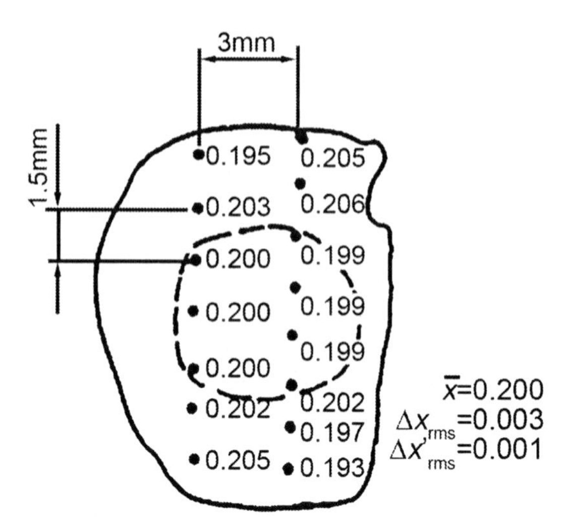

Fig. 4.53. The transverse compositional distribution of a $Hg_{1-x}Cd_xTe$ ($x = 0.200$) sample gotten by an electronic probe measurement. In the beam region outlined by the *dashed curve* with $\varphi = 3\sim4$ mm, $\Delta x_i \approxeq 0.001$

small beam can be scanned to map the transverse compositional uniformity of the sample (Chu et al. 1985). But because of its complexity, this method may be not convenient. There is a more convenient method to determine the uniformity. Based on the absorption edge law, we can determine quantitatively the composition x and its mean-square deviation Δx using software to analyze the absorption spectra of large beams.

The absorption edge law for $Hg_{1-x}Cd_xTe$ is the theoretical basis of the method. By analyzing the measurement of the absorption edge of thin $Hg_{1-x}Cd_xTe$ ($x = 0.17$–0.443) samples at different temperatures (T=4.2–300 K), a quantitative expression for the Urbach exponential absorption edge can be extracted (Chu et al. 1991, 1992a). The thicknesses of samples used varied from several μm to 20 μm. The samples are thinned bulk crystals, so their longitudinal compositional distributions are homogeneous. Using an electronic probe to determine the transverse uniformity of the samples, yields $\Delta x_i \approx 0.001$ in the region covered by a light beam with a diameter of $\phi = 3$–4 mm. Figure 4.53 shows the compositional distribution of a $x = 0.200$ $Hg_{1-x}Cd_xTe$ sample. The mean square deviation of the composition is $\Delta x_i = 0.001$ (Chu et al. 1985). Because for good homogeneous samples, the mean concentration deviation of macroscopic regions is small compared to the average concentration, it is concluded that the absorption tail data taken on these samples are dominated by lattice disorder, impurity defects, etc., rather than compositional inhomogeneities.

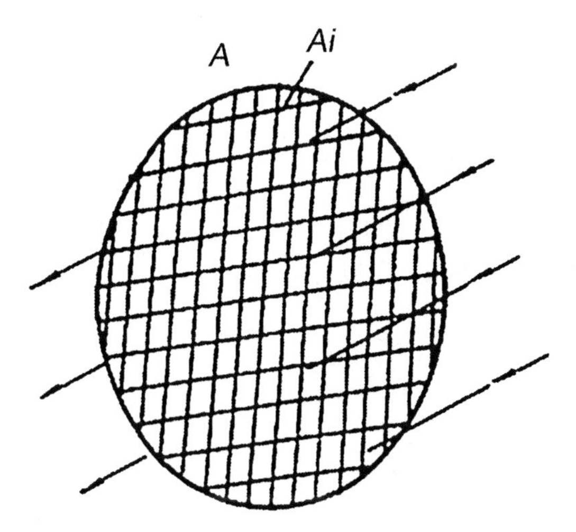

Fig. 4.54. The transmittance of a large surface area region of a sample is the sum of transmittances of many small regions

These results not only serve as a standard to judge theoretical results, but also can be used to judge the homogeneity of other tested samples. If we suppose that the tested samples are homogeneous along the longitudinal direction, any increase of the band tail in samples with well controlled defects can be ascribed to transverse concentration inhomogeneities.

In order to describe quantitatively the inhomogeneities of $Hg_{1-x}Cd_xTe$ samples, we need to employ a compositional distribution function, and then we can determine the average and mean deviation of the composition in different cases. Through analyzing the electron probe measurement data of many $Hg_{1-x}Cd_xTe$ samples, it was found that transverse compositional distributions could be represented by a logarithmic normal distribution defined as:

$$f(x) \equiv \frac{1}{\sqrt{2\pi}\sigma\left[\dfrac{1+\operatorname{erf}\left(\dfrac{-\mu}{\sqrt{2}\sigma}\right)}{2}\right]}\frac{1}{x}\exp\left\{\frac{-(\ln x-\mu)^2}{2\sigma^2}\right\}$$

$$\cong \frac{1}{\sqrt{2\pi}\sigma}\frac{1}{x}\exp\left\{\frac{-(\ln x-\mu)^2}{2\sigma^2}\right\} \tag{4.288}$$

where σ and μ are two parameters determining mean square deviation $\Delta x_c \cong x_o\sigma$, and the average composition $x_o = e^{\mu}$. The last equality in (4.288) and that for Δx_c are valid for cases where μ/σ are large compared to unity.

The transmittance of a large surface area region can be gotten by measuring the sum of the transmittances of many small regions (Fig. 4.54). If we regard the composition of each small region of area A_i as homogeneous, we can get the average total transmittance of the large area A over a composition range x_1 to x_2 as follows:

$$\overline{T}(E) = \frac{1}{A}\sum_i A_i \int_{x1}^{x2} f(x_i) T_i(E, x_i) dx_i \tag{4.289}$$

where

$$T_i(E,x) = \frac{\left[1 - R(x)\right]^2 \exp\left[-\alpha(E,x)\cdot d\right]}{1 - R(x)^2 \exp\left[-2\alpha(E,x)\cdot d\right]} \tag{4.290}$$

is the ith small region transmittance. In the equation, d is the thickness of the sample, E is the photon energy, and the absorption coefficient and the reflectivity R are both functions of composition x. Note that after the integration is done the argument of the sum over i no longer depends on i so 1/A times the sum is unity. Here, we are interested in the edge of transmission. In the indicated spectral range the absorption coefficient can be calculated from an equation in the literature (Chu et al. 1991, 1992a). The reflectivity R can be calculated from its relation to the high frequency dielectric function. Through (4.229)–(4.231), the transmittance edge can be calculated.

Based on the above model, by adjusting the parameters of the compositional distribution μ and σ, to fit the experimental transmission edge of a sample, we can determine its average $\bar{x} = \exp\left(\mu + \frac{\sigma^2}{2}\right)$, and its mean square deviation composition Δx. During the calculation, the upper and lower limits of the integral x_1 and x_2 can be set so $f(x_{1,2})$ are the same fraction $\eta = f(x_{1,2})/f(\bar{x})$ of the compositional distribution around its value at $\bar{x}$. Typically η is about 1%. The values of x_1 or x_2 are given by the expression:

$$x_{1,2} = \bar{x}\exp\left[-\frac{3}{2}\sigma^2\left(1 \mp \frac{8}{9}\ln\eta\right)\right] \cong \bar{x}\left(1 \pm \sigma\sqrt{-2\ln\eta}\right) \tag{4.291}$$

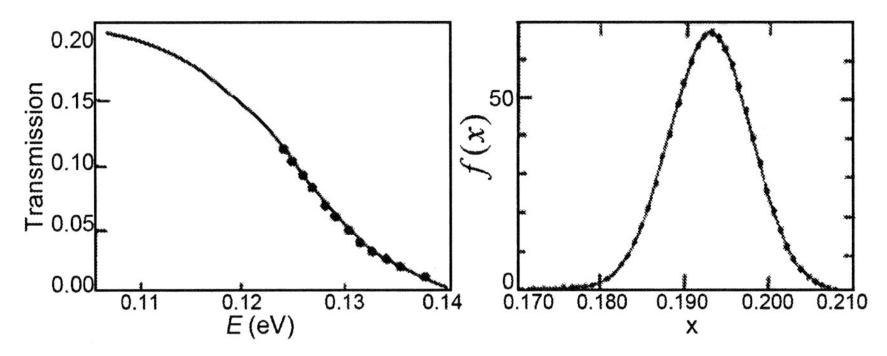

Fig. 4.55. (**a**) The transmission spectrum (*solid curve*) and fits (*dots*) of a $Hg_{1-x}Cd_xTe$ sample, (**b**) The deduced probability distribution of the sample's composition

where, the last approximate equality holds for small σ. In principle, x_1, x_2 can be any two numbers separated from $\bar{x}$. Then the average transmittance can be calculated by numerical methods. Figure 4.55a is the transmission spectrum of a sample, in which the solid line is the experimental curve and the dots are the result of fitting. Based on this data fitting, the compositional distribution parameters μ and σ are obtained as shown in Fig. 4.55b. From the figure the average value and mean square deviation of the composition are $x_0 = 0.193$, and $\Delta x_c = 0.0049$, respectively.

From the absorption edge formula adopted in the calculation the inhomogeneous contribution deduced from experimental data is $\Delta x_i = 0.001$; the root mean square (rms) concentration variation is not just that due to a homogeneous sample. Thus, the rms composition variation obtained through the above method is not the actual rms deviation of the sample. The upper limit of actual rms deviation is the quadrature sum of Δx_c and Δx_i,

$$\Delta x \leq \sqrt{\Delta x_c^{\ 2} + \Delta x_i^{\ 2}} \tag{4.292}$$

For the sample treated in Fig. 4.55, the real mean square deviation of composition is $\Delta x \approx 0.0050$, which is consistent with, but slightly smaller than the electronic probe data.

In the above discussion, we assumed that the longitudinal composition of samples is homogeneous. The assumption is a good approximation for a bulk material. The longitudinal compositional inhomogeneity affects the transmission spectrum differently, so the effect was not considered in the above discussion. However, for the $Hg_{1-x}Cd_xTe$ epitaxial film materials, such as those grown by LPE, MBE, and MOCVD, the effect should not be neglected.

4.6.3 The Longitudinal Compositional Distribution of $Hg_{1-x}Cd_xTe$ Epilayers

$Hg_{1-x}Cd_xTe$ epilayers have become the primary materials in the manufacture of infrared detectors. Compared with bulk single crystal material, epilayers have more low-density of dislocations, but at the same time have better transverse compositional uniformity. However, because of the unbalanced chemical potential between HgCdTe epilayers and CdTe substrates, the inter diffusion between the epilayer and the substrate can not be overcome, so there are compositional gradients near the interface. This can have an adverse affect on device performance (Edwall et al. 1984). Thus it is necessary to find a simple but effective method to determine the longitudinal as well as the transverse compositional distributions of HgCdTe layered materials.

Some common methods to determine sample compositions, such as SIMS (secondary ion mass spectroscopy) (Bubulac et al. 1992), or electronic probes, require cleavage, or mesa etch of samples (Price and Boyd 1993). These destructive measurement methods can not be applied to materials intended for device fabrication. A method using infrared transmission spectra at room temperature to determine the homogeneity of samples is nondestructive, simple and convenient (Price and Boyd 1993; Chu 1983a; Hougen 1989; Gopal et al. 1992). The method can also be used to determine the compositional distribution of epilayers.

The intrinsic absorption edge law forms the theoretical basis of the method. If a sample is non-uniform in the area covered by an incident beam, the slope and shape of the absorption edge in the intrinsic absorption range is changed. This non-uniformity can be deduced by analyzing the absorption edge spectra.

For a two-sheet HgCdTe/CdTe epilayer sample, when a beam impinges on the front surface, there are multiple reflections from the front air/epilayer (subscript 1) interface, the epilayer/CdTe substrate (subscript 2) interface, and the back substrate/air (subscript 3) interface (Hougen 1989). The total transmittance $T_{1,3}$ is

$$T_{1,3}=\frac{(1-R_1)(1-H)T_{2,3}a_1}{1-R_1(1-H)R_{2,3}(a_1)^2} \tag{4.293}$$

where a_1 is the absorption of the epilayer, and T is the transmittance, R is the reflectivity both of which can be calculated using the refractivity formula (Liu et al. 1994b).

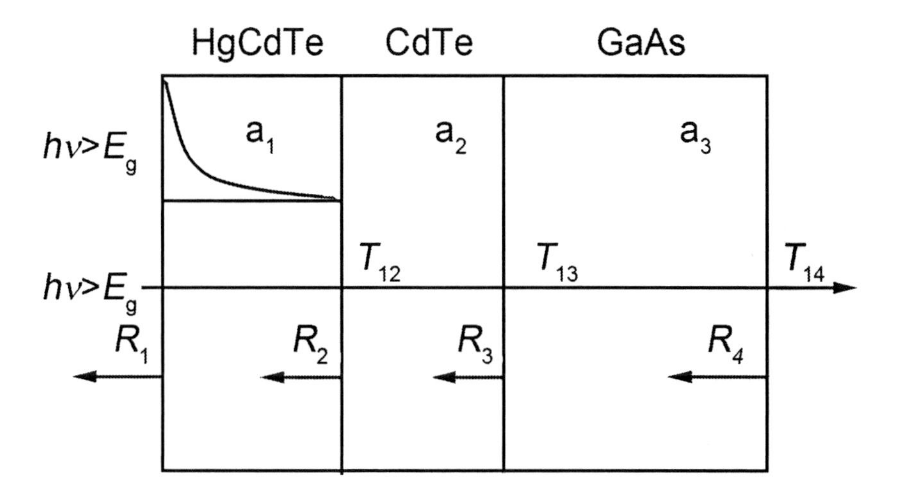

Fig. 4.56. A schematic of the absorption, reflection and transmission of a three-sheet sample consisting of HgCdTe/CdTe/GaAs

For a three-sheet HgCdTe/CdTe/GaAs epilayer sample, such as those grown by MBE, or MOCVD, we must consider multiple reflections from each interface; the front air/MCT epilayer (subscript 1) interface, the epilayer/CdTe buffer layer (subscript 2) interface, the buffer layer /GaAs substrate (subscript 3) interface, and the substrate/ air (subscript 4) interface, shown in Fig. 4.56. The total transmittance $T_{1,4}$, assuming there is no absorption in the buffer or substrate layers, is:

$$T_{1,4} = \frac{(1-R_1)(1-H)T_{2,4}a_1}{1-R_1(1-H)R_{2,4}(a_1)^2} \tag{4.294}$$

The parameter H in (4.234) and (4.235) accounts for the intensity loss in the surface of the front epilayer, and is regarded as a variable parameter in calculations. We adjust H to make the maximum of calculated curve equal the maximum of the experimental data.

Due to the longitudinal inhomogeneity of a $Hg_{1-x}Cd_xTe$ epilayer, the absorption of the layer varies with the composition $x(z)$ at different depths, that is:

$$\alpha_{MCT}(k,T)d = \int_0^d \alpha[k,T,x(z)]dz \tag{4.295}$$

where α is the absorption coefficient of the $Hg_{1-x}Cd_xTe$, d is the thickness of epilayer, k is the wave number, and T is temperature. As we know, the absorption curve is composed of the Urbach absorption tail and the Kane intrinsic absorption range. The intrinsic absorption edge of $Hg_{1-x}Cd_xTe$ follows the Urbach absorption law (Chu et al. 1992a):

$$\alpha = \alpha_0 \exp\left[\delta(E - E_0)/k_B T\right] \tag{4.296}$$

where the definition of the parameters is the same as those in (4.226). An approximation to the intrinsic absorption coefficient valid in the initial part of the Kane range also depends on energy exponentially (Chu et al. 1994):

$$\alpha = \alpha_g \exp\left[\beta(E - E_g)\right]^{1/2} \tag{4.297}$$

where $\beta(T, x) = -1 + 0.083T + (21 - 0.13T)x$.

If the transverse compositional inhomogeneity can be neglected, the longitudinal compositional distribution can be written (Hougen 1989):

$$x(z) = \frac{1 - (x_s + sd)}{1 + 4(z/\Delta z)^2} + (x_s + sd) - sz \tag{4.298}$$

where z is the distance in the epilayer from substrate, and x_s, s, Δz are three not entirely independent fitting parameters. x_s depends on the composition. Δz depends on the thickness of the buffer layer, which mainly affects the transmission spectrum near 0–10% of T_{max}. s depends on rate of change of the longitudinal compositional distribution, and is determined by the slope of the transmittance spectrum curve near 30–80% of T_{max}.

Substituting (4.295) to (4.298) into (4.293) or (4.294), the transmittance of an epilayer can be calculated. By adjusting the parameters x_s, s, Δz in the (4.298), and the parameter H in (4.291) and (4.293), the calculated curve can be fitted to the measured curve. If the two curves coincide, the parameters x_s, s, Δz are considered to be determined and the longitudinal compositional distribution can be calculated using (4.298). The whole fitting process can be carried out using the appropriate software.

In a fitting process the calculated curve and the measured curve may not entirely overlap. This can be induced by a transverse compositional inhomogeneity (Chu et al. 1992a; Gopal et al. 1992; Li et al. 1995a). In the Hougen model, the effects of a transverse compositional inhomogeneity were not included. However, in a full calculation, both effects should be considered simultaneously. If we assume that compositional distribution along the transverse direction has a lognormal form (Chu et al. 1992a), the composition of the epilayer at depth z from the surface is:

$$f(x)=\frac{1}{\sqrt{2\pi}\sigma x(z)}\exp\left\{\frac{-\left\{\ln\left[x(z)\right]-\ln x_o\right\}^2}{2\sigma^2}\right\} \tag{4.299}$$

where $f(x)$ is probability distribution function of the composition $x(z)$ at z, $x_o=\bar{x}e^{-\sigma^2/2}$ is related to the average composition $\bar{x}$, and σ is the mean square deviation of the logarithm the compositional distribution taken to be independent of z. Equation (4.299) indicates that the compositional probability distribution has most of its area (~95%), confined to a range from $x(z)-2x_0\sigma$ to $x(z)+2x_0\sigma$, and this range can be used as the integration limits when calculating averages.

The samples examined next were grown by LPE, MBE and MOCVD to thicknesses varying between 4–20 μm. A CdTe substrate was used for the LPE growth. The substrates used for the MBE and MOCVD growths were GaAs. A 5 μm CdTe buffer layer was grown on the GaAs, and then the $Hg_{1-x}Cd_xTe$ epilayer was added. In order determine the longitudinal compositional distribution, usually samples were chosen that had uniform transverse compositions. The method used to determine whether or not the transverse compositional distribution was homogeneous, was to scan a small sized beam over the surface and measure the transmission spectra at a series of points.

Before measurements, samples needed to be cleaned using an organic solvent, etched using bromine carbinol, and washed using de-ionized water. The transmissivity of the samples were measured using an infrared spectrophotometer, whose precision is $\sim 10^{-3}$, and whose beam diameter is ϕ = 3 mm. SIMS measurements were made using an IMS-3f type instrument produced by CAMECA. In order to insure the measurement precision (particularly, to avoid Hg desorption caused by heating), a primary O_2^+ ion beam with low energy (10–12.5 keV), and a low ion flow density ($\approx$150 μA cm^{-2}) was used to bombard the surface of the samples. This produced an etch speed of the $Hg_{1-x}Cd_xTe$ layer of about 0.8 μm min^{-1}.

Figures 4.57a, b, and c show the transmission spectra of MBE, LPE and MOCVD epilayers, at room temperature. In the figure, the dots are experimental data; the solid curves are the fits from (4.293) to (4.298), and the dash curve is a calculated result from (4.240). The parameters deduced from the fitting procedure are in Table 4.13. The interference fringes in the transmission spectra of the MBE and the MOCVD samples effect the determination of the parameter H. However, the value of H does not affect the actual compositional distribution. From the figure, we can see that for

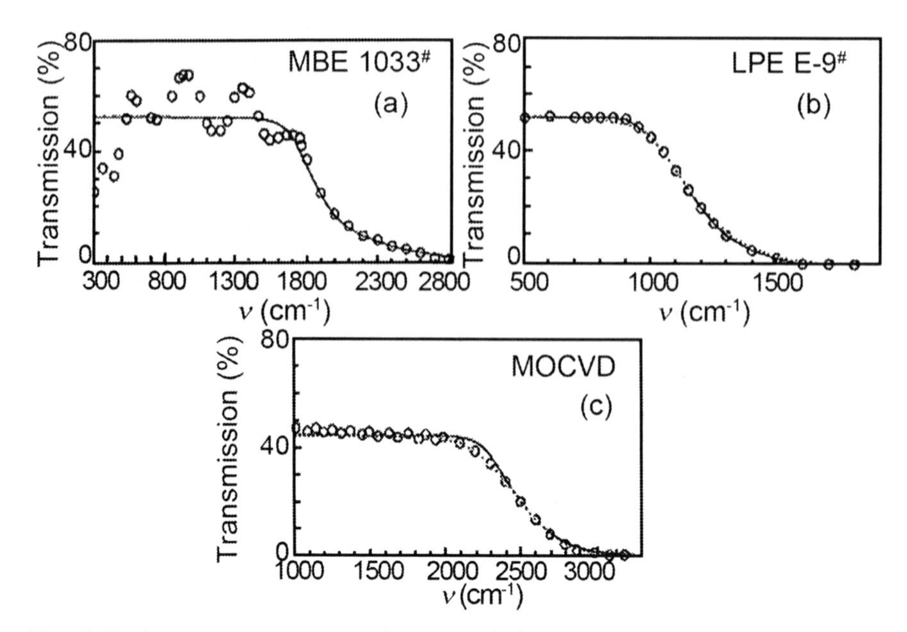

Fig. 4.57. At room temperature, the transmission spectra of (**a**) an MBE, (**b**) an LPE, (**c**) an MOCVD sample

the MBE sample the deduced curve fits the data well. However, for the LPE and MOCVD samples, if only the effect of the longitudinal compositional distribution was considered, the experimental transmission spectra could not be interpreted. It was necessary to add the effect of transverse compositional distribution (the dash lines) to cause the calculated curves to consistently fit the experimental data. To confirm this result the transmission spectra of a small scanned beam was conducted on different samples. For the MBE samples, the shapes of the transmission spectra at different locations were almost identical. However, the LPE and the MOCVD samples' transmission spectra shapes were slightly different across their surface areas (Chu et al. 1992a; Gopal et al. 1992).

Through fitting the transmission spectra at room temperature, the compositional distribution of samples can be obtained. Figure 4.58 shows the longitudinal compositional distribution of the MBE sample studied in Fig. 4.57a. Figure 4.58a is the calculated result from (4.300), and the solid curve is the composition profile of Cd, and the dash curve is that of Hg. Figure 4.58b are SIMS profiles of the compositions of Hg and Te. N is the number of secondary ions detected, and t is the time of the etch exposure. The etch speed of $Hg_{1-x}Cd_xTe$ epilayer in the ion beam to is 0.8 μm min^{-1}. The surface compositional distribution of a sample also can be determined

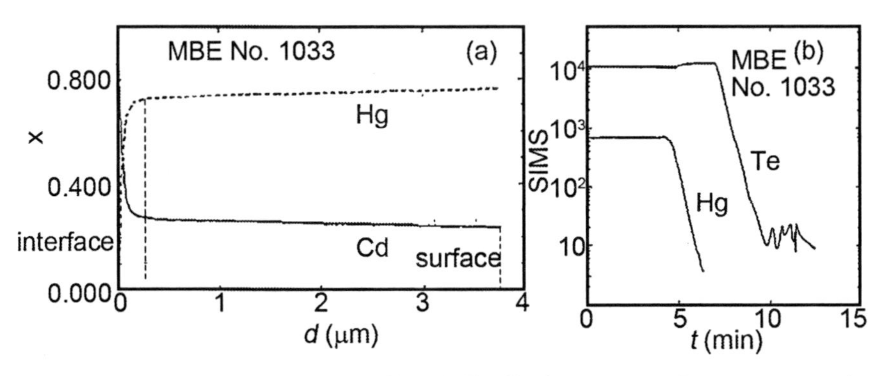

Fig. 4.58. The longitudinal compositional distribution curves of an MBE sample, (**a**) a calculated result (**b**) SIMS spectra of Hg, and Te concentrations

by an electron probe. Analyzing the compositional distribution of Te through its SIMS spectra, allows a determination of the distance between the CdTe/HgCdTe interface and the surface to be about 4 μm. This is the same as the result deduced from interference fringes. The effect of the sample surface on the transmission spectrum is substantial. The surface composition (0.239) deduced from (4.300) is close to the result (0.2378) deduced from an electron probe measurement. Figure 4.58 shows the longitudinal variation of the Cd concentration from the CdTe interface to the sample surface (the solid curve), and the Hg concentration (the dashed curve). The Cd concentration increases from the surface to the CdTe/HgCdTe interface where it increases rapidly in the transition region up to 1. The width of transition region obtained by fitting the room temperature transmission spectra is appreciably narrower than the result of this SIMS measurement.

Figure 4.59 shows the longitudinal spatial compositional distribution of the LPE sample whose absorption spectrum is displayed in Fig. 4.57b, where both the transverse and longitudinal compositions were considered. Figure 4.59a is a result calculated from (4.299). In Fig. 4.59b, the solid line is a SIMS curve, the dashed curve is the longitudinal compositional distribution calculated from (4.298). From the SIMS profile, we deduce a thickness of the LPE epilayer of about 16 μm. There is little difference between the surface compositions (0.186±0.0015) calculated from (4.299) and that (0.1854) measured by an electron probe. From Fig. 4.59, we see that the calculated longitudinal composition distribution closely follows the SIMS profile.

Using the same method, the longitudinal spatial compositional distribution of MOCVD samples can be obtained. From Table 4.13, we see that the width of transition region in which the composition changes from 1 to 0.466 at the HgCdTe/CdTe interface of one sample is $\Delta z = 0.4$ μm. Along the grown direction outside the transition region, the composition changes from 0.466 to 0.289. The extracted transverse compositional inhomogeneity is σ=0.005. In order to check this result, the room temperature transmission was measured at different locations in the big beam area using a small beam (ϕ=1 mm), and the fitting procedure was carried out for four spots. The

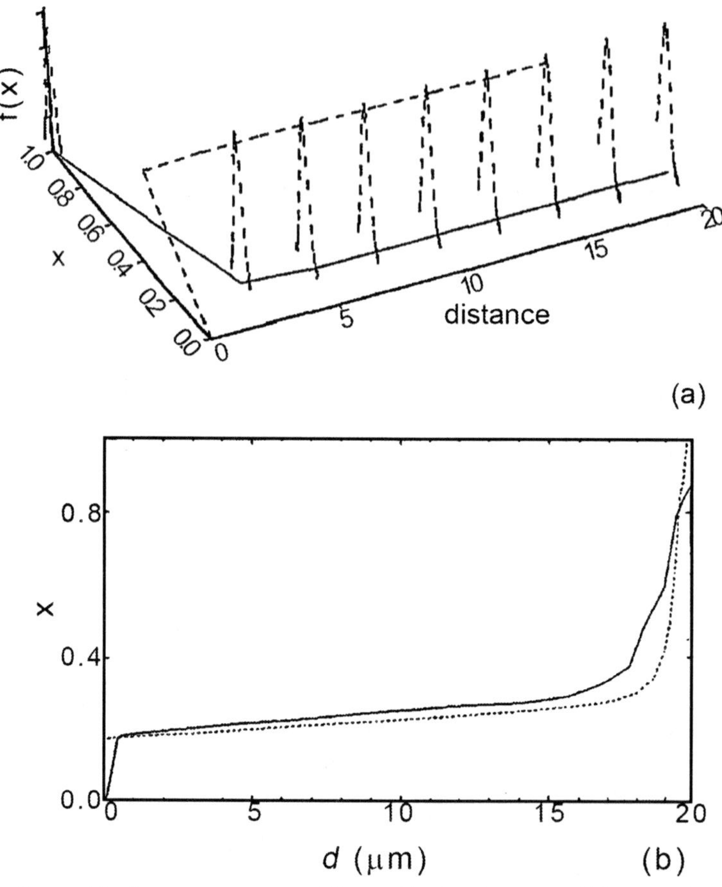

Fig. 4.59. The spatial compositional distribution of an LPE sample

Table 4.13. The fitting parameters deduced for MBE, LPE, and MOCVD samples

Grown method	Surface composition	Slope	The width of transition region	H	Root mean square deviation of the transverse composition
	x_s	S(1 cm^{-1})	Δz(μm)		σ
MBE	0.2366	86	0.05	0	
LPE	0.176	51	0.08	0	0.0015
MOCVD	0.289	104	0.4	0.15	0.005

surface compositions found were 0.2892, 0.2901, 0.2884, and 0.2896. It is obvious that the degree of the dispersion of the transverse composition gotten by experiment is comparable to that predicted by the theoretical model.

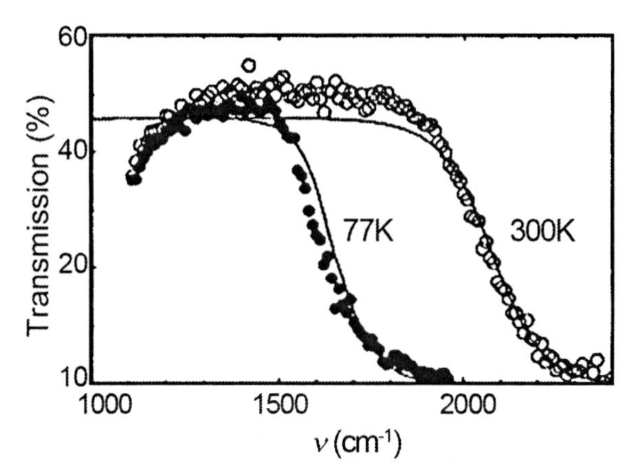

Fig. 4.60. The transmission spectra of an epilayer at room temperature and at liquid helium temperature

The above fitting model can also be used to fit spectra taken at low temperature. Figure 4.60 is the transmission spectra of an epilayer sample taken at room temperature and at liquid helium temperature. The dots are the experimental data, and the solid curves are the fitted calculated results. In the fitting process, the values of x, s, Δz in (4.298) were first determined based on the room temperature transmission spectra, and then these values were substituted, along with $T = 77$ K, into (4.293)–(4.298), to deduce the transmissivity at 77 K. The resulting curve also fits the low temperature experimental data, proving that the above fitting model is also applicable at low temperature.

4.6.4 Using Infrared Transmission Spectra to Determine the Parameters of a HgCdTe/CdTe/GaAs Multilayer Structure Grown by MBE

The above fitting method used mostly data from the intrinsic absorption and the absorption edge spectral ranges, and while it was mentioned, did not explicitly treat the interference effects seen in the transparent spectral range. The theoretical model is relatively simple for a single layer grown on a substrate. If there are two layers on the substrate, the more complex effects of the multilayer structure must be considered in choosing parameters to fit the entire spectra.

We first examine the theory of the transmissivity of a HgCdTe/CdTe/GaAs structure. An MCT epilayer grown by MBE at low temperature, 180 K, experiences little inter-diffusion between layers, and retains a rather good interfacial integrity. This can be proven by the use of the elliptical polarization technique. Thus we can regard the heterojunction as abrupt and the interface as ideal in transmission calculations.

The most common methods to calculate the transmissivity of a multilayer dielectric layer uses a recursion method, or the interference matrix method. The recursion method traces a beam incident at a specified angle to learn its net reflection and transmission through the multilayer film (See Sect. 4.4). Then by choosing several representative incident beam angles, the full angular dependence of the transmissivity, reflectivity and absorptivity of the film can be deduced. When the beam direction changes the complex refractive index entering into the calculation of the transmissivity and the reflectivity must be changed accordingly. The interference matrix method, starts from the tangential components of the electric and magnetic field intensities of a light wave written in matrix form, and doing matrix multiplications with the various materials dielectric and magnetic tensors to obtain the tangential components of electric and magnetic field intensities at the back side of the film. There is no essential difference between the two methods but the calculation for the interference matrix method is easier to program. Further discussion can be found in the literature (Wang 1997) and textbooks on optical thin films.

The precision of the parameter determination for a two-epilayer structure deduced from transmission spectra measurements lies in the sensitivity to the variation of parameters and the accuracy of the optical constants used. In theoretical calculations, the primary factors affecting the transmissivity are the optical constants of the material. The effect of the optical constants on the calculated result can be seen from the formula for the complex phase $\delta = (n-ik)2\pi \ell / \lambda$, where ℓ is the path length. When the path length is the thickness of the epilayer d and it is fixed, the change of the complex phase depends on the change of the complex refractive index $(n-ik)$, and correspondingly, the interference peak positions in the transmission spectrum change. In the intrinsic absorption region ($E > E_g$), the absorption coefficient of MCT is large and the transmissivity is small. Then the interference peaks are suppressed by absorption, so here, the effect of the refraction is far smaller than the effect of the absorption. However, in the free carrier absorption region at longer wavelengths, the absorption coefficient is small, so the interference peaks are evident, and in this spectral range the transmission is dominated by refraction. Near E_g at the absorption edge, the effects of refraction and absorption on the phase difference are comparable, and the both effects must be included in any analysis.

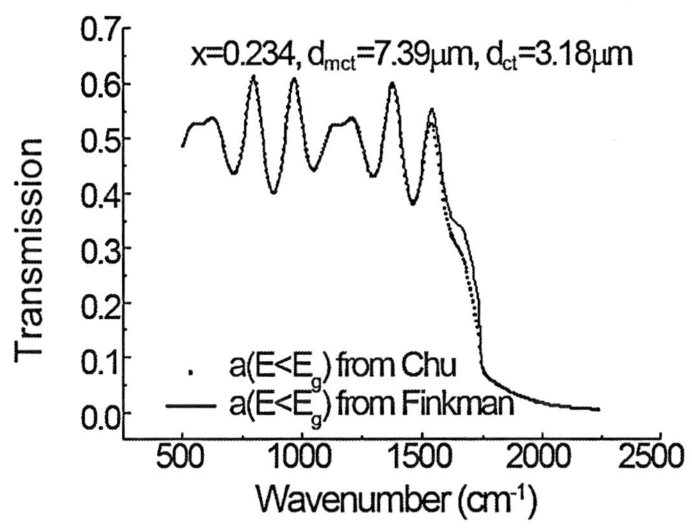

Fig. 4.61. Two transmittance curves for an MBE sample calculated from two different absorption formulas

For an actual calculation, the absorption of the CdTe in the infrared wave band is small and can be neglected, and its refractive index n changes little with wavelength. In the 2–15 μm range, the index of the CdTe is about n= 2.67, and the index of the GaAs is n = 3.2. Once these two parameters have been determined, the primary parameters affecting the transmission of a HgCdTe/CdTe/GaAs sample is mostly due to the optical constants of the HgCdTe.

As reported above Chu et al. (1992a) carried out a detailed investigation of the MCT absorption spectra leading to two formulas, one for the Urbach tail range and the second for the Kane intrinsic range. Other absorption formulas were obtained by Schacham and Finkman et al. (Finkman and Nemirovsky 1979; Finkman and Schacham 1984), which apply to the concentrations $x = 0.215$, 0.29, and 1.0. Yang et al. (1996) tested the results deduced from both the Chu formulas and the Finkman formulas by comparing their results to the transmission spectrum of a HgCdTe (x = 0.234) sample grown by MBE. Figure 4.61 shows the absorption calculated from both formulas. The two results have differences only near E_g. The absorption coefficient deduced from the Finkman formula is smaller.

The Finkman and Nemirovsky (1979) formula for the index of refraction applicable in the transparent range has the form, $n(\lambda,T)^2 = A + B/[1-(C/\lambda)^2] + D\lambda^2$. Its precision is about 2%. Kucera (1987) carried out a more detail study of the index of MCT. They measured MCT samples whose compositions varied from 0.18 to 1.0 at room temperature and at 77 K, and deduced the variation of the index of refraction. They analyzed experimental data through theoretical fits to more complex formula than those used by Finkman et al. Liu (1994) measured the refractive index of MCT samples with compositions in the range x=0.17 to 0.443, and arrived at a series of experimental formulas that are similar to those of Finkman et al.:

$$n(\lambda,T)^2 = A + B/[1-(C/\lambda)^2] + D\lambda^2 \tag{4.300}$$

Where

$$\begin{aligned} A &= 13.173 - 9.852x + 2.909x^2 + 10^{-3}(300-T), \\ B &= 0.83 - 0.246x - 0.0961x^2 + 8\times10^{-4}(300-T), \\ C &= 6.706 - 14.437x + 8.531x^2 + 7\times10^{-4}(300-T), \\ D &= 1.953\times10^{-4} - 0.00128x + 1.853\times10^{-4}x^2. \end{aligned} \tag{4.301}$$

These expressions have small differences with those of Finkman et al., but they are based on rather more data points. Yang et al. (1996) compared the transmission spectra calculated with the K. Liu formula and the Kucera formula with experimental data. He found the both formulas yield results consistent with his data in the longer wave band. However, near the absorption edge, the Kucera formula has a rather large difference with the data, because the refractivity given by the formula increases sharply near the energy band gap and suppresses the interference peak near the absorption edge. The Liu formula provides a better fit. Thus we adopt the Liu formula of refractivity in further discussions.

The thickness of a monolayer film can be calculated from the intervals between interference peaks in the interference spectrum. For positive interference the phase difference between two beams (the direct beam and the one reflected from the back surface and then the front surface so the path length is 2d) passing through a monolayer film, in which there is only one optical constant, is $\Delta\delta = 2\pi = \Delta\left(2\pi\frac{\ell}{\lambda}n\right) = \Delta k 2dn$. Once the wave-number difference between two peaks, Δk, is measured, the thickness of the film, is $d = \frac{\pi}{n\Delta k}$, is easily deduced. However, for a multilayer film, the beam will pass through each layer multiple times, and the phase differences depend on the optical constants and thicknesses of every layer, so the analysis to extract the layer thicknesses is more complex.

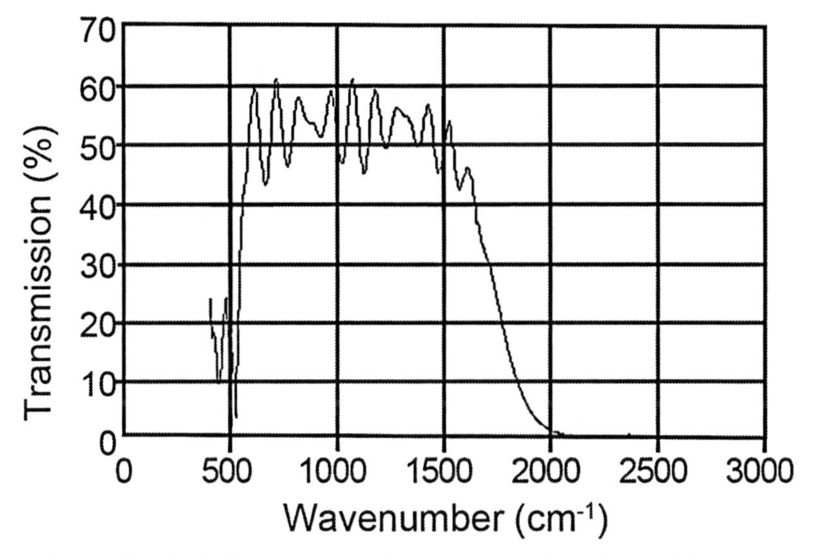

Fig. 4.62. The infrared transmission spectrum of a MCT/CT/GaAs sample

Figure 4.62 is a typical experimental infrared transmission spectrum of an MCT/CT/GaAs sample. From the makeup of interference peaks, we can see that the spectrum is a superposition of two wave modes, one with a shorter period than the other. A cursory view of the interference spectrum suggests that the shorter period corresponds to MCT with a larger optical thickness n_1d_1, and the longer period to CdTe with a smaller optical thickness n_2d_2. Actually, when the phase difference between the two layers is $n\pi$, the spectrum is the common result of interference of both layers. As a first approximation without regard to the effect of other beams, the approximate thicknesses of the MCT and the CT layers can be deduced. When the spectrum is analyzed more carefully, we find that the spacing of the interference peaks with the shorter period is not equidistant. Figure 4.63a–c constitutes a study of the effects on the transmission spectra of varying the HgCdTe epilayer thickness and composition, and the CdTe buffer layer thickness while fixing some parameters. The resulting fits to the data indicates the epilayer thickness is ~0.2 μm, and other effects can be distinguished from the spectrum. This study shows that the HgCdTe epilayer thickness can be determined by examining the interference peaks, and the CdTe buffer layer thickness can be determined by the change of the shape of interference nodes.

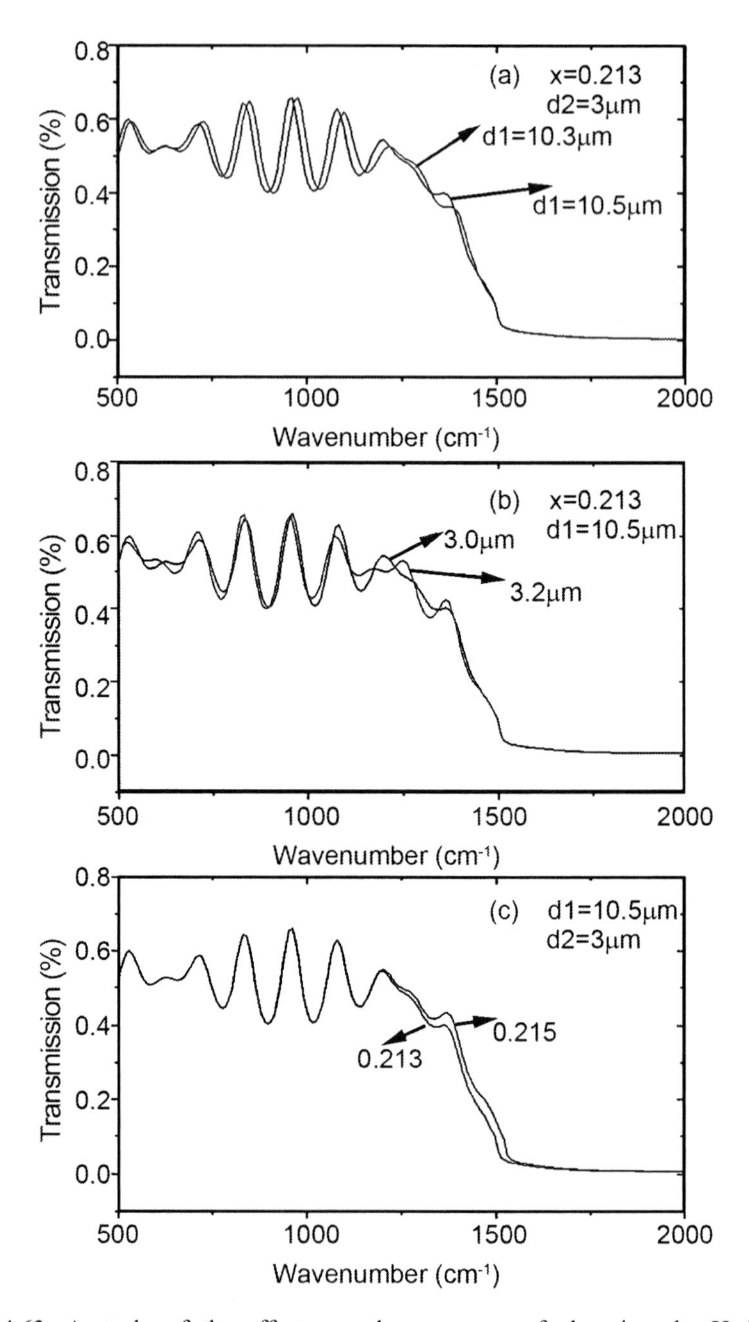

Fig. 4.63. A study of the effects on the spectrum of changing the HgCdTe/CdTe/GaAs alloy concentrations, and epilayer thicknesses while holding some parameters fixed

The FTS-7 infrared spectrometer used in the experiment reported in Fig. 4.62 has a 2 cm^{-1} resolution and the actual error of reading its spectrum is less than 10 cm^{-1}. Correspondingly, the error of the calculated thickness is less than ±0.1 μm. Including uncertainties in the indices, the absolute error of the HgCdTe epilayer thickness obtained by the spectrum method is less than ±0.2 μm, and is better than that of common optical microscope measurements. From the point of view of infrared focal plane device applications, this method satisfies common specifications for choosing film thicknesses.

The accurate determination of the MCT composition from infrared spectra is a very important problem and many contributions have been published. (Hougen 1989; Natarajan et al.1986; Micklethwaite 1988; Ariel et al. 1995, 1996, 1997; Anandan et al. 1991; Jeoung et al. 1996; Yang et al. 1996; Gong et al. 1996)

References

Alterovitz SA, Bu-Abbud GH, Woollam JA, et al. (1983) J Appl Phys 54:1559

Anandan M, Kapoor AK, Bhaduri A, Warrier AVR (1991) Optical characterisation of $Hg_{1-x}Cd_xTe$ bulk samples. Infrared Phys 31:485–491

Anderson WW (1980) Infrared Phys 20:363

Anderson PL, et al. (1982) J Vac Sci Technol 21:125

Ariel V, Garber V, Roserfeld D, et al. (1995) Appl Phys Letts 66:2101

Ariel V, Garber V, Bahir G, et al. (1996) Appl Phys Letts 69:1864

Ariel V, Garber V, Bahir G, et al. (1997) Appl Phys Letts 70:1849

Arwin H, Aspnes DE (1984) J Vac Sci Technol A 2:1316

Aspnes DE (1974) J Opt Soc Am 64:639

Aspnes DE, Studna AA (1975) Appl Opt 14:220

Aspnes DE, Studna AA (1983) Phys Rev B 27:985

Aspnes DE, Kelso SM, Logan, RA, et al. (1986) J Appl Phys 60:754

Azzam RMA, Bashara NM (1977) Ellipsometry and Polarized Light. North-Holland Publishing Company, Amsterdam

Baars J, Sorger F (1972) Solid State Commun 10:875

Baltz R Von, Escher W (1972) Phys Stat Sol B 51:499

Baranskii PI, Gorodonichii OP, Schevchenko NV (1990) Infrared Phys 30:259

Bartoli FJ, et al. (1986) J Vac Sci Technol A 4:2047

Bassani F, Parravicini GP (1975) Electronic States and Optical Transitions in Solid. Pergamon Press, Oxford, pp 149–167

Belyaev AE, Schevchenko NV, Demidenko ZA (1991) Chinese J Millim Waves 10:241

Benson JD, Cornfeld AB, Martinka M, et al. (1996) Electr Mater 25:1406

Bevan MJ, Duncan WM, Weatphal GH, et al. (1996) J Electr Mater 25:1371

Bewley WW, Lindle JR, Vurgaftman I, Meyer JR, Johnson JL, Thomas ML, Tennant WE (2003) Negative luminescence with 93% efficiency from midwave infrared HgCdTe diode arrays. Appl Phys Letts 83:3254–3256
Blair J, Newnham R (1961) In: Metallurgy of Elemental and Compound Semiconductors Vol. 12. Wiley (Inter science), New York, p 393
Blakemore JS (1985) Semiconductor Statistics. Dover Publications, Mineola, New York
Blue MD (1964) Phys Rev 134:A226
Brice JC, Capper P (1987) Properties of Mercury Cadmium Telluride. INSPEC, Institution of Electrical Engineers
Brossat T, Raymond F (1985) J Crystal Growth 72:280
Bubulac LO, Edwall DD, Cheung JT (1992) J Vac Sci Technol B 10:1633
Burstein E (1954) Phys Rev 93:632
Caldwell RS, Fan HY (1959) Phys Rev 114:664
Cahan BD, Spanier RF (1969) Surf Sci 16:166
Callen H (1949) Phys Rev 76:1394
Chen AB, Sher A (1978) Phys Rev Lett 40:900
Chen AB, Sher A (1982) CPA band calculation for (Hg, Cd) Te. J Vac Sci Technol 21:138–141
Chen AB, Sher A (1995) Semiconductor Alloys. Plenum Press, New York and London
Chen LY, Feng XW, Ma HZ, et al. (1994) Appl Opt 33:1299
Chu JH (1983a) Infrared Res 2:25
Chu JH (1983b) Chin J IR Res 2:89
Chu JH (1983c) Infrared Res 2:439
Chu JH (1983d) AD-A13550419, USA
Chu JH (1984) Ph.D. thesis, Shanghai Institute of Technical Physics
Chu JH, Mi ZY (1987) Prog Phys 7:311
Chu JH, Xu SQ, Tang DY (1982) Chin Sci Bull 27:403
Chu JH, Wang RX, Tang DY (1983a) Infrared Res 2:241
Chu JH, Xu SQ, Tang DY (1983b) Appl Phys Lett 43:1064
Chu JH, Xu SQ, Ji HM, et al. (1985) Infrared Res A 4:255
Chu JH, Mi ZY, Tang DY (1991) Infrared Phys 32:195–211
Chu JH, Miao JW, Shi Q, et al. (1992a) J Infrared Milli Waves 11:411
Chu JH, Mi ZY, Tang DY (1992b) J Appl Phys 71:3955
Chu JH, Li B, Liu K, Tang DY (1994) J Appl Phys 75:1234
Demay Y, Gailliard JP, Medina P (1987) J Crystal Growth 81:97
Djurisic AB, Li EH (1999) J Appl Phys 85:2854
Djurisic AB, Y Chang, Li EH (2002) Mater Sci Eng R 38:237
Dornhaus R, Nimtz G (1976) In: Solid State Phys Springer Tracts in Modern Phys Vol 78. Springer Berlin Heidelberg New York
Edwall DD, Gertner ER, Tennant WE (1984) J Appl Phys 55:1453
Edward DP (1991) Handbook of Optical Constants of Solids II. Academic Press, New York
Ehrenreich H (1957) J Phys Chem Solids 2:131
Faraday M (1846) Phil Trans Roy Soc, p 1

Finkman E, Nemirovsky Y (1979) J Appl Phys 50:4356
Finkman E, Schacham SE (1984) J Appl Phys 56:2896
Franz W (1958) Z Naturf 13a:484
Frohlich H (1937) Proc R Soc A 160:230
Gering JM, Crim DA, Morgan DG, Coleman PD, Kopp W, Morkoç H (1987) A small-signal equivalent-circuit model for GaAs-$Al_xGa_{1-x}As$ resonant tunneling heterostructures at microwave frequencies. J Appl Phys 61:271–276
Gong HM, Hu XN, Li YJ, Shen J, et al. (1996) In: Basic applications studies of HgCdTe materials and devices, 95 collection of the theses. Shanghai Institute of Technical Physics, p 276
Gopal V, Ashokan R, Dhar V (1992) Infrared Phys 33:39
Gurauskas E, Kavaliauskas J, Krivaite G, et al. (1983) Phys Stat Sol B 115:771
Hartley RH, Folkard MA, et al. (1992) J Crystal Growth 117:166
Hauge PS, Dill FH (1973) J Res Dev 17: 472
He L, Becker CR, Bicknell-Tassius RN, et al. (1993) J Appl Phys 73:3305
Heinz B (1991) Optische Konstanten von Halblerter-Mehrschicht-systemen Dissertation. RWTH, Azchen
Herrmann KH, Melzer V (1996) Infrared Phys Technol 37:753
Herrmann KH, Happ M, Kissel H, et al. (1993a) J Appl Phys 73:3486
Herrmann KH, Melzer V, Muller U (1993b) Infrared Phys 34:117
Hougen CA (1989) J Appl Phys 66:3763
Huang CH (1990) Ph.D. thesis, Shanghai Institute of Technical Physics
Huang ZM, Chu JH (2000) Appl Opt 39:6390
Huang ZM, Chu JH (2001) Infrared Phys Technol 42:77–80
Huang G, Yang J, Chen X, et al. (2000) Proc SPIE 4086:270
Huga E, Kimura H (1963) J Phys Soc Jpn 18:777
Jensen B, Torabi A (1983) J Appl Phys 54:5945
Jeoung Y, Lee T, Kim H, Kim J, Han M, Kang T (1996) New method for the estimation of bulk HgCdTe composition by infrared transmission. Infrared Phys and Technol 37:445–450
Johnson EJ (1967) Semicond Semimet 3:153–258
Kane EO (1957) J Phys Chem Solids 1:249
Kane EO (1960) J Phys Chem Solids 2:181
Keldysh LV (1958a) Sov Phys JETP 34:788
Keldysh LV (1958b) Sov Phys JETP 6:763
Kerker M (1969) The Scattering of Light. Academic Press, New York, London
Kim JS, et al. (1994) Semicond Sci Technol 9:1696
Kireev PS (1978) Semicond Physics (English Translation). Mir Publishers, Moscow, p 570
Kucera Z (1987) Phys Status Solidi A 100:659
Landau LD, Lifshitz EM (1969) Statistical Physics. Addison-Wesley, Reading, Massachusetts
Lax B, Zeiger HJ, Dexter RV (1954) Physica 20:818
Li B, Chu JH, Liu K, et al. (1995a) J Phys C 6:23
Li B, Chu JH, Liu K, et al. (1995b) J Phys C 7:29

Lindle JR, Bewley WW, Vurgaftman I, Meyer JR, Varesi, JB, Johnson SM (2003a) Negative luminescence from MWIR HgCdTe/Si devices. Optoelectronics, IEE Proceedings 150:365–370
Lindle JR, Bewley WW, Vurgaftman I, Meyer JR, Varesi JB, Johnson SM (2003b) Efficient 3–5 μm negative luminescence from HgCdTe/Si photodiodes. Appl Phys Letts 82:2002–2004
Lindle JR, Bewley WW, Vurgaftman I, Meyer JR, Johnson JL, Thomas ML and Tennant WE (2004) Negative luminescence from mid-wave infrared HgCdTe diode arrays. Physica E: Low-dimensional Systems and Nanostructures 20:558–562
Liu K (1994) Ph.D. thesis, Shanghai Institute of Technical Physics
Liu K, Chu JH, Tang DY (1994a) J Appl Phys 75:4176
Liu K, Chu JH, Li B, et al. (1994b) Appl Phys Lett 64:2818
Lorentz HA (1906) Theory of Electrons. Teubner, Leipzig
Mao DH, Syllaios AJ, Robinson HG, et al. (1998) J Electron Mater 27:703
Maracas GN, Edwards JL,Shiralagi K, et al. (1992) J Vac Sci Technol A 10:1832
Marple DTF (1966) Phys Rev 112:785
Martienssen W (1957) J Phys Chem Solids 2:257
Merkel KG, Snyder PG, Woollam JA, et al. (1989) Jpn J Appl Phys 28:133
Micklethwaite WFH (1988) J Appl Phys 63:2382
Mollmann KP, Kissel H (1991) Semicond Sci Technol 6:1167
Moss TS (1952) Photoconductivity in the Elements. Butterworth, London, p 61
Moss TS (1954) Proc Phys Soc B 67:775
Moss TS (1959) Optical Properties of Semiconductors. Buttorworth, London, p 48
Moss TS, Burrell GJ, Ellis B (1973) Semiconductor Opto-Electronics. Butterworth, London, pp 59–88
Mroczkowski JA, Nelson DA (1983) J Appl Phys 54:2041
Muller MW, Sher A (1999) Appl Phys Lett 74:2343
Natarajan V, Taskar NR, Bhat IB, et al. (1986) J Electron Mater 17:479
Nathan V (1998) Optical absorption in $Hg_{1-x}Cd_xTe$. J Appl Phys 83:2812–2814
Orioff GJ, et al. (1994) J Vac Sci Technol A 12:1252
Packard RD (1969) Appl Opt 8:1901
Philipp HR (1972) J Appl Phys 43:2836
Price SL, Boyd PR (1993) Semicond Sci Technol 8:842
Qian DR (1986) Phys Stat Sol A 94:573
Redfield D (1963) Phys Rev 130:916
Schaake HF, Tregilgas JH (1983) J Electron Mater 12:931
Schacham SE, Finkman E (1985) J Appl Phys 57:2001
Schmit JL, Stelzer EL (1969) J Appl Phys 40:4865
Scott MW (1969) J Appl Phys 40:4977
Seeger K (1989) Semiconductor Physics (Fourth Edition). Springer, Berlin Heidelberg New York p 319
Sharma RK, Verma D, Sharma BB (1994) Infrared PhysTechnol 35:673
Snyder PG, Woollam JA, et al. (1990) J Appl Phys 68:5925
Sze SM (1981) Physics of Semiconductor Devices. Wiley, New York
Tian JG, Zhang CP, Zhang GY (1991) Appl Phys Lett 59:2591

Toyazawa Y (1959) Progr Theoret Phys (Kyoto) Suppl 12:111
Urbach F (1953) Phys Rev 92:1324
Viňa L, Umbach C, Cardona M, et al. (1984) Phys Rev B 29:6752
Wang SL (1997) Ph.D. thesis, Shanghai Institute of Technical Physics
Wiley JD, Dexter RN (1969) Phys Rev 181:1181
Wiley JD, Peerey PS, Dexter RN (1969) Phys Rev 181:1173
Yadava RDS, Sundersheshu BS, Anandan M, et al. (1994) J Electron Mater 23:1349
Yang JR (1988) Ph.D. thesis, Shanghai Institute of Technical Physics
Yang JR, Wang SL, Guo SP, He L (1996) J Infrared Milli Waves 15:328
Yuan S, Krenn H, Springholz G, et al. (1993) Appl Phys Lett 62:885
Yuan S, Springholz G, Bauer G, et al. (1994) Phys Rev B 49:5476
Zawadzki W, Szymanska W (1971) J Phys Chem Solids 32:1151
Zhong GY, Tang WG, Qian TL (1983) Infrared Res 2:45

5 Transport Properties

5.1 Carrier Concentration and the Fermi Level

5.1.1 Carrier Statistical Laws

Electrons are spin 1/2 particles and therefore obey the Pauli exclusion principle which states that no two spin 1/2 particles can occupy the same one particle quantum state. Consequently their populations are arranged in a Fermi–Dirac distribution. According to Fermi–Dirac statistics at zero temperature $T = 0$ K, the electrons in a metal fill all possible states up to the Fermi level. The density of states can be calculated from a band structure model and in this highly degenerate case, an energy integration from the bottom of the conduction band to an energy for which all the electrons are accommodated determines the Fermi level. In intrinsic semiconductors, at $T = 0$ K electrons fill the valence bands and none occupy the conduction bands. Semi-metals and highly doped semi-conductors at finite but low temperatures are also degenerate but their population distributions are somewhat different from those of a metal. In this chapter we will discuss the statistics and transport properties of narrow gap semiconductors.

For an electron gas, at temperature T and in thermal equilibrium, the electron occupation probability $f(E)$ at energy E is given by the Fermi–Dirac distribution:

$$f(E) = \frac{1}{\exp[(E - E_F)/k_B T] + 1} \tag{5.1}$$

where E_f is the Fermi level. At temperature 0 K, the electrons fill in the lowest possible states. At $E = E_f$, $f(E)$ drops suddenly from 1 to 0. When the temperature increases the change of $f(E)$ near E_f becomes smooth, i.e. in the energy range of $E_f \pm kT$, $f(E)$ decrease from 1 to 0, as show in Fig. 5.1a.

Equation (5.1) shows that when E is several k_BT greater than E_F, the denominator is >>1, so $f(E) \cong 0$; and when E is several k_BT smaller than E_F, the denominator is $\cong 1$, so $f(E) \cong 1$. It is obvious that the energy difference E_2–E_1 from $f(E_1) \cong 1$ to $f(E_2) \to 0$ is only several k_BT. For

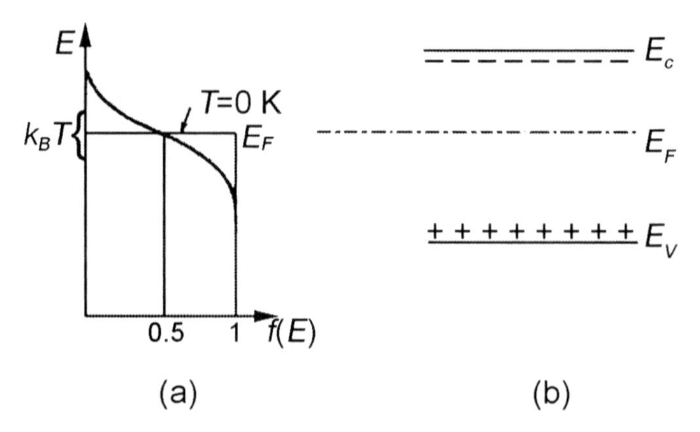

Fig. 5.1. (**a**) A sketch of the Fermi–Dirac distribution function; (**b**) The conduction band edge, valence band edge, and the Fermi level

free electrons in a box, the density of energy states in the range of $E \rightarrow E+dE$, excluding spin, is:

$$D(E)dE = \frac{(2m_0)^{3/2} E^{1/2}}{4\pi^2 \hbar^3} dE \tag{5.2}$$

where m_0 is the free electron mass, and $\hbar = h/2\pi$, is Plank's constant. This expression is derived assuming that the electrons are restricted to a box with infinite walls, which from the Schrödinger equation and periodic boundary conditions leads to the momentum $\boldsymbol{p} = \hbar\,\boldsymbol{k}$, and the energy dispersion $E = p^2/2m_0$. When the spin degeneracy of electrons is taken into account, the density of energy states doubles. The electron concentration energy density at energy E is given by the product of the density of energy states (5.2) and the occupation probability (5.1). Then the electron concentration is:

$$n = \int_0^\infty f(E)D(E)dE \tag{5.3}$$

or

$$n = \frac{(2m_0)^{3/2}}{2\pi^2 \hbar^3} \int_0^\infty \frac{E^{1/2}}{\exp[(E - E_F)/k_B T] + 1} dE \tag{5.4}$$

When $T \rightarrow 0$ K the integral reduces to $\int_0^{E_F} E^{1/2} dE$. From (5.4), we get the relationship between n and E_F:

$$E_F = \frac{\hbar^2 (3\pi^3 n)^{2/3}}{2m_0} \tag{5.5}$$

For a typical metal the electron concentration is $n \approx 5\times10^{28}$ m^{-3}, so $E_F \approx 6$ eV. The characteristic Fermi temperature is $T_F = E_F/k_B \approx 70{,}000$ K. This indicates that for all metals below their melting point, Fermi–Dirac statistics must be applied. However, for high energy states, where $E - E_F \gg k_B T$, (5.1) reduces to $f(E) = \exp\left[\left(E_F - E\right)/k_B T\right]$, or:

$$f(E) = A\exp(-E/k_B T) \tag{5.6}$$

The normalization coefficient is $A \approx \exp(E_F/k_B T)$. Equation (5.6) is the Maxwell–Boltzmann distribution function.

In the same way, we can calculate a momentum distribution. The momentum distribution is obtained from the energy distribution by using the relations $E = p^2/2m_0$ and $m_0 dE = pdp$. The expression for the momentum distribution is:

$$\mathbb{N}(p)dp = \frac{1}{\pi^2\hbar^3}\frac{p^2 dp}{\exp[(E - E_F)/k_B T] + 1} \tag{5.7}$$

It is obvious that the shape of momentum distribution curve differs from that of energy distribution. The electron concentration distribution along a given momentum component can be calculated. It cuts off at the Fermi momentum when $T = 0$ K, but broadens when the temperature is finite.

5.1.2 Intrinsic Carrier Concentration *n*

From the Fermi distribution and the conduction and valence bands density of states functions, the intrinsic carrier concentration of HgCdTe can be deduced. A proper treatment requires that we consider the degeneracy and the non-parabolic character of the energy bands. The electron concentration in the conduction band is:

$$n = \int_{E_C}^{\infty} f(E)D(E)dE \tag{5.8}$$

in which

$$f(E) = \left[1 + \exp\left(\frac{E - E_F}{k_B T}\right)\right]^{-1} \tag{5.9}$$

and

$$D(E)=\frac{k^2}{\pi^2}\frac{dk}{dE} \tag{5.10}$$

are the Fermi–Dirac distribution and the density of states functions, respectively. E_C is the energy at the bottom of the conduction band and E_F is the Fermi energy. Note that the Fermi energy, the conduction band edge, and the band gap are all slightly temperature dependent.

The conduction band is called degenerate if the Fermi level lies above the conduction band edge enough at temperature T so that $\exp[(E_c-E_F)/k_BT]\ll 1$. In the other extreme it is non-degenerate. A similar statement applies to the valence band.

The energy band structure of $Hg_{1-x}Cd_xTe$, in the concentration range where $x \leq 0.4$, is a narrow gap semiconductor where the spin orbit splitting $\Delta >> E_g$. As we have demonstrated in Chap. 3, applying the ***k·p*** perturbation of Kane in this approximation leads to a non-parabolic conduction band.

$$E-E_C=\frac{\hbar^2k^2}{2m_0}-\frac{E_g}{2}+\frac{1}{2}(E_g^2+\frac{8}{3}P^2k^2)^{\frac{1}{2}} \tag{5.11}$$

From (5.8)–(5.11), the electron concentration function can be deduced (Harman 1961):

$$n=\frac{3}{4\pi^2}\left(\frac{3}{2}\right)^{\frac{1}{2}}\left(\frac{k_BT}{P}\right)^3\int_0^\infty\frac{\varepsilon^{\frac{1}{2}}(\varepsilon+\phi)^{\frac{1}{2}}(2\varepsilon+\phi)}{1+\exp(\varepsilon-\eta)}d\varepsilon \tag{5.12}$$

in which $\phi=\frac{E_g}{k_BT}$ is the reduced bandgap, $\eta=\frac{(E_F-E_C)}{k_BT}$ is the reduced Fermi energy level, and $\varepsilon=\frac{E-E_C}{k_BT}$ is a reduced energy. All are dimensionless parameters. There is a similar equation for the hole concentration p. The Fermi level is set by the neutrality condition, which for an intrinsic semiconductor is $n_i = p_i$. We note this leads to a temperature dependent Fermi level. It has not been emphasized in this discussion but remember that 1_g is temperature dependent, and so is E_F.

Here interesting cases are for lightly doped $Hg_{1-x}Cd_xTe$ alloys in the ranges $x > 0.17$ and $\sim 100 < T < 300$ K. These ranges correspond to those of a narrow gap semiconductor, for which the carrier concentration is nearly intrinsic. In an intrinsic case, because the conduction band mass is so small, the Fermi energy level can approach the conduction band edge. If the material is doped, and is at a low enough temperature so it is extrinsic, the Fermi level can lie above the conduction band edge. If $\phi \gg \varepsilon$, then we have:

$$(\varepsilon+\phi)^{\frac{1}{2}} \approx \phi^{\frac{1}{2}} + \frac{1}{2}\varepsilon\phi^{-\frac{1}{2}} - \frac{1}{8}\varepsilon^2\phi^{-\frac{3}{2}} \tag{5.13}$$

In (5.12), the effective conduction band mass at the bottom of the conduction band m_c* is defined in terms of the matrix element P as:

$$m_c^* \equiv \frac{2\hbar^2 E_g}{4P^2}. \tag{5.14}$$

Suppose

$$N_C = 2(2\pi m_c^* k_B T / h^2)^{\frac{3}{2}} \tag{5.15}$$

is the effective density of states in the conduction band. Adopt the Fermi-Dirac integral expression defined as:

$$F_j(\eta) \equiv \frac{1}{\Gamma(j+1)} \int_0^\infty \frac{\varepsilon^j d\varepsilon}{1+\exp(\varepsilon-\eta)}, \tag{5.16}$$

where $\Gamma(x)$ is a gamma function of argument x.

Then, we can obtain the expression:

$$n = N_C \left[F_{\frac{1}{2}}(\eta) + \frac{15}{4\phi} F_{\frac{3}{2}}(\eta) + \frac{105}{32\phi^2} F_{\frac{3}{2}}(\eta) - \frac{105}{128\phi^3} F_{\frac{7}{2}}(\eta) \right] \tag{5.17}$$

With the parameters defined as:

$$\alpha \equiv F_{\frac{3}{2}}(\eta) / F_{\frac{1}{2}}(\eta)$$

$$\beta \equiv F_{\frac{5}{2}}(\eta) / F_{\frac{1}{2}}(\eta),$$

$$\gamma \equiv F_{\frac{7}{2}}(\eta) / F_{\frac{1}{2}}(\eta)$$

Equation (5.17) can be rewritten:

$$n = A \cdot N_C F_{\frac{1}{2}}(\eta), \tag{5.18}$$

where

$$A \equiv 1 + \frac{15\alpha}{4\phi} + \frac{105\beta}{32\phi^2} - \frac{105\gamma}{128\phi^3}. \quad (5.19)$$

From evaluations of the Fermi–Dirac integral, when η is in the range –4–1.3, α lies in the range of 1.003–1.33, β lies in the range 1.004–1.56, and γ lies in the range 1.005–1.71. Thus using typical φ values found in the experiments, A is a number slightly larger than 1. $N_c F_{1/2}(\eta)$ is the expression for an electron concentration in a parabolic conduction band, so A is a correction factor that accounts for the non-parabolic nature of the conduction band, and therefore depends on composition and temperature.

The near parabolic heavy-hole band, leads to a hole concentration expression:

$$p_h = N_{hV} F_{1/2}(-\phi - \eta),$$

in which $N_{hV} = 2(2\pi m_h^* k_B T / h^2)^{\frac{3}{2}}$ is the effective density of states in the heavy hole valence band. Because the effect mass of the light-hole m_{lh} is much smaller than the effect mass of the heavy-hole m_{hh}, the contribution of the light-holes to the density of states in the valence band can be neglected, $m_h^* \approx m_{hh}$. For an n-type semiconductor or a weak p-type semiconductor, for which the valence band hole population is non-degenerate, the net hole population reduces to a classical approximation (Blakemore 1965):

$$p = N_V \exp(-\phi - \eta) \quad (5.20)$$

with $N_V = N_{hV}$.

The neutrality condition for an intrinsic semiconductor is $n = p = n_i$, thereby yielding the expression:

$$A \cdot N_C F_{\frac{1}{2}}(\eta) = N_V \exp(-\phi - \eta). \quad (5.21)$$

If the bandgap is wide enough and the temperature low enough so the conduction band is far from degenerate, then $F_{1/2} = e^{+\eta}$, and from (5.18), (5.20) and (5.21) along with the neutrality condition $n_i = p_i$ we get:

$$n_i = \sqrt{A N_C N_V} e^{-\frac{\phi}{2}}, \quad (5.22a)$$

and

$$E_{Fi} - E_V = \frac{1}{2}\left(E_g + \ln\frac{N_V}{AN_C}\right) = \frac{1}{2}E_g + k_B T\left(\frac{3}{4}\ln\frac{m_h^*}{m_c^*} - \frac{1}{2}\ln A\right). \quad (5.22b)$$

Compared with the usual expression for a parabolic conduction band, this concentration has a $\sqrt{A}$ correction factor.

In commonly encountered situations, including the non-degenerate and the weakly degenerate conduction band cases, the approximate expression for $F_{1/2}(\eta)$ is of the form:

$$F_{\frac{1}{2}}(\eta) = [B + \exp(-\eta)]^{-1}. \quad (5.23)$$

Based on an analysis by Blakemore (1965), when $\eta >> 1.3$, then B = 0.27, the error between (5.23) and the true value of $F_{1/2}(\eta)$ is in the ±3% range. Substituting (5.23) into (5.21), we get:

$$\exp(-\eta) = [0.0182 + A(\frac{m_c^*}{m_h^*})^{\frac{3}{2}} \exp(\frac{E_g}{k_B T})]^{\frac{1}{2}} - 0.135. \quad (5.24)$$

Equations and (5.22) together with (5.18) yields:

$$n_i = A \cdot N_C \left\{ 0.135 + \left[0.0182 + A\left(\frac{m_c^*}{m_h^*}\right)^{\frac{3}{2}} \exp\left(\frac{E_g}{k_B T}\right) \right]^{\frac{1}{2}} \right\}^{-1}. \quad (5.25)$$

Equations (5.14) and (5.25) with $P = 8\times10^{-8}$ eV cm, and $m_h^* = 0.55m_0$, then become:

$$n_i = \frac{A_1 \times 9.56(10^{14}) E_g^{3/2} T^{3/2}}{1 + \sqrt{1 + 3.6 A_1 \cdot E_g^{3/2} \exp(\frac{E_g}{k_B T})}}, \quad (5.26)$$

in which

$$A_1 \equiv 1 + \frac{15\alpha}{4}\frac{k_B T}{E_g} + \frac{105\beta}{32}(\frac{k_B T}{E_g})^2 - \frac{105\gamma}{128}(\frac{k_B T}{E_g})^3. \quad (5.27)$$

Equation (5.26) is a good approximate expression for intrinsic carrier concentration of HgCdTe alloys. For a specific example take $x = 0.17$, and 77 K $\le T \le$ 300 K, then the value of E_g/kT lies between 5 and 10, so $3.6A_1E_g^{3/2}\exp(E_g/kT) >> 1$. Thus the 1 in the square root in the denominator

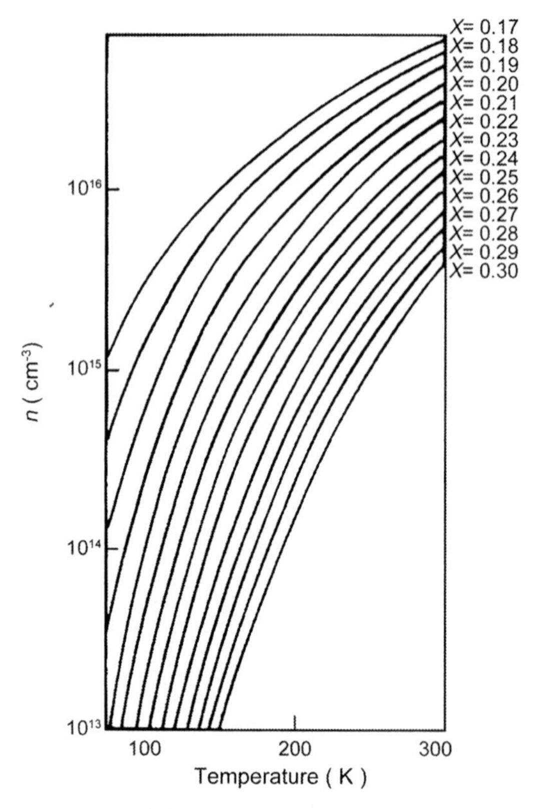

Fig. 5.2. $Hg_{1-x}Cd_xTe$ intrinsic carrier concentration n_i as a function of temperature T for different x

of (5.26) can be neglected, but the 1 in front of the square root sign must be retained. At the same time the third term and the fourth term in (5.27) are also far less than land can be neglected. Based on experience calculating the intrinsic carrier concentration the expression actually used can be written:

$$n_i = (1+3.25k_BT/E_g)9.56(10^{14})E_g^{3/2}T^{3/2}[1+1.9E_g^{3/4}\exp\left(\frac{E_g}{2k_BT}\right)]^{-1} \quad (5.28)$$

The results of such calculations are nearly identical to those obtained from numerical evaluations of (5.25). When $E_g \gg 2k_BT$, (5.28) can be reduced still further to the expression:

$$n_i = (1+3.25k_BT/E_g)5.03\times10^{14}E_g^{3/4}T^{3/2}\cdot\exp\left(\frac{-E_g}{2k_BT}\right) \quad (5.29)$$

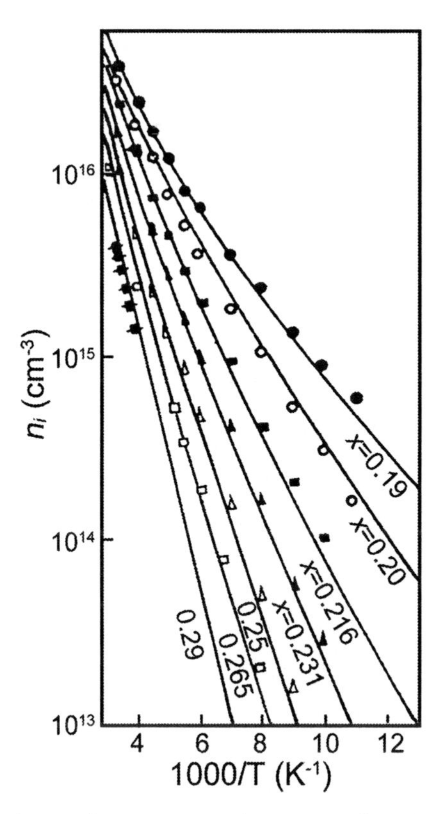

Fig. 5.3. The intrinsic carrier concentration as a function of the reciprocal temperature for $Hg_{1-x}Cd_xTe$ samples with different compositions

Figure 5.2 is a plot of intrinsic carrier concentration n_i as a function of temperature T for different HgCdTe alloy compositions.

In Fig. 5.3 experimental n_i data published by various authors is compared with predictions calculated from the formula (5.28) (Chu 1983a). The figure illustrates the excellence of the predictions. In the figure the curves are calculated for samples with x = 0.19, 0.20, 0.216, 0.231, 0.25, 0.265, and 0.29. Data are the result of experiment for samples with x = 0.19, 0.20, 0.216, and 0.251 taken by Chu (1983a), with x = 0.265 taken by Elliott (1971), and with x = 0.290 taken by Nemirovsky and Finkman (1979).

Schmit (1970) deduced an expression for $n_i(x,T)$ in the parabolic band approximation, and used his empirical formula for $E_g(x,T)$ to calculate $n_i(x,T)$. There was a large difference between the results when compared to experimental data. In 1982, he put forward an amended formula for $n_i(x,T)$, which was considerably more accurate:

$$n_i = \left[9.908 - 5.21x + 3.07\left(10^{-4}\right)T + 5.94\left(10^{-3}\right)Tx\right] \cdot \left(10^{14}\right)T^{3/2}E_g^{\ 3/4}\exp\left(-E_g/2k_BT\right) \quad (5.30)$$

when E_g was taken from the Hansen et al. (1982) expression.

$$E_g(eV) = -0.302 + 1.93x + (1-2x)5.35\left(10^{-4}\right)T - 0.810x^2 + 0.832x^3 \quad (5.31)$$

In deducing (5.30), $m_{hh} = 0.55m_0$ and $P = 9\times10^{-8}$ eV cm were used.

Later Hansen and Schmit suggested a new expression for n_i (Hansen and Schmit 1983).

$$n_i = \left[5.585 - 3.820x - 1.753\left(10^{-3}\right)T - 1.364\left(10^{-3}\right)Tx\right] \cdot \left(10^{14}\right)T^{3/2}E_g^{\ 3/4}\exp\left(-E_g/2k_BT\right) \quad (5.32)$$

in which $E_g(x,T)$ was given once again by (5.31). In the fitting procedure, $m_{hh} = 0.443m_0$ and $P = 8.49\times10^{-8}$ eV cm were used.

Based on experimental data for x = 0.205–0.220 and x = 0.290 samples, taking account of degeneracy but ignoring the conduction band non-parabolic correction, Nemirovsky and Finkman (1979), using m_{hh} = $0.55m_0$, $P = 1.953\times10^{-8}(18+3x)$eV cm, deduced an expression:

$$n_i = \left[1.265\times10^{16}T^{3/2}(6+x)^{-3/2}E_g^{\ 3/2}\right] \left\{1+\left[1+22.72(6+x)^{-3/2}E_g^{\ 3/2}\exp\left(E_g/2k_BT\right)\right]^{1/2}\right\}^{-1}, \quad (5.33)$$

in which $E_g(x,T)$ is given by the Finkman and Nemirovsky (1979) formula:

$$E_g(eV) = -0.337 + 1.948x + (1-1.89x)\times6.006\left(10^{-4}\right)T\,. \quad (5.34)$$

Comparisons of these formulas with experimental results follow. The samples used were single crystals of HgCdTe mainly grown by the solid state re-crystallization technique (a few were grown by the tellurium solvent technique). After cutting the crystals into slices and being subjected to a low temperature thermo-treatment, n-type samples with rather good electrical properties and low electron concentrations were

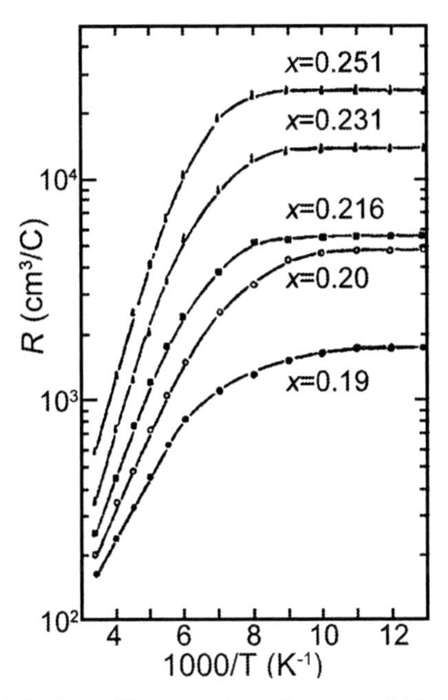

Fig. 5.4. Measured Hall coefficient values for several $Hg_{1-x}Cd_xTe$ samples

obtained. Effective donor concentrations N_D* were determined from the plateau of the low temperature part in the Hall curve $N_D^* = 1/eR$ (Fig. 5.4). Sample compositions were determined from the spectral cut-off response wavelength of photo-devices fabricated on the materials.

The parameters of the samples are listed in Table 2.2 in the second chapter. In the condition where all the effective donors are ionized, there is a useful formula:

$$n_i^2 = n(n - N_D^*)\,. \tag{5.35}$$

For each temperature, from the Hall coefficient the electron concentration n in the conduction band can be determined, and in combination with the N_D^* value obtained at 77 K the intrinsic carrier concentration n_i at this temperature can be deduced from (5.35).

The carrier concentration can be calculated by using the intrinsic carrier concentration formula and (5.35). Then the result can be compared with experiments. Bajaj et al. (1982) measured the carrier concentration of HgCdTe samples grown by using liquid phase epitaxy. The experimental dots in Fig. 5.5 are the carrier concentrations of samples with x = 0.227 and x = 0.192. In the low temperature saturation range, the carrier concentration is the effective donor concentration N_D^*. From the experimental

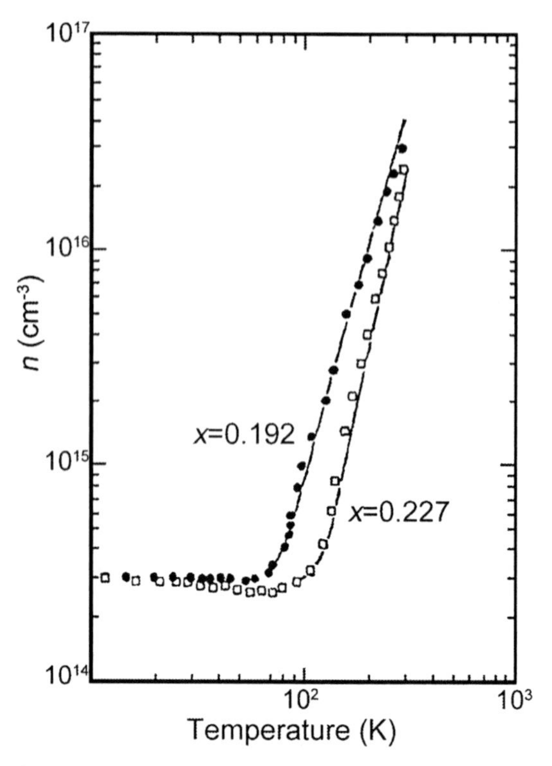

Fig. 5.5. The carrier concentrations of $Hg_{1-x}Cd_xTe$ (x = 0.192 and x = 0.227) samples as functions of temperature

result of Bajaj et al. (1982), for the sample with x = 0.227, $N_D^* = 2.95\times10^{14}$ cm^{-3}. The solution of (5.35) for n is:

$$n = \frac{1}{2}\left(N_D^* + \sqrt{N_D^* + 4n_i^2}\right). \tag{5.36}$$

For x = 0.192 and x = 0.227, first calculate the intrinsic carrier concentration n_i from (5.29) and then calculate the carrier concentration from (5.36). The results are the solid curves in Fig. 5.5, which are in good agreement with the experiments.

Figure 5.6 shows the experimental value of the intrinsic carrier concentration of the sample with x = 0.205, and the result calculated from (5.29) (solid curve), from the formula of Schmit, (5.30) (dashed curve), and from the formula of Nemirovsky and Finkman, (5.33) (dash–dot curve).

From Figs. 5.5–5.7, we see that in the ranges x = 0.19–0.29, and T = 77–300 K, the intrinsic carrier concentration formula, (5.29), accurately predicts the experimental results.

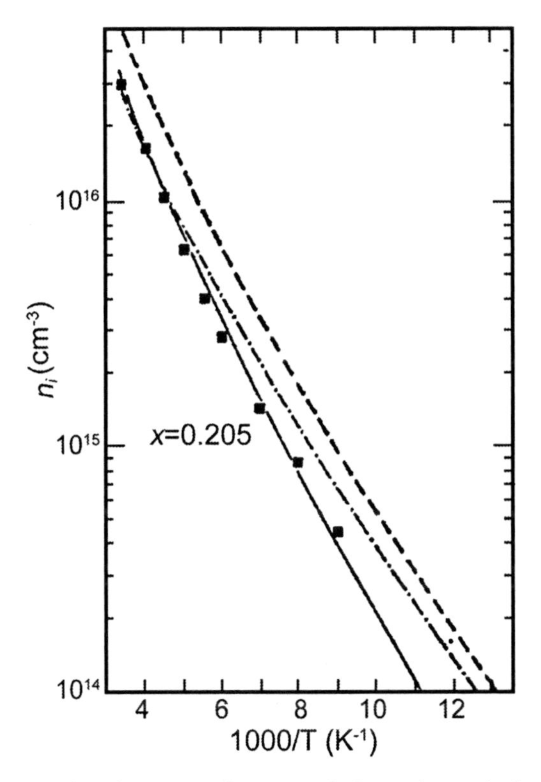

Fig. 5.6. The result of an experiment and three theoretical calculations of the intrinsic carrier concentration of a $Hg_{1-x}Cd_xTe$ ($x = 0.205$) sample

To compare the intrinsic carrier concentrations predicted by various authors, their results for three concentrations, $x = 0.20$, 0.24, and 0.30 were calculated and plotted verses temperature in Fig. 5.7. The curves derived from (5.29), (5.30), and (5.33), are plotted as solid, dashed and dash–dot curves, respectively. In all cases, the result calculated from (5.30) is higher than the others. At $x = 0.24$, the calculated results from (5.29) lies close to that from (5.33). With increasing composition, the curve calculated from (5.29) lies slightly higher than the one from (5.33). When $x = 0.24$ and for temperatures near room temperature, the curve from (5.29) is higher than the one from (5.33), but at lower temperatures the curves cross and their magnitudes reverse.

Based on the comparison with experimental results, (5.29) does accurately predict the relation between the intrinsic carrier concentration and temperature in HgCdTe crystals, especially in the high temperature range. From this fact, we can estimate roughly the composition and the quality of crystal from the shape of the log(R)–$1/T$ curve obtained in Hall

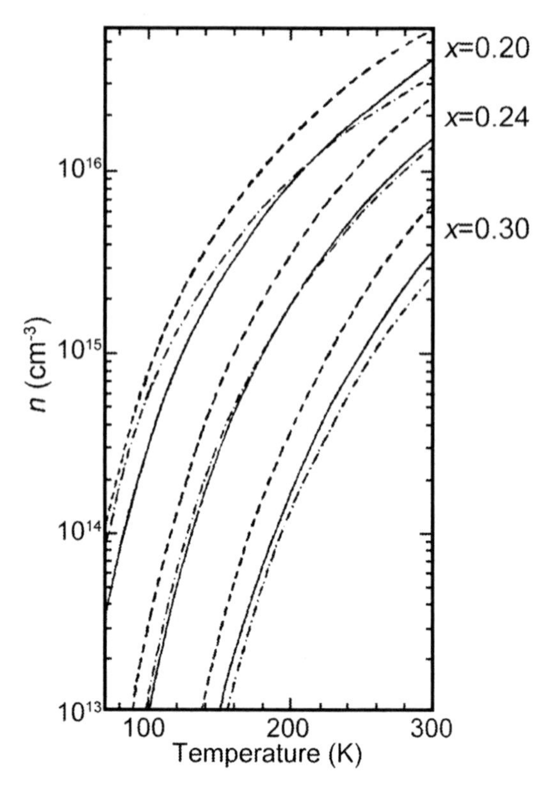

Fig. 5.7. Comparisons of intrinsic carrier concentration curve predictions verses temperature from (5.29) (*solid*), (5.30) (*dashed*), and (5.33) (*dot–dash*)

measurements. According to (5.29) and (5.36), we calculate the log(R)–$1/T$ curve for samples with different compositions in the range x = 0.18–0.26 (Fig. 5.8), supposing the effective donor concentration is N_D^* = 1/20 n_i (at 300 K). The reason of the supposition is: (1) For this condition, N_D^* has a negligible effect on R near 300 K. (2) For samples with x = 0.20, which are of greatest interest, this value of N_D^* is the upper limit for acceptable device material. Then draw Fig. 5.8 on transparent paper and place it on the log R verses $1/T$ experimental curve (using the same scale). If the sample qualifies, the ascending part of the log R–$1/T$ curve should nearly overlap one of the theoretical curves, while the flat part will be higher than the theoretical curve. Then the x value of the theoretical curve is the composition of the sample. If the effective donor concentration is too high, and/or the composition is inhomogeneous the slope of the curve will be too low and the experimental curve will not fit any of the theoretical ones. In this case, it shows that the quality of the material is not good.

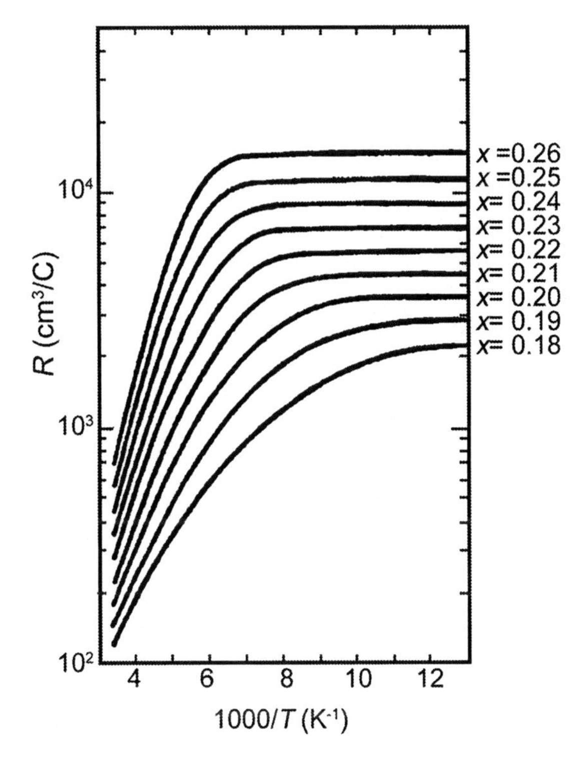

Fig. 5.8. The Hall coefficient of $Hg_{1-x}Cd_xTe$ samples verses $1/T$ with effective donor concentrations $N_D{}^* = 1/20\ n_i$ (at 300 K)

5.1.3 The Carrier Concentration and the Fermi Level for Compensated Semiconductors

The usual method for obtaining the Fermi level uses (5.24). The Fermi level E_F is often obtained from the carrier concentration n by numerical methods. To establish insight into trends it is valuable to establish an approximate analytical relationship between the carrier concentration n, the energy levels of the donors and the acceptors, and the Fermi level.

There is a long history of research into calculations of the carrier concentration and the Fermi level (Kittle 1976; Shockley 1955; Коренбдит 1956; Blakemore 1965; Roy 1977; Huang 1958). Kittle pointed out (Kittle 1976) those in cases with high compensation, because the charge neutrality equation is a transcendental equation, only numerical treatments are possible. Thus, over a broad temperature range and for cases involving a collection of impurities, a graphical method put forward by Shockley in

1995 (Shockley 1955) is often used to calculate carrier concentrations and Fermi levels. In 1956, Коренбдит also put forward a graphical method (Коренбдит 1956). However, his graphical values did not agree with experimental values for impurities with low ionization energies. In an analytical treatment, Blakemore produced a rather detailed discussion of the carrier concentration and the Fermi level in compensated semiconductors using the law of mass action (Blakemore 1965). But to do this he had to simplify the analysis by neglecting intrinsic excitations. In 1977, Roy (1977) treated impurity excitations to free carriers, but once again in this part of the analysis neglected intrinsic excitations. This approach applies mainly to low temperature and high impurity concentration cases. He did consider intrinsic excitations and obtained cubic equations for the carrier concentration and the Fermi level. This form of the calculation was intended to apply up to 300 K, whether or not the intrinsic excitations were included. However, the predicted locations of the Fermi levels deviate from experiments by over a 0.1 eV. Also in these calculations Roy considered only the case of a single impurity type and did not include the possibility of compensating impurities. In what follows a uniform analytical treatment of the carrier concentration and the Fermi level is developed and the result is compared to that of Shockley's graphical method and experiments.

(1) The n^4 Type Formula

The charge neutrality condition is:

$$N_C e^{-\frac{E_c - E_F}{k_B T}} + N_A \left(\frac{1}{e^{\frac{E_F - E_A}{k_B T}} + 1} - 1 \right) = N_D \left(\frac{1}{e^{\frac{E_D - E_F}{k_B T}} + 1} - 1 \right) + N_V e^{-\frac{E_F - E_v}{k_B T}} \tag{5.37}$$

where the left side is the net negative charge made up of the number of electrons in the conduction band plus the number of ionized acceptors, and the right side is the net positive charge made up of the number of ionized donors plus the number of holes (Huang 1958). Define:

$$\begin{aligned} E_F' &\equiv E_F - E_V, \\ E_i &\equiv E_C - E_D, \\ &\text{and} \\ E_i' &\equiv E_A - E_V, \end{aligned} \tag{5.38}$$

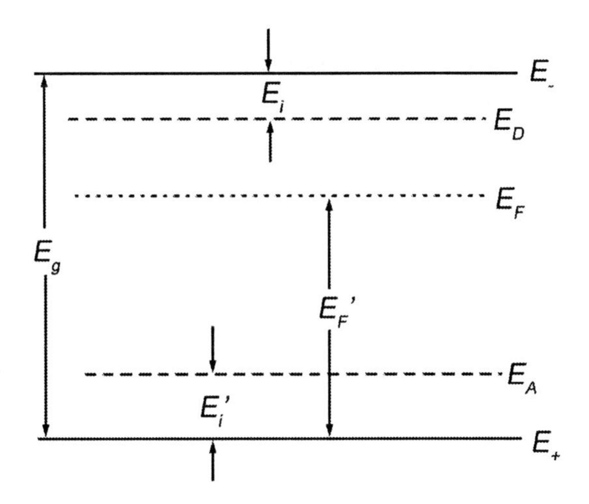

Fig. 5.9. A sketch of the energy level structure

where $E_F^{'}$ is the Fermi energy referenced to the valence band edge, E_i is the donor ionization energy and $E_i^{'}$ is the acceptor ionization energy.

From Fig. 5.9, other relations are:

$$\begin{cases} E_C - E_F = E_g - E_F^{'} \\ E_D - E_F = E_g - E_i - E_F^{'} \\ E_F - E_A = E_F^{'} - E_i^{'} \end{cases} . \tag{5.39}$$

Let:

$$\begin{cases} n = N_C e^{-\frac{E_C - E_F}{k_B T}} \\ p = N_V e^{-\frac{E_F - E_V}{k_B T}} \\ n_i^2 = N_V N_C e^{-\frac{E_g}{k_B T}} = np \\ n_0^2 \equiv N_C N_D e^{-\frac{E_i}{k_B T}} \\ p_0^2 \equiv N_V N_A e^{-\frac{E_i^{'}}{k_B T}} \end{cases} , \tag{5.40}$$

where n is the electron concentration in the conduction band, p is the hole concentration in the valence band and n_i is the intrinsic carrier concentration. The new quantities defined in (5.40) are related to the number of electrons excited from donors, n_o, and holes excited from acceptors, p_o. Substituting (5.38)–(5.40) into (5.37), produces a transcendental algebraic equation:

$$n+\frac{N_A}{1+\frac{pN_A}{p_0^2}}=\frac{N_D}{1+\frac{nN_D}{n_0^2}}+p\,. \tag{5.41}$$

In this formula, the second term on the left is the ionized acceptor concentration, N_A^-, the first term on the right is the ionized donor concentration N_D^+. If there are i kinds of donors with concentrations N_{Di}, and energy levels E_{Di}, and j kinds of acceptors with concentrations N_{Aj} and energy levels E_{Aj}, then (5.41) can be generalized to the more usual form:

$$n+\sum_j N_{Aj}^- =\sum_i N_{Di}^+ + p \tag{5.42}$$

Equation (5.41) can be expressed as a quadratic equation in n:

$$\sum_{l=0}^{4} A_l n^l = 0 \tag{5.43}$$

in which

$$\begin{aligned}
A_0 &= -n_0^2 N_A n_i^4\\
A_1 &= -[p_0^2 n_0^2 + N_A N_D (n_0^2 + n_i^2)]n_i^2\\
A_2 &= N_A n_0^2 (p_0^2 + n_i^2) - N_D p_0^2 (n_0^2 + n_i^2)\,.\\
A_3 &= p_0^2 n_0^2 + N_A N_D (p_0^2 + n_i^2)\\
A_4 &= p_0^2 N_D
\end{aligned}$$

If we substitute p_0, n_0, N_A, N_D for n_0, p_0, N_D, N_A, respectively and substitute p for n, we can get the same a quadratic equation for p. The value of n is determined from (5.43) and then the Fermi level is:

$$E_F = E_C - k_B T \ln \frac{N_C}{n} \tag{5.44}$$

An analytic solution to quadratic equation is complicated. At lower temperatures where the intrinsic excitation can be neglected, $n_i \to 0$, then (5.43) reduces to:

$$n^2 + (\frac{n_0^2}{N_D} + N_A)n + (\frac{N_A}{N_D} - 1)n_0^2 = 0. \tag{5.45}$$

This expression has been deduced from the law of mass action (Blakemore 1965) and discussed in detail in the temperature range where the intrinsic excitation can be neglected and only the impurity excitation considered. In higher temperature ranges, where both mechanisms must be considered (5.45) is not valid, and another the way of simplifying (5.43) must be used.

(2) Simplified n^3 Type Formulas Based on the Degree of the Donor Ionization

If we know the donor's (or acceptor's) degree of ionization, the n^4 type formula (5.43) can be simplified to an n^3 type formula. In the literature (Huang 1958), $\chi_D^2 \equiv \frac{N_C}{N_D} e^{-E_i/k_B T}$ is defined to be the donor's ionization degree. Substituting the n_0^2 expression obtained from (5.41) it becomes:

$$\chi_D^2 = \frac{n_0^2}{N_D^2}. \tag{5.46}$$

The same procedure for the acceptors yields:

$$\chi_A^2 = \frac{p_0^2}{N_A^2}, \tag{5.47}$$

for the acceptor's degree of ionization. The weak ionization case occurs when $\chi^2 \gg \frac{1}{2}$, when $\chi^2 = \frac{1}{2}$, it is the half ionization case, and when $\chi^2 \to \infty$, it is the complete ionization case. Equation (5.41) can be rewritten:

$$n + \frac{N_A + \gamma' \frac{pN_A^2}{p_0^2}}{\alpha' + \beta' \frac{pN_A}{p_0^2}} = \frac{N_D + \gamma \frac{nN_D^2}{n_0^2}}{\alpha + \beta \frac{nN_D}{n_0^2}} + p \tag{5.48}$$

in which $\alpha(\alpha')$, $\beta(\beta')$, $\gamma(\gamma')$ are parameters corresponding to different ionization degrees. The usual case corresponds to $\alpha(\alpha')=1$, $\beta(\beta')=1$, $\gamma(\gamma')=0$. The weak ionization cases correspond to $\alpha(\alpha')=0$, $\beta(\beta')=1$, $\gamma(\gamma')=0$. The half ionization cases correspond to $\alpha(\alpha')=2$, $\beta(\beta')=0$, $\gamma(\gamma')=0$. The heavy ionization cases correspond to $\alpha(\alpha')=1$, $\beta(\beta')=0$, $\gamma(\gamma')=-1$. The complete ionization cases correspond to $\alpha(\alpha')=1$, $\beta(\beta')=0$, $\gamma(\gamma')=0$.

Now if we do not allow the ionization of acceptors, the usual case, and we consider donors with four different degrees of ionization, we get four different cubic equations. The donor ionization cases are:

- Weak ionization: ($\alpha=0$, $\beta=1$, $\gamma=0$, $\alpha'=1$, $\beta'=1$, $\gamma'=0$)

$$n^3+N_A\left(1+\frac{n_i^2}{p_0^2}\right)n^2-(n_0^2+n_i^2)n-\frac{N_A}{p_0^2}(n_0^2+n_i^2)n_i^2=0. \tag{5.49}$$

- Half ionization: ($\alpha=2$, $\beta=0$, $\gamma=0$, $\alpha'=1$, $\beta'=1$, $\gamma'=0$)

$$n^3+\left[\left(N_A-\frac{N_D}{2}\right)+\frac{N_A n_i^2}{p_0^2}\right]n^2-n_i^2\left(1+\frac{N_A N_D}{2p_0^2}\right)n-\frac{n_i^4 N_A}{p_0^2}=0. \tag{5.50}$$

- Heavy ionization: ($\alpha=1$, $\beta=0$, $\gamma=-1$, $\alpha'=1$, $\beta'=1$, $\gamma'=0$)

$$n^3+\left[\frac{N_A-N_D}{1+\frac{N_D^2}{n_0^2}}+\frac{N_A n_i^2}{p_0^2}\right]n^2-\frac{n_i^2\left(1+\frac{N_A N_D}{p_0^2}\right)}{1+\frac{N_D^2}{n_0^2}}n-\frac{1}{1+\frac{N_D^2}{n_0^2}}\frac{n_i^4 N_A}{p_0^2}=0. \tag{5.51}$$

- Complete ionization: ($\alpha=1$, $\beta=0$, $\gamma=0$, $\alpha'=1$, $\beta'=1$, $\gamma'=0$)

$$n^3+\left[(N_A-N_D)+\frac{N_A n_i^2}{p_0^2}\right]n^2-n_i^2\left(1+\frac{N_A N_D}{p_0^2}\right)n-\frac{n_i^4 N_A}{p_0^2}=0. \tag{5.52}$$

The above cubic equations can be written collectively as:

$$n^3+b_1 n^2+b_2 n+b_3=0. \tag{5.53}$$

Let P, Q and R be defined as: $P \equiv b_2 - \frac{b_1^2}{3}$, $Q \equiv b_3 - \frac{b_2 b_3}{3} + \frac{2b_1^3}{27}$, $R \equiv \frac{P^3}{27} + \frac{q^2}{4}$. Because n must be a positive real number, n>0, then the relevant solutions are:

- When $R > 0$,

$$n = (-\frac{Q}{2} + \sqrt{R})^{1/3} + (-\frac{Q}{2} - \sqrt{R})^{1/3}, \tag{5.54}$$

- When $R < 0$,

$$n = 2\sqrt{\frac{-P}{3}} \cos\frac{\phi}{3} - \frac{b_1}{3}, \tag{5.55}$$

in which $\cos\phi = \frac{\sqrt{27}}{2} \frac{Q}{P\sqrt{-P}}$.

Equations (5.49)–(5.52), are four cubic equations, that can be simplified still further if in the range of temperature being studied is such that the intrinsic excitation can be neglected, $n_i \to 0$. Solutions to these equations and (5.45) in different approximations are listed in Table 5.1. We emphasize that the n^3 type formula also apply to situations where there are several donor impurities. In such cases substitute $\sum_i (\frac{N_{Di}^2}{n_{0i}^2})$ for $\left(\frac{N_D^2}{n_0^2}\right)$, $\sum_i n_{0i}^2$ for n_0^2, and $\sum_i N_{Di}$ for N_D.

Table 5.1. Solutions to (5.49)–(5.52) and (5.45) in different approximations

n^4 type	The usual formula	$n = \frac{1}{2}[-(\frac{n_0^2}{N_D} + N_A) + \sqrt{(\frac{n_0^2}{N_D} + N_A)^2 - 4(\frac{N_A}{N_D} - 1)n_0^2}]$
n^3 type	Weak donor ionization	$n = \frac{1}{2}[N_A + \sqrt{N_A^{\ 2} + 4n_0^2}]$
	Half donor ionization	$n = \frac{N_D}{2} - N_A$
	Heavy donor ionization	$n = \frac{N_D - N_A}{1 + N_D^2 / n_0^2}$
	Complete donor ionization	$n = N_D - N_A$

Table 5.2. The result of the n^2 type equation of carrier concentration

	Weak donor ionization	Half donor ionization	Heavy donor ionization	Complete donor ionization
Weak acceptor ionization	$\frac{\sqrt{4n_i^2[1+\frac{n_0^2}{n_i^2}][1+\frac{p_0^2}{n_i^2}]}}{2[1+\frac{p_0^2}{n_i^2}]}$	$\frac{\frac{N_D}{2}+\sqrt{\frac{N_D^2}{4}+4n_i^2[1+(\frac{p_0^2}{n_i^2})]}}{2[1+(\frac{p_0^2}{n_i^2})]}$	$\frac{N_D+\sqrt{N_D^2+4n_i^2[1+\frac{p_0^2}{n_i^2}+\frac{N_D^2}{n_0^2}]}}{2[1+\frac{p_0^2}{n_i^2}+\frac{N_D^2}{n_0^2}]}$	$\frac{N_D+\sqrt{N_D^2+4n_i^2[1+\frac{p_0^2}{n_i^2}]}}{2[1+\frac{p_0^2}{n_i^2}]}$
Half acceptor ionization	$\frac{-\frac{N_A}{2}+\sqrt{\frac{N_A^2}{4}+4n_i^2[1+\frac{n_0^2}{n_i^2}]}}{2}$	$\frac{(\frac{N_D}{2}-\frac{N_A}{2})+\sqrt{\frac{(N_A-N_D^2)}{4}+4n_i^2}}{2}$	$\frac{(N_D-\frac{N_A}{2})+\sqrt{(N_D-\frac{N_A}{2})^2+4n_i^2[1+\frac{n_0^2}{n_i^2}]}}{2(1+\frac{N_D^2}{n_i^2})}$	$\frac{(N_D-\frac{N_A}{2})+\sqrt{(N_D-\frac{N_A}{2})^2+4n_i^2}}{2}$
Heavy acceptor ionization	$\frac{-N_A+\sqrt{N_A^2+4n_i^2[1+\frac{n_0^2}{n_i^2}+\frac{N_A^2}{p_0^2}]}}{2}$	$\frac{(\frac{N_D}{2}-N_A)+\sqrt{(\frac{N_D}{2}-N_A)^2+4n_i^2(1+\frac{N_A^2}{p_0^2})}}{2}$	$\frac{(N_D-N_A)+\sqrt{(N_D-N_A)^2+4n_i^2[1+\frac{N_A^2}{p_0^2}][1+\frac{n_0^2}{n_i^2}]}}{2(1+\frac{N_D^2}{n_i^2})}$	$\frac{(N_D-N_A)+\sqrt{(N_D-N_A)^2+4n_i^2[1+\frac{N_A^2}{p_0^2}]}}{2}$
Complete acceptor ionization	$\frac{-N_A+\sqrt{N_A^2+4n_i^2[1+\frac{n_0^2}{n_i^2}]}}{2}$	$\frac{(\frac{N_D}{2}-N_A)+\sqrt{(\frac{N_D}{2}-N_A)^2+4n_i^2}}{2}$	$\frac{(N_D-N_A)+\sqrt{(N_D-N_A)^2+4n_i^2[1+\frac{N_D^2}{n_0^2}]}}{2}$	$\frac{(N_D-N_A)+\sqrt{(N_D-N_A)^2+4n_i^2}}{2}$

Note 1: The seven formulas for complete ionization are special cases of the five formulas for heavy ionization. Substituting $\frac{N_D^2}{n_0^2} = \frac{1}{\chi_D^2} \to 0$, and $\frac{N_A^2}{p_0^2} = \frac{1}{\chi_A^2} \to 0$, the complete ionization conditions, into the formulas for heavy ionization, reduces them to the formulas for complete ionization.
Note 2: The n^2 type formulas can be generalized to cases where there are several donors and several acceptors by making the following substitutions: $n_0^2 \to \sum_i n_{0i}^2$,

$$N_D \to \sum_i N_{Di},\ \frac{N_D^2}{n_0^2} \to \sum_i \frac{N_{Di}^2}{n_{0i}^2},\ p_0^2 \to \sum_j p_{0j}^2,\ N_A \to \sum_j N_{Aj},\ \frac{N_A^2}{p_0^2} \to \sum_j \frac{N_{Aj}^2}{p_{0j}^2}.$$

(3) Further Simplified n^2 Type Formulas Based on the Degree of Donor and Acceptor Ionization

If we consider four kinds of donor and acceptor ionization degrees, there are sixteen different cases. From (5.48), when $n_i^2 \neq 0$ there are 16 equations for n. The solutions of these equations are listed in Table 5.2. From Table 5.2, we can deduce a generalized analytic expression:

$$n = \{\sqrt{(\xi N_D - \eta N_A)^2 + 4n_i^2[1 + (\frac{n_0^2}{n_i^2})_{DW} + (\frac{N_A^2}{p_0^2})_{AS}][1 + (\frac{p_0^2}{n_i^2})_{AW} + (\frac{N_D^2}{n_0^2})_{DS}]} + (\xi N_D - \eta N_A)\} / \{2[1 + (\frac{p_0^2}{n_i^2})_{AW} + (\frac{N_D^2}{n_0^2})_{DS}]\} \quad (5.56)$$

The terms with the subscripts *DW, DS, AW,* and *AS* take on the indicated form in weak donor ionization, heavy donor ionization, weak acceptor ionization, and heavy acceptor ionization, respectively, or else they are 0. ξ (or η) is 0, 1/2, 1 in the case of weak, half and heavy (or complete) donor ionization (or acceptor ionization), respectively.

(4) Calculational Results

The applicable scope and conditions for the n^4 type, the n^3 type, and the n^2 type formulas and the circumstances for their use are indicated in Table 5.3.

When we engage in practical calculations we first determine n_0^2, p_0^2, n_i^2 based on N_C, N_V, N_D, N_A, E_i, E_i', E_g and T, and then estimate the ionization degrees χ_D and χ_A. Based on these ionization degrees, we choose the proper formulas from Table 5.1 or Table 5.2, or calculate the value of n by using (5.49)–(5.52) and then evaluate E_F. Table 5.4 is an example of a calculation using Shockley's graphical method (Shockley 1955). In this evaluation the parameters chosen were $N_A = 10^{14}\,\mathrm{cm}^{-3}$, $N_D = 10^{15}\,\mathrm{cm}^{-3}$, $E_i = E_i^{'} = 0.04\,\mathrm{eV}$, and $E_g = 0.72\,\mathrm{eV}$. From the free electron approximation we have $N_C = N_V = 4.831 \times 10^{15} T^{3/2}$ (Blakemore 1965), and the value of k_B used was $k_B = 8.625 \times 10^{-5}\,\mathrm{eV/K}$.

The graphical values presented in the table come from Shockley's original paper (Shockley 1955). From the table, we see the analytic values are consistent with the graphical values at all temperatures and have a higher precision. The law of mass action, in which one considers only the impurity excitation and takes no account of the intrinsic excitation, is consistent with the graphical values below 300 K and do not consistently agree with the graphical values above 300 K. At 600 K, the discrepancy between the Fermi levels is up to 0.22 eV. It is obvious that we cannot neglect the intrinsic excitation at elevated temperatures. From these calculations, we see that above 600 K the carrier concentration n is dominated by the intrinsic carrier concentration n_i, but below 600 K the discrepancy between n_i and n increase gradually. At 400 K, n is 1.5 times larger than n_i and by 300 K, n is 40 times larger than n_i. It is obvious that below 600 K the intrinsic excitation n_i and the impurity excitation must both be considered. This indicates that in the cases from 300 to 600 K, we can

Table 5.3. The applicable scope and conditions for the n^4 type, the n^3 type, and the n^2 type formulas and the circumstances for their use

	The applicable scope	The applicable condition	The formula chosen	
n^4 type	One donor One acceptor		$n_0 \neq 0$	Quadratic equation (5.43)
			$n_0 \to 0$	(5.43)
n^3 type	Several donors and/or several acceptors	Known donor ionization degree	$n_0 \neq 0$	Cubic equation (5.49)–(5.52)
			$n_0 \to 0$	Table 5.1
n^2 type	Several donors and/or several acceptors	Known ionization degrees of donors and acceptors	$n_0 \neq 0$	Quadratic equation (5.56) or Table 5.2

neither calculate n using the law of mass action nor simply regard it as an intrinsic excitation. We can calculate n following the procedure presented in the proceeding paragraphs, Shockley's graphical method, or a complete numerical evaluation.

Judging from the above comparisons, the analytical treatment validated.

Table 5.4. An example of a calculation using Shockley's graphical method

T (K)	n_0^2 (cm^{-6})	p_0^2 (cm^{-6})	χ_D^2	χ_A^2	n_i^2 (cm^{-6})	n	$E_- - E_F$		
						Calculated via the above analysis (cm^{-6})	Calculated via the above analysis	Calculated by Shockley	Calculated by the law of mass action (5″)
600	3.227×10^{34}	3.227×10^{33}	$\sim10^4$	$\sim10^5$ (complete)	4.5726×10^{33}	6.8080×10^{16}	0.3596	0.36	0.5835
450	1.6453×10^{34}	1.6453×10^{33}	$\sim10^4$	$\sim10^5$ (complete)	1.8652×10^{31}	4.7921×10^{15}	0.3560		0.4207
400	1.2123×10^{34}	1.2123×10^{33}	$\sim10^4$	$\sim10^5$ (complete)	1.2904×10^{30}	1.6718×10^{15}	0.3467		0.3681
300	5.3506×10^{34}	5.3506×10^{33}	$\sim10^3$	$\sim10^4$ (complete)	5.1861×10^{26}	9.0014×10^{14}	0.2649	0.27	0.2649
200	1.3444×10^{33}	1.3444×10^{32}	$\sim10^3$	$\sim10^4$ (complete)	1.3934×10^{20}	9.0000×10^{14}	0.1661		0.1661
150	4.0287×10^{32}	4.0287×10^{31}	$\sim10^2$	1.6 (heavy)	3.4126×10^{15}	9.0000×10^{14}	0.1190	0.12	0.1190
50	1.6005×10^{29}	1.6005×10^{28}	0.16	0.16 (weak)	$\to0$	3.5300×10^{14}	0.03658	0.038	0.03773
37.5	4.7201×10^{27}	4.7201×10^{26}	0.0047	0.0047 (weak)	$\to0$	3.4970×10^{13}	0.03352	0.034	0.03389
20	3.6727×10^{22}	3.6727×10^{21}	$\to0$	$\to0$ (weak)	$\to0$	3.6720×10^{8}	0.03603	0.036	0.03621

In the following two examples, $N_A = 10^{12}$ cm^{-3}, and all the other parameters are unchanged.

37.5	4.7201×10^{27}	4.7201×10^{24}	0.0047 (weak)	4.7 (heavy)	$\to0$	6.8204×10^{13}	0.03136	0.031	0.03147
20	3.6727×10^{22}	3.6727×10^{19}	$\to0$ (weak)	$\to0$ (weak)	$\to0$	3.5469×10^{10}	0.02814	0.028	0.02841

Table 5.5. From a Ge sample example in the literature (Hannay 1959): N_A=10^{14} cm^{-3}, N_D=10^{15} cm^{-3}, $E_g = 0.66\,\text{eV}$, $E_i = E_i^{'} = 0.01\,\text{eV}$, $m_e = 0.25m_0$, $m_h = 0.3m_0$

T (K)	N_-	N_+	n_0^2	p_0^2	χ_D^2	χ_A^2
300	3.13×10^{18}	4.115×10^{18}	2.1268×10^{33}	2.7961×10^{32}	$\sim 10^3$ (complete)	$\sim 10^4$ (complete)

T (K)	n_i^2	n	$E_- - E_F$	
			Calculated	Graphic
300	1.0824×10^{26}	9.0012×10^{14}	0.2110	0.21

Table 5.6. Comparison with the date of Morin and Maita (1963) on a Si sample: N_A = 74×10^{14} cm^{-3}, N_D = 10^{11} cm^{-3}, $E_i^{'} = 0.046\,\text{eV}$, $m_h = 0.3m_0$. The calculated value of p_0 is obtained based on the formula in Table 5.1

T (K)	N_V	p_0^2	χ_A^2	p (cm^{-6})	
				Calculated	Graphic
20	7.1001×10^{16}	1.6071×10^{20}	$\to 0$ (weak)	1.58×10^{9}	1.6×10^{9}
25	9.9227×10^{16}	3.9914×10^{22}	$\to 0$ (weak)	1.56×10^{11}	1.8×10^{11}
33.5	1.5280×10^{17}	1.2725×10^{25}	$\sim 10^{-4}$ (weak)	3.52×10^{12}	4×10^{12}
50	2.8066×10^{17}	4.8410×10^{27}	$\sim 10^{-2}$ (weak)	6.95×10^{13}	7×10^{13}
80	4.115×10^{17}	5.3490×10^{29}	1 (half)	3.70×10^{14}	3.7×10^{14}

5.2 Conductivity and Mobility

5.2.1 The Boltzmann Equation and Conductivity

Under quite general conditions that apply to nearly all practical situations encountered in device structures, the equation governing the change in populations of Fermion single particle states is:

$$\frac{df(u,t)}{dt} = \sum_{u'}\left\{\left[1-f(u,t)\right]W(u/u')f(u,t)-\left[1-f(u',t)\right]W(u'/u)f(u,t)\right\} \tag{5.57}$$

where

- $f(u,t)$ is the Fermi function for the state u at time t,
- A single particle state is designated by $u \Rightarrow \alpha, \mathbf{k}, s$ with the band index α, the wave vector $\boldsymbol{k}$, and the spin index s, and
- $W(u'/u)$ is the transition probability per unit time from an initial state u to a final state u' which is given by the golden rule.

This equation is called the single Fermion particle "Master Equation." From the derivation of this equation, which starts from the Schrodinger equation, it has been shown that (Sher and Primakof 1960):

$$\frac{W(u/u')}{W(u'/u)} = \frac{e^{-E(u)/k_BT}}{e^{-E(u')/k_BT}} \tag{5.58}$$

where $E(u)$ is the energy of state u. A consequence of (5.58) is that in equilibrium where the time derivative vanishes the solution to (5.57) is the Fermi distribution:

$$f_0(u) = \frac{1}{e^{(E-E_F)/k_BT}+1}. \tag{5.59}$$

E_F is determined from the condition that the sum of $f_0(u)$ over all states u of the system must equal the total number of free electrons N in the solid.

For simplicity restrict the following discussion to a single band case where only interband scattering occurs (suppress the band index α), and to non-degenerate situations where the Fermi distribution reduces to a Boltzmann distribution,

$$f_o(\mathbf{k},s) = \frac{e^{-E/k_BT}}{Z}, \tag{5.60}$$

independent of s, and the partition function is:

$$Z = \sum_{\mathbf{k}} e^{-E/k_BT}. \tag{5.61}$$

Then (5.57) reduces to its simpler form often called the Boltzmann gain–loss equation:

$$\frac{df(u,t)}{dt} = \sum_{u'} \{W(u/u')f(u,t) - W(u'/u)f(u,t)\}. \tag{5.62}$$

To solve transport problems it is necessary to be able to treat cases where the material is spatially inhomogeneous. The above quantum formalism does not simultaneously have electron spatial $\boldsymbol{r}$ coordinates and wave vectors $\boldsymbol{k}$ because of the uncertainty principle. The way this is overcome is to schematically decompose the sample into a collection of mesoscopic regions, each homogeneous and large enough so (5.57) applies, but these regions may have varying properties from one location $\boldsymbol{r}$ to another. Then a new occupation number is defined with an equilibrium functional form:

$$F_o(\mathbf{r},\mathbf{k},s) = \frac{nV}{2} e^{-E(\mathbf{r},\mathbf{k})/k_B T(\mathbf{r})}, \tag{5.63}$$

where n/2 is the average density of electrons per spin and V is the volume of the sample. Then the total time derivative of F(**r**, **k**, s, t) can be expressed as:

$$\begin{aligned}
\frac{dF(\mathbf{r},\mathbf{k},s,t)}{dt} &= \frac{\partial F}{\partial t} + \frac{\partial \mathbf{k}}{\partial t} \bullet \nabla_{\mathbf{k}} F + \mathbf{v} \bullet \nabla_{\mathbf{r}} F + \mathbf{v} \cdot \nabla_{\mathbf{r}} T \frac{\partial F}{\partial T} \\
&= \frac{\partial F}{\partial t} - \frac{e}{\hbar}(\mathrm{E} + \mathbf{v} \times \mathbf{B}) \bullet \nabla_{\mathbf{k}} F + \mathbf{v} \bullet \left(\nabla_{\mathbf{r}} F + \nabla_{\mathbf{r}} T \frac{\partial F}{\partial T} \right) \\
&= \left(\frac{dF}{dt} \right)_{coll} = \sum_{u'} \{ W(u/u') F(u,t) - W(u'/u) F(u,t) \}
\end{aligned} \tag{5.64}$$

where u now stands for $\boldsymbol{r},\boldsymbol{k},s$, $\boldsymbol{E}$ and $\boldsymbol{B}$ are externally applied electric and magnetic fields respectively, and $\mathbf{v} = \frac{\partial \mathbf{r}}{\partial t}$ is an electron velocity. The first equality stems from the properties of a total derivative, the second comes from the Lorentz force law, and the third is the collision rate due to various scattering mechanisms. Equation (5.64) is also called the Boltzmann equation.

Let us simplify the treatment once more so the focus is on the properties most often encountered. While generalizations can be done it complicates the discussions. Suppose the system is spatially homogeneous so $\nabla_{\mathbf{r}} F = 0$, $\nabla_{\mathbf{r}} T = 0$, and $W(u'/u) = W(\mathbf{k}',s'/\mathbf{k},s)\delta_{\mathbf{r}',\mathbf{r}}$, and in steady state so $\frac{\partial F}{\partial t} = 0$. Then (5.64), suppressing the index $\boldsymbol{r}$, reduces to the form:

$$\frac{-e}{\hbar} \mathrm{E} \bullet \nabla_{\mathbf{k}} F(\mathbf{k},s) = \sum_{\mathbf{k}',s'} [W(\mathbf{k},s/\mathbf{k}',s') F(\mathbf{k}',s') - W(\mathbf{k}',s'/\mathbf{k},s) F(\mathbf{k},s)] \tag{5.65}$$

Next multiply both sides of (5.64) by **k** and sum them over all **k** to obtain:

$$
\begin{aligned}
\frac{-eE}{\hbar}\sum_{\mathbf{k}}\mathbf{k}\left(\hat{z}\bullet\nabla_{\mathbf{k}}F\right) &= 2\sum_{\mathbf{k},\mathbf{k}'}\mathbf{k}\left[W(\mathbf{k}/\mathbf{k}')F(\mathbf{k}')-W(\mathbf{k}'/\mathbf{k})F(\mathbf{k})\right] \\
&= 2\sum_{\mathbf{k},\mathbf{k}'}\left(\mathbf{k}\left[W(\mathbf{k}/\mathbf{k}')\delta F(\mathbf{k}')-W(\mathbf{k}'/\mathbf{k})\delta F(\mathbf{k})\right]\right) \\
&= -2\sum_{\mathbf{k},\mathbf{k}'}(\mathbf{k}-\mathbf{k}')W(\mathbf{k}'/\mathbf{k})\delta F(\mathbf{k}) \\
&\equiv -2\sum_{\mathbf{k}}\frac{\mathbf{k}}{\tau_m(\mathbf{k})}\delta F(\mathbf{k})
\end{aligned}
\tag{5.66}
$$

A number of steps have been taken in (5.66). The electric field $\mathrm{E}=\mathrm{E}\hat{\mathbf{z}}$ has been set to lie in the z direction. In the first equality the sum over s' has been done yielding the factor of 2, and then the spin index suppressed because the quantities involved do not depend on spin for most mechanism of interest in this discussion. In the second equality $\delta F(\mathbf{k})\equiv F(\mathbf{k})-F_0(\mathbf{k})$ and the equality stems from the principal of detailed balance derived from (5.58) and the form of $F_0(\mathbf{k})$ in (5.63). The third equality is obtained by exchanging $\mathbf{k}$ and $\mathbf{k}'$ in the first term in the brackets of the second equality which is permitted because the sums run over all $\mathbf{k}$ and $\mathbf{k}'$. The third equality results from the definition of $\tau_m(\mathbf{k})$ which is:

$$
\frac{\mathbf{k}}{\tau_m(\mathbf{k})}\equiv\sum_{\mathbf{k}'}(\mathbf{k}-\mathbf{k}')W(\mathbf{k}'/\mathbf{k}) \tag{5.67}
$$

Take the dot product of both sides of (5.67) with $\hat{\mathbf{k}}$ and divide by k to get:

$$
\frac{1}{\tau_m(\mathbf{k})}=\sum(1-\cos\theta)W(\mathbf{k}'/\mathbf{k}) \tag{5.68}
$$

where $\hat{\mathbf{k}}'\bullet\hat{\mathbf{k}}=\cos\theta$. This is the momentum relaxation time which differs from the total scattering relaxation time defined as:

$$
\frac{1}{\tau_R}\equiv\sum_{\mathbf{k}'}W(\mathbf{k}'/\mathbf{k}) \tag{5.69}
$$

The momentum relaxation time is always greater than or equal to the total scattering relaxation time, $\tau_m\geq\tau_R$, because the contributions from forward scattering events where θ is small are suppressed.

Actually what should enter into the problem is the velocity relaxation time rather than the momentum relaxation time (Chen and Sher 1995). This makes a difference in a non-parabolic conduction band, but when the band is parabolic these two relaxation times are equal.

Equation (5.66) can be rewritten in the form:

$$\sum_{\mathbf{k}} \mathbf{k}\left[\frac{e\mathrm{E}}{\hbar}\hat{\mathbf{z}}\cdot\nabla_{\mathbf{k}}F_0(\mathbf{k})-\frac{1}{\tau_m(\mathbf{k})}\delta F(\mathbf{k})\right]\cong 0 \tag{5.70}$$

Where another approximation has been added in which E is small allowing F to be replaced by F_0 in the first term of the bracket. A sufficient solution to (5.70) sets all the brackets in the sum to zero. Then we have:

$$\begin{aligned}\delta F(\mathbf{k}) &\cong \frac{e\mathrm{E}\tau_m(\mathbf{k})}{\hbar}\hat{\mathbf{z}}\cdot\nabla_{\mathbf{k}}F_0(\mathbf{k})\\ &\cong -\frac{e\hbar\tau_m(\mathbf{k})k_z}{m^* k_B T}F_0(\mathbf{k})\mathrm{E}\end{aligned} \tag{5.71}$$

Next study the steady state current density $\mathbf{j}$ defined for electron conduction as:

$$\mathbf{j} \equiv -e\sum_{\mathbf{k},s}\frac{\mathbf{v}F(\mathbf{k})}{V} \tag{5.72}$$

Here we have $F(\mathbf{k}) = F_0(\mathbf{k}) + \delta F(\mathbf{k})$, and by symmetry $\sum_{\mathbf{k}} \mathbf{v}F_o(\mathbf{k}) = 0$. Moreover in view of (5.71) the x and y components of $\boldsymbol{j}$ vanish, $j_x = j_y = 0$. Thus (5.72) reduces to the expression:

$$\begin{aligned}j_z &= -\frac{2e}{V}\sum_{\mathbf{k}} v_z\delta F(\mathbf{k}) = \frac{2e^2\hbar}{m^* k_B TV}\sum_{\mathbf{k}} v_z k_z \tau_m(\mathbf{k})F_0(\mathbf{k})\\ &= \frac{2e^2}{k_B TV}\sum_{\mathbf{k}} v_z^2\tau_m(\mathbf{k})F_0(\mathbf{k})\mathrm{E} = \frac{e^2 n}{k_B T}\sum_{\mathbf{k}}\frac{v^2}{3}\tau_m(|\mathbf{k}|)\frac{e^{-E/k_BT}}{\sum_{\mathbf{k}} e^{-E/k_BT}}\mathrm{E}\\ &= \frac{2e^2 n}{3m^* k_B T}\frac{\int_0^\infty dE E^{3/2}\tau_m(|\mathbf{k}|)e^{-E/k_BT}}{\int_0^\infty dE E^{1/2}e^{-E/k_BT}}\mathrm{E} \equiv \sigma\mathrm{E}\end{aligned} \tag{5.73}$$

The first equality results from substituting for δF from (5.71), the second used the relation $m^* v_z = \hbar k_z$, the third is valid only for mechanisms for which τ_m depends only on the magnitude of $\boldsymbol{k}$, i.e. on the electron energy, so by symmetry in the sum $v_z^2 \Rightarrow \dfrac{v^2}{3}$, the forth equality results form substituting for F_0 from (5.63), and the next equality uses $E = \dfrac{1}{2}m^* v^2$ and

$$\sum_{\mathbf{k}} G(|\mathbf{k}|) \Rightarrow \frac{V}{(2\pi)^3}\int dk^3 G(|\mathbf{k}|) = \frac{V}{(2\pi)^3}\frac{4\pi\sqrt{2}m^{*3/2}}{\hbar^3}\int_0^\infty dE E^{1/2} G(|\mathbf{k}|),$$

and finally the last equality is the definition of the conductivity σ.

$$\sigma = \frac{2e^2 n}{3m^* k_B T}\frac{\int_0^\infty dE E^{3/2}\tau_m(|\mathbf{k}|)e^{-E/k_BT}}{\int_0^\infty dE E^{1/2}e^{-E/k_BT}} \equiv e\mu n. \tag{5.74}$$

The last identity is the definition of the mobility μ:

$$\mu = \frac{2e}{3m^* k_B T}\frac{\int_0^\infty dE E^{3/2}\tau_m(|\mathbf{k}|)e^{-E/k_BT}}{\int_0^\infty dE E^{1/2}e^{-E/k_BT}} \tag{5.75}$$

If τ_m is independent of $\mathbf{k}$ then it can be taken outside the integral and the expressions for μ and σ assume classical results often quoted in texts:

$$\mu = \frac{2e\tau_m}{3m^*}\frac{\Gamma(5/2)}{\Gamma(3/2)} = \frac{e\tau_m}{m^*},$$

and (5.76)

$$\sigma = \frac{e^2\tau_m n}{m^*}$$

In realistic cases τ_m is never independent of $\mathbf{k}$ and the simple results are inaccurate.

It has been shown that for several important scattering mechanisms (Seeger 1989):

$$\tau_m(\mathbf{k}) \cong \tau_0\left(\frac{E}{k_BT}\right)^r \tag{5.77}$$

For deformation potential coupling of acoustic phonons to the electrons $r = -1/2$, while for impurity scattering $r = 3/2$. The quantity τ_0 differs from one mechanism on another since it depends on the inverse square of the coupling constant, the density of accessible final states, and the temperature. Inserting (5.77) into (5.75) and integrating yields:

$$\mu = \frac{4e\tau_0}{3\sqrt{\pi m^*}}\Gamma\left(\frac{5}{2}+r\right) = \frac{4e\tau_0}{\sqrt{\pi m^*}}\begin{cases}\frac{1}{3}; \text{ acoustic phonon} \\ 2; \text{ ionized impurity}\end{cases} \tag{5.78}$$

A detailed treatment of scattering due to ionized impurity and polar longitudinal optical phonon mechanisms has been done (Chen and Sher 1995; Krishnamurthy and Sher 1994, 1995).

It is useful to introduce the mean free path per electron defined as:

$$\ell \equiv \frac{\langle v\tau_m \rangle}{nV} = \frac{1}{2}\sum_{\mathbf{k},s} v\tau_m(\mathbf{k})F_0(\mathbf{k}) = 2\sqrt{\frac{2k_BT}{\pi m^*}}\tau_0\Gamma(2+r) \tag{5.79}$$

The ratio of the mobility to the mean free path is:

$$\frac{\mu}{\ell} = \frac{e}{3}\sqrt{\frac{2}{m^*k_BT}}\frac{\Gamma(5/2+r)}{\Gamma(2+r)} \tag{5.80}$$

which is independent of τ_0.

5.2.2 Experimental Results of the Electron Mobility of $Hg_{1-x}Cd_xTe$

For $Hg_{1-x}Cd_xTe$ alloys, the Hall mobility has been obtained following the normal procedure of measuring the Hall coefficient and the conductivity. Because the mobility of electrons is much larger than that of holes, the electron mobility is much easier to measure.

Tang (1974) examined the electron mobility of HgCdTe. Figure 5.10 shows experimental results for the electron mobility of $Hg_{1-x}Cd_xTe$ (x = 0–1) samples from 4.2 to 300 K. For a variety of compositions, the shape of the relation between the temperature and the mobility is almost the same. At high temperatures, from ~100 to ~300 K, because the scattering is mainly due to the crystal lattice vibrations, the mobility increases when the temperature decreases, $\mu \propto T^{-n}$. The value of n changes little with composition x.

At low concentrations x, even at high temperatures, scattering due to impunities often plays the main role, causing the mobility to decrease when the temperature decreases. Figure 5.11 shows the variation of the average electron mobility with composition at 300 K. At the concentration where $E_g \cong 0$, the mobility reaches a maximum. Scott (1972) analyzed the intrinsic scattering mechanisms that affect the electron mobility at room temperature. These include acoustics phonon scattering, piezoelectric scattering, optical phonon scattering, and electron–hole scattering. For a parabolic energy band and in a non-degenerate case, the electron mobility determined by acoustic phonon scattering is (Scott 1972):

$$\mu_{ac}\left(\mathrm{cm}^2/\mathrm{V\,s}\right)=\frac{3\times10^{-5}C_l}{\left(m^*/m_0\right)^{5/2}T^{3/2}E_c^2} \tag{5.81}$$

where, C_l is the bulk modulus. E_c is the electron deformation potential energy. These two parameters change little over the full range of compositions $x = 0$ to 1 because the elastic constants of HgTe and CdTe are nearly the same (Chen and Sher 1995). For this reason the change of μ_{ac} with composition primarily comes from the change of (m^*/m_0). μ_{ac} deduced from (5.81) is plotted as a dotted curve in Fig. 5.10. It is over 100 times larger than the experimental curve. At the end of the x range near 0, the influence of degeneracy and the non-parabolic character of the energy band must be considered. But even taking these influences into account, the values of μ_{ac} does not come within an order of magnitude of the experimental curve. Therefore, in $Hg_{1-x}Cd_xTe$, acoustic phonon scattering is negligible.

In binary Zincblende structured crystals, piezoelectric coupled scattering is induced by polarization resulting from lattice distortions associated with acoustics phonons. The calculated mobility caused by piezoelectric scattering μ_{ac}, can be a few times greater than μ_{ac} (Scott 1972). Therefore this mechanism also has little effect to the room temperature electron mobility of $Hg_{1-x}Cd_xTe$.

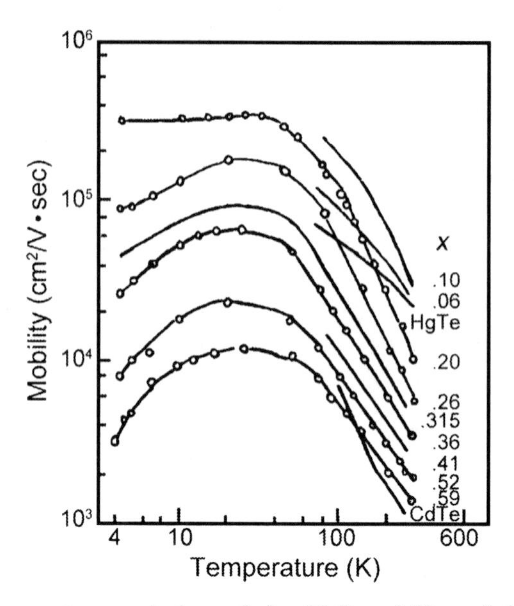

Fig. 5.10. The temperature variation of the Hall mobility of $Hg_{1-x}Cd_xTe$ for different *x*

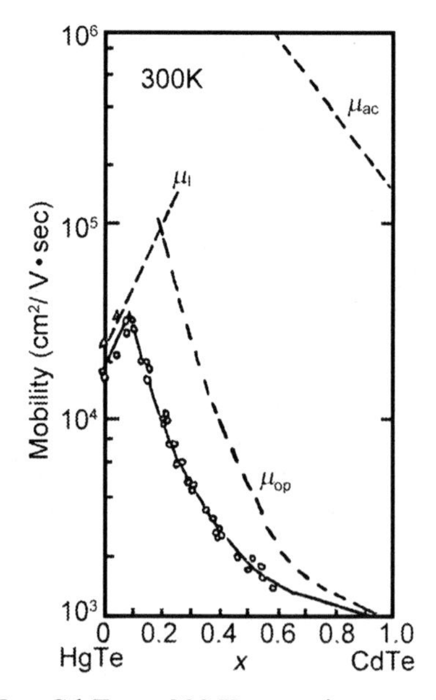

Fig. 5.11. Mobility of $Hg_{1-x}Cd_xTe$ at 300 K, experiment and theory: the *dots* are experimental data, the *dashed* curve is the theory at low concentrations for ionized impurity scattering with the impurity content chosen to fit the data, and the *dashed curve* at higher concentrations is the prediction of the effect of optical phonon scattering

The scattering due to longitudinal optical phonons coupled through the Fröhlich polar interaction (Chen and Sher 1995) is what mainly dominates the room temperature electron mobility of $Hg_{1-x}Cd_xTe$. Over the whole composition range from HgTe to CdTe, a weak electron–phonon coupled theory is applicable. For a parabolic energy band and a non-degenerate semiconductor, the electron mobility due to optical phonon scattering is (Scott 1972; Ehrenreich 1959):

$$\mu_{op} = 2.5\times10^{32}\left(\frac{T}{300}\right)^{1/2}\frac{e_0^2\varepsilon_\infty^2\omega_l\left(e^z-1\right)\left[e^{-\xi}G^{(1)}\right]}{\left(\varepsilon_s-\varepsilon_\infty\right)\omega_t^2\left(m_0^*/m_0\right)^{3/2}} \tag{5.82}$$

Where, e_0 is the electron charge, using electrostatic units to scale it. ε_s and ε_∞ respectively are the static and high frequency dielectric constants. $e^{-\xi}G^{(1)}$ is a function defined by Ehrenreich (1959), which to fit CdTe data was taken to be $e^{-\xi}G^{(1)}$ = 0.7. The expression in (5.82) is only valid for temperatures T such that $T \gg T_l \equiv \hbar\omega_l / k_B$. Equation (5.82) was derived for a semiconductor compound, and to predict the value for the $Hg_{1-x}Cd_xTe$ alloy the concentration weighted average for HgTe and CdTe is used. $z = E_f/kT$ is a reduced Fermi energy. ω_l, ω_t respectively are the longitudinal and transverse optical photon vibration frequencies. Equation (5.82) only includes longitudinal optical phonon scattering coupled through the Fröhlich interaction. Transverse optical phonon scattering introduces only a small correction since these phonons are coupled through deformation potentials. Before doing the alloy average the contribution from the transverse phonons should be added. The result is depicted in Fig. 5.11. Comparing the prediction with an experimental result, as x decreases from 1 the deviation between them becomes progressively greater. At x = 0.3, the theoretical value is three times greater than the experimental value. This suggests that either the approximations employed are leading to an inaccurate prediction, or another mechanism is active.

Chattopadhyay and Nag (1974) re-examined the room temperature electron mobility of $Hg_{1-x}Cd_xTe$ from a solution to the Boltzmann equation that accounted for both optical and acoustic phonon types. Adopting some new experiment data, and considering the non-parabola character of the conduction band, they got the result in Table 5.7.

The result shows that for x with low or high values (x = 0.2, 0.8, 1.0), the theoretical mobility values agree with the experimental values. But for the x values in the mid range (x = 0.44, 0.6), the difference is large, almost double the experimental values. This indicates that in this temperature range, ionized impurity scattering is a key mechanism that determines the

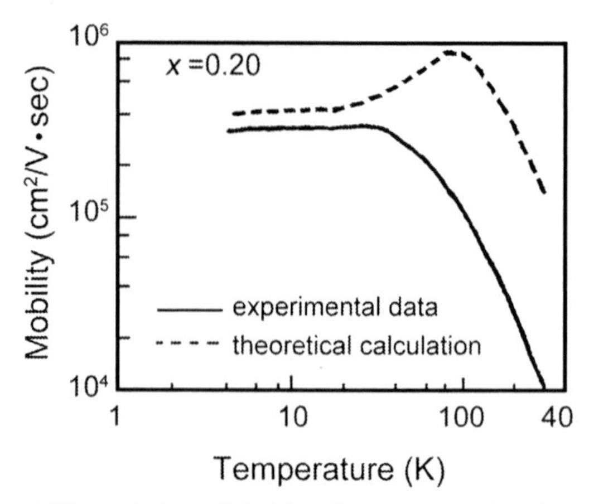

Fig. 5.12. The mobility of $Hg_{0.8}Cd_{0.2}Te$, the computational curve is on the assumption that only ionized impurity scattering and electron–hole scattering exist

Table 5.7. The room temperature electron mobility of $Hg_{1-x}Cd_xTe$ from a solution to the Boltzmann equation

x	Mobility including non-parabola energy band and optical phonon scattering (cm^2/V s)	Mobility including non-parabola energy band, optical and acoustic phonons, impurity, and electron–hole scattering (cm^2/V s)	Experimental (cm^2/V s)
0.2	1.65×10^4	1.44×10^4	1.03×10^4
0.4	5.6×10^3	5.6×10^3	2.74×10^3
0.6	2.73×10^3	2.49×10^3	1.55×10^3
0.8	1.6×10^3	1.44×10^3	1.26×10^3
1.0	1.04×10^3	9.22×10^2	$1.\times10^3$

electron mobility. When the temperature is increased, the mobility due to ionized impurity scattering increases. Because of the increase of the carrier density of narrow gap material, at about 100 K, the intrinsic carrier density becomes very high, causing electron–hole scattering to enter and the mobility to decrease. However, even at 300 K, based on this theory the mobility is still several times higher than the experimental value. So, for a $x = 0.20$ sample, the effect of the electron–hole scattering mechanism on the electron mobility at 300 K can be ignored.

We see in Fig. 5.12 (Scott 1972), that for a sample with $x = 0.2$ in the high temperature region, the experimental curve approximates a straight line $\mu_n \propto T^{-2.3}$. While the carrier concentration is not quoted in the paper, judging from the shape of the experimental curve around 40 K, for reasons that will be explained later the concentration has to be a few times 10^{15} cm^{-3}.

This high temperature behavior was interpreted to be the result of the net effect of longitudinal optical phonon and electron–hole scattering mechanisms. When the data is taken from high temperature to 77 K, the mobility determined by this mechanism rises to $\mu_{op} \approx 2.5\times10^5$ cm^2/V s. At this temperature, the mobility predicted from this theory, arising from the combination of ionized impurity scattering and electron–hole scattering, is $\mu_I \approx 8\times10^5$ cm^2/V s. The joint action of these two mechanisms at still lower temperatures lowers the mobility to 1.9×10^5 cm^2/V s. The value determined in the experiment 1.5×10^5 cm^2/V s approaches the theoretical value quite well.

A computationally intensive study has also been conducted and compared to temperature dependent data ranging from 4.2 to 300 K, on samples with $x = 0.22$ (Krishnamurthy and Sher 1994). The experiments were conducted on samples with low doping concentrations, $\sim 1\times10^{14}$ cm^{-3}. This study attempts to calculate the mobility including the most important scattering mechanisms with no parameters adjusted to fit the experimental data. The features of this calculation are:

- Accurate band structures and wave functions derived from the HPTB (hybrid psudopotential-tight-binding)/CPA method are used. Just to test the sensitivity of the final result a Kane band structure was also tried.
- Full Fermi statistics are included. The temperature dependence of the band gap taken from the literature is used. It is essential to be able to evaluate accurate Fermi energies.
- The transition probabilities in (5.57) are calculated for ionized impurity and longitudinal optical phonon scattering from the Golden rule using the calculated wave functions. Just to see how sensitive the answers are to the doping concentration a range from 10^{14} to 10^{18} cm^{-3} is examined. Also the donor ionization energy is varied from $E_D = 0$ to 30 meV is included to see how this changes the mobility at low temperatures where carrier freeze out can occur. The donor ionization energy is rarely treated as an important parameter in theories, partially because it is so poorly known for the so called "residual impurities" responsible for low donor concentrations.
- The full one particle master equation (5.57) is solved by expanding the functions $f(\mathbf{k},s,t)$ in a series of Hermite polynomials. Only a finite number of polynomials, 8, are needed to reach convergence when the applied electric field is small. Once again to test the sensitivity of the answers to approximations use in the literature, a relaxation time calculation was also done.

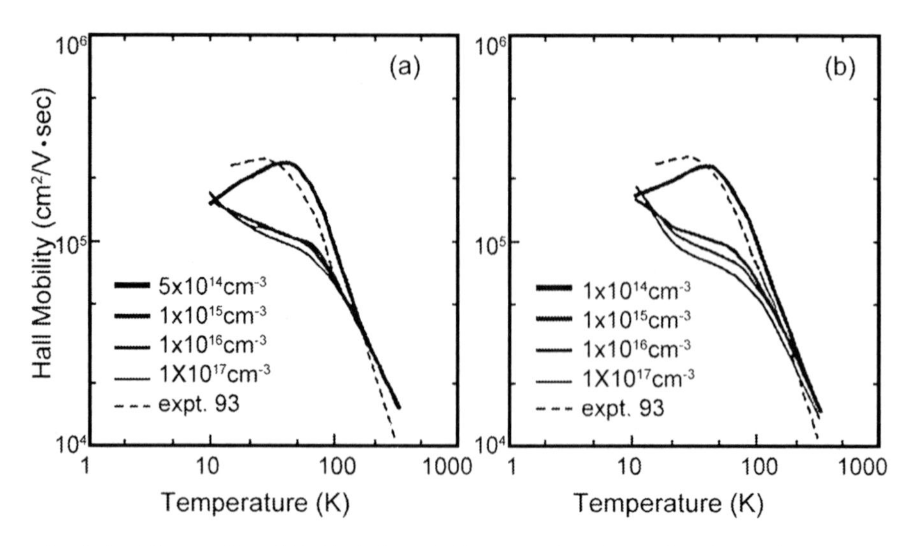

Fig. 5.13. The predicted Hall drift mobility as a function of T for various donor concentrations: (**a**) $E_D = 0$ eV and (**b**) $E_D = 30$ meV, compared to data for two $x = 0.22$ samples

The results are collected in Figs. 5.13 and 5.14. In Fig. 5.13 the mobility as a function of T for different doping concentrations is depicted in two cases: (a) where the donor ionization energy is $E_D = 0$ eV, and (b) where $E_D = 30$ meV. The dashed curve is the data for a sample collected by J. Bajaj (private communication). For $N_D = 1\times10^{14}$ cm^{-3} and at temperatures above ~40 K, the scattering is dominated by polar coupled optical phonon scattering. Below ~40 K it becomes dominated by impurity scattering whose rate increases as the temperature decreases. Surprisingly the scattering rate increases for $E_D = 30$ meV relative to that for $E_D = 0$ eV. At the higher donor ionization energy at low temperature there is some carrier freeze out, yet the scattering rate increases. The reason is that while there are fewer ionized impurities their screening is less efficient so each of them is a more effective scatterer. In any case the theory agrees with experiment to within ~25% over the entire temperature range. Once the other minor mechanisms, deformation potential and piezoelectric coupled acoustic and optical phonon, electron–hole, electron–electron, neutral impurity, and mesoscopic concentration fluctuation scattering are added to this theoretical method the agreement is expected to improve. Notice that for higher donor concentrations the peak in the mobility around 40 K disappears. This accounts for the lack of a peak seen in the Scott data presented in Fig. 5.12.

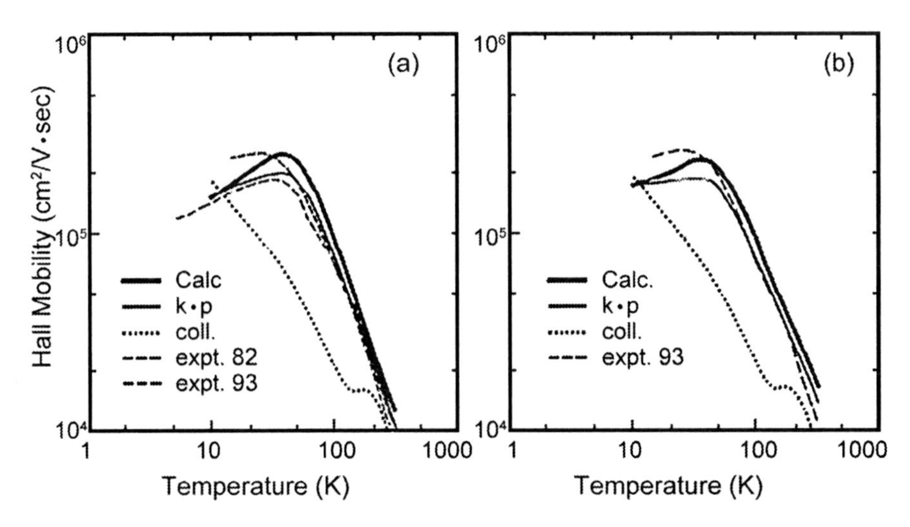

Fig. 5.14. The Hall drift mobility of an $x = 0.22$ sample, and calculated results for several different approximations while using the same coupling constants; (**a**) $E_D = 0$ eV and (**b**) $E_D = 30$ meV

There are noticeable differences in Fig. 5.14 at low temperatures between the mobility curves predicted by the HPTB and Kane band theories. Once the calculated results are as good as those presented here these differences become important. The relaxation time approximation over estimates the scattering rate over the entire temperature range. To bring that prediction into agreement with experiment in a fitting process, one would have to choose unphysical coupling constants.

The relationship at 4.2 K between the electron mobility determined by ionized impurity scattering at different alloy concentrations have been reported (Scott 1972; Long 1968; Stankiewicz et al. 1971). Both experimental tests and theoretical computations have been done. At the concentration where $E_g = 0$, the mobility reaches a maximum. The slope in the negative gap energy range is small, while it is larger in the composition range where the gap opens. The declining rate approaching the negative forbid energy is slow, while it speeds up toward the positive one. The source of the asymmetry is caused by the relationship between the dielectric constant, i.e. the screening of the ionized impurity scattering potential, and the electron density, which acts differently on the two sides of the peak. Taking this relationship into account, the theoretical curve has the same shape as the experimental curve. Figure 5.15 is a comparison between the theory and the experiment. In this experiment data was also taken at high pressure, which changes the forbidden energy gap. When $x =$ 0.07, the sample was studied at pressures ranging from 0 to 8,300

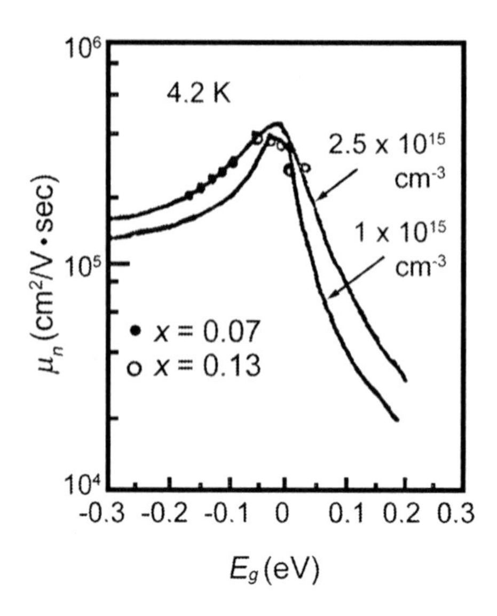

Fig. 5.15. The mobility limited by ionized impurity scattering versus band gap

atmospheres, causing the gap to vary from –0.17 eV to –0.091 eV. For a sample that had $x = 0.13$, in the same pressure range, the gap varied from –0.060 eV to +0.021 eV. The band gap of this sample became 0 at a pressure of about 6,250 atmospheres.

In most samples the Hall effect is dominated by the electron conduction because the electron mobility is much larger than the hole mobility. This makes accurate hole mobility measurements more difficult. Here we first list the data that Elliott (1971) obtained and then that of Kinch et al. (1973) as examples.

Elliott

At $x = 0.24$, and $T = 83.3–133$ K

$\mu_n \approx 120–90$ cm^2/V s and

$b = \mu_n / \mu_h$, the ratio of the electron mobility to the hole mobility, $b =$ 225–380.

At $x = 0.25$, and $T = 125–180$ K

$\mu_n \approx 350–284$ cm^2/V s; $b = 52–99$

At $x = 0.265$, and $T = 122–185$ K

$\mu_n \approx 210–156$ cm^2/V s; $b = 71–130$

Kinch et al.

At $x = 0.20$, N_A-$N_D \approx 2\times10^{15}$ cm^{-3}, and $T = 77$ K

$\mu_p = 600$ cm^2/V s

This value is independent of N_A-N_D and was regard as the limiting mobility due to phonon scattering.

The carrier mobility is an important parameter to measure the quality of a semiconductor material. Often the trend of the mobility variation with temperature is used to study scattering mechanisms. It is generally thought that in n-$Hg_{1-x}Cd_xTe$ impurity scattering will dominate at low temperatures, while phonon scattering dominates at high temperatures Some minor roll is also played by alloy, and carrier–carrier scattering (Szymanska and Dietl 1978; Subowski et al. 1981; Chen and Colombo 1993; Yoo and Kwack 1997). As noted earlier other minor contributions may arise in some samples.

5.2.3 An Approximate Analytic Expression for the Electron Mobility of n-type $Hg_{1-x}Cd_xTe$

In according to the Kane model the $Hg_{1-x}Cd_xTe$ conduction band satisfies following formula:

$$\frac{\hbar^2 k^2}{2m_e^*} = E(1+\frac{E}{E_g}) \tag{5.83}$$

where m_e* is the electron effective mass and the relationship between E_g and x follows the Chu CXT formula (Chu et al. 1993, 1982, 1985).

$$E_g = -0.295 + 1.87x - 0.28x^2 + (1-14x+3x^2)10^{-4}T + 0.35x^4 \tag{5.84}$$

Given the density of free electrons N_D, the Fermi level is deduced from the solution to the equation:

$$N_D = \frac{2}{\sqrt{\pi}} N_c \int_0^\infty \frac{y^{1/2}(1+\beta y)^{1/2}(1+2\beta y)}{1+\exp(y-\eta)} dy \tag{5.85}$$

The sum over all states of the Fermi–Dirac distribution function is converted into an integral in (5.85) following the usual prescription where N_c is the density of conduction band states,

$$y = E/k_B T, \beta = k_B T/E_g, \text{and}\, \eta = k_B T/E_F .$$

The contribution to the mobility of an electron with wave number $\boldsymbol{k}$ from ionized impurity scattering and calculated by Zawadzki (1982) in a first order Born approximation for the relaxation time is:

$$\mu_i(\mathbf{k}) = \frac{1}{2\pi} \frac{(4\pi\varepsilon_0\varepsilon_s)^2}{q^3\hbar N_I} \frac{1}{F_I} \left(\frac{dE}{dk}\right)^2 k \tag{5.86}$$

where $F_{\rm I}$ = $2kL_D$, L_D is the Debye length, $N_{\rm I}$ is the total impurity density $N_{\rm I}$=N_A+N_D, ε_s and ε_∞ are respectively the static permittivity and the high frequency permittivity.

$$\varepsilon_s = 20.8 - 16.8x + 10.6x^2 - 9.4x^3 + 5.3x^4$$

and (5.87)

$$\varepsilon_\infty = 15.1 - 10.3x - 2.6x^2 + 10.2x^3 - 5.2x^4$$

Summing (5.85) over all states yields:

$$\mu_I = 2.91 \times 10^{57} \frac{\varepsilon_s{}^2 T^3 m^*}{N_I N_D} \int_0^\infty \frac{\exp(y-\eta) y^3 (1+\beta y)^3}{[1+\exp(y-\eta)]^2 F_I (1+2\beta y)^2} dy \tag{5.88}$$

Electron–electron scattering, including screening, only adds a factor F_c to the argument of (5.88):

$$F_c = \frac{3[1+\exp(-\eta) F_{1/2}^2(\eta)]}{4F_2(\eta)} \tag{5.89}$$

where $F_s(\eta) = \frac{2}{\sqrt{\pi}} \int_0^\infty \frac{\xi^s}{1+\exp(\eta-\xi)} d\xi$ is the Fermi–Dirac integral. In the classical limit, $F_c = 0.647$, but in the extreme degenerate limit, $F_c = 1$.

Optical phonon scattering is effective only at higher temperatures, here, we will use the Bate et al. approximation (Bate et al. 1965):

$$\mu_{op}\left[\mathrm{cm}^2/\mathrm{V\,s}\right] = \frac{1.74 m_0}{\alpha\hbar\omega_l m_e^*} \frac{\exp(z)-1}{z^{1/2}} \frac{F_{1/2}(\eta)}{D_{00}(\eta,z)} \tag{5.90}$$

where z = θ/T, θ is the Debye Temperature, $\theta \equiv \hbar\omega_l / k_B$ and ω_l is the longitudinal optical phonon frequency. For a $Hg_{1-x}Cd_xTe$ alloy the phonon frequencies are weighted combinations of those of HgTe and CdTe, in particular an often used approximation is, $\omega_l \cong (1-x)(\omega_l)_{HgTe} + x(\omega_l)_{CdTe}$, so:

$$\left.\frac{1}{\mu_{op}}\right|_{Hg_{1-x}Cd_xTe} = \left.\frac{1-x}{\mu_{op}}\right|_{HgTe} + \left.\frac{x}{\mu_{op}}\right|_{CdTe} \tag{5.91}$$

and

$$\hbar\omega_l = (17.36 - 1.24x)\times 10^{-3}\text{ eV; HgTe}$$
$$\hbar\omega_l = (18.60 + 3.72x)\times 10^{-3}\text{ eV; CdTe} \tag{5.92}$$

$D_{00}(\eta, z)$ is:

$$D_{00}(\eta,z) = 2\int_0^\infty \frac{y^{1/2}(y+z)^{1/2}}{[\exp(\eta - y)+1][\exp(y-\eta)+\exp(-z)]}dy \tag{5.93}$$

The polar constant, α which has the physical significance that $\alpha/2$ is the average number of phonons (lattice distortions) attached to and moving with the electron, i.e. dressing the electron (Kittle 1963), is defined as:

$$\alpha = \frac{e^2}{4\pi\varepsilon_0\hbar}\left(\frac{m_e^*}{2\hbar\omega_l}\right)^{1/2}\left(\frac{1}{\varepsilon_\infty}-\frac{1}{\varepsilon_s}\right) = 17.76\left(\frac{m_e^*}{m_0}\right)^{1/2}\left(\frac{500}{\theta}\right)^{1/2}\left(\frac{1}{\varepsilon_\infty}-\frac{1}{\varepsilon_s}\right) \tag{5.94}$$

Dressed electrons are called "polarons." The dressing increases the mass: $m_{polaron}^* = \left(1+\frac{\alpha}{6}\right)m_e^*$ because the moving electron has to drag the lattice distortion with it.

Alloy scattering, for a degenerate electron gas calculated in a second order perturbation theory yields (Dubowski 1978):

$$\mu_{dis} = \frac{1.93\times 10^{-32}}{m_e^* x(1-x)N_D\Omega_0 V^2}\int_0^\infty \frac{\exp(y-\eta)y(1+\beta y)}{[1+\exp(y-\eta)]^2(1+2\beta y)^2 F_d}dy \tag{5.95}$$

Here, Ω_0 is the volume of a unit cell, V represents the crystal potential energy s-state matrix element difference between the two constituent compounds of the alloy , generally V is ~9×10^{-23} eV/cm^3, F_d is an another constant accounting for other matrix elements that are considered. This equation is a rough approximation to alloy scattering in $Hg_{1-x}Cd_xTe$ because the effective V is too large for second order perturbation theory to hold. While this equation predicts trends, a CPA calculation is needed to get accurate results (Chen and Sher 1995). Since alloy scattering near the band edges is not a major contributor to the mobility of low x material, the inaccuracy causes little error in mobility calculations.

Taking multiple scattering mechanisms into account, the net mobility can be written:

$$\mu = [(F_c\mu_l)^{-1} + (\mu_{op})^{-1} + (\mu_{dis})^{-1}] \tag{5.96}$$

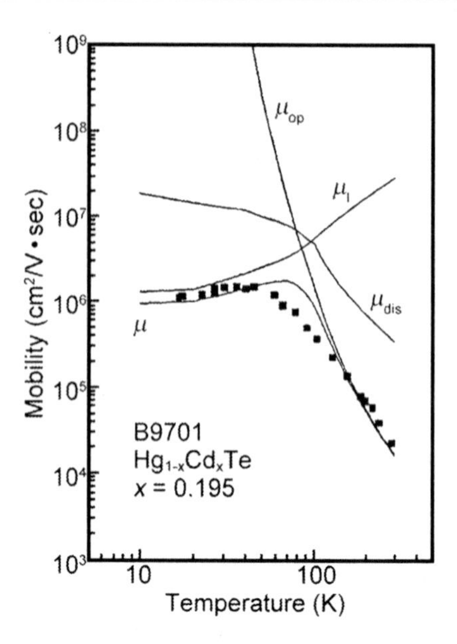

Fig. 5.16. The Hall mobility of a sample with $x \approx 0.200$ and the contributions to the temperature variation of the mobility caused by different scattering mechanisms

Figure 5.16 provides mobility temperature dependence rules for an $x \approx 0.200$ alloy (Gui 1998). In this figure the dots are experimental values, and the solid curves are the contributions to the mobility deduced from impurity scattering, alloy scattering, and optical phonon scattering. Different scattering mechanisms dominate different temperature ranges, impurities play an important role in the low temperature range (below 50 K), but in the high temperature range (<100 K), polarized optical phonon scattering dominates. Alloy scattering in this formalism makes little contribution to the mobility over the whole temperature range.

5.2.4 An Expression for the Hole Mobility of p-$Hg_{1-x}Cd_xTe$

To manufacture photo-voltaic detectors of good capability, p-$Hg_{1-x}Cd_xTe$ material with a high hole mobility is needed. The problem differs from that of the electron case. There are both heavy and light holes at the top of the valance band, and both hole bands are anisotropic. There are more heavy holes than light holes, because of the density of states for the heavy hole band is much higher than that of light hole band. In this treatment only the heavy hole contribution is included. In a p-$Hg_{1-x}Cd_xTe$ ($x \sim 0.22$) material, the average density of states for the heavy hole band is $\sim 10^{18}\,cm^{-3}$ at 77 K, therefore the Boltzmann distribution is a good approximation for this case. The work done (Chen and Dodge 1986; Chen and Tregilgas 1987; Meyer and Bartoli 1987; Yadave et al. 1994) indicated that the hole mobility is mainly determined by ionized impurity scattering. For ionized impurity scattering, the Brooks–Herring theory predicts (Wiley 1975):

$$\mu_i = 3.284 \times 10^{15} \frac{\varepsilon_s T^{3/2}}{N_I (m_h^* / m_0)^{1/2}} \left(\ln(1+b) - \frac{b}{1+b} \right)^{1/2} \tag{5.97}$$

where

$$b = 1.294 \times 10^{14} \frac{m_h^* T^2 \varepsilon_s}{m_0 p_1} \tag{5.98}$$

$$p_1 = p + (p + N_D)[1 - (p + N_D) / N_A] \tag{5.99}$$

and m_h^* is an effective average heavy hole mass for the top of the valance band is about 0.28–0.33m_0, p is the density of free holes, and $N_1 = N_A + N_D$.

The hole mobility expression for polar coupled longitudinal optical phonon scattering, (5.90) becomes (Bate et al. 1965):

$$\mu_{op} = \frac{1.74 m_0 \pi^{1/2}}{\alpha \hbar \omega_l m_h^*} \frac{\exp(z) - 1}{2z^{3/2} \exp(z/2) K_1(z/2)} \tag{5.100}$$

where, K_l is a Bessel function.

The contribution to the hole mobility from alloy scattering, is approximately (Makowski and Glickman 1973):

$$\mu_{dis} = \frac{32.8}{(m_h^* / m_0)^{5/2} T^{1/2} \Delta E_v^2 x(1-x)}, \tag{5.101}$$

where ΔE_v is band mismatch energy between HgTe and CdTe, and was estimated to be about 0.30 eV (Kowalczyk et al. 1986; Sporken et al. 1989).

Here we only examine the relationship between the hole mobility and the carrier density at 77 K. For simplicity, suppose that all the acceptors are ionized. In Fig. 5.17, the contributions to the hole mobility from three different scattering mechanisms in a $Hg_{0.78}Cd_{0.22}Te$ material are presented. When the carrier density is less than 10^{17} cm^{-3}, the hole mobility is independent of the carrier concentration and is mainly determined by polar optical phonon and alloy scattering. Above a carrier concentration of 10^{17} cm^{-3} the mobility becomes dominated by impurity scattering.

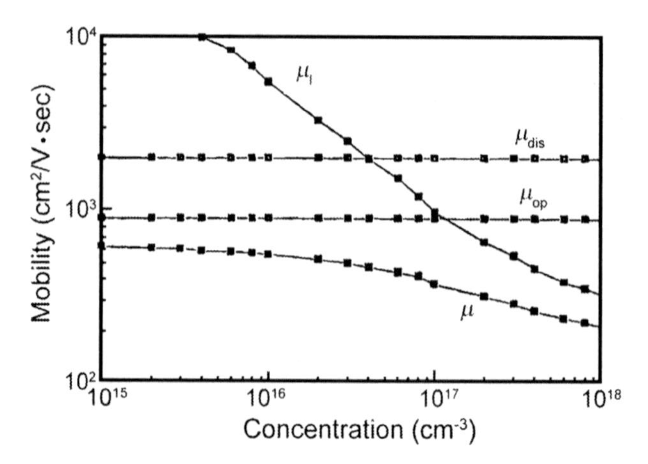

Fig. 5.17. The predicted hole mobility contributions verses carrier concentration for p-$Hg_{0.78}Cd_{0.22}Te$ at 77 K from three important mechanisms, optical phonons, impurities, and alloy scattering

5.3 Transport Properties in Magnetic Field

5.3.1 Conductivity Tensor

Supposing the electric field lies along the z direction $E = (0, 0, E_z)$ and the magnetic field is zero, the steady state Boltzmann equation in the momentum relaxation time approximation and assuming the sample is homogeneous is:

$$\frac{\partial f}{\partial v_z}\frac{e}{m^*}E_z + \frac{f - f_0}{\tau_m} = 0 \tag{5.102}$$

or

$$f = f_0 - \frac{e}{m^*}\tau_m \frac{\partial f}{\partial v_z}E_z \tag{5.103}$$

In this section $f(\mathbf{r}, \mathbf{k}, s, t)$ is the state population distribution (or occupation number) at $\boldsymbol{r}$, and under the homogeneous condition, $\nabla_{\mathbf{r}} f(\mathbf{r},\mathbf{k},s,T,t) = 0$; $\varepsilon(\mathbf{k})$ is the electron energy; and $\boldsymbol{E}$ is the applied electric field. For completeness a spatially and time dependent temperature $T(\boldsymbol{r},t)$ has been added to the list of variables in the functional dependence of f. Note that discussing a time dependent temperature implies that the non-equilibrium population distribution at $\boldsymbol{r}$ can be characterized by a temperature (Sher and Primakof 1960). However, (5.102) is only valid if the temperature is spatially and time independent, if not there is an additional term as noted in (5.63) valid if all the time dependence of T arises through its dependence on $\boldsymbol{r}(t)$. If there is a direct time dependence of T then there is still another term, $\frac{\partial T}{\partial t}\frac{\partial f}{\partial T}$.

A population distribution in the electric field could be obtained by solving the differential equation, (5.103). The second term on the right side of (5.103) is the deviation from the equilibrium distribution due to the applied electric field. The homogeneous assumption when there is only an electric field applied assumes the contacts to the device are ohmic so there is no space charge build up at the electrodes. In the low electric field case the deviation of f from f_0 is small and thus the disturbed distribution function f in the second term on the right can be replaced by the equilibrium distribution f_0. Remembering that $v = \hbar^{-1}\nabla_k\varepsilon$, we have:

$$\frac{eE_z}{m^*}\frac{\partial f_0}{\partial v_x} = \frac{e}{\hbar}\frac{\partial f_0}{\partial \varepsilon}\cdot\nabla_k\varepsilon\cdot\boldsymbol{E},$$

so (5.103) becomes:

$$f = f_0 - \frac{\partial f_0}{\partial \varepsilon}\frac{e}{\hbar}\tau_m\left(\nabla_k\varepsilon\cdot\boldsymbol{E}\right), \tag{5.104}$$

which is the same as (5.71),

When a magnetic field $\boldsymbol{B}$ is present in addition to the electric field $\boldsymbol{E}$ the homogeneous condition no longer applies, and then one can show (5.103) generalizes to:

$$f = f_0 - \frac{\partial f_0}{\partial \varepsilon}\frac{e}{\hbar}\tau_m\left(\nabla_k\varepsilon\cdot\boldsymbol{G}\right), \tag{5.105}$$

where $\boldsymbol{G}$ is (Seeger 1985):

$$\boldsymbol{G} \equiv \frac{\boldsymbol{F} - e\left[\boldsymbol{B}\times m^{*-1}\tau_m\boldsymbol{F}\right] + \tilde{\alpha}\boldsymbol{B}\cdot\left(\boldsymbol{F}\cdot\boldsymbol{B}\right)}{1+\boldsymbol{B}\cdot\tilde{\alpha}\boldsymbol{B}}, \tag{5.106}$$

and the effective force is:

$$e\boldsymbol{F} = e\boldsymbol{E} + T\nabla_{\mathrm{r}}\left[(\varepsilon-\varepsilon_f)/T\right] = e\boldsymbol{E} + \nabla_{\mathrm{r}}(\varepsilon-\varepsilon_f) - (\varepsilon-\varepsilon_f)\frac{\nabla_{\mathrm{r}}T}{T} \tag{5.107}$$

The contribution of a temperature gradient is included in the second term on the right of (5.107). ε_f is the Fermi level, and the tensor $\tilde{\alpha}$ is a constant times the identity matrix $\tilde{I}$ in the case of spherical energy band:

$$\tilde{\alpha} = \left(\frac{e\tau_m}{m^*} \right)^2 \tilde{I} . \tag{5.108}$$

In the case of a non-spherical energy band $\tilde{\alpha}$ is:

$$\tilde{\alpha} = e^2 \begin{pmatrix} \frac{\tau_y \tau_z}{m_y m_z} & 0 & 0 \\ 0 & \frac{\tau_x \tau_z}{m_x m_z} & 0 \\ 0 & 0 & \frac{\tau_x \tau_y}{m_x m_y} \end{pmatrix} . \tag{5.109}$$

The general population distribution function in electric, magnetic and temperature gradient fields can be expressed in the form:

$$f = f_0 - \frac{e}{\hbar} \frac{\partial f_0}{\partial \varepsilon} \tau_m \nabla_k \varepsilon \cdot \mathbf{G} = f_0 + \frac{e\tau_m}{k_B T} \mathbf{v} \cdot \mathbf{G} f_0 , \tag{5.110}$$

where we have assumed there is a parabolic band so $\nabla_{\mathbf{k}} \varepsilon = \hbar \mathbf{v}$, and f_0 was taken to be a Boltzmann distribution. If the system is sufficiently degenerate so f_0 is a sharp Fermi distribution, then $\frac{\partial f_0}{\partial \varepsilon} = -\delta(\varepsilon - \varepsilon_F)$ and the calculation proceeds differently.

We next discuss the contribution of each term in (5.110) to the distribution function. When no magnetic field and no temperature gradient is present, i.e., only the electric field exists ($\boldsymbol{F} = \boldsymbol{E}$), then we have:

$$f = f_0 + \frac{e\tau_m}{k_B T} \mathbf{v} \cdot \mathbf{E} f_0 \tag{5.111}$$

In this version of the formalism, the current density is $\mathbf{j} = en \int_0^\infty \mathbf{v} f(\varepsilon) \rho(\varepsilon) d\varepsilon$, and with $\mathbf{j} = \tilde{\sigma} \boldsymbol{E}$, the conductivity tensor for an isotropic band becomes:

$$\tilde{\sigma}_I = \begin{pmatrix} \sigma_0 & 0 & 0 \\ 0 & \sigma_0 & 0 \\ 0 & 0 & \sigma_0 \end{pmatrix}, \tag{5.112}$$

where

$$\sigma_0 = \left(ne^2 / m^*\right)\langle \tau_{mw} \rangle. \tag{5.113}$$

The weighted average momentum relaxation time:

$$\langle \tau_{mw} \rangle = \left\langle \frac{\varepsilon}{k_B T} \tau_m \right\rangle, \tag{5.114}$$

is τ_m averaged over a Maxwell–Boltzmann distribution function after being weighted by $\frac{\varepsilon}{k_B T}$ to get:

$$\langle \tau_{mw} \rangle = \frac{4}{3\sqrt{\pi}} \int_0^\infty \tau_m \left(\frac{\varepsilon}{k_B T}\right)^{3/2} \exp\left(-\frac{\varepsilon}{k_B T}\right) d\left(\frac{\varepsilon}{k_B T}\right). \tag{5.115}$$

Using (5.113), the current density is $\mathbf{j} = \sigma_0 \mathbf{E}$.

Note that $\langle \tau_{mw} \rangle$ is not the same as the average relaxation time, $\langle \tau_m \rangle^{-1} = \int_0^\infty \frac{1}{\tau_m} f_0 \rho(\varepsilon) d\varepsilon$. Because it is the transition rates from different $\boldsymbol{k}$ contributions that add to get a total transition rate, the relaxation times add reciprocally. Also if the system is degenerate so f_0 is a Fermi distribution then $\langle \tau_{mFw} \rangle = \frac{\varepsilon_F}{k_B T} \tau_{mF}(\varepsilon_F)$ instead of (5.113). Also the relaxation times τ_{mF} differ from τ_m; for details see Krishnamurthy and Sher (1993).

In a weak magnetic field, neglecting the second term of the denominator and the first and third terms of numerator in (5.106), the second term contributing to the conductivity varies linearly with magnetic field:

$$\tilde{\sigma}_{II} = \begin{pmatrix} 0 & \gamma B_z & -\gamma B_y \\ -\gamma B_z & 0 & \gamma B_x \\ \gamma B_y & -\gamma B_x & 0 \end{pmatrix} \tag{5.116}$$

Here we have:

$$\gamma = \gamma_0 \equiv \left(ne^3 / m^{*2}\right)\left\langle (\tau_m^2)_w \right\rangle, \tag{5.117}$$

where $\left\langle (\tau_m^2)_w \right\rangle$ is the same as (5.114) with τ_m in the integrand replaced by τ_m^2. Notice that all terms like $v_x v_y$ vanish in the integral over ε as long as τ_m is symmetrical in $\boldsymbol{k}$, i.e. depends only on ε. Also in the integral v_x^2, v_y^2, and v_z^2 can each be replaced by $\frac{v^2}{3}$, and in the parabolic band approximation, $\frac{v^2}{3} = \frac{2}{3}\frac{\varepsilon}{m^*}$. Then, to this level of approximation the current density can be obtained by a matrix multiplication of the conductivity tensor $\tilde{\sigma} = \tilde{\sigma}_I + \tilde{\sigma}_{II}$ given in (5.112) and (5.116), times the electric field vector $\boldsymbol{E}$.

More generally giving up the low frequency approximation we find:

$$\gamma \equiv \left(\frac{ne^3}{m^{*2}}\right)\left\langle \left(\frac{\tau_m^2}{1+\omega_c^2\tau_m^2}\right)_w \right\rangle, \tag{5.118}$$

where ω_c is cyclotron frequency:

$$\omega_c = \frac{eB}{m^*}. \tag{5.119}$$

If the second term in denominator of (5.110) is not neglected, then there is an addition $\tilde{\sigma}_{III}$ added to the conductivity matrix. This correction term contribution to the conductivity is:

$$\tilde{\sigma}_{III} = \begin{pmatrix} -\beta B_x^2 & -\beta B_x B_y & -\beta B_x B_z \\ -\beta B_x B_y & -\beta B_y^2 & -\beta B_y B_z \\ -\beta B_x B_z & -\beta B_y B_z & -\beta B_z^2 \end{pmatrix}, \tag{5.120}$$

where

$$\beta \equiv -\left(\frac{ne^4}{m^3}\right)\left\langle \left(\frac{\tau_m^3}{1+\omega_c^2\tau_m^2}\right)_w \right\rangle \tag{5.121}$$

When the magnetic field is weak enough, then:

$$\frac{1}{1+\omega_c^2\tau_m^2} \cong 1-\omega_c^2\tau_m^2 . \tag{5.122}$$

Since $\omega_c^2 \propto B^2 = B_x^2 + B_y^2 + B_z^2$, the matrix in (5.120) can be expressed as:

$$\tilde{\sigma}_{III} = \begin{pmatrix} \beta_0(B_y^2+B_z^2) & -\beta_0 B_x B_y & -\beta_0 B_x B_z \\ -\beta_0 B_x B_y & \beta_0(B_x^2+B_z^2) & -\beta_0 B_y B_z \\ -\beta_0 B_x B_z & -\beta_0 B_y B_z & \beta_0(B_x^2+B_y^2) \end{pmatrix} . \tag{5.123}$$

where

$$\beta_0 \equiv -\left(\frac{ne^4}{m^{*3}}\right)\left\langle\left(\tau_m^3\right)_w\right\rangle . \tag{5.124}$$

Hence, the current density becomes $\mathbf{j} = (\tilde{\sigma}_I + \tilde{\sigma}_{II} + \tilde{\sigma}_{III})\mathbf{E}$.

In the case of a week magnetic field, the conductivity tensor reduces to:

$$\tilde{\sigma} \cong \begin{pmatrix} \sigma_0+\beta_0(B_y^2+B_z^2) & \gamma_0 B_z - \beta_0 B_x B_y & -\gamma_0 B_y - \beta_0 B_x B_z \\ -\gamma_0 B_z - \beta_0 B_x B_y & \sigma_0+\beta_0(B_x^2+B_z^2) & \gamma_0 B_x - \beta_0 B_y B_z \\ \gamma_0 B_y - \beta_0 B_x B_z & -\gamma_0 B_x - \beta_0 B_y B_z & \sigma_0+\beta_0(B_x^2+B_y^2) \end{pmatrix} \tag{5.125}$$

Choosing a coordinate system and letting $\boldsymbol{B}$ lie along the z direction, $B_x = B_y = 0$, then:

$$\tilde{\sigma} \cong \begin{pmatrix} \sigma_0+\beta_0 B_z^2 & \gamma_0 B_z & 0 \\ -\gamma_0 B_z & \sigma_0+\beta_0 B_z^2 & 0 \\ 0 & 0 & \sigma_0 \end{pmatrix} , \tag{5.126}$$

or:

$$\sigma = \begin{pmatrix} \sigma_{xx} & \sigma_{xy} & 0 \\ \sigma_{yx} & \sigma_{yy} & 0 \\ 0 & 0 & \sigma_{zz} \end{pmatrix} , \tag{5.127}$$

with:

$$\sigma_{xx} = \sigma_{yy}\text{, and } \sigma_{xy} = -\sigma_{yx} . \tag{5.128}$$

5.3.2 Hall Effect

Let us consider a current in the x direction flowing along the longest side of a rectangular sample and having a magnetic field B_z perpendicular to the current flow direction. Besides drifting along x axis under the influence of an electric field, the carriers also experience a force in y direction according to the Lorentz's law, and if the contacts on the y surfaces are blocking, or feed into an infinite impedance volt meter, a space charge leading to a Hall voltage V_y is established between the two sides of the sample in the y direction given by:

$$V_y = \frac{R_H I_x B_z}{d}. \tag{5.129}$$

Here d is the width of the sample, and R_H is the Hall coefficient. The Hall electric field E_y is, $E_y = R_H j_x B_z$ and j_x is drift current density. According to $\boldsymbol{j} = \boldsymbol{\sigma} \cdot \boldsymbol{E}$ and (5.127) and (5.128), $j_x = \sigma_{xx} E_x + \sigma_{xy} E_y$ and $j_y = \sigma_{yx} E_x + \sigma_{yy} E_y$. In the Hall experiment, since no current flows between the Hall electrodes on the y surfaces we have,

$$j_y = \sigma_{yx} E_x + \sigma_{yy} E_y = 0 \tag{5.130}$$

That results in the expression, $E_y = \frac{-\sigma_{yx}}{\sigma_{yy}} E_x = \frac{\sigma_{xy}}{\sigma_{xx}} E_x$. Then the drift current density j_x becomes:

$$j_x = \frac{\sigma_{xx}^2 + \sigma_{xy}^2}{\sigma_{xx}} E_x \tag{5.131}$$

The Hall coefficient and the Hall resistivity can be obtained from their definitions, respectively as:

$$R_H(B) = \frac{V_H}{dj_x B} = \frac{E_y}{j_x B} = \frac{\sigma_{xy}}{\sigma_{xx}^2 + \sigma_{xy}^2} \frac{1}{B}, \tag{5.132}$$

$$\rho(B) = \frac{V_x}{Lj_x} = \frac{E_x}{j_x} = \frac{\sigma_{xx}}{\sigma_{xx}^2 + \sigma_{xy}^2}. \tag{5.133}$$

V_H is the Hall voltage and V_x is the voltage cause by the drift current through the magnetic field modified resistance; d and L are the width and the length of the sample, respectively.

The components of conductivity tensor in the low magnetic field approximation allow the Hall coefficient to be calculated. From (5.126) and (5.127) and $\boldsymbol{j} \cong \tilde{\sigma}\boldsymbol{E}$ the current density components become:

$$\begin{cases} j_x = (\sigma_0 + \beta_0 B_z^2)E_x + \gamma_0 B_z E_y \\ j_y = -\gamma_0 B_z E_x + \left(\sigma_0 + \beta_0 B_z^2\right)E_y \end{cases}. \tag{5.134}$$

Then, Hall electric field and Hall coefficient are:

$$E_y = \frac{\gamma_0}{(\sigma_0 + \beta_0 B_z^2)^2 + \gamma_0^2 B_z^2} j_x B_z, \tag{5.135}$$

and

$$R_H = \frac{\gamma_0}{(\sigma_0 + \beta_0 B_z^2)^2 + \gamma_0^2 B_z^2} \tag{5.136}$$

Since B_z is weak this reduces to:

$$R_H \approx \frac{\gamma_0}{\sigma_0^2} = \frac{1}{ne} \frac{\left\langle (\tau_m^2)_w \right\rangle}{\left\langle (\tau_m)_w \right\rangle^2}. \tag{5.137}$$

This can also be expressed as:

$$R_H = \frac{r_H}{ne}, \tag{5.138a}$$

or

$$r_H = \frac{\left\langle (\tau_m^2)_w \right\rangle}{\left\langle (\tau_m)_w \right\rangle^2} \tag{5.138b}$$

r_H is called the Hall factor.

Supposing $\tau_m = \tau_0 \left(\dfrac{E}{k_B T} \right)^r$ then we have:

$$\begin{aligned} \left\langle (\tau_m)_w \right\rangle^2 &= \left\{ \frac{4}{3\sqrt{\pi}} \tau_0 \left(\frac{3}{2} + r \right)! \right\}^2, \\ \left\langle (\tau_m)_w^2 \right\rangle &= \frac{4}{3\sqrt{\pi}} \tau_0^2 \left(\frac{3}{2} + 2r \right)! \end{aligned} \tag{5.139}$$

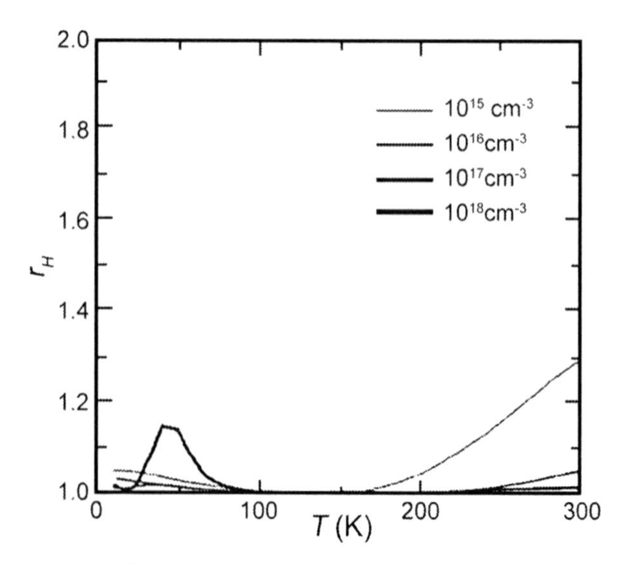

Fig. 5.18. The Hall coefficient as a function of temperature for several different carrier concentrations

so:

$$r_H = \frac{3\sqrt{\pi}}{4} \frac{\left(2r + \frac{3}{2}\right)}{\left\{\left(r + \frac{3}{2}\right)!\right\}^2}. \tag{5.140}$$

The magnitude of the Hall factor depends on which scattering mechanisms are active. For acoustic deformation potential scattering where $r = -1/2$, then $r_H = 1.18$. For impurity ion scattering where $r = +3/2$, then $r_H = 1.93$. So according to this collection of approximations, r_H is usually of the order of 1 but may be as large as 2. For HgCdTe materials standard theories place r_H is in the range $0.9<r_H<1.4$. However when the calculations for a $Hg_{0.78}Cd_{0.22}Te$ alloy are done with complete band structures, as functions of temperature and donor density a somewhat different result emerges (Krishnamurthy and Sher 1993) as shown in Fig. 5.18. The calculation includes polar optical phonon and ionized impurity scattering. Now except for a low temperature peak in the 10^{18} cm^{-3} case, and an elevated value at high temperature in the low carrier concentration case, all the other values lie close to unity. When this calculation is repeated using a parabolic band the Hall coefficient has temperatures where its value approaches 2. Thus approximating r_H by unity is a better approximation for many cases than previously thought. However it does exhibit some temperature and carrier concentration dependence.

The Hall coefficient is negative for an n-type semiconductor because of the negative charge of electrons, while it is positive for holes in a p-type semiconductor. Thus the Hall affect is an important technique in determining both carrier concentrations and the sign of the charge of the dominate carriers through:

$$n(p) = -(+)\frac{r_H}{R_H e}. \tag{5.141}$$

In (5.141), r_H is allowed to carry a sign, however in most equations its absolute value is used.

If the Hall coefficient of a material is known, the magnetic field B_z can be determined by measuring the Hall voltage. The Hall mobility is defined as $R_H\sigma_0$. There is a factor of r_H between the Hall mobility and the drift mobility, μ, that is defined in (5.75).

$$\mu_H \equiv R_H\sigma_0 = \left(\frac{r_H}{ne}\right)ne\mu = r_H\mu. \tag{5.142}$$

In the case where both electrons and holes contribute to the current, then the current density is the sum of both contributions:

$$j_y = \sigma_0 E_y - er_H\left(p\mu_p^2 - n\mu_n^2\right)E_x B_z \tag{5.143}$$

In the limit of a weak magnetic field:

$$E_y = R_H j_x B_z \cong R_H\sigma_0 E_x B_z, \tag{5.144}$$

and then, we obtain:

$$R_H = \frac{er_H}{\sigma_0^2}\left(p\mu_p^2 - n\mu_n^2\right) = \frac{r_H}{e}\frac{p\mu_p^2 - n\mu_n^2}{\left(p\mu_p + n\mu_n\right)^2} = \frac{r_H}{e}\frac{p - nb^2}{\left(p + nb\right)^2}. \tag{5.145}$$

Here $b = \mu_n/\mu_p$ is the ratio of the electron to the hole mobility. The sign of the Hall coefficient changes at $p = nb^2$.

In the limit of a high magnetic field, $\omega_c\tau_m >> 1$, and σ_0, β_0, and γ_0 are replaced with σ, β and γ, respectively, which in this case are:

$$\sigma = \left(\frac{ne^2}{m^*}\right)\left\langle\left(\frac{\tau_m}{1+\omega_c^2\tau_m^2}\right)_w\right\rangle \cong \left(\frac{ne^2}{m^*\omega_c^2}\right)\left\langle(\tau_m^{-1})_w\right\rangle$$

$$\beta = -\left(\frac{ne^4}{m^{*3}}\right)\left\langle\left(\frac{\tau_m^3}{1+\omega_c^2\tau_m^2}\right)_w\right\rangle \cong -\left(\frac{ne^4}{m^{*3}\omega_c^2}\right)\left\langle(\tau_m)_w\right\rangle \tag{5.146}$$

$$\gamma = \left(\frac{ne^3}{m^{*2}}\right)\left\langle\left(\frac{\tau_m^2}{1+\omega_c^2\tau_m^2}\right)_w\right\rangle \cong \left(\frac{ne^3}{m^{*2}\omega_c^2}\right)$$

With $\boldsymbol{B}$ lying along z direction, the conductivity tensor (5.125) becomes:

$$\sigma_w = \begin{pmatrix} \sigma - \beta B_x^2 & -\beta B_x B_y + \gamma B_z & -\beta B_x B_z - \gamma B_y \\ -\beta B_x B_y - \gamma B_z & \sigma - \beta B_y^2 & -\beta B_y B_z + \gamma B_x \\ -\beta B_x B_z + \gamma B_y & -\beta B_y B_z - \gamma B_x & \sigma - \beta B_z^2 \end{pmatrix} \tag{5.147}$$

With $B_y = B_x = 0$ and $E_z = 0$, the current components j_x and j_y are:

$$\begin{cases} j_x = \sigma E_x + \gamma B_z E_y \\ j_y = -\gamma B_z E_x + \sigma E_y \end{cases} . \tag{5.148}$$

From $j_y = 0$, we get $E_x = \left(\dfrac{\sigma}{\gamma B_z}\right) E_y$, and then:

$$E_y = \frac{\gamma}{\sigma^2 + (\gamma B_z)^2} j_x B_z = R_H j_x B_z . \tag{5.149}$$

Hence we find:

$$R_H = \frac{ne^3 / m^{*2}\omega_c^2}{\left(ne^2 / m^*\omega_c^2\right)^2 \left\langle \tau_m^{-1} \right\rangle^2 + \left(ne^3 B_z / m^{*2}\omega_c^2\right)^2} \approx \frac{1}{ne} . \tag{5.150}$$

Note that the strong magnetic field condition $(\mu_H B_z)^2 >> 1$ is required for the validity of (5.150). If $(\mu_H B_z)^2 = 9$ and $\mu_H = 10^4$ cm^2/V s, then $B \approx 3$ T. For narrow gap semiconductors, the strong magnetic field condition can easily be reached because of the high electron mobility in these materials, and thus (5.150) is often a good approximation.

5.3.3 Magneto-Resistance Effect

From the analysis of the last Sect. 5.3.2, the transverse magnetic field B_z also contributes to limiting the current along external electric field direction E_x; see for example (5.148). The conductivity and resistivity therefore depend on the magnetic field.

From $\boldsymbol{j} \cong \tilde{\sigma}\boldsymbol{E}$ in the weak magnetic field limit, and (5.125), we have:

$$\begin{cases} j_x = (\sigma_0 + \beta_0 B_z^2)E_x + \gamma_0 B_z E_y \\ j_y = -\gamma_0 B_z E_x + (\sigma_0 + \beta_0 B_z^2)E_y \end{cases}. \tag{5.151}$$

If $j_y = 0$, i.e., steady state has been established in y direction, then we have:

$$j_x = \left\{ \sigma_0 + \beta_0 B_z^2 + \frac{(\gamma_0 B_z)^2}{\sigma_0 + \beta_0 B_z^2} \right\} E_x. \tag{5.152}$$

Equation (5.152) also can be expressed by defining a magneto-resistivity ρ_B, i.e. $j_x \equiv E_x / \rho_B$, with $\rho_0 \equiv 1/\sigma_0$. The relative change of the resistivity due to the applied magnetic field is defined as:

$$\frac{\Delta\rho}{\rho_B} \equiv \frac{\rho_B - \rho_0}{\rho_B}. \tag{5.153}$$

From (5.152), we obtain:

$$\frac{\Delta\rho}{\rho_B} = -B_Z^2 \left[\frac{\rho_0}{\sigma_0} + \left(\frac{\gamma_0}{\sigma_0} \right)^2 \right] = T_M \left(\frac{e\langle(\tau_m)_w\rangle B_z}{m} \right)^2. \tag{5.154}$$

Where T_M is magneto-resistance scattering coefficient, that is defined as $T_M \equiv \dfrac{\langle(\tau_m^3)_w\rangle\langle(\tau_m)_w\rangle - \langle(\tau_m^2)_w\rangle^2}{\langle(\tau_m)_w\rangle^4}$.

If $\tau_m(E)$ obeys an exponential law like (5.77), we have:

$$T_M = \frac{9\pi}{16} \frac{\left(3r + \frac{3}{2}\right)!\left(r + \frac{3}{2}\right)! - \left\{\left(2r + \frac{3}{2}\right)!\right\}^2}{\left\{\left(r + \frac{3}{2}\right)!\right\}^4}. \tag{5.155}$$

For acoustic deformation potential and ion impurity scattering, T_M = 0.38 (r = −1/2) and T_M = 2.15 (r = +3/2), respectively. Considering $\mu = e\langle(\tau_m)_w\rangle / m$, (5.154) also can be written:

$$\frac{\Delta\rho}{\rho_B} = T_M \left(\mu B_z\right)^2 . \tag{5.156}$$

Note that the above discussion applies to the weak magnetic field limit, i.e., $(\mu_H B_z)^2 << 1$. In this limit, the relative change caused by the magneto-resistance is proportional to B_z^2, and $\Delta\rho << \rho_B$.

For intrinsic semiconductors, both electrons and holes have to be taken into account in calculating the magneto-resistance because both contribute to the conductivity. The contributions from electrons and holes are denoted by the subscripts n and p, respectively, and σ, γ, and β are replaced by $\sigma_n + \sigma_p$, $\gamma_n + \gamma_p$ and $\beta_n + \beta_p$, respectively. Then rewrite (5.154) as:

$$\frac{\Delta\rho}{\rho_B B_z^2} = -\frac{\beta_n + \beta_p}{\sigma_n + \sigma_p} - \left(\frac{\gamma_n + \gamma_p}{\sigma_n + \sigma_p}\right)^2 . \tag{5.157}$$

Note that γ_n and γ_p have opposite signs due to the opposite signs of the charge for electrons and holes, while both σ and β are positive because they are proportional to even powers of their charges. Supposing that the scattering mechanisms are the same for electrons and holes, then from (5.157), we obtain:

$$\frac{\Delta\rho}{\rho_B B_z^2} = \frac{9\pi}{16}\left\{\frac{\left(3r+\frac{3}{2}\right)!}{\left\{\left(r+\frac{3}{2}\right)!\right\}^3}\cdot\frac{p\mu_p^3 + n\mu_n^3}{p\mu_p + n\mu_n} - \left(\frac{\left(2r+\frac{3}{2}\right)!}{\left\{\left(r+\frac{3}{2}\right)!\right\}^2}\cdot\frac{p\mu_p^2 - n\mu_n^2}{p\mu_p + n\mu_n}\right)^2\right\} . \tag{5.158}$$

For an intrinsic semiconductor, n = p, and in the case of acoustic deformation potential scattering, $r = -1/2$, and letting $b = \mu_n / \mu_p$, we have:

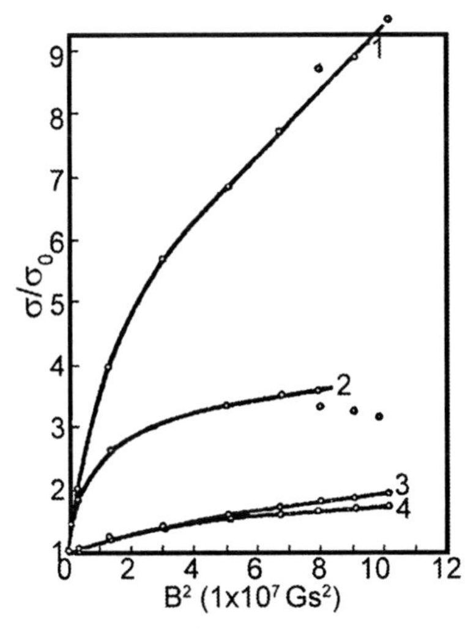

Fig. 5.19. The magneto-resistance coefficient of PbTe as a function of B^2: *Line 1*: Transverse magneto-resistance at 20 K; *Line 2*: Longitudinal magneto-resistance at 20 K; *Line 3*: Transverse magneto-resistance at 77 K; *Line 4*: Longitudinal magneto-resistance at 77 K

$$\begin{aligned}\frac{\Delta\rho}{\rho_B B_z^2} &= \frac{9\pi}{16}\left(1-\frac{\pi}{4}\right)\mu_n^2\left(1+\frac{\frac{\pi}{2}-1}{1-\frac{\pi}{4}}\cdot\frac{1}{b}+\frac{1}{b^2}\right)\\ &= \frac{9\pi}{16}\left(1-\frac{\pi}{4}\right)\mu_p^2\left(1+\frac{\frac{\pi}{2}-1}{1-\frac{\pi}{4}}\cdot b+b^2\right)\end{aligned} \tag{5.159}$$

For narrow gap semiconductors, b is large ($b \approx 100$), so the electrons are the major contributor to the magneto-resistance, and (5.159) reduces to:

$$\frac{\Delta\rho}{\rho_B B_z^2} = \frac{9\pi}{16}\mu_p^2\left(\frac{1+b^3}{1+b}-\frac{\pi}{4}(1-b)^2\right). \tag{5.160}$$

Figure 5.19 shows the transverse and longitudinal magneto-resistance of PbTe at 20 and 77 K (Putley 1955). It can be seen that $\Delta\rho/\rho_0$ is proportional to B^2 at high magnetic fields.

For semimetals, the averages in the calculations of σ, γ, and β are not necessary because of the sharp cutoff of the Fermi function; and the 9π/6 and the π/4 are replaced with 1. Then from (5.160) we obtain:

$$\frac{\Delta\rho}{\rho_B B_z^2} = \mu_n \mu_p . \tag{5.161}$$

For p-type semiconductors, both heavy and light holes must be taken into account. If $r = -\frac{1}{2}$ in the $\tau_m(E)$ power law, then from (5.160), we get:

$$\frac{\Delta\rho}{\rho_B B_z^2} = \frac{9\pi}{16}\mu_h^2 \left(\frac{1+\eta b_1^3}{1+\eta b_1} - \frac{\pi}{4}\left(\frac{1+\eta b_1^2}{1+\eta b_1} \right)^2 \right), \tag{5.162}$$

where μ_h and μ_l are the mobilities of the heavy and the light holes, respectively, $b_1 = \mu_l / \mu_h$, and $\eta = p_l / p_h$ is the ratio of concentration of light holes to heavy holes.

If the transverse magnetic field is strong enough the current density is:

$$j_x = \left(\sigma + \frac{\gamma^2 B_z^2}{\sigma} \right) E_x . \tag{5.163}$$

The relative change of magneto-resistance is:

$$\frac{\Delta\rho}{\rho_B} = 1 - \frac{\left(\sigma + \frac{\gamma^2 B_z^2}{\sigma} \right)}{\sigma_0} \tag{5.164}$$

In the limit of a strong magnetic field, $\sigma << \frac{\gamma^2 B_z^2}{\sigma}$, so σ can be neglected, and (5.164) reduces to:

$$\frac{\Delta\rho}{\rho_B} = 1 - \frac{\gamma^2}{\sigma\sigma_0} B_z^2 \tag{5.165}$$

Combining (5.137) and (5.165), we obtain:

$$\frac{\Delta\rho}{\rho_B} = 1 - \left\{ \langle (\tau_m)_w \rangle \langle (\tau_m^{-1})_w \rangle \right\}^{-1} . \tag{5.166}$$

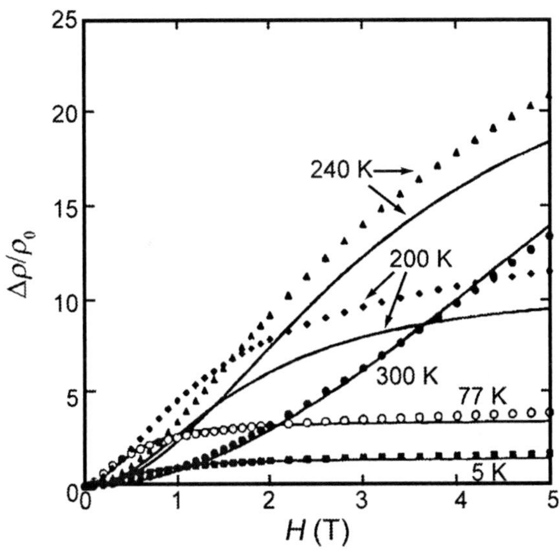

Fig. 5.20. The magneto-resistance as function of magnetic field at different temperatures. The *solid curves* are calculated from a normal magnetoresistance theory based on a two-layer model (Song et al. 1993)

If $\tau_m(\varepsilon)$ obeys a power law, then we have:

$$\frac{\Delta\rho}{\rho_B}=1-\frac{9\pi}{16\left(r+\frac{3}{2}\right)!\left(\frac{3}{2}-r\right)!}. \tag{5.167}$$

For acoustic deformation scattering, r = –1/2, so $\Delta\rho/\rho_B=0.116$, and for impurity ion scattering, r = +3/2, so $\Delta\rho/\rho_B=0.706$. In metals, only the electrons near Fermi level contribute to the conductivity. The average carrier energy is approximately ε_f; the average τ_m is its value at the Fermi energy, and is nearly a constant, i.e., r = 0. From (5.159), $\Delta\rho/\rho_B=0$, so there is no magneto-resistance effect in metals.

Figure 5.20 (Song et al. 1993) shows experimental results for InSb.

5.3.4 Magneto-Transport Experimental Methods

The magneto-transport is one of very important and direct methods in the study of semiconductor physics and devices. Many of the physical properties of materials can be investigated by investigating the longitudinal current, the temperature and the magnetic field dependence of the magnetic resistance voltage $V_\rho(B,T,I)$ and the Hall voltage $V_H(B,T,I)$.

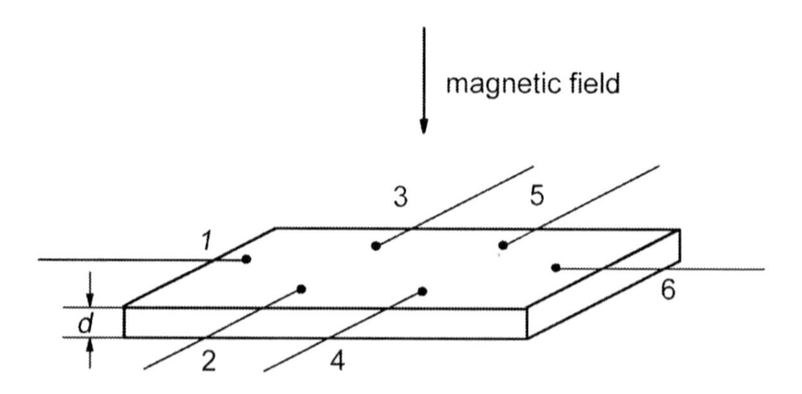

Fig. 5.21. The sample geometry for performing a standard Hall measurement

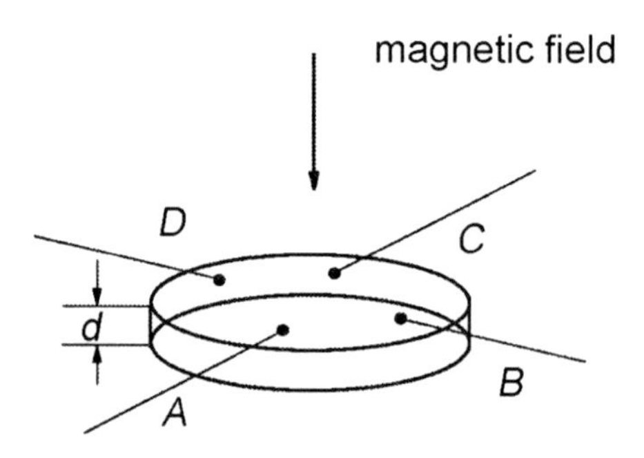

Fig. 5.22. A sample configuration for the Van der Pauw method

There are two ways to perform Hall measurements, i.e., the Van der Pauw method (Van der Pauw 1958) and the standard method (Putley 1960). In the standard method, the sample configuration must be in the shape of a rectangular bar as indicated in Fig. 5.21. A constant current flows along the x-axis between contacts labeled 1 and 6. The magneto-resistance voltage is obtained between contacts 2 and 4 or 3 and 5, and the Hall voltage is between contacts 2 and 3 or 4 and 5. Since the sample shape has a strong effect on the current distribution in the sample, to ease the data interpretation the ratio of the length to the width should be large, e.g., larger than 3. To minimize the errors induced by thermal effects and contact resistance, one usually takes the measurements by changing the directions of the current and the magnetic field.

In 1958, Van der Pauw pointed out that Hall effects also can be measured in irregular shaped samples as long as they have a uniform thickness and are without cavities, if the electrodes are located at the edges

Table 5.8. The value of shape factor *f* for different resistance ratios

$R_{AB,CD}/R_{BC,DA}$	1	2	5	10	20	50	100	200	500	1,000
F	1	0.96	0.82	0.70	0.59	0.47	0.40	0.35	0.30	0.26

From Seeger (1989)

of the sample and the size of the electrodes is small enough. A typical arrangement is showed in Fig. 5.22, where points A, B, C, and D are the locations of electrodes, and d is the thickness of the sample. This measurement technique is called the Van der Pauw method.

As shown in Fig. 5.22, the magnetic field is perpendicular to the sample surface. In the Van der Pauw method, first induce a current to flow from A to C. Then the Hall coefficient is obtained by measuring the voltage between B and D:

$$R_H = \frac{\left[V_{BD}(B) - V_{BD}(0)\right]d}{I_{AC}B}. \tag{5.168}$$

Then, let current flow from B to D, measuring the Hall voltage between A and C:

$$R_H^{'} = \frac{\left[V_{AC}(B) - V_{AC}(0)\right]d}{I_{BD}B} \tag{5.169}$$

The average value of R_H and R'_H is taken to be the Hall coefficient of the sample.

When current I_{AB} flows from A to B, there is a voltage V_{CD} across C and D, and a "resistance" is defined as:

$$R_{AB,CD} = |V_{CD}|/I_{AB}. \tag{5.170}$$

When current I_{BC} flows from B to C, there is a voltage V_{DA} across D and A, and again another "resistance" is also defined as:

$$R_{BC,DA} = |V_{DA}|/I_{BC}. \tag{5.171}$$

The resistivity is then calculated from the expression:

$$\rho = \frac{\pi d}{\ln 2} \frac{R_{AB,CD} + R_{BC,DA}}{2} f \tag{5.172}$$

Where f is a shape factor, experimentally determined from the ratio $R_{AB,CD}/R_{BC,DA}$. The value of f for different resistance ratios is presented in Table 5.8.

In addition to the standard and Van der Pauw methods, another method is used for a Robinson sample. In this case the current flows from one electrode at the center of round sample to a circular electrode at out edge of the sample. For a detailed description, readers are referred to a paper by Seeger (1989).

The magnetic transport measurement system consists of a direct current system, a sample support, a cryostat, a superconducting magnet, and a temperature control system. The current detection system includes a signal source with a multiple independent switch unit, a measuring and controlling switch box, a programmable current source, a high precision volt meter, a computer, and ports. Typically the lower limit of the current source is 1 pA, and that of the high precision volt meter measures differences of less than that 2 μV under stable external conditions. The sample support, which can be rotated in an angular range of 0–90° to change the angle between the sample and the magnetic field, includes controllable thermal couples, a heater, electrode wires, etc. The power supply and superconducting magnet can produce a current of up to 100 A with a corresponding magnetic field up to 12–17 T. In the constant current source there is a standard resistor and the voltage across it is fed into a computer to control the magnetic field. The current fed into the superconducting magnet needs to be stable, otherwise, the magnet can drop out of its superconducting state, especially at high magnetic fields. The temperature control system can vary the temperature in the range between 0.3–300 K, with temperature fluctuation of $<1\%$.

Magnetic transport measurements are important methods to obtain the electrical parameters of materials. The measurement techniques are developing rapidly even though the physical principles remain the same. The improvements in the measurement techniques lies in increasing the magnitude of the fields generated by superconductor magnets, improving the computer control of the magnetic fields and the temperatures, and improvements in the stabilization of current and voltage sources. These improvements have greatly enhanced the measurement accuracies of the electrical properties of materials.

5.4 Mobility Spectrum in a Multi-Carrier System

5.4.1 The Conductivity Tensor of a Multi-Carrier System

The effect of multiple carrier species on semiconductor magneto-transport must be taken into account. This is particularly crucial for narrow gap semiconductors, in which the coexistence of multiple species tends to be the rule rather than the exception. Species competing with the majority carriers in narrow gap materials and detectors can be due to surface inversion or accumulation layers caused by passivation, thermally generated minority carriers, multiple p and n doping layers, and heavy and light holes in the valence bands. The experimental Hall and resistivity data are very sensitive to the concentrations and mobilities of multiple carrier species. These effects can most easily be understood from the conductivity tensor derived from the semi-classical Bolzmann equation:

$$-e\boldsymbol{E}\cdot\boldsymbol{v}_k\frac{\partial f_0^j}{\partial\varepsilon}+\frac{S_j e(\boldsymbol{v}_k\times\boldsymbol{B})}{\hbar}\cdot\frac{\partial\left[f^j(k)-f_0^j(k)\right]}{\partial\boldsymbol{k}}=\frac{\left[f^j(k)-f_0^j(k)\right]}{\tau_{\mathbf{k}}} \tag{5.173}$$

where $\boldsymbol{B}$ is the magnetic field taken to lie along the z axis, $\boldsymbol{E}$ the electric field in the x, direction, and for a wavevector $\boldsymbol{k}$, $\boldsymbol{v}_{\mathbf{k}}$ is its velocity, $\varepsilon(\boldsymbol{k})$ is its energy, $f_0(\boldsymbol{k})$ is the equilibrium Fermi distribution function, $f^j(\mathbf{k})-f_0^j(\mathbf{k})$ is the perturbation of the distribution function induced by the electric and magnetic fields, $\tau_{\mathbf{k}}$ is the momentum relaxation time, and $S_j = +1$ for a hole and -1 for an electron. In (5.173) all the dynamical variables have a super script j on them to indicate their species. The solution is:

$$f^j(\mathbf{k})-f_0^j(\mathbf{k})=\frac{f_0^j(1-f_0^j)e\tau_{\mathbf{k}}\boldsymbol{v}_{\mathbf{k}}}{k_B T}\cdot\left(\frac{\boldsymbol{E}+(S_j e\tau_{\mathbf{k}}/m_k)\boldsymbol{B}\times\boldsymbol{E}}{1+(e\tau_{\mathbf{k}}/m_{\mathbf{k}})^2B^2}\right) \tag{5.174}$$

where T is the temperature and $m_{\mathbf{k}}=\hbar^2(\partial^2\varepsilon/\partial k^2)^{-1}$ is the carrier's 'optical' effective mass at wave vector $\boldsymbol{k}$. With this result for a given perturbed state distribution function, one can evaluate the net current density by integrating over all states:

$$\mathbf{j}=\tilde{\boldsymbol{\sigma}}\cdot\boldsymbol{E}=\sum_j\int dk\left[f^j(\mathbf{k})-f_o^j(\mathbf{k})\right]e\boldsymbol{v}_{\mathbf{k}}^j \tag{5.175}$$

Here $\tilde{\sigma}$ is the conductivity tensor and the sum $\sum_j$ is over all carrier species j. Then a substitution of $f^j(k)-f_0^j(k)$ from (5.174) and noting the dependences of j_x and j_y on E_x, one obtains the tensor components:

$$\sigma_{xx}=\sum_j\int dk\left(\frac{e^2}{k_B T}\right)\frac{f_0^j(1-f_0^j)(v_{x\mathbf{k}}^j)^2\tau_{\mathbf{k}}^j}{1+(e\tau_{\mathbf{k}}^j/m_{\mathbf{k}}^j)^2B^2}, \tag{5.176}$$

$$\sigma_{xy} = \sum_j \int d\mathbf{k} S_j \left(\frac{e^2}{k_B T} \right) \frac{(e\tau_{\mathbf{k}}^j / m_{\mathbf{k}}^j) B f_0^j (1 - f_0^j)(v_{x\mathbf{k}}^j)^2 \tau_{\mathbf{k}}^j}{1 + (e\tau_{\mathbf{k}}^j / m_{\mathbf{k}}^j)^2 B^2}. \tag{5.177}$$

The experimental determination of conductivity tensor is obtained from the definitions of the transport coefficients:

$$R_H(B) \equiv \frac{\sigma_{xy} / B}{\sigma_{xx}^2 + \sigma_{yy}^2},$$

and (5.178)

$$\rho(B) \equiv \frac{\sigma_{xx}}{\sigma_{xx}^2 + \sigma_{yy}^2}.$$

Then we find:

$$\sigma_{xx} = \frac{1}{\rho(B)[(R_H(B)B / \rho(B))^2 + 1]}, \tag{5.179}$$

$$\sigma_{xy} = \frac{R_H(B)B}{\rho^2(B)[(R_H(B)B / \rho(B))^2 + 1]}. \tag{5.180}$$

Mathematically, the parameters for each species can be fitted according to (5.176) and (5.177) from the experimental data for σ_{xx} and σ_{xy}. However, each term in the sum over j and integral over dk will have its own relaxation time, velocity and effective mass. Clearly, one cannot extract all of these quantities from the experimental dependences of σ_{xx} and σ_{xy} on magnetic field. A more modest and achievable goal is to obtain the net concentration n_j and μ_j for each of the species contributing to the transport. Equations (5.176) and (5.177) simplify considerably if we take the relaxation time to be independent k and use $\mu_j \approx e\tau^j / m^j$. This rough approximation leads to the expressions:

$$\sigma_{xx} = \sum_j \frac{n_j e \mu_j}{1 + \mu_j^2 B^2}, \tag{5.181}$$

$$\sigma_{xy} = \sum_j S_j \frac{n_j e \mu_j^2 B}{1 + \mu_j^2 B^2}. \tag{5.182}$$

If there is only a single carrier species, substitution of these approximations into (5.178) yields the familiar result, $R_H \to S_j / n_j e$. A more detailed treatment, in which one accounts for the wave vector dependence of the relaxation time for the appropriate scattering mechanism, but still treats $m_{\mathbf{k}}^j$ as being independent of $\boldsymbol{k}$ (i.e. parabolic bands), leads to the slightly modified relation $R_H = S_j r_H^j / n_j e$ at low magnetic fields, where $r_H = \left\langle (\tau_{\mathbf{k}}^2)_w \right\rangle / \left\langle \tau_{\mathbf{k}w} \right\rangle^2$ is the ratio of energy weighted averages over the occupied states in the band. The Hall factor, r_H, rarely falls outside the range $0.9<r_H<1.4$ in $Hg_{1-x}Cd_xTe$ alloys. In many cases of practical importance to the characterization of infrared detector materials, the minor correction due to the Hall factor is of much less interest than determining how many carrier species contribute to the conduction, and obtaining the temperature dependences of n_j and μ_j for each.

It should be pointed out that the derivation leading to (5.181) and (5.182) is equally applicable to either a three-dimensional (3D) or a two-dimensional (2D) carrier population, since in both cases there is no current out of the plane (j_Z = 0). The dimensionality of the integral $\int d\boldsymbol{k}$ thus depends on the dimensionality of the system, and the units of some of the quantities, such as n, σ, $\boldsymbol{j}$, R_H and ρ, will vary accordingly. The mixed-conduction analysis of the Hall mobility and resistivity data gives only net area densities, with no direct indication of whether the carriers populate a thick 3D region or are confined to a 2D plane. Tiled measurements can be used to distinguish 2D from 3D properties. In principle, the mixed-conduction analysis can be used to study the carrier properties and any inhomogeneity along the z axis, e.g. if there is a doping or compositional gradient.

The validity of (5.181) and (5.182) depends on two assumptions. First, the carrier concentrations and mobilities are independent of the magnetic field, which is never strictly true. Particular care is required when a magnetic field-dependent shift of the energy gap leads to a significant variation of the intrinsic carrier concentration with field (Meyer et al. 1988), or when the effective mobility strongly decreases with increasing magnetic field due to the magnetic freeze-out phenomena (Aronzon and Tsidilkovskii 1990). Second, whereas (5.181) and (5.182) were derived from a semi-classical treatment, the discreteness of the Landau levels causes Shubnikov-de Haas (SdH) (and sometimes quantum Hall) oscillations to be superimposed on the classical conductivities. These are insignificant under many conditions of interest, since they are washed out by either collision broadening at low fields, or by thermal broadening at high temperatures. They often dominate the transport in the opposite limits of high B and low T. Hence data from the quantum regime are unsuitable for the mixed-conduction analysis unless one can somehow ‘remove’ the oscillatory component. On the other hand, SdH and quantum Hall effects are in themselves valuable sources of additional information, which complement that available from the semi-classical mixed-conduction analysis.

The main drawback of the multi-carrier fitting procedure (MCF) (Gold and Nelson 1986) based on the conventional mixed-conduction approximation is its arbitrariness, i.e. one must decide in advance how many carrier species to assume in the fitting and the approximate mobility for each carrier species. As first discussed by Beck and Anderson (1987), this shortcoming can be avoided by employing the ‘mobility spectrum’ technique. We will discuss this technique in detail in Sect. 5.4.3.

5.4.2 Multi-Carrier Fitting Procedure

The MCF is based on the conventional mixed-conduction approximation. In this ‘conventional’ treatment, one simply performs a least-square fit (perhaps with some form of weighting) to the experimental dependences of σ_{xx} and σ_{xy} on magnetic field given by (5.179) and (5.180), using n_j and μ_j as adjustable parameters.

Only if one knows in advance how many carrier species to assume in the fit and the range of carrier mobilities, can the fitted results approach the true situation. Figure 5.23 shows the fitting result for an LPE grown sample n-$Hg_{1-x}Cd_xTe$ (Gui et al. 1997) using a MCF assuming there are 2–4 kinds of carriers. Figure 5.24 displays the results of a two-carrier fitting procedure if different initial values of mobilities are used. Mathematically, all the fits are in good agreement with the experimental data. It clearly demonstrates the shortcoming of this fitting procedure, i.e. its arbitrariness.

There is still another procedure named the reduced-conductivity-tensor scheme, RCT (Kim et al. 1993, 1994), that can be used to analysis such a multiple carrier system. At zero magnetic field, the conductivity tensor defined in (5.181) and (5.182) can be simplified to:

$$\sigma_0 \equiv \sigma_{xx}(0) = \sum_j e n_j \mu_j \ , \tag{5.183}$$

$$\sigma_{xy}(0) = 0 \, . \tag{5.184}$$

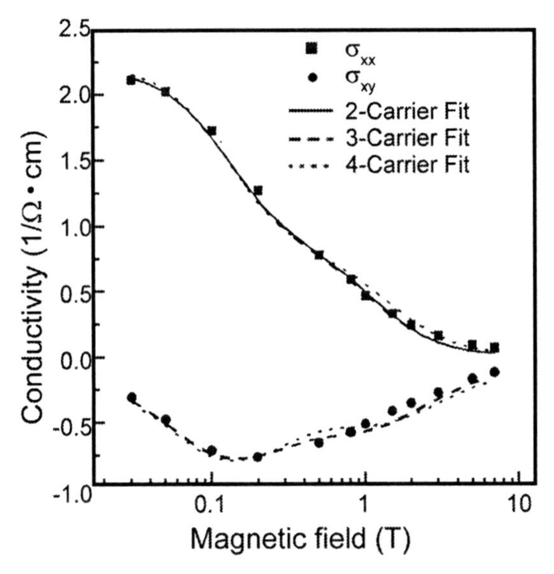

Fig. 5.23. Comparison between the experimental conductivity tensor as a function of magnetic field (*solid symbols*) and multi-carrier fitting results (*curves*) at 100 K

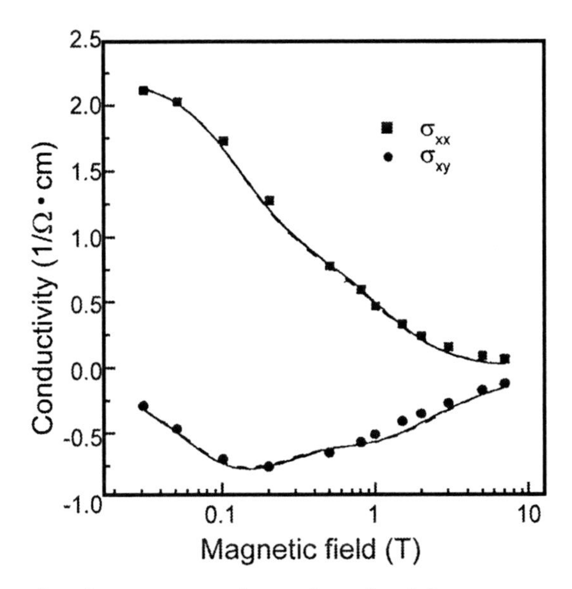

Fig. 5.24. Comparison between experimental conductivity tensor components as a function of magnetic field (*solid symbols*) and two-carrier fitting results (*curves*) with different initial values of mobility at 100 K

Table 5.9. Multi-carrier fitting results of an n-$Hg_{1-x}Cd_xTe$ LPE sample

	Fig. 5.23			Fig. 5.24	
n_1 (cm^{-3})	1.07×10^{14}	1.01×10^{14}	1.15×10^{14}	1.07×10^{14}	5.92×10^{13}
μ_1 (cm^2/V s)	8.08×10^4	7.81×10^4	8.0×10^4	8.08×10^4	1.1×10^5
n_2 (cm^{-3})	5.92×10^{14}	5.22×10^{14}	6.95×10^{14}	5.92×10^{14}	8.11×10^{14}
μ_2 (cm^2/V s)	8.84×10^3	1.07×10^4	6.89×10^3	8.84×10^3	9.21×10^3
n_3 (cm^{-3})		1.12×10^{14}	5.28×10^{13}		
μ_3 (cm^2/V s)		4.95×10^2	6.05×10^2		
n_4 (cm^{-3})			9.8×10^{12}		
μ_4 (cm^2/V s)			3.87×10^1		

Then the reduced conductivity tensor components are defined as:

$$X(B)=\frac{\sigma_{xx}(B)}{\sigma_0}, \tag{5.185}$$

$$Y(B)=\frac{2\sigma_{xy}(B)}{\sigma_0}. \tag{5.186}$$

The factor of 2 in (5.186) causes $Y(B)$ to have a maximum value of 1 rather than 1/2, i.e. $|Y|\le 1.0$.

A parameter f_j, which is a weighting parameter in the fitting procedure for the jth carrier contribution to the total conductivity at zero magnetic fields, is defined as:

$$f_j \equiv |n_j\mu_j| \Big/ \sum_j |n_j\mu_j| \tag{5.187}$$

Substitution of f_j into (5.185) and (5.186) produces:

$$X(B,\{\mu\},\vec{f})=\sum_j \frac{f_j}{1+\mu_j^2B^2}, \tag{5.188}$$

$$Y(B,\{\mu\},\vec{f})=\sum_j \frac{2Bf_j\mu_j}{1+\mu_j^2B^2}. \tag{5.189}$$

where $\{\mu\}$ and $\{f\}$ represent the sets of all μ_j, and f_j, and the value of f_j are by definition limited to:

$$\begin{cases} 0\le f_j \le 1 \\ \sum_j f_j = 1 \end{cases} \tag{5.190}$$

A conventional least-square fit to the experimental data is performed, using μ_j and f_j as adjustable parameters:

$$\chi^2(\vec{\mu},\vec{f}) = \frac{1}{2(L+1)}\sum_{n=0}^{L}\{[X(B_n,\vec{\mu},\vec{f}) - X_{\exp}(B_n)]^2 + [Y(B_n,\vec{\mu},\vec{f}) - Y_{\exp}(B_n)]^2\}, \quad (5.191)$$

where L is the number of experimental magnetic field points, and X_{exp} and Y_{exp} are the experimental results.

The RCT method yields a slight improvement in the carrier parameters deduced from the dependence of the resistivity on magnetic field. The magneto-resistance is also a function of magnetic field B:

$$M(B) = \frac{\rho(B) - \rho(0)}{\rho(0)}, \quad (5.192)$$

where $\rho(B)$ is the resistivity at field B. This can be used to provide extra information to aid the extraction of the carrier parameters. The resistivity is related to the conductivity tensor components σ_{xx} and σ_{xy}:

$$\rho(B) = \frac{\sigma_{xx}(B)}{\sigma_{xx}^2(B) + \sigma_{xy}^2(B)}. \quad (5.193)$$

From (5.185), (5.186), (5.192), and (5.193), it is easy to get:

$$M(B) = \frac{X}{X^2 + Y^2} - 1. \quad (5.194)$$

For one-carrier system, $M(B) = 0$, i.e. the magneto-resistance is independent of the magnetic field. In a two-carrier system we find:

$$M(B) = \frac{(\alpha\Delta B)^2}{1 + (\beta\Delta B)^2}, \quad (5.195)$$

where $\alpha, \beta, \text{and}\ \Delta$ are defined as:

$$\alpha \equiv \sqrt{f_1(1-f_1)},\ \beta \equiv \mu_1/\Delta - f_1,\ \text{and}\ \Delta \equiv \mu_1 - \mu_2 \quad (5.196)$$

At a low magnetic field we have, $M(B) = (\alpha\Delta B)^2 \propto B^2$. At high magnetic fields, $M(B) = (\alpha/\beta)^2$ and $M(B)$ is a constant independent of the magnetic field, and only depends on the carrier mobilities and concentrations.

The RCT scheme provides a criterion for distinguishing between one-carrier and multi-carrier conduction, and has an advantage in the analysis of two-terminal devices (photo-conductive detectors). However, it still does not overcome the arbitrariness in the assignments associated with the conventional mixed-conduction approach, i.e. one must decide in advance how many carrier species to assume in the fitting procedure and the approximate mobility of each carrier species.

5.4.3 Mobility Spectrum Analysis

Mobility spectrum analysis (MSA) that has appeared in recent years, overcomes some defects of the conventional method. This method obtains the conductivity spectrum as a function of mobility from the relationship between the conductivity tensor and the magnetic field strength. In the spectrum, each a peak value in the spectrum corresponds to one of the carrier species, and the sign of the mobility indicates the type of the carrier (Beck and Anderson 1987).

During the MSA, we first assume the mobility of electrons and holes in samples is continuously distributed. Then (5.181) and (5.182) can be written as the integral forms:

$$\sigma_{xx}(B) = \int_0^\infty \frac{s^p(\mu) + s^n(\mu)}{1+\mu^2 B^2} d\mu, \tag{5.197}$$

$$\sigma_{xy}(B) = \int_0^\infty \frac{[s^p(\mu) - s^n(\mu)]\mu B}{1+\mu^2 B^2} d\mu. \tag{5.198}$$

Here the conductivity/unit mobility functions for electrons and holes (that is the mobility spectrum of conductivity/unit mobility), $s^p(\mu)$ and $s^n(\mu)$, are defined as:

$$s^p(\mu) = ep(\mu)\mu \sum_{j=1}^{j_h} \delta\left(\mu - |\mu_j|\right), \tag{5.199}$$

$$s^n(\mu) = en(\mu)\mu \sum_{j=1}^{j_{el}} \delta\left(\mu - |\mu_j|\right) \tag{5.200}$$

where $p(\mu)$ and $n(\mu)$ are the concentrations of holes and electrons as a functions of mobility, and j_{el}, and j_h are the numbers of electron and hole species contributing to the conductivity. The aim of mobility spectrum

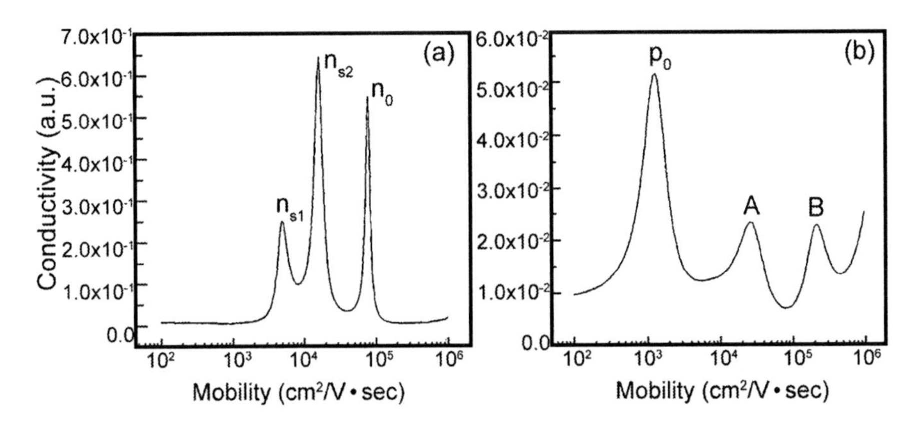

Fig. 5.25. At 100 K, the mobility spectrum of n-$Hg_{1-x}Cd_xTe$ sample L9701-1 (**a**) electrons; (**b**) holes

analysis is to obtain the value of $s^p(\mu)$ and $s^n(\mu)$ by a series of transforms. If the approximations under which (5.181) and (5.182) were derived are lifted, so one must start form (5.176) and (5.177) instead, then the generalizations of (5.197) and (5.198) become infinite sums (Beck and Anderson 1987). These sums are over Fourier coefficients related to energy derivatives along constant energy surfaces. Beck and Anderson developed an accurate mathematical procedure (see Appendix A of their paper) to obtain unique envelope functions for $s^p(\mu)$ and $s^n(\mu)$, based on the constraint that no carrier species makes a negative contribution to the conductivity. The ultimate purpose of the MSA is to determine, the number of carrier species contributing to the conductivity, and their approximate mobilities and concentrations.

Figure 5.25 is, at 100 K, the mobility spectrum of n-$Hg_{1-x}Cd_xTe$ sample L9701-1 discussed earlier. From the figure, we see the electronic mobility spectrum of sample L9701-1 has three sharp peaks, corresponding to three species of electrons: bulk electrons n_0, and interface electrons n_{s1} and n_{s2}, respectively. The hole mobility spectrum also has three peaks; p_0 corresponding to minority bulk carrier, but the A and B peaks are artifacts mirroring the peaks in the electronic mobility spectrum. With the mobility spectrum in Fig. 5.25 normalized to the conductivity, the electron and hole concentration distributions verses mobility can be deduced, and is shown in Fig. 5.26. From these figures, the concentrations and mobilities of the various carriers can be found. The concentration of the bulk electrons lies in the range ~10^{13}–10^{14} cm^{-3}, and their mobility is about ~7–8×10^4 cm^2/V s. There are two kinds of interface electrons, whose concentrations are both in the range ~10^{13}–10^{14} cm^{-3}, and whose mobilities are about

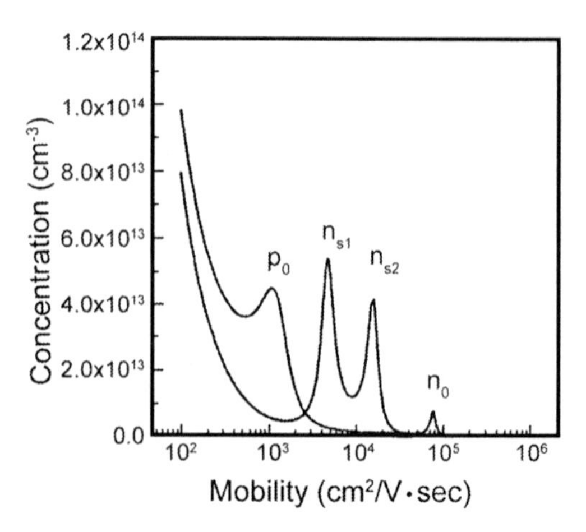

Fig. 5.26. The carrier concentration distribution as a function of mobility of sample L9701-1 (n-$Hg_{1-x}Cd_xTe$) obtained by the mobility spectrum method

$\sim$1–2 × 10^4 cm^2/V s, and $\sim$1–5 × 10^3 cm^2/V s. From the mobility spectrum, intrinsic holes are found with a concentration in the range $\sim 10^{13}$–10^{14} cm^{-3}, which is the same as that of the bulk electrons, and their mobility is about $\sim 1\times 10^3$ cm^2/V s. Because the mobility of holes is far less than that of the electrons, in low fields, the hole contribution to the conductivity tensor can not be separated out. Even so the mobility spectrum does separately determine the contribution of the holes.

The above example demonstrates how a mobility spectrum enables the determination of not only the species of carriers in a sample, but also their approximate mobilities and concentrations. Thus it is an effective tool for the analysis of multi-carrier systems.

In still another method, Dziuba and Gorsk (1992) adopted an iterative approximation method to obtain the mobility spectrum of samples. This method differs from the envelope function of Beck and Anderson. In this method, (5.197) and (5.198) are replaced by sums that course grain the integrals:

$$\sigma_{xx}(B) = \sum_i^m \frac{[s^p(\mu_i) + s^n(\mu_i)]\Delta\mu_i}{1+\mu_i^2 B^2} = \sum_i^m \frac{S_i^{xx}\Delta\mu_i}{1+\mu^2 B^2}, \tag{5.201}$$

$$\sigma_{xy}(B) = \sum_i^m \frac{[s^p(\mu_i) - s^n(\mu_i)]\mu_i B\Delta\mu_i}{1+\mu_i^2 B^2} = \sum_i^m \frac{S_i^{xy}\mu_i B\Delta\mu_i}{1+\mu^2 B^2}. \tag{5.202}$$

The parameter m is the number of discrete mobilities into which the mobility spectrum is subdivided. The functions S_i^{xx} and S_i^{xy} are defined as:

$$S_i^{xx} = s^p(\mu_i) + s^n(\mu_i), \tag{5.203}$$

and

$$S_i^{xy} = s^p(\mu_i) - s^n(\mu_i). \tag{5.204}$$

The method postulates an initial spectrum, and uses a Jacobi iterative procedure to solve the system of (5.201) and (5.202) for the mobility spectrum. In this method, the range of mobilities spanned depends on the magnetic field intensity employed in measurement. The upper and lower bounds satisfies the conditions: $1/B_{\max}^{\exp} = \mu_{\min} \le \mu \le \mu_{\max} = 1/B_{\min}^{\exp}$, where $B_{\min}^{\exp}$ and $B_{\max}^{\exp}$ are the minimum and maximal values of the magnetic field intensity. For some semiconductor materials, the mobility spanned is too small for practical magnetic fields. For example, for the holes in $Hg_{1-x}Cd_xTe$ alloys, whose mobility falls in the range ~10^2–10^3 cm^2/V s, several tens of Tesla magnetic field intensities are needed. In addition, although $s^p(\mu_i)$ and $s^n(\mu_i)$ deduced from this method are reasonable, other extraneous results also often appear, e.g. negative values of $s^p(\mu_i)$ and $s^n(\mu_i)$. So while their starting point was useful it was left to others to improve it into a method having an explicit physical meaning.

5.4.4 Quantitative Mobility Spectrum Analysis

Although the mobility spectrum has a good many advantages, its results presented so far were only qualitative or semi-quantitative. To improve the accuracy, Meyer et al. (1993) developed a more quantitative mobility spectrum analysis (QMSA). The method, named the Hybrid Mixed Conduction Analysis (HMCA), regards the prior mobility spectrum results as initial values, then treats the experimental data by a MCF and thereby extracts unique electrical parameters reflecting the true behavior of the material. Figure 5.27 is the fitted result for the n-$Hg_{1-x}Cd_xTe$ ($x = 0.214$) sample L9701–1 using the result depicted in Fig. 5.26 for initial values. The dots are the experimental data and the curves are the fitted results. From HMCA we find:

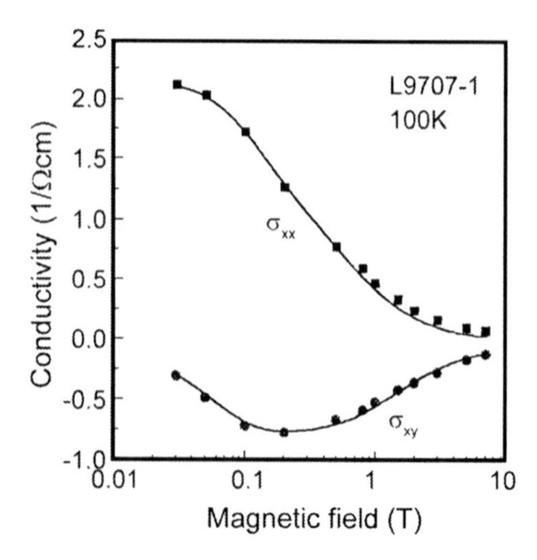

Fig. 5.27. At 100 K, the relationship between the conductivity tensor components and the magnetic field of sample L9701-1 (The *dots* are experimental data, and the *curves* are the results deduced by the HMCA method)

- The concentration of bulk electrons is 9.7×10^{13} cm^{-3} and their mobility is 7.7×10^4 cm^2/V s,
- There are two kinds of interface electrons in the material with concentrations of 2.4×10^{14} cm^{-3} and 4.1×10^{14} cm^{-3} and the corresponding mobilities are 1.62×10^4 cm^2/V s and 5.6×10^3 cm^2/V s, respectively, and
- The concentration of intrinsic ionized holes is 1×10^{14} cm^{-3} and their mobility is 3×10^2 cm^2/V s.

We also get $(n_0p_0)^{1/2} = 1 \times 10^{14}$ cm^{-3}, which is near the intrinsic carrier concentration ($n_i = 9.5 \times 10^{13}$ cm^{-3}) at 100 K. So the result of the HMCA analysis provides a clearer physical meaning than that of the MCF method.

The second kind of QMSA was developed by Gui et al. To get an accurate mobility spectrum having an explicit physical meaning, Gui et al. (1998a,b; Gui 1998), combined the mobility spectrum of Beck and Anderson and an iterative approximation method. In the analysis, the mobility spectrum of Beck and Anderson is taken as initial values, then they use their iterative algorithm to get accurate solutions under the constraint that $s^p(\mu_i)$ and $s^n(\mu_i)$ are positive at each iterative step. This assures the efficiency of algorithm, the precision of the result, and that it has an explicit physical meaning. In QMSA, the range of mobilities covered expands from $1/B_{\max}^{\exp} \le \mu \le 1/B_{\min}^{\exp}$ to 10^2–10^6 cm^2/V s, to encompass the most interesting mobility range.

In (5.181) and (5.182), if the contributions to the conductivity components of the *i*th carrier species are extracted from the sums on the right sides of the equations, and the net conductivities on the left sides of the equations are replaced by their corresponding experimental values, then (5.202) and (5.203) can be transformed as follows:

$$S_i^{xx} = (1+\mu_i^2 B_i^2)\left[\sigma_{xx}^{\exp}(B_i) - \sum_{j=0}^{i-1}\frac{S_j^{xx}}{1+\mu_j^2 B_i^2} - \sum_{j=i+1}^{m}\frac{S_j^{xx}}{1+\mu_j^2 B_i^2}\right], \tag{5.205}$$

$$S_i^{xy} = \frac{(1+\mu_i^2 B_i^2)}{\mu_i B_i}\left[\sigma_{xy}^{\exp}(B_i) - \sum_{j=0}^{i-1}\frac{S_j^{xy}\mu_j B_i}{1+\mu_j^2 B_i^2} - \sum_{j=i+1}^{m}\frac{S_j^{xy}\mu_j B_i}{1+\mu_j^2 B_i^2}\right]. \tag{5.206}$$

To speed the computations a super relaxation method was adopted to solve the linear equations, (5.205) and (5.206):

$$S_i^{xx}(k+1) = (1-\omega_{xx})S_i^{xx}(k) + \omega_{xx}(1+\mu_i^2 B_i^2) \cdot\left[\sigma_{xx}^{\exp}(B_i) - \sum_{j=0}^{i-1}\frac{S_j^{xx}(k+1)}{1+\mu_j^2 B_i^2} - \sum_{j=i+1}^{m}\frac{S_j^{xx}(k)}{1+\mu_j^2 B_i^2}\right], \tag{5.207}$$

$$S_i^{xy}(k+1) = (1-\omega_{xy})S_i^{xy}(k) + \omega_{xy}\cdot\frac{(1+\mu_i^2 B_i^2)}{\mu_i B_i} \cdot\left[\sigma_{xy}^{\exp}(B_i) - \sum_{j=0}^{i-1}\frac{S_j^{xy}(k+1)\mu_j B_i}{1+\mu_j^2 B_i^2} - \sum_{j=i+1}^{m}\frac{S_j^{xy}(k)\mu_j B_i}{1+\mu_j^2 B_i^2}\right]. \tag{5.208}$$

where $S_i^{xx}(k)$ and $S_i^{xy}(k)$ are the results of the kth step of iteration for S_i^{xx} and S_i^{xy}, respectively. In the formulas, ω_{xx} and ω_{xy} set the convergence rate of the iterative procedure. When $\omega_{xx} = \omega_{xy} = 1$, the convergence rate is maximized, but the initial shape of mobility spectrum is rapidly destroyed and the result often diverges. When $\omega_{xx} = \omega_{xy} = 0$, the convergence rate is the slowest, and the final result remains the initial mobility spectrum. After extensive calculation tests, Gui found $\omega_{xx} = 0.05$, $\omega_{xy} = 0.01$ are the optimal choices. In addition, to improve the accuracy as far as possible the data was smoothed and extra points were interpolated between the actual data to increase the sampling density used in the iterative procedure to 100

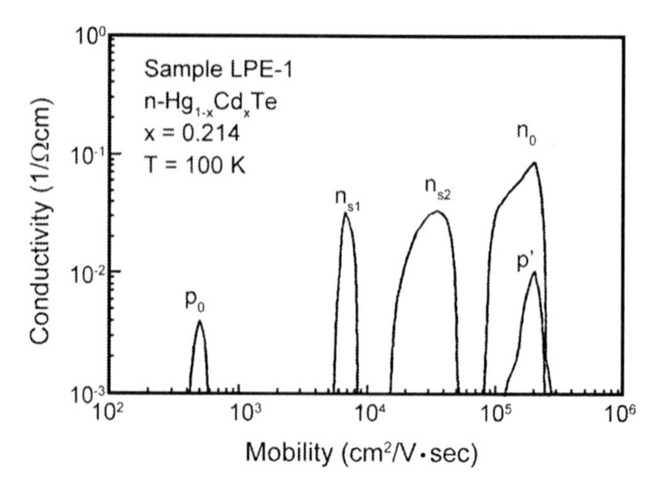

Fig. 5.28. At 100 K, the QMSA spectrum of the n-$Hg_{1-x}Cd_xTe$ sample L9701-1

points per order of magnitude. This compares favorably against the iterative approximation of Dziuba and Gorsk who used only 20 points, with some of their results being unphysical.

Figure 5.28 is the QMSA at 100 K of the n-$Hg_{1-x}Cd_xTe$ sample L9701-1. In the spectrum, the concentration of each carrier species j is: $n_j = \sum_{i=1}^{i_j}(\sigma_i - \sigma_{i-1})/(e\mu_i)$, a weighted sum of all carriers within the jth peak, where i_j is the total number of mobility steps taken in the peak. From the QMSA spectrum of this sample, the bulk electron concentration is determined to be $9.7\times10^{13}\,cm^{-3}$ and its mobility is $1.2\times10^5\,cm^2/V$ s. There are two kinds of interface electrons whose concentrations are $6\times10^{14}\,cm^{-3}$, and $4.8\times10^{14}\,cm^{-3}$, with corresponding mobilities of $2.4\times10^4\,cm^2/V$ s, and $6.5\times10^3\,cm^2/V$ s, respectively. Finally, the intrinsic hole concentration is $6\times10^{14}\,cm^{-3}$ and their mobility is $5.0\times10^2\,cm^2/V$ s.

$Hg_{1-x}Cd_xTe$ samples grown by MBE are usually capped with a very thin CdTe layer. Usually a two-dimension electron gas forms at the interface between the $Hg_{1-x}Cd_xTe$ and CdTe cap layers. If a conventional fixed magnetic measurement is adopted, usually the real electric parameters of materials are not obtained. To deduce true electric information, samples were first measured at various magnetic fields ranging from 100 to 4,000 Gs, then the experimental data was analyze using HMCA (Gui et al. 1997). The samples were cut into 5×5 mm^2 squares and four indium Ohmic contacts were fabricated on the samples in a Van der Pauw configuration. Hall measurements were performed using the Van der Pauw method.

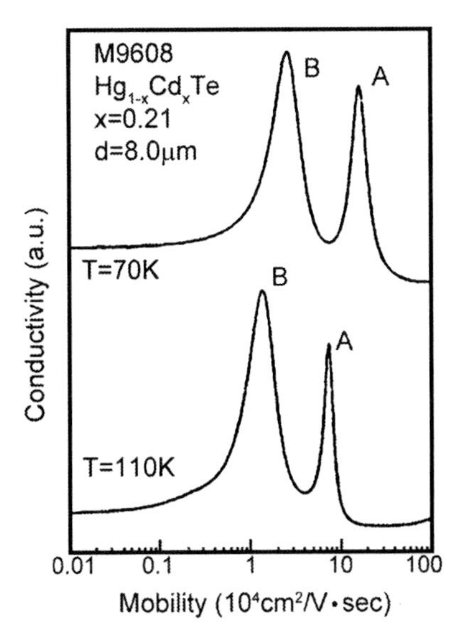

Fig. 5.29. At 77 and 110 K, the mobility spectra of sample M9608

Sample M9608 is $Hg_{1-x}Cd_xTe$ (x = 0.21) grown by MBE on a $Zn_{0.04}Cd_{0.96}Te$ substrate (Gui et al. 1997). The thickness of its epitaxial layer is 5.0 μm, and its CdTe cap layer thickness is 30 nm. Figure 5.29 displays the mobility spectrum of sample M9602 taken at 70 and 110 K. At both temperatures, the mobility spectra have two peaks. Thus the sample has two kinds of carriers, A and B. The mobility spectrum results were used as the initial values and a MCF was used to refine these results. At 70 K, the A species carrier concentration and mobility are $2.46 \times 10^{14}\ cm^{-3}$, and $1.52 \times 10^5\ cm^2/V$ s; and the B species carrier concentration and mobility are $1.85 \times 10^{15}\ cm^{-3}$, and $1.92 \times 10^4\ cm^2/V$ s. At 110 K, the A species carrier concentration and mobility are $3.82 \times 10^{14}\ cm^{-3}$, and $5.71 \times 10^4\ cm^2/V$ s; and the B species carrier concentration and mobility are $2.37 \times 10^{15}\ cm^{-3}$, and $1.07 \times 10^4\ cm^2/V$ s. If we were to assume there are three or four carrier species during the MCF, a good fit to the experimental data is still obtained. But the result is not a true reflection of the real behavior of the sample. An essential prerequisite to a quantitative evaluation of the behavior of samples is the information obtained from a mobility spectrum. Otherwise there is no assurance the results reflect truly the character of samples.

Between 15 and 280 K, the transport character at various magnetic fields were measured every 5 K, and the HMCA method was used to deuce the variation of the carrier concentration and mobility with temperature of sample M9608 (Gui et al. 1997). From Fig. 5.30a,b, we see: the

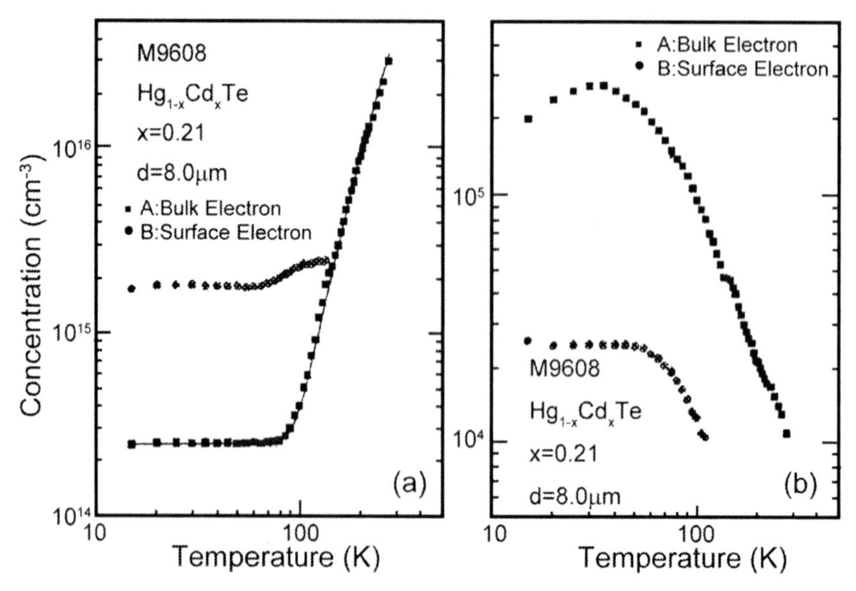

Fig. 5.30. The HMCA relations between: (**a**) the carrier concentrations, and (**b**) the mobilities of the temperature variation of sample M9608

concentration of the A carriers is nearly constant between 15 and 80 K, and then rapidly ascends with increasing temperature above 80 K. The mobility of the A carriers ascends slowly with increasing temperature at low temperatures, attains a maximum at 35 K, and then descends rapidly as the temperature again increases. The concentration of the B carriers is independent of temperature. The mobility of the B carriers is almost independent of temperature between 15 and 70 K, and then descends rapidly as the temperature increases above 70 K. This behavior is consistent with the character of surface electrons in n-type $Hg_{1-x}Cd_xTe$ samples reported in the literature (Reine et al. 1993). Above 120 K, the contribution to the conductivity from B carriers decreases, and is no longer seen in the mobility spectrum.

From this characterization, we conclude that the A carriers are bulk electrons in the epitaxial layer. For narrow band gap $Hg_{1-x}Cd_xTe$ alloys, the bulk electron concentration is dominated by ionized donor impurities at low temperature, and the intrinsic carrier contribution enters at about 100 K, and then dominates the bulk electron concentration at higher temperatures. In Fig. 5.28b, the variation of the A carrier mobility is consistent with that expected of bulk electrons. Impurity scattering and polarized optical phonon scattering limit the bulk electron mobility at low, and high temperature, respectively. The temperature dependence of the

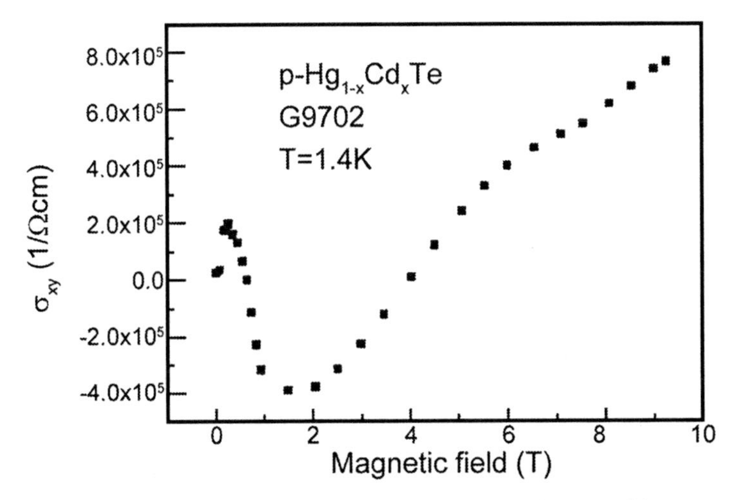

Fig. 5.31. The conductivity tensor component σ_{xy} variation with magnetic field at 1.4 K of a p-$Hg_{1-x}Cd_xTe$ sample G9702 (between 0 and 0.5 T, the sample is p-type; between 0.5 and 4 T it is n-type; and above 4 T it reverts to p-type)

mobility determined by impurity scattering is: $\mu_N \propto T^{3/2}$, and the mobility determined by polarized optical phonon scattering is: $\mu_{op} \propto \exp(T_p / T)$. One commonly finds the electronic concentration at the surface or interface of $Hg_{1-x}Cd_xTe$ to be almost independent of temperature (Parat et al. 1990), so the B carriers are identified as the electrons at the interface between the $Hg_{1-x}Cd_xTe$ and the CdTe.

Samples of p-$Hg_{1-x}Cd_xTe$ were grown using MBE and characterized at The Shanghai Institute of Technical Physics (Gui et al. 1997). The substrate materials were $Cd_{0.96}Zn_{0.04}Te$ or GaAs. The samples were cut into 5×5 mm^2 squares and four indium Ohmic contacts, with diameters of the spot welds being smaller than 0.5 mm, were made in a Van der Pauw configuration. Hall measurements were performed using the Van der Pauw method for 15 different temperatures ranging between 1.4 and 175 K at various magnetic fields between 0 and 9 T. First, a high density, 400 field point, magnetic scan between 0 and 9 T was performed at 1.4 K, to determine whether or not quantum oscillations were excited at either the interfacial two-dimension electron or hole gas. No quantum oscillations were found. Then measurements at 20–30 temperatures were preformed at fixed magnetic fields and the data analyzed by the quantitative mobility spectrum method.

Figure 5.31 is the measured functional relation at 1.4 K of the conductivity tensor component σ_{xy} versus the magnetic field of a p-$Hg_{1-x}Cd_xTe$ sample G9702 (Gui et al. 1997). From the figure, one can infer that the

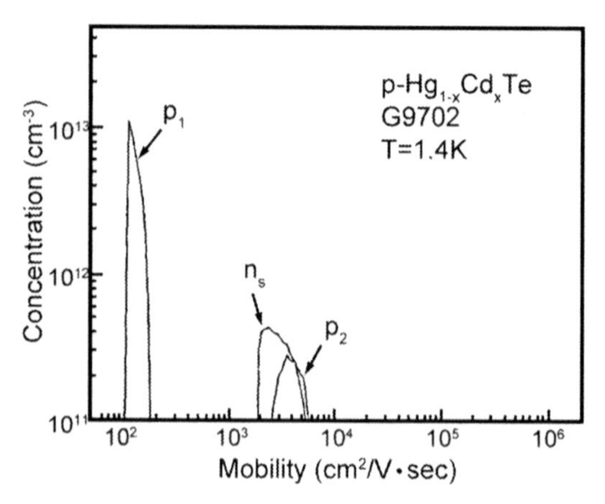

Fig. 5.32. The QMSA spectrum at 1.4 K of the p-$Hg_{1-x}Cd_xTe$ sample G9702. There are two kinds of holes, p_1 with a low mobility, p_2 with a high mobility, and an electron species n

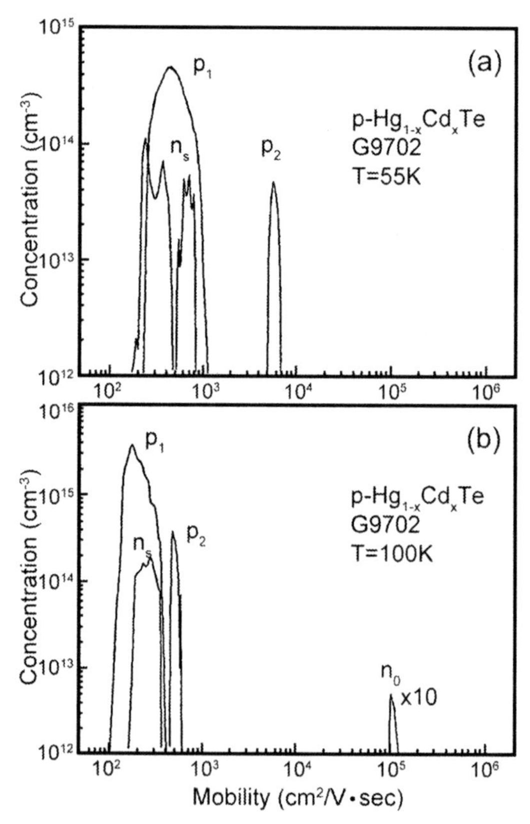

Fig. 5.33. (**a**) At 55 K and (**b**) 100 K, the QMSA spectrum of the p-$Hg_{1-x}Cd_xTe$ sample G9702

conductive type of the sample first converts from p-type to n-type as the magnetic field increases, and then back again from n-type to p-type, a double conversion process. To research the cause of these conversions, the experimental data at this temperature was subjected to a QMSA, and the result is shown in Fig. 5.31.

From Fig. 5.32 there are two hole and one electron species in sample G9702. Each species carrier concentration and mobility can be obtained accurately from the spectra. The concentration and mobility of the p_1 holes are $3.7 \times 10^{13}\,cm^{-3}$, and $1.1 \times 10^2\,cm^2/V\,s$; that of the p_2 holes are $7.0 \times 10^{12}\,cm^{-3}$, and $3.5 \times 10^3\,cm^2/V\,s$; and that of the surface electrons are $1.0 \times 10^{13}\,cm^{-3}$, and $2.2 \times 10^3\,cm^2/V\,s$. In the low magnetic field range, the contributions to σ_{xy} from p_2 and p_1 holes are larger than that of the surface electrons n_s, so the sample is p-type. Because the mobility of holes p_2 is rather high, and its effect decreases gradually as the magnetic field increases the p to n conversion occurs. Then over a large field range, n_s dominates and the sample is n-type. When the magnetic field is increased again, the p_1 holes despite their low mobility gradually dominate σ_{xy}.

Figure 5.33 is the quantitative mobility spectra of sample G9702 at 55 K and 100 K. At 55 K, the surface electrons in the spectrum appear to be in a complex state because its peak is split and highly broadened. At the elevated temperature, the concentration of holes greatly increases. In the QMSA spectrum, there are two hole peaks and the concentration of p_1 is almost 30 times larger than that of p_2. The mobility of the holes at 55 K is slightly larger than is the case at 1.4 K. The mobility of p_1 is $5.1 \times 10^2\,cm^2/V\,s$ and that of p_2 is $7.6 \times 10^3\,cm^2/V\,s$. At 100 K, the peak corresponding with the bulk electrons n_0 appears and its concentration is $2.1 \times 10^{12}\,cm^{-3}$.

In Figs. 5.32 and 5.33, the bewildering p_2 peak is present. This hole species mobility is far higher than the bulk holes, up to $10^4\,cm^2/V\,s$ (33 K). Holes with this high a mobility may come from two origins. One possibility is the light holes. Their effective mass is far smaller than that of the heavy holes, and based on mobility theory one expects that $\mu \propto (m^*)^{-1}$ and the concentration of these carriers is $n \propto (m^*)^{3/2}$. Thus the light hole mobility is far higher and its concentration is far smaller than those of the heavy holes. The experimental data agrees qualitatively with this theory. Another possible origin for the p_2 hole peak is from two-dimension interface holes. To pin down the origin of the p_2 peak, additional experiments and theoretic calculations are required. Here, the available experimental data is presented along with a call for additional research.

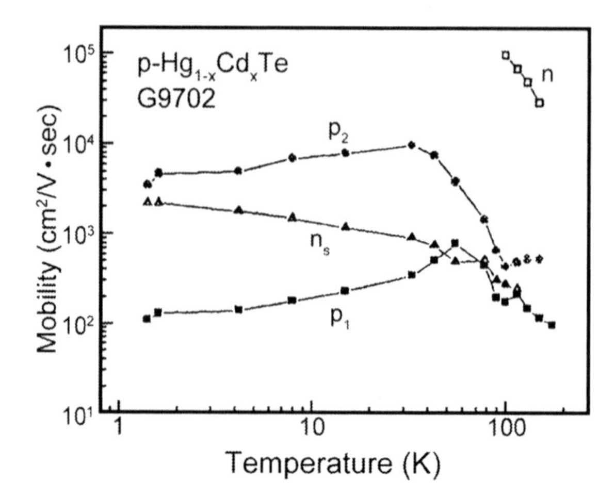

Fig. 5.34. In sample G9702, the variation of the carrier mobilities of the different species with temperature are collected. The behavior of the different species are identified: the *solid squares* are the majority hole carriers, the *solid circles* are the holes with a high mobility, the *solid triangles* are the surface electrons, and the *open squares* are the bulk electrons

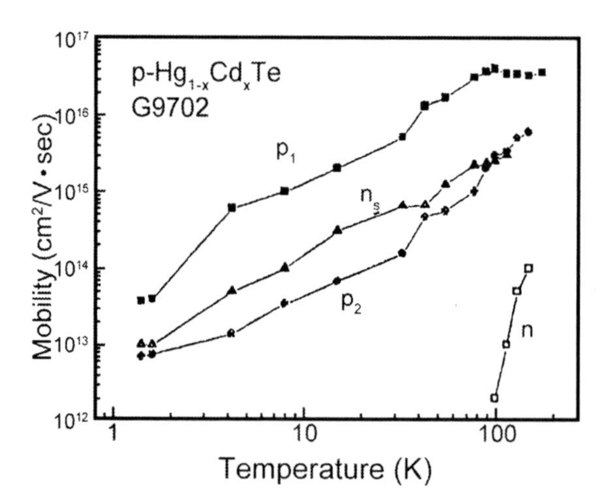

Fig. 5.35. In sample G9702, the variation of the different species carriers concentrations with temperature (The symbols in this figure are the same as those defined Fig. 5.32)

In materials, the different carrier species' mobilities and concentrations change with temperature. These changes are shown in Figs. 5.34 and 5.35 for sample G9702.

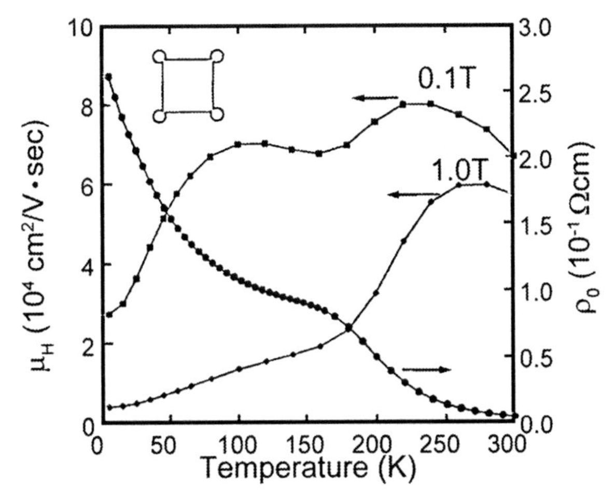

Fig. 5.36. The resistivity (left axis label), and the Hall mobility (right axis label) verses temperature measured at two magnetic fields: 0.1 and 1 T

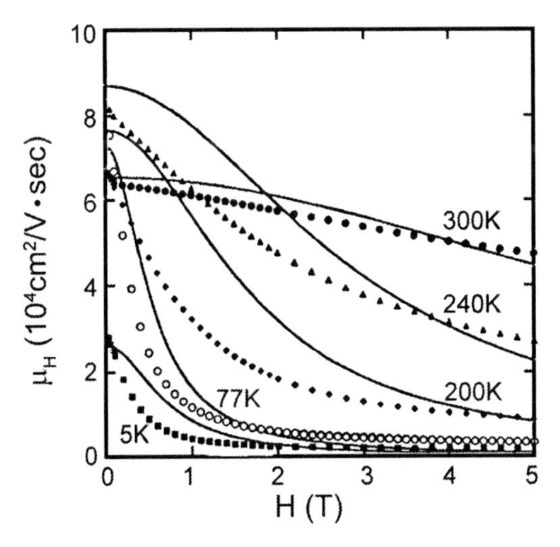

Fig. 5.37. At difference temperatures, the relationship between the Hall mobility and the magnetic field (The *solid curves* are calculated from a normal magneto-resistance theory based on a two-layer model, see Song et al. 1993)

Song et al. (1993) researched the transport characteristics of an n-type InSb film grown by MOCVD. The 3 μm thick InSb film was grow on a GaAs(100) substrate. Using the four-point probe Van der Pauw method, he measured the dependence of the Hall mobility on the temperature and the magnetic field. Figure 5.36 is the relation between the Hall mobility and the temperature, at two different fields, $B = 0.1$ and 10 T.

At 0.1 T, the mobility has two peaks, one is at T = 100 K where $\mu_H = 6.7\times10^4$ cm²/V s, and the other is at T = 240 K where μ_H = 5.1×10^4 cm²/V s. The location of these two peaks is near the bottom of valleys of the resistivity curve. As shown as the figure, the measured mobility depends sensitively on the magnetic field. At different temperatures, the dependence of the Hall mobility on the magnetic field is shown in Fig. 5.37.

5.5 Quantum Effects

5.5.1 Magneto-Resistance Oscillation

As mentioned above, one of the approximate conditions under which the conductivity expressions, (2.10) and (2.11), are correct is that the magnetic field is relatively small so the quantum effects do not exist. However, quantum effects will occur even at low magnetic fields in HgCdTe materials due to the small effective masses of their electrons. This results in the magneto-resistance oscillations. A magneto-resistance oscillation is produced as successive Landau sub-band energy levels cross the Fermi energy level as the magnetic field increases. The conditions for which oscillations generally occur are:

1. Low temperatures $\hbar\omega_c \gg k_B T$,
2. High magnetic fields $\omega_c\tau \gg 1$, and
3. Highly degenerate carrier distributions, $\varepsilon_F > \hbar\omega_c$ (where $\varepsilon_F = E_F - E_c$, $\omega_c = eB/m^*$, and $\mu = \dfrac{e\tau}{m^*}$).

In the absence of a magnetic field the energy level distribution in the conduction band is continuous. It splits into a series of Landau energy bands when a magnetic field is applied. As a result, the density of energy states varies periodically and reaches maxima at energies for which: $\varepsilon = \left(N + \dfrac{1}{2}\right)\hbar\omega_c$, where N = 0, 1, 2,... are integers that labels the Landau levels. This model assumes the conduction band is highly degenerate, so initially the Fermi energy $\varepsilon_F > \hbar\omega_c$, and that ε_F is independent of the magnetic field, a prediction by Wamplar and Springford (1972) for $N \geq 3$. When ε_F falls at the position of a density of states maximum, the electron motion suffers its most intense scattering, and the magneto-resistance reaches a maximum. Therefore, the magneto-resistance of a sample changes periodically as the magnetic field increases, called SdH effect.

When a magnetic field is applied to a semiconductor sample, the electron motion perpendicular to the applied magnetic field is quantized, while that parallel to the applied magnetic field is unaffected. If the conduction band is parabolic, the energy of an electron moving in the z-direction is:

$$\varepsilon = \left(N + \frac{1}{2}\right)\hbar\omega_c + \frac{\hbar k_z^2}{2m^*}. \qquad (5.209)$$

One should note that the spin of the electron is not considered in (5.209).

If the spin of the electron is considered, then in a uniform magnetic field in the *z*-direction, and for a non-parabolic conduction band, the energies of the electrons can be written (Nimtz et al. 1983):

$$\varepsilon_N = -\frac{\varepsilon_g}{2} + \left\{\left(\frac{\varepsilon_g}{2}\right)^2 + \left[\frac{\hbar^2 k_z^2}{2m_0^*} + \left(N + \frac{1}{2}\right)\hbar\omega_c \pm \frac{1}{2} g^* \mu_B B\right]\varepsilon_g\right\}^{1/2}, \qquad (5.210)$$

where ε_g is the energy gap, $\omega_c = eB/m_0^*$ is the cyclotron angular frequency, μ_B the Bohr magneton $\mu_B = e\hbar/2m_0$, g^* the effective Lande factor, and $N = 0, 1, 2, 3\ldots$. The term $\hbar^2 k_z^2/2m^*$ can be ignored in the case of a high magnetic field. Hence, (5.210) can be written:

$$\varepsilon_N \left(1 + \frac{\varepsilon_N}{\varepsilon_g}\right) = \left(N + \frac{1}{2} \pm \frac{1}{2}\delta\right)\hbar\omega_c, \qquad (5.211)$$

where

$$\delta = g^* \mu_B B/\hbar\omega_c = \frac{g^*}{2}\frac{m^*}{m_0}, \qquad (5.212)$$

is the ratio of the spin-splitting to the Landau energy band splitting, and m_0 is the free electron mass.

Equations (5.210) and (5.211) indicate that conduction band is split by a magnetic field into many sub-bands, the Landau sub-bands. Then including spin, and if the Fermi level does not change with magnetic field, the energy $\left(N \pm \frac{1}{2}\delta\right)\hbar\omega_c$ equaling ε_F now sets the conditions for the periodic maxima of the resistivity.

The theories of SdH effect were studied widely toward the end of the 1950s (Roth and Argyres 1966). Since there are both transverse and longitudinal SdH oscillations, and their scattering mechanisms differ, the theoretical formulas deduced by various authors are also different. However, the periodicity and the magnitude of the SdH oscillations as a function of the temperature are the same for the different formulas. Even though the initial theoretical results were generalized (Roth and Argyres 1966) the new formulations were still only correct for the parabolic energy band model. For n-InSb, the non-parabolic shape of the conduction band must be considered (Лифшиц and Косевиц 1955). As long as the non-parabolic energy band maintains a spherical energy surface (InSb satisfies this condition), the theoretical formula for a parabolic energy band is also suitable for a non-parabolic one if F/B is substituted for $\varepsilon_f / \hbar\omega_c$, where F is the SdH field-period separating oscillation peaks.

For an isotropic non-parabolic energy band, degenerate electrons, elastic scattering, and collision-broadening, the oscillating parts of the longitudinal and the transverse magneto-resistance are respectively:

$$\rho_{//} \approx \rho_0 \left\{ 1 + \sum_{r=1}^{\infty} b_r \cos\left(\frac{2\pi F}{B} r - \frac{\pi}{4} \right) \right\}, \tag{5.213}$$

$$\rho_{\perp} \approx \rho_0 \left\{ 1 + \frac{5}{2} \sum_{r=1}^{\infty} b_r \cos\left(\frac{2\pi F}{B} r - \frac{\pi}{4} \right) + R \right\}, \tag{5.214}$$

or

$$\frac{\Delta\rho_{//}}{\rho_0} \approx \sum_{r=1}^{\infty} b_r \cos\left(\frac{2\pi F}{B} r - \frac{\pi}{4} \right), \tag{5.215}$$

$$\frac{\Delta\rho_{\perp}}{\rho_0} \approx \frac{5}{2} \sum_{r=1}^{\infty} b_r \cos\left(\frac{2\pi F}{B} r - \frac{\pi}{4} \right) + R, \tag{5.216}$$

where b_r denotes the amplitude of the oscillations and has the form:

$$b_r = \frac{(-1)^r}{\sqrt{r}} \left(\frac{B}{2F} \right)^{1/2} \cos\left(\pi r \frac{g^*}{2} \frac{m^*}{m_0} \right) \frac{r\beta T m^* / m_0 B}{\sinh(r\beta T m^* / m_0 B)} e^{-r\beta T_D m^* / m_0 B}, \tag{5.217}$$

$\rho_{||}$ and $\rho_{\perp}$ are the longitudinal and transverse resistivity respectively, ρ_0 is the resistivity at zero magnetic field, $\Delta\rho = \rho(B) - \rho_0$, T is the temperature, $m' = m^* / m_0$, m_0 is the mass of the free electron, and

$$\beta = \frac{2\pi^2 k_B m_0}{\hbar e} = 14.70\,\mathrm{T/K}. \tag{5.218}$$

In practice, because of the frequency of the collisions, and the index factor in (5.217) causing the magnitude of higher r terms to decay rapidly, the series of terms for $r \geq 2$ can be ignored. As a result, only the terms with $r = 1$ need be retained in (5.215) and (5.216). R is a complex factor, but R can be ignored here. Therefore, the longitudinal and transverse magneto-resistances are both cosine functions of $1/B$ and both of them have the same phase. T_D is the Dingle temperature (Ye 1987), which is a measure of collision-broadening. Both of the temperature factors T and T_D lead to oscillation amplitude decay with $1/B$, however, the influence of T_D can be separated out from the formula. Because the slopes of the resistivity curves are zero at the peaks of the oscillation, we have (retaining only the $r = 1$ term):

$$\ln\left[\frac{\Delta\rho}{\rho_0}\frac{\sinh\chi}{\chi}(\varepsilon_F/\hbar\omega_c)^{1/2}\right] = -2\pi^2 k_B T_D/\hbar\omega_c + \ln(\cos(\pi\nu)/\sqrt{2}), \tag{5.219}$$

where $\nu = m'g^*/2$ is a parameter, and χ is:

$$\chi = 2\pi^2 k_B T/\hbar\omega_c. \tag{5.220}$$

T_D can be determined by finding the slope of $\ln\left[\frac{\Delta\rho}{\rho_0}\frac{\sinh\chi}{\chi}(\varepsilon_F/\hbar\omega_c)^{1/2}\right]$ as a function of $1/B$. The intercept at the vertical axis approximates a straight line from which $\cos(\pi\nu)/\sqrt{2}$ can be determined, and hence also ν and g^*. The carrier concentration n can be obtained in terms of the field oscillation effective frequency $P = \frac{1}{F}$:

- For carriers in three dimensions:

$$n = (1/3\pi^2)(2e/\hbar)^{3/2} P^{3/2} = 5.66\times 10^{15} P^{3/2}(\mathrm{cm}^{-3}), \tag{5.221}$$

- And for carriers constrained to two dimensions:

$$n = eP/\pi\hbar = 4.82\times 10^{10} P(\mathrm{cm}^{-2}). \tag{5.222}$$

The oscillating terms of the longitudinal and the transverse magneto-resistance are almost the same if the R term is ignored. Note that the oscillation amplitude of the transverse term is 2.5 times greater than that of the longitudinal, but this predicted ratio has not been proven by experiment. The oscillation curve peaks of ρ as a function of B occur at:

$$\varepsilon_f = \left(N + \frac{1}{2}\right)\hbar e B_N / m^*;\; N = 0,1,2\cdots. \tag{5.223}$$

Assuming that ε_f does not vary with the magnetic field, the differences between the reciprocals of the magnetic field corresponding to two adjacent peaks is a constant F^{-1}. The reciprocal of the oscillation field-frequency F is from (5.223):

$$F^{-1} = \Delta\left(\frac{1}{B}\right) = \frac{1}{B_N} - \frac{1}{B_{N+1}} = \frac{\hbar e}{m^* \varepsilon_f}. \tag{5.224}$$

Thus we have:

$$F = \left[\Delta\left(\frac{1}{B}\right)\right]^{-1} = \frac{m^*}{\hbar e}\varepsilon_f. \tag{5.225}$$

For a degenerate, non-parabolic energy band, ε_f can be expressed as a function of the electron concentration n, and the electron effective mass m_f^* at the at Fermi level. The relations are complicated because m_f^* is also a function of n. In the case where the energy gap ε_g is far smaller than the spin–orbit splitting (InSb satisfies this condition), the Kane model simplifies. As a result, Zawadzki and Szmanska (1971) obtained:

$$\varepsilon_f = \frac{1}{2}\varepsilon_g(\sqrt{\Delta} - 1), \tag{5.226}$$

where

$$\Delta = 1 + 2\pi^2\left(\frac{3}{\pi}\right)^{2/3}\left(\frac{\hbar^2}{m_0^* \varepsilon_g}\right) n^{2/3}, \tag{5.227}$$

and

$$m^*(\varepsilon_f) = m_0^* \sqrt{\Delta}. \tag{5.228}$$

Here m_0^* is the electron effective mass at the conduction band edge. For an n-Insb sample with the electron concentration being $10^{15}\,\text{cm}^{-3}$, the second term of Δ is far smaller than unity. From a binomial expansion of

$\sqrt{\Delta}$, one can get $m^*(\varepsilon_f)$ as a function of ε_f. Substituting $m^*(\varepsilon_f)$ into (5.225), we obtain:

$$F = \frac{\hbar}{2e}(3\pi^2 n)^{2/3}, \tag{5.229}$$

and in this approximation F is only a function of n. Therefore, one can deduce the electron concentration n from the measured F.

It was mentioned above that field-frequency and amplitude as functions of the temperature are the same for both the transverse and the longitudinal magneto-resistances. Strictly speaking, this is just an approximate result. In fact, the physical mechanism responsible for the transverse magneto-resistance is more complex than that for the longitudinal one. The longitudinal magneto-resistance mainly derives from elastic ionized impurity induced electron scattering that only couples states within Landau sub-bands. The transverse magneto-resistance, derives from both collision that couple electrons within sub-bands, but also from collisions causing the electrons to transition between sub-bands, which is the reason for the appearance of R in (5.216). According to the theory (Roth and Argyres 1966), there is a phase difference of $\pi/4$ between R and the other term. R will influence the waveform of the magneto-resistance oscillation if it cannot be ignored. Using the R term to determine physical parameters often results in errors. But this kind of error can be avoided if the physical parameters are deduced from the longitudinal magneto-resistance oscillation curve.

In the case where spin splitting occurs, there are two kinds of magneto-resistance peaks, namely:

$$\begin{aligned} E_f &= \left(N + \frac{1}{2}\delta\right)\hbar\omega_c, \text{ and} \\ E_f &= \left(N - \frac{1}{2}\delta\right)\hbar\omega_c, \end{aligned} \tag{5.230}$$

where N = 0,1,2,..., and B_N^+, and B_N^- denote the magnetic fields at the peaks in the two different cases. The field-frequency of the magneto-resistance oscillation is the same for the different electron spin orientations and is given by:

$$\frac{1}{F} = \frac{1}{B_N^+} - \frac{1}{B_{N-1}^+} = \frac{1}{B_N^-} - \frac{1}{B_{N-1}^-} \equiv \Delta\left(\frac{1}{B}\right). \tag{5.231}$$

The theory proved (Fu et al. 1983) that the relations between the oscillation field-frequency F and the electron concentration n and the Fermi level is the same as given in (5.229):

$$F = \frac{\hbar}{2e}(3\pi^2 n)^{2/3} = \frac{m^*}{\hbar e}\varepsilon_f . \tag{5.232}$$

The electron concentration n, as well as the parameter $m^* \varepsilon_f$, can be obtained from a measurement of F. In addition, the following result can be obtained experimentally:

$$\delta_{\exp} = \frac{\left(\frac{1}{B_N^+}\right) - \left(\frac{1}{B_N^-}\right)}{\Delta\left(\frac{1}{B}\right)} . \tag{5.233}$$

If m^* is known, g^* can be obtained from (5.212).

When the spin splitting and non-parabolic energy band (Adams and Holstein 1959; Roth and Argyres 1966), are included, the magneto-resistance as a function of magnetic field becomes:

$$\frac{\Delta\rho}{\rho_0} = \sum_{m'=0}^{\infty} \frac{5}{2}\left(\frac{RP}{2B}\right)^{1/2} \frac{\beta T m' \cos(R\pi\nu)}{\sinh(R\beta T m'/B)} e^{-R\beta T_D m'/B} \cos 2\pi(R/PB - 1/8 - R\gamma) \tag{5.234}$$

In n-InSb (Stephens et al. 1975), in n-AsGa (Stephens et al. 1978) and in HgTe (Justice et al. 1988) a great deal of experimental data was analyzed using (5.234) and the results are quite consistent with those from other experimental methods. It indicates that starting from SdH measurements, (5.234) is a useful way to extract two-dimensional and three dimensional carrier concentrations, effective masses, g^* factors, and Dingle temperatures.

For n-type narrow energy gap semiconductors, their Fermi levels often satisfy the condition $\varepsilon_F > kT$ at low temperature. Many quantum energy levels will be occupied if $\hbar\omega_c < \varepsilon_F$, on the other hand, only the lowest quantum energy level will be occupied if $\hbar\omega_c > \varepsilon_F$. This condition is so called "quantum limit." In the case of the non-quantum limit, the N*th* energy level will cross the Fermi level, and if the magnetic field is

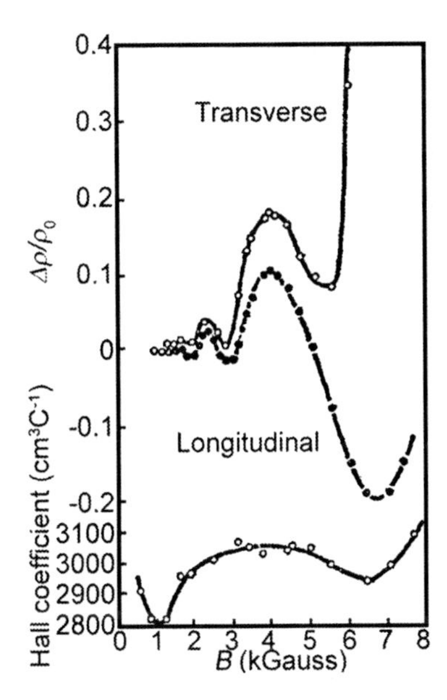

Fig. 5.38. The magneto-resistance oscillations and Hall coefficient of an n-type InSb sample at 1.7 K

relatively weak, it leads to the condition: $\varepsilon_F \approx \left(N+\frac{1}{2}\right)\hbar\omega_c$. As the magnetic field increases, the *N*th energy level will cross over the Fermi level which is pinned to the metal contacts. Therefore, this energy level will empty. The electrons will transfer to occupy the lower energy states consisting of:

- The number lower Landau levels that have been raised higher in the conduction band where the density of states is higher so they have some empty states,
- Transfer into the metal contacts to modify the space charge distributions at the interfaces, and
- Recombine with donors (carrier freeze-out). When only the lowest quantum energy level is occupied, that is "quantum limit."

As a result, the conductance will vary greatly as successive energy levels cross the Fermi level, resulting in the magneto-resistance oscillations.

Frederikse et al. (1957) observed the magneto-resistance oscillation for InSb and measured the Hall coefficient as a function of the magnetic field at 1.7 K, shown in Fig. 5.38. The oscillations of the transverse and longitudinal magneto-resistances with an increase of magnetic field can be seen in Fig. 5.38. The periods of the oscillations are $1/B$.

5.5.2 The Longitudinal Magneto-Resistance Oscillations of n-InSb

The SdH oscillation of the magneto-resistance in high magnetic fields is an effective tool used to study the energy bands of semiconductors. Especially for narrow energy gap semiconductors, the condition for a high magnetic field is easy to achieve in an ordinary laboratory since the electron effective mass is small. So, many experimentalists have endeavored to study the effect for several narrow energy gap semiconductors, InSb single crystals (Frederikse et al. 1957; Broom 1958; Isaacon and Weger 1964; Атирханов et al. 1963, 1964; Атирханов and Баширов 1966; Anteliffe and Stradling 1966; Павлов et al. 1965; Бреелер et al. 1966; Stephens et al. 1975; Глуэман et al. 1979). The early work validated the existence of this effect in several materials (Frederikse et al. 1957; Broom 1958) and clarified the relationship between the effect and some experimental conditions. After the appearance of a fine quantum theory (Roth and Argyres 1966), the breath of the experiments expanded to include investigations of the strength of the magnetic field, the measurement temperature, and the electron concentration of samples. Most of these efforts were aimed at validation of the theories, and then to obtain important parameters such as the carrier concentration, the electron effective mass, and the spectrum splitting factor. In order to get reliable parameters, the influence on the peak positions of the Landau sub-band induced magneto-resistance oscillations from the electron spin-splitting (Атирханов et al. 1963, 1964; Атирханов and Баширов 1966; Павлов et al. 1965; Бреелер et al. 1966), the temperature broadening (Павлов et al. 1965) and the collision broadening (Глуэман et al. 1979), had to be studied carefully.

As mentioned above, there are two kinds of magneto-resistances. One is the transverse magneto-resistance (the electric field is perpendicular to the magnetic field), and the other is the longitudinal magneto-resistance (the electric field is parallel to the magnetic field). The phenomena driving the two kinds of magneto-resistance oscillations are similar. Research on the transverse magneto-resistance is more prevalent, though many authors have reported measured results for both the transverse and the longitudinal ones. The reason the transverse resistance has been studied more extensively is because its amplitude is several times larger than that of the longitudinal one. In fact, the same parameters are obtained from an analysis of the longitudinal as the transverse magneto-resistance.

The electron concentration is the order of $10^{15}\,\mathrm{cm}^{-3}$ for an n-Insb single crystal sample examined by Fu et al. (1983). For this kind of sample (5.232) predicts a small oscillation field-frequency. Therefore, the oscillation can be observed in lower magnetic fields. At the same time, condition (3) is also satisfied. If a sample with a high electron mobility is selected,

Table 5.10. The main measured parameters of four samples

Sample number	Orientation index	Electron concentration (10^{15} cm^{-3})	Hall mobility (cm^2/V s)	The Fermi level/meV at 4.2 K (calculated from the Kane model)
I	[112]	6.7	1.68×10^5	9.05
II	[112]	5.4	2.07×10^5	7.88
III	[110]	3.9	2.06×10^5	6.38
IV		5.5	1.55×10^5	10.55

condition (2) is satisfied. Condition (1) is also satisfied for the sample measured at 4.2 K.

Some 10 × 1 × 1 (mm^3) parallelepiped samples were cut from a single-crystal ingot, and processed by grinding, polishing, and welding on electrodes. Then the samples were measured at 77 K to get their Hall coefficients and resistivities. Samples were chosen for the continuing study depending on the results from the previous measurements. Table 5.10 lists the main parameters of four samples chosen for further measurements.

The electron concentrations shown in Table 5.10 are the mean values of results obtained from the measurements on two pairs of Hall electrodes spaced 3 mm from each other. The deviations between the mean values and those from each pair of Hall electrodes are all less than ±3% for sample numbers I, II, and III. The bulk of the fourth sample is somewhat larger, so the deviation reached ±6%. It has been proven experimentally that Hall coefficients do not change from 77 to 4.2 K, therefore, the electron concentrations shown in Table 5.10 are results taken at 4.2 K. The last column shown in Table 5.10 are the Fermi levels at 4.2 K calculated from the Kane model using $\varepsilon_g = 0.2335$ eV.

Measurements were performed at liquid helium temperature. The sample holder was designed so the orientation of the sample in the magnetic field could be determined from outside of the Dewar bottle. One can judge whether the sample is parallel or perpendicular to the magnetic field using the criterion that the Hall voltage of the sample is either zero or a maximum.

Either direct current or magnetic field modulation methods are generally performed to measure the magneto-resistance. The most straightforward method is to record the resistance variation as a function of the magnetic field. A weakness of this method is the modulation of resistance is not obvious when the carrier concentration of the sample is small. When the second magnetic field modulation method is used so $B = B_0 + (\Delta B)_0 \sin \omega t$

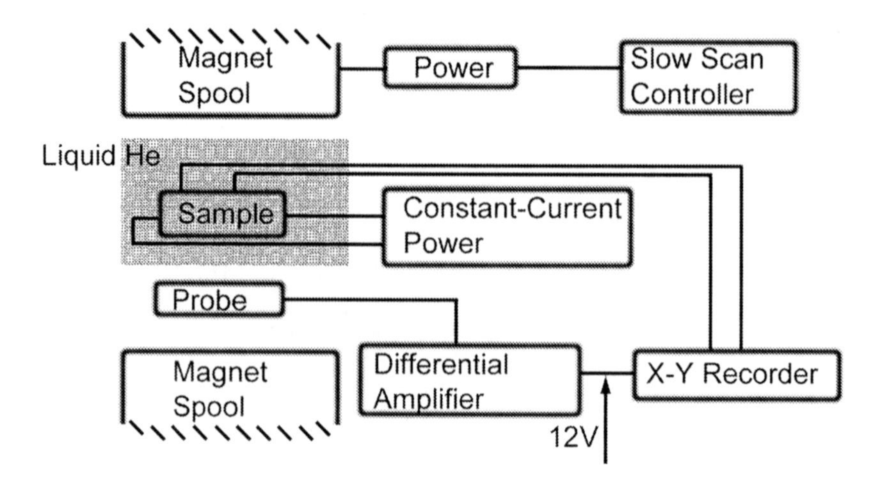

Fig. 5.39. A sketch of the measurement equipment

(α is the modulation frequency, B_0 is the dc component, and $(\nabla B)_0$ is the amplitude of the ac component of the magnetic field). Then the oscillation of the magneto-resistance stands out because, the highly sensitive, phase-sensitive demodulation technology can be used to get $d\rho/dB$.

To obtain the results reported next, a dc magnetic field was used accompanied by differential amplifier technique designed to improve the signal to noise ratio. The oscillating part of the magneto-resistance as a function of the magnetic field given by (5.213) and (5.214) is always superposed on a background curve that is also a function of magnetic field. The background curve of the magneto-resistance is much smaller in a longitudinal arrangement than that in transverse one. Moreover, the background magneto-resistance in a longitudinal configuration can be treated as a nearly zero constant independent of the magnetic field. Thus the oscillating waveform of the curve stands out against the background when a difference channel is employed to amplify $\rho_B - \rho_0$. In the transverse configuration the background is large and varies with magnetic field, limiting the applicability of the differential amplifier technique. As a result, it is easy to introduce errors.

Figure 5.39 is a sketch of the measurement equipment. The electromagnet generated magnetic field can be varied in the range from 0.02 to 0.73 T. The Hall probes of a CT5 Gauss meter are fixed on the poles of the magnetic. The Hall voltage obtained is input into the x axis of an x-y recorder after being amplified. The Gauss meter is calibrated using nuclear magnetic resonance technology.

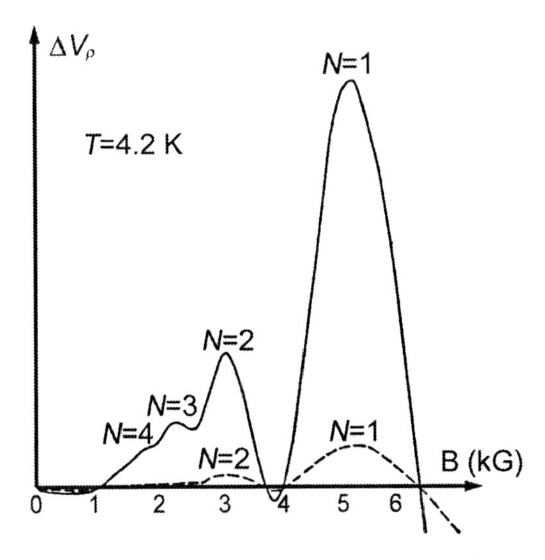

Fig. 5.40. For the same sample under the same conditions, the contrast between results obtained by the ordinary direct-current method and that plus differential amplifier method (The *doted curve* shows the result from the direct-current plus differential amplifier method, and the *dashed curve* is that from the ordinary direct-current method)

The current going through the sample was kept stable using a constant-current source. The output voltage across the resistance could be varied from $V_\rho = 1, 2, 5$ or 10 mV by adjusting the current at $B = 0$. In this way, it is easy to read out the magneto-resistance directly. The voltage that is proportional to the magneto-resistance is recorded on the y axis of the x-y recorder.

The sample is located at the inside of a single-layer red copper envelope. In order to reduce the temperature gradient in the sample, the envelope is entirely immersed in the cooling liquid so that good thermal contact is established between the sample and the cold liquid. Also, in order to reduce the magnetic effect of a secondary current, two different measurements are performed with the magnetic field reversed between them and the mean value is recorded. Figure 5.40 shows the contrast between results obtained under the same conditions by the direct-current and the direct-current plus difference amplifier methods. Two extra oscillation peaks can be identified when the direct-current plus differential amplifier method is employed. Owing to the superiority of the direct-current plus differential amplifier method, even the SdH effect in a sample with a low electron concentration can be studied in relatively low magnetic fields.

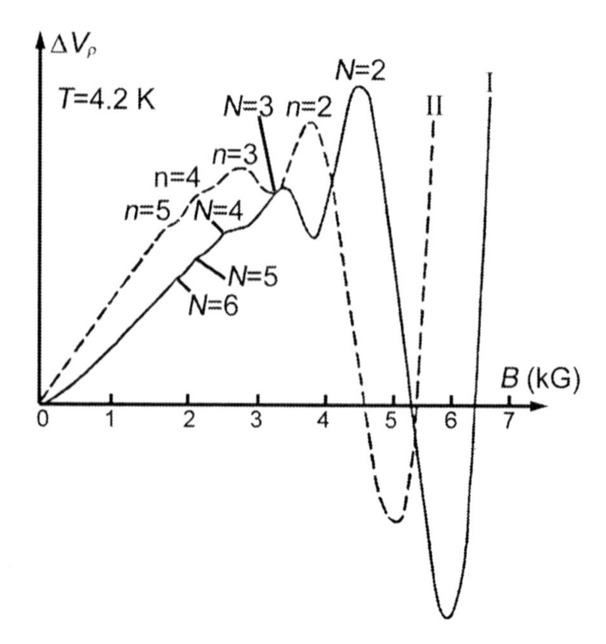

Fig. 5.41. The results of magneto-resistance measurements at liquid helium temperature using a differential amplifier; the *doted curve* is for sample I, and the *dashed curve* is for sample II

The carrier concentration can be deduced from the measured field-frequency. Figure 5.41 shows the oscillation curves of the longitudinal magneto-resistance of samples I and II. N is the Landau quantum number in Fig. 5.41. Table 5.11 collects the reciprocal of the magnetic fields corresponding to the oscillation peaks of four samples, as well as the oscillation field-frequency F. The electron concentrations calculated from the mean F values according to the expression in (5.225), and from the Hall coefficient are in the last two columns. Comparing the two columns, the largest difference is 4.2×10^{15} cm^{-3}. The results for samples I and II tend to agree with each other.

The effective mass of the participating electrons can be estimated from the oscillation amplitude as a function of the temperature. In the measurement procedure the applied electric field must be small enough so the sample temperature caused by Joule heating does not increase. This restriction places an extra burden on the required sensitivity of the measurement equipment. The oscillation amplitude will decrease as the temperature is raised as shown in Fig. 5.42, up to the point where at last, no signal can be extracted from the noise.

Table 5.11. The reciprocal of magnetic field $1/B_N$ corresponding to the oscillation peaks, and the oscillation field-frequency F

Sample	Landau level quantum number N	The reciprocal of magnetic field corresponding to the oscillation peak $1/B_N$ (T^{-1})	Frequency /T	The mean value of frequency F/Tesla	Carrier concentration/c	
					Calculated by F m^{-3}	Calculated by Hall coefficient
I	N=2	2.19				
	N=3	3.00				
	N=4	3.78	1.28	1.25	7.9×10^{15}	6.7×10^{15}
	N=5	4.59	1.23			
II	N=1	1.46				
	N=2	2.53				
	N=3	3.55				
	N=4	4.59	0.96			
	N=5	5.62	0.97	0.96	5.3×10^{15}	5.4×10^{15}
III	N=1	1.91				
	N=2	3.27				
	N=3	4.71	0.69			
	N=4	6.16	0.69	0.69	3.3×10^{15}	3.9×10^{15}
IV	N=2	1.63				
	N=3	2.33				
	N=4	3.05	1.39	1.43	9.7×10^{15}	5.5×10^{15}
	N=5	3.73	1.47			

The effective mass of the participating electrons can be calculated based on these curves from (5.225). Assuming that the electron scattering mechanism does not change in the temperature range of the experiments (4.2–10 K), the Dingle temperature T_D is unchanged, and the high-order terms with $r \geq 2$ can be ignored (these two assumptions will be discussed in Sect. 5.5.3), the oscillation amplitude predicted for a given peak as a function of the temperature from (5.227) is mainly determined by the following equation:

$$\frac{\beta T m'/B}{\sinh(\beta T m'/B)} = \frac{Tx}{\sinh(Tx)}, \tag{5.235}$$

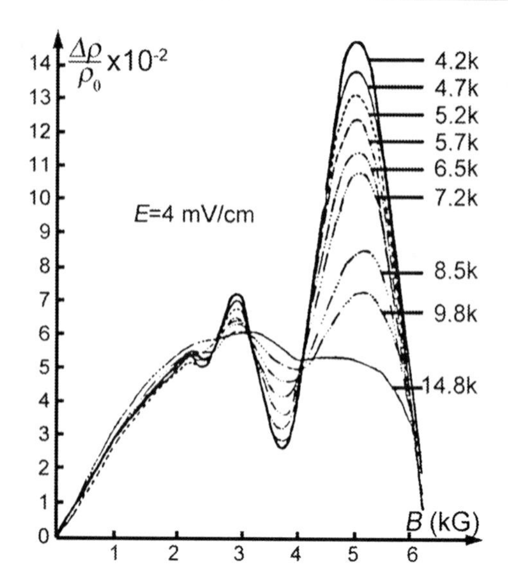

Fig. 5.42. The magneto-resistance as a function of the magnetic field obtained from the second pair of electrodes of sample III in a weak electrical field at different temperatures (The crystal lattice temperature is marked on every curve)

Table 5.12. The effective masses and Fermi levels of the participating electrons for four samples

Sample	I	II	III	IV
Effective mass m^*/m_0	0.015	0.014	0.013	0.017
Fermi level (meV)	9.05	7.88	6.38	10.55

where $x = \beta m'/B$, the relation between x and χ in (5.220) is $\chi = Tx$. For the oscillation amplitudes $A(T)$ at two different temperature T_1 and T_2, the ratio of $A(T_1)$ and $A(T_2)$ is:

$$\frac{A(T_1)}{A(T_2)} = \frac{T_1}{T_2}\frac{\sinh(T_2 x)}{\sinh(T_1 x)}. \qquad (5.236)$$

Therefore, x can be obtained from (5.236), and then m' also can be deduced. For example, for the two peaks of sample I, we have at B = 4.57 KG, A (4.2 K) = 17.5 and A (7.8 K) = 5.6; and at B = 3.33 KG, A (4.2 K) = 4.5, and A (7.8 K) = 0.76. From this information inserted into (5.244) $m^* = 0.015\ m_0$ is obtained. The results for the other three samples are collected in Table 5.12. The values shown in Table 5.12 are the effective masses at the varying Fermi level, which are not constant in these experiments. Removing Δ from (5.234) and (5.236), we have:

$$m^*(\varepsilon_f) = m_0^*\left(1 + \frac{2\varepsilon_f}{\varepsilon_g}\right). \tag{5.237}$$

Here we choose ε_g = 0.235 eV to get m_0^* using the data from Table 5.12. The mean of the four samples is, $m^* = 0.0137m_0$, which is quite consistent with the recognized value, $m^* = 0.0136m_0$. However, the deviation of the values among the different sample reaches ±14%, which is relatively large.

The Dingle temperature T_D also can be obtained from the experimental results. Still assuming the terms with $r \geq 2$ can be ignored in (5.217), the amplitude of oscillation curve can be written:

$$A(T,B) = \text{constant}\cdot\sqrt{B}\frac{\beta Tm'/B}{\sinh(r\beta Tm'/B)}e^{-\beta T_D m'/B}. \tag{5.238}$$

m' and T have been obtained in prior arguments. T_D can be obtained from the ratio of amplitudes of two adjacent peaks at the same temperature. The results are T_D = 5.4, 5.1 and 5.7 K respectively for samples numbers II, III and IV, with errors varying between ±0.5 K. For sample III, the results at different temperatures are: T_D= 5.09 K at 4.2 K, T_D=5.11 K at 7.2 K.

These results, when inserted back into (5.215)–(5.217) validate the two assumptions used to calculate the effective mass of the participating electrons in the previous section. Namely, (1) the electron scattering mechanism is unchanged in the range of the experimental temperatures, and (2) the higher order terms, $r \geq 2$, can be ignored. This is because the ratio of the two oscillation amplitude contributions from the $r = 2$ term and $r = 1$ term is: $\sqrt{2}\frac{\sinh(\beta Tm'/B)}{\sinh(2\beta Tm'/B)}e^{-\beta T_D m'/B} \approx 0.03$, for β= 0.5 T, T = 4.2 K, and T_D = 5.1 K. As a result, the neglect of higher-order terms will not significantly deform the predicted waveform.

It is found experimentally that the oscillation amplitude is related to the applied electrical field. The oscillating curves at 4.2 K have been observed when different electrical fields are applied. The temperature of the sample was monitored in order to insure that the temperature remained at 4.2 K. The results are shown in Fig. 5.43. The oscillation amplitude of curves remains nearly unchanged as long as the applied electrical field is small, E < 0.0052 V/cm. Beyond that field, the oscillation amplitude decreases as the applied electrical field increases, and in the end it cannot be observed.

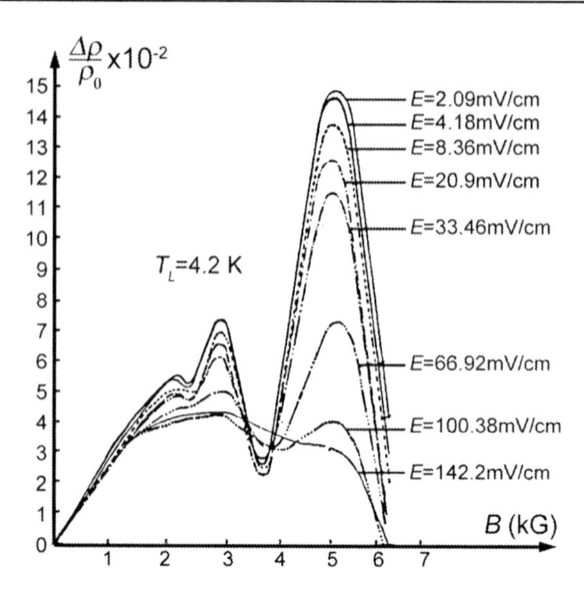

Fig. 5.43. The longitudinal magneto-resistance as a function of magnetic field obtained by measuring the voltage across the second pair of electrodes of sample III. The measured results are obtained using different electrical field while holding the lattice temperature constant. The values of electrical fields employed are marked on every curve

The effect of increasing the applied electrical field is similar to that of increasing the temperature of sample. Since the lattice temperature T_L of the sample is keep at 4.2 K, the effect of the electrical field is to increase the temperature of electron system. Hot electrons at a temperature $T_e > T_L$ arise because their principal scattering mechanism is elastic, so transfer of the electron energy gained from the electric field to the crystal lattice is slow. The hot electron temperature as a function of the applied electrical field can be obtained by comparing the oscillation amplitudes of the peak at $B = 5.23$ kG in Fig. 5.43, with that in Fig. 5.42. The hot electron temperature curve verses electric field deduced from this comparison is shown in Fig. 5.44. This analysis assumes that the non-equilibrium electron distribution can be characterized by a Fermi distribution with temperature T_e. This assumption is likely to be valid because the rapid elastic scattering is just what is required to drive a non-equilibrium distribution into one with a temperature (Sher and Primakof 1963). The method generally used to study hot electron distributions is to measure the mobility of carriers in a pulsed electrical field, but this method fails for degenerate semiconductors. The SdH effect provides an alternative that can be used to study hot carrier distributions of degenerate semiconductors.

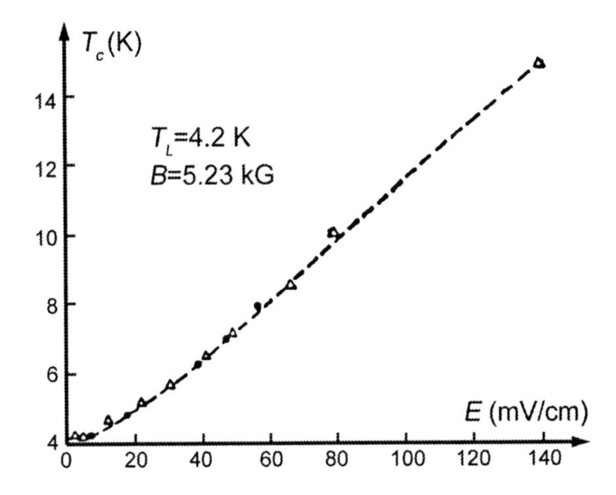

Fig. 5.44. The hot electron temperature curve as a function of the applied electrical field for sample III at B = 5.23 kG and T = 4.2 K. The *asterisks* denote the results obtained from the first pair of electrodes, and the *filled-triangle* denotes those obtained from the second pair of electrodes

As can be seen from the experimental results discussed above, the electron concentration was obtained from the frequency of the magneto-resistance oscillation for this InSb sample. The electron effective mass, the Dingle temperature and the hot electron temperature were obtained from the amplitude of the oscillation peak as functions of the temperature and applied magnetic field. All the results are consistent with those obtained from other methods. The Dingle temperature result indicates that the principal scattering mechanism remains unchanged in the range from 4 to 8 K. So the same value of the Dingle temperature is used in this temperature range. The hot electron temperature reached 14 K at the highest applied electrical field used in these experiments.

5.5.3 The Magneto-Resistance Oscillations in n-$Hg_{1-x}Cd_xTe$

The magneto-resistance oscillation is an important experimental tool for studying degenerate semiconductors, semimetals, and two-dimensional electron gases. It is especially useful in studies of narrow-gap semiconductors in which the effective masses are small the mobilities are large and consequentially the magnetic field needed to reach a measurable magnetic quantum condition is small.

Many papers on the narrow-gap semiconductor InSb (Fu 1983 and references therein) and the semimetal HgTe (Nimtz et al. 1983) have been

Table 5.13. Parameters of the samples selected for further study

Sample number	X	$\varepsilon_{g\ 4.2K}$ (meV)	μ_{77K} (cm^2/V s)	n_{77K} (cm^{-3})
TA1	0.172	20.21	2.4×10^5	7.6×10^{15}
TA2	0.162	2.43	3.5×10^5	9.2×10^{15}
TA3	0.169	14.88	2.7×10^5	1.05×10^{15}

reported. The $Hg_{1-x}Cd_xTe$ alloy with a band gap that is adjustable from semiconductor to semimetal has been studied extensively. The early researches on the SdH effect include numerous papers (Nimtz et al. 1983; Averous et al. 1980; Гуревичет and ЭФрос 1962; Antcliff 1970; Suizu and Narita 1972). Next we present some results gleaned from studies of the SdH effect in $Hg_{1-x}Cd_xTe$.

SdH oscillations in $Hg_{1-x}Cd_xTe$ samples with the compositions x = 0.162–0.172 at the temperatures 2.3–24 K were measured by Zheng et al. (1987) using an electromagnet and a superconductor magnet. The x values were obtained by the density method and a scanning electron microscope. The samples were etched in a bromine–alcohol solution, and subsequently indium balls were used to solder the electrodes. The samples were screened at liquid nitrogen temperature, where qualification required that the electron concentrations were larger than 10^{15} cm^{-3}, the mobility was typically large, and the samples were uniform. The parameters of the selected samples are listed in Table 5.13.

The ε_g values in Table 5.13 were calculated from the following equation (Chu et al. 1983b):

$$E_g = -0.295 + 1.87x - 0.28x^2 + (6.0 - 14x + 3.0x^2)10^{-4}T + 0.35x^4 . \quad (5.239)$$

The DC difference method was used in the measurement, and the signals were recorded with an x-y recorder after being amplified by the data amplifier. Either the longitudinal resistance or the transverse resistance can be measured when using the electromagnet by adjusting the orientation of the sample support from outside the dewar. And when using the superconductor magnet, it is more convenient to remove the sample support fixture when measuring at low temperature and sequentially repositioning the sample.

The measured longitudinal magneto-resistance is used to calculate the energy parameters. Although the amplitude is smaller than the transverse one, the influence of R in (5.216) is avoided, the background magneto-resistance is far smaller, and consequently the resulting parameters are more accurate. A representative relationship between the magneto-resistance and the magnet field is shown in Fig. 5.45. The longitudinal

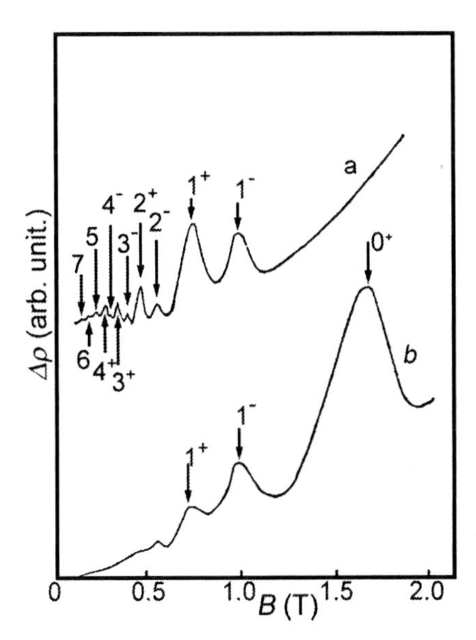

Fig. 5.45. The magneto-resistance oscillation peaks of $Hg_{1-x}Cd_xTe$ sample TA1 at T = 2.3 K, where a is for the longitudinal magneto-resistance oscillation peaks and b the transverse magneto-resistance oscillation peaks

Table 5.14. The electron concentrations calculated from (5.232)

Sample	Composition x	E_g (meV)	$\Delta(1/B)$ (T^{-1})	n (10^{15} cm^{-3})	$m^*(E_f)/m_0$ (10^{-3})	δ	g^*	E_f (meV)	m_0^*/m_0 (10^{-3})
TA1	0.172	20.21	0.96	6.02	7.05	0.44	124.8	20.0	2.37
TA2	0.162	2.43	0.86	7.10	7.60	0.45	115.4	21.5	0.40
TA3	0.169	14.88	0.81	5.00	7.90	0.44	111.4	22.5	1.96

magneto-resistance curve clearly shows the spin splittings of four Landau levels.

The relationship between the oscillation field-frequency F and the electron concentration n can be determined from the experimental results. Figure 5.46 displays the relationship between the Landau state quantum number and the reciprocal of the magnetic field corresponding to the maxima and minima of the oscillation curve, and from the slope of the curve the oscillation frequency F is obtained. The electron concentrations calculated from (5.232) are listed in Table 5.14. These electron concentrations are 20% smaller than those deduced from the Hall measurement data in Table 5.10. The variance is attributed to the difference between the temperatures at which the two data sets were taken. The SdH method is practically unaffected by electrode contact impedance induced errors, therefore it should yield more accurate results than the Hall measurement.

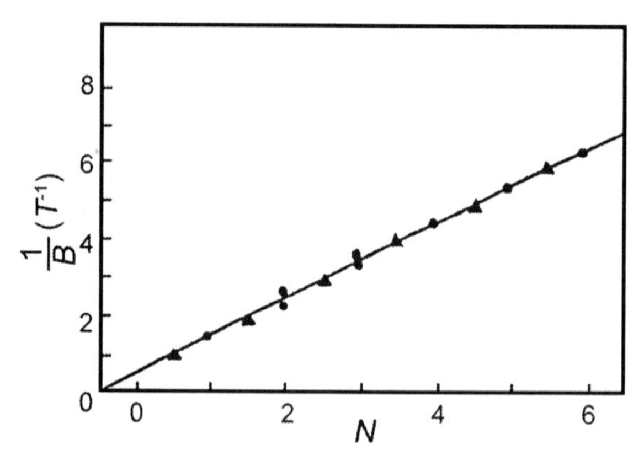

Fig. 5.46. The relationship between the quantum number N and the reciprocal of the magnetic field ($1/B$), $\Delta(1/B)$=0.96 T^{-1}, where the points correspond to the maxima of the peaks and the triangle the minima of the valleys

As we have seen the experimental results for the amplitude ratio for one magneto-resistance oscillation peak at two temperatures can be analyzed to yield the effective mass m^* at the Fermi level expressed in (5.235). Since there are two temperatures appeared in (5.235), therefore, the effective mass obtained is an average value of those at the two temperatures; so the real effective mass m^* at 4.2 K is a little smaller than that listed in Table 5.14. Also the Fermi level can be obtained if m^* is known from (5.226), and the results are listed in Table 5.14. By using the expression (5.240) (from Chap. 3):

$$\frac{m_0^*}{m_0} = 0.05966 \frac{E_g(E_g + 1)}{E_g + 0.667}, \tag{5.240}$$

the results of m_0^*/m_0 are 1.79, 0.22, 1.32, respectively for the samples of TA1,TA2,TA3.

The spin splitting of the Landau levels can also be deduced from the experimental results. The experimental value for the spin splitting parameter, δ_{exp}, can be obtained from (4.233), and for the sample TA1, $\delta_{\text{exp}(N=1)}$=0.48, $\delta_{\text{exp}(N=2)}$=0.45, $\delta_{\text{exp}(N=3)}$=0.43. There are systematic errors in these three values. Assuming the Fermi level $\varepsilon_f(0)$ at zero magnetic field, and is $\varepsilon_f(B)$ at a finite magnetic field, then using the relationship, given by Kichigin et al. (1981), between the two values for the parabolic energy band:

$$\varepsilon_f(B) = \varepsilon_f(0)\left[1 + \frac{1}{12}\left(\frac{\hbar\omega_c}{2\varepsilon_f(0)}\right)^2\right], \tag{5.241}$$

and replacing the aforementioned $\delta_{\exp}$(B) with $\delta_{\exp}(0)$, one obtains 0.46, 0.44, 0.43, respectively. The correction could be larger for a non-parabolic energy band evaluation at N=1. Therefore, the average value $\overline{\delta}_{\exp}(0)$ =0.44 is recommended.

A 0^+ peak was found in the transverse magneto-resistance curve (b) in Fig. 5.45, while the longitudinal magneto-resistance, curve (a), does not have this feature. This happens in all the measured samples. Similarly to InSb, this occurs because all the electrons are in two spin split energy bands corresponding to $N = 0$ when a large enough magnetic field is applied. For the 0^+ peak to appear, it requires that all the electron spins are in the same direction. The probability of spin reversal due to scattering is very small when $\boldsymbol{E} \parallel \boldsymbol{B}$, while in the case of $\boldsymbol{E} \perp \boldsymbol{B}$ the perturbation of the electric field is large enough to reverse the spin. Therefore, the 0^+ peak appears in the transverse magneto-resistance, while it does not in the longitudinal magneto-resistance.

The position of the 0^+ peak provides another method to obtain δ (Гуревичет and ЭФрос 1962). Taking the temperature into account, one finds:

$$B_0^+ = \frac{2\hbar c}{e}\left(\frac{\pi^2 n}{2}\right)^{2/3}\left[\delta^{1/2} + 0.536\left(\frac{k_B T}{\hbar\omega_c}\right)^{1/2}\right]^{-\frac{2}{3}}. \tag{5.242}$$

Using the real B_0^+, from this equation a value for δ of sample TA1 is obtained once the correction to the Fermi level is taken into account. The aforementioned average value, $\delta = 0.44$, is quite accurate.

The experimental results can also be used to extract the effective g factor g^*. The g^* value can be obtained from (5.220) knowing m^* and δ, and is listed in Table 5.14. This is the effective g factor of electrons at the Fermi level. A g_0^* value for electrons at the bottom of the conduction band can be obtained from the following equation:

$$g_0^* = \frac{m_0}{m_0^*}\frac{\Delta}{\Delta + \frac{3}{2}\varepsilon_g}, \tag{5.243}$$

where Δ is the valence band spin splitting energy. It is well known that Δ is about 0.8–1.0 eV for all the chemical compounds of the $Hg_{1-x}Cd_xTe$ series ($0 \le x \le 1$), and $\Delta >> 3.5\varepsilon_g$ for all the samples treated here. Therefore g_0^* is proportional to the reciprocal of m^*. g_0^* is about 500 for sample TA1 and TA3, while it is still larger for sample TA2.

5.6 Hot Electron Effects

5.6.1 Hot Electrons

If there is no temperature gradient in the crystal and only an electrical field along the z direction, i.e. $E_x = E_y = 0$, the Boltzmann equation in the momentum relaxation time approximation becomes:

$$\frac{\partial f}{\partial v_z}\frac{e}{m}E_z + \frac{f - f_0}{\tau_m} = 0. \tag{5.244}$$

The $\frac{\partial f}{\partial v_z}$ function can be expanded in terms of $\frac{\partial f_0}{\partial v_z}$, and only the linear term retained if the electric field is small. Then $\frac{\partial f}{\partial v_z}$ in the aforementioned equation can be replaced by $\frac{\partial f_0}{\partial v_z}$, and the last equation can be rewritten as:

$$f = f_0 - \frac{e}{m^*}\tau_m \frac{\partial f_0}{\partial v_z} E_z. \tag{5.245}$$

Including the higher order terms, the equation becomes:

$$f = f_0 - \left(\frac{e\tau_m}{m^*}E_z\right)\frac{\partial f_0}{\partial v_z}E_z + \left(\frac{e\tau_m}{m^*}E_z\right)^2 \frac{\partial^2 f_0}{\partial v_z^2} + \cdots. \tag{5.246}$$

The series converges when the drift velocity of the carriers is far less than their thermal motion velocities, i.e., E_z is not too large. If τ_m doesn't depend on the direction of v_z, the drift velocity is defined by:

$$v_d\big|_z = \frac{\int_{-\infty}^{\infty} v_z f d^3 v}{\int_{-\infty}^{\infty} f_0 d^3 v}, \tag{5.247}$$

where $d^3v = dv_x dv_y dv_z$. Inserting the expression for f into this expression causes it to become:

$$v_d\big|_z = \frac{\int_{-\infty}^{\infty} v_z \left(-\frac{e\tau_m}{m^*} E_z \frac{\partial f}{\partial v_z} \right) d^3v}{\int_{-\infty}^{\infty} f_0 d^3v}. \tag{5.248}$$

In this expression for $v_d\big|_z$ only the odd power terms of E_z are non-zero. Comparing this expression with $v_{dz} = \mu E_z$, one obtains:

$$v_{dz} = \mu E_z = \mu_0 \left(E_z + \beta E_z^3 + \cdots \right) \tag{5.249}$$

and

$$\mu = \mu_0 \left(1 + \beta E_z^2 + \cdots \right) \tag{5.250}$$

where β is a coefficient, μ_0 is the mobility at zero electrical field, and βE_z^3 is a deviation from Ohm's law. The carriers are called "hot carriers" when the mobility is expressed by $\mu_0 \left(1 + \beta E_z^2 \right)$, or by an expression with higher order terms. Then as will be seen below, if the electron distribution can be characterized by a temperature T_e, this temperature is higher than the lattice temperature. However at very high fields, electrons begin to transfer to higher bands, and the distribution becomes bimodal and cannot be assigned a temperature. Then more sophisticated methods must be employed (Chen and Sher 1995).

The coefficient β can be calculated from scattering theory or an approximate empirical assignment. Assuming that the electrons follows a Maxwell–Boltzmann distribution with a hot temperature, i.e., the distribution function is: $f \propto \exp\left(-\varepsilon / k_B T_e \right)$, where the electron temperature T_e is larger than the lattice temperature T_L. In steady state the energy gained by a carrier from the electric field per unit time (the power gain) exactly equals that lost by scattering. The power gain is the product of the electric force eE and the carrier drift velocity. Steady state is reached when:

$$\left(\frac{\partial \varepsilon}{\partial t} \right)_{\text{collision}} = \mu e E^2. \tag{5.251}$$

The power lost by inelastic scattering equals the difference between the average hot carrier energy $\frac{3}{2}k_BT_e$ and their average lattice temperature energy $\frac{3}{2}k_BT_L$, divided by the relaxation time τ_ε:

$$\left(\frac{\partial \varepsilon}{\partial t}\right)_{\text{energy collision}} = \frac{\frac{3}{2}k_B\left(T_e - T_L\right)}{\tau_\varepsilon}, \tag{5.252}$$

where the relaxation time τ_ε that enters this equation is the energy relaxation time which often differs somewhat from the momentum relaxation time $\tau_m \le \tau_\varepsilon$ (Chen and Sher 1995).

Then we have:

$$\mu e E^2 = \frac{\frac{3}{2}k_B\left(T_e - T_L\right)}{\tau_\varepsilon}, \tag{5.253}$$

which shows that the carrier mobility is a function of the electron temperature T_e, which is now denoted $\mu(T_e)$. The zero field (or the Ohm) mobility μ_0 is a function of the lattice temperature T_L, denoted by $\mu_o = \mu(T_L)$. Then expanding $\mu(T_e)$ in a Taylor series about T_L yields the expression:

$$\frac{\mu}{\mu_0} = 1 + \frac{T_e - T_L}{\mu_o}\frac{\partial \mu(T_e)}{\partial T_e}\bigg|_{T_L} + \cdots = 1 + \left(T_e - T_L\right)\frac{d\ln\mu(T_e)}{dT_e}\bigg|_{T_L} \tag{5.254}$$

Furthermore, from (5.250), one obtains:

$$\frac{\mu - \mu_0}{\mu_0} = \beta E^2. \tag{5.255}$$

Consequently,

$$\left(T_e - T_L\right)\frac{d\ln\mu(T_e)}{dT_e}\bigg|_{T_L} = \beta E^2. \tag{5.256}$$

Putting this expression into (5.253), one obtains:

$$\mu e E^2 = \frac{\frac{3}{2}k_B \left. \frac{d \ln \mu(T_e)}{dT_e} \right|_{T_L}^{-1} \beta E^2}{\tau_\varepsilon}. \tag{5.257}$$

Therefore the relaxation time is given approximately by:

$$\tau_\varepsilon \cong \frac{\frac{3}{2}\left(\frac{k_B T_L}{e}\right)\beta}{\left[\mu_0 \left(\frac{d \ln \mu(T_e)}{d \ln T_e}\right)_{T_e}\right]}, \tag{5.258}$$

where the approximation in this equation stems from replacing μ by μ_o.

Since we have shown that $\mu = \frac{e}{m}\langle \tau_m \rangle_w$, for $\tau_m \propto \varepsilon^r$, $\frac{d \ln \mu}{d \ln T_e} \sim \frac{1}{r}$, which is of order of magnitude 1. If μ_0 is known, β can be determined from calculations or obtained by measuring μ. Thus the energy relaxation time τ_ε can be obtained approximately from (5.258) once β is determined.

For a degenerate semiconductor, (5.253) is replaced by:

$$\mu e E^2 = \frac{\left\{\langle \varepsilon(T_e) \rangle - \langle \varepsilon(T_L) \rangle\right\}}{\tau_\varepsilon}, \tag{5.259}$$

where

$$\langle \varepsilon(T_e) \rangle = \frac{\frac{3}{2}k_B T_e F_{3/2}\left(\frac{\zeta_n}{k_B T_e}\right) N_c(T_e)}{n}, \tag{5.260}$$

and $F_{3/2}(x)$ is the Fermi–Dirac integral for j=3/2. The energy relaxation time τ_ε can be calculated by putting (5.260) into (5.259). Assuming the mobility is about 10^4 cm^2/V s and $|\beta|$ is about 10^{-4} cm^2/V^2, the order of magnitude of τ_ε will be 10^{-10} s.

5.6.2 Hot Electron Effects in HgCdTe

Assuming that the magnetic field B is along z direction the dispersion relationship for the electron energy can be written in the following form:

$$E = \frac{\hbar^2 k_z^2}{2m^* a_0} + (n+1)\hbar\omega_c + \frac{E_g a_0}{2} - \frac{E_g}{2}, \tag{5.261}$$

where E is the energy referenced to the bottom of the energy band of the lowest spin split Landau level, n is the Landau level index, k_z is the z component of the electron wave vector, m^* is the effective mass at the bottom of the energy band, and a_0 is:

$$a_0 = \left[1 + \frac{2\hbar\omega_C}{E_g}\left(1 - |g|\frac{m^*}{2m_0}\right)\right]^{\frac{1}{2}}, \tag{5.262}$$

and g is the spin splitting g factor, m_0 is the free electron mass, and $\omega_c = \frac{eB}{m^*}$. Actually in the experiments reported here the electrons occupy only the lowest Landau level so n =0 will be taken. Assume that the electric field is parallel to $\boldsymbol{B}$, and it drives a non-equilibrium hot electron gas into a Maxwell–Boltzmann distribution with a characteristic temperature T_e. The power loss per electron caused by emitting (or absorbing) acoustic phonons is expressed by:

$$\begin{aligned} P_{ac} = & \frac{(m^* a_0)^{1/2}\omega_0}{\pi(2\pi k_B T_e)^{1/2}\hbar}\exp\left(-\frac{m^* a_0 s^2}{2k_B T_e}\right)\int_0^\infty q_\perp dq_\perp \int_0^\infty \frac{dq_z}{q_z} C|f(q)|^2 \\ & \times\exp\left(-\frac{l^2 q_\perp^2}{2} - \frac{\hbar^2 q_z^2}{8m^* a_0 k_B T_e} - \frac{m^* a_0 s^2}{2k_B T_e}\cdot\frac{q_\perp^2}{q_z^2}\right)\left[(N_A+1)\exp(-\gamma_e) - N_A\exp(\gamma_e)\right] \end{aligned} \text{'}\tag{5.263}$$

where $\hbar\omega_0$ is the phonon energy, s is the velocity of sound for an acoustic phonon, $l = (\hbar/eB)^{1/2}$ is the Landau cyclotron radius, N_R is the average phonon occupation number, $C|f(q)|^2$ is the electron–phonon coupling term, $q_\perp$ and q_z are respectively the transverse and the longitudinal components of the phonon wave vector $\mathbf{q}$, and $\gamma_e = \hbar\omega_0/(2k_B T_e)$. Details about features of this expression can be found in the literature (Kahlert and Bauer 1973; Bauer et al. 1975; Pinchuk 1977). The energy of an acoustic phonon can be written as:

$$\hbar\omega_0 = \hbar s q = \hbar s q_\perp\left(1 + \frac{q_z^2}{q_\perp^2}\right)^{1/2} \approx \hbar s/l, \tag{5.264}$$

where, in the magnetic quantum limit case, the phonons contributing to the electron scattering has a characteristic length which corresponds to the Landau cyclotron radius, i.e. $q_{\perp} \sim \frac{1}{l} = \left(\frac{\hbar}{eB}\right)^{-\frac{1}{2}}$, and q_z denotes the z component of a hot phonon wave vector. The electron–phonon coupling term for the deformation potential acoustic phonon scattering is:

$$C|f(q)|^2 = C_{ac}\left[\frac{q^5}{\left(q^2+q_s^2\right)^2}\right], \tag{5.265}$$

where $C_{ac} = \frac{E_1^2\hbar}{2\rho s}$, E_1 is the acoustic deformation potential energy parameter, ρ is the mass density, and q_s the reciprocal of the screening length.

For the piezoelectric acoustic phonon coupling case:

$$C|f(q)|^2 = C_{pz}\left[\frac{q^3}{q^2+q_s^2}\right], \tag{5.266}$$

where $C_{pz} = \hbar^2 e^2 e_{14}^2 / \left(2\rho u_p \varepsilon_s^2\right)$, and e_{14} is piezoelectric modulus, u_p is the piezoelectric velocity, and ε_s is the material's permittivity.

Assuming the phonons are in thermal equilibrium and follow Bose–Einstein statistics, the energy loss transition rate for an electron due to deformation potential scattering can be written in the following forms (Basu et al. 1988):

$$P_{ac} = w_{ac}\int_0^{\infty} f_{ac}(v)\exp(-v)\frac{dv}{v}, \tag{5.267}$$

where

$$w_{ac} = \frac{\left(m^* a_0\right)^{\frac{1}{2}} E_1^2 \omega_0 N_0}{8\pi\left(k_B T_e\right)^{\frac{1}{2}} l^3 \rho s}\left[\exp\left(\frac{\hbar\omega_0}{k_B T_L}\right) - \exp\left(\frac{\hbar\omega_0}{k_B T_e}\right)\right]\exp\left(-\frac{\hbar\omega_0 + m^* a_0 s^2}{2k_B T_e}\right) \tag{5.268}$$

$$f_{ac}(v) = \left(1+\frac{\rho v}{u_0}\right)^{\frac{5}{2}}\left(1+\frac{\gamma}{v}\right)^{-\frac{3}{2}}\left(1+\frac{\beta v + \alpha_s}{u_0}\right), \tag{5.269}$$

and $\beta \equiv 8m^* a_0 l^2 k_B T_e / \hbar^2$, $\gamma \equiv \dfrac{\hbar^2 s^2}{8(k_B T_e l)^2}$, $u_0 \equiv (lq_\perp)^2$, $\alpha_s \equiv q_s^2 l^2$, and $v \equiv \hbar^2 q_z^2 / (8m^* a_0 k_B T_e)$.

The transition rate due to piezoelectric scattering is:

$$P_{pz} = w_{pz} \int_0^\infty f_{pz}(v) \exp(-v) \frac{dv}{v}, \tag{5.270}$$

where

$$w_{pz} = \frac{(m^* a_0)^{\frac{1}{2}} e^2 e_{12}^2 \omega_0 N_0}{8\pi\rho u_p \varepsilon_s^2 (k_B T_e)^{\frac{1}{2}} l} \times \left[\exp\left(\frac{\hbar\omega_0}{k_B T_L}\right) - \exp\left(\frac{\hbar\omega_0}{k_B T_e}\right) \right] \exp\left(-\frac{\hbar\omega_0 + m^* a_0 u_p^2}{2k_B T_e} \right), \tag{5.271}$$

and

$$f_{pz}(v) = \left(1 + \beta \frac{v}{u_0}\right)^{\frac{5}{2}} \left(1 + \frac{\gamma}{v}\right)^{-\frac{1}{2}} \left(1 + \frac{\beta v + \alpha_s}{u_0}\right)^{-2}. \tag{5.272}$$

In (5.271) and (5.272), $f_{ac}(v)$ and $f_{pz}(v)$ are both slowly varying functions of v, so to a good approximation they can be pulled out of the integral. Furthermore v can be set unity, $v = 1$, corresponding to $\hbar^2 q_z^2 / (2m^* a_0) = 4k_B T_e$, in the upper limit of the integral. The lower limit of the integral is $E_c / 4k_B T_e$, where E_c is the cut-off energy and varies proportional to $B^{2/3}$. Therefore, we find:

$$P_{ac} \cong w_{ac} f_{ac}(v = 1) \ln[(4k_B T_e / E_c) \exp(-C_e)], \tag{5.273}$$

and

$$P_{pz} = w_{pz} f_{pz}(v = 1) \ln[(4k_B T_e / E_c) \exp(-C_e)], \tag{5.274}$$

where f_{ac} and f_{pz} are values at $v = 1$, and C_e is the Euler constant.

Since $P_{ac} + P_{pz}$ is the net power loss of an electron caused by the acoustic phonon and piezoelectric scattering, and the procedure takes place in a duration of the energy relaxation time τ_e, one obtains:

$$P_{ac} + P_{pz} = \frac{1}{2} k_B \left(T_e - T_L \right) / \tau_e , \tag{5.275}$$

where the $\frac{1}{2}$ reflects a limit on the degrees of freedom due to the Landau quantization, and τ_e the energy relaxation time of the hot electron.

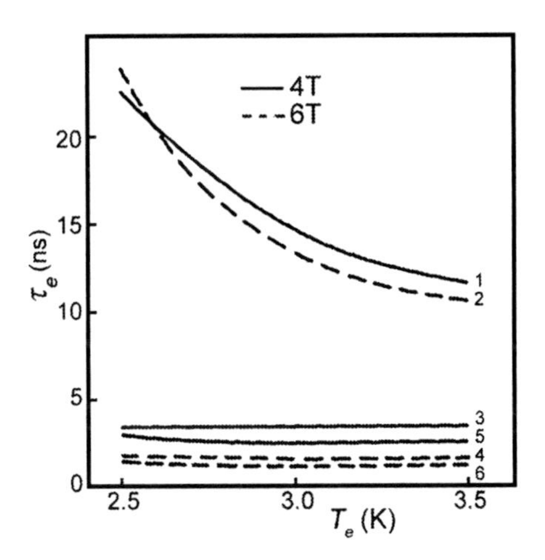

Fig. 5.47. The relationship between the energy relaxation time τ_e and T_e at T_L = 1.5 K (*Curves 1* and *2* are experimental results, and the others are the results of theoretical calculations)

The calculation for $Hg_{0.8}Cd_{0.2}Te$ ($n = 10^{14}$ cm^{-3}) was carried out by Basu et al. (1988), where the parameters used were: T = 1.5 K, B = 4 T, 6 T, m^* = 0.0065 m_0, $E_g = 70\,\text{meV}$, the density $\rho = 7.654 \times 10^3\,\text{kg/m}^3$, the Landau g factor $|g| = 90 (at\, B = 4T)$, $|g| = 80$ (at $B = 6T$), the acoustic wave velocity $s = 3.017 \times 10^3$ m/s, the acoustic deformation potential $E_1 = 9.0\,\text{eV}$, the piezoelectric modulus $e_{14} = 0.0335\,\text{c/m}^2$, the piezoelectric velocity $u_p = 1.948 \times 10^3$ m/s, the dielectric constant $k_0 = 18$, and the cut-off energy, which depends on the energy broadening due to thermal motion, was set to $E_c = 0.1\,\text{meV}$ at $B = 4T$. With the temperature at $T_e = 2.5 \sim 3.5\circ\text{K}$, and the Fermi energy located at $2k_B T_e$ below the band edge, the calculated τ_e is in the range of 3–5 ns, which is about one sixth to one eighth of the experimental result, shown in Fig. 5.47.

It was further suggested by Basu that the phonon distribution would heat as they interact with the hot electrons. Assuming that the occupation number of the hot phonon distribution is N_R, and their number in equilibrium with the external heat bath is N_0, then a rough approximation to the rate at which N_R changes is:

$$\left(\frac{\partial N_R}{\partial t}\right)_e = -\frac{N_R - N_0}{\tau_p}, \tag{5.276}$$

where τ_p is the phonon relaxation time. Consequently in this model the power transfer is:

$$\begin{aligned} P_{ac} &= -\frac{1}{nV}\sum_q \hbar\omega_0 \left(\frac{\partial N_R}{\partial t}\right)_e \\ &= \frac{\hbar\omega_0}{4\pi^2 n}\left(\frac{\partial N_R}{\partial t}\right)_e \int_0^{q_{\perp m}} q_\perp dq_\perp \int_o^{q_{zm}} dq_z \end{aligned}, \tag{5.277}$$

where n is the carrier concentration, V is the crystal volume, $q_{\perp m}^2 = l^{-2}$, and $\hbar^2 q_{zm}^2 / (2m^* a_0) = 4k_B T_e$. Replacing ω_0 with s/l for simplicity, one obtains:

$$\left(\frac{\partial N_R}{\partial t}\right)_e = \frac{4\sqrt{2}\pi^2 n l^2}{\omega_0 \left(m^* a_0 k_B T_e\right)^{\frac{1}{2}}} P_{ac} = a \cdot P_{ac} \tag{5.278}$$

Putting (5.263), (5.276) and (5.278) together, one obtains:

$$N_R = \frac{a\alpha \exp(-\gamma_e) + N_0/\tau_p}{a\alpha\left[\exp(\gamma_e) - \exp(-\gamma_e)\right] + \tau_p^{-1}}, \tag{5.279}$$

where

$$\alpha = w_{ac1} f_{ac} \ln\left[\left(4k_B T_e / E_c\right)\exp(-C_e)\right], \tag{5.280}$$

and

$$w_{ac1} = \frac{\left(m^* a_0\right)^{\frac{1}{2}} E_1^2 \omega_0}{8\pi \left(k_B T\right)^{\frac{1}{2}} l^3 \rho s} \exp\left(-\frac{m^* a_0 s^2}{2k_B T_e}\right). \tag{5.281}$$

Consequently, (5.278) can be re-written as:

$$a\alpha\left[\left(N_R+1\right)\exp\left(-\gamma_e\right)-N_R\exp\left(\gamma_e\right)\right]=-\frac{N_R-N_0}{\tau_p}=\left(\frac{\partial N_R}{\partial t}\right)_e. \tag{5.282}$$

The hot phonons tend toward their equilibrium state by their interaction with the electrons, therefore am approximation to (5.276) can be written as:

$$\left(\frac{\partial N_R}{\partial t}\right)_e=-\frac{1}{2}k_B\left(T_e-T_L\right)/\tau_e, \tag{5.283}$$

where T_L is the lattice temperature, T_e is the electron temperature, and τ_e the hot electron relaxation time. Consequently, we find:

$$a\alpha\left[\left(N_R+1\right)\exp\left(-\gamma_e\right)-N_R\exp\left(\gamma_e\right)\right]=\frac{0.5k_B\left(T_e-T_L\right)}{\tau_p}. \tag{5.284}$$

In the all aforementioned equations, τ_p must be known to calculate τ_e. The exact value of τ_p is unknown, because the phonon relaxation time mechanism is very complicated, but τ_p can be deduced by using it as a fitting parameter to the τ_e experimental results. Doing so leads to the

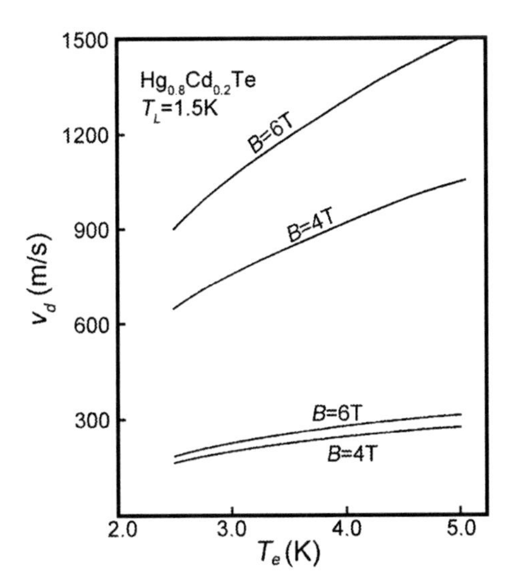

Fig. 5.48. The relationship between the drift mobility of hot electrons v_d, and the electron temperature T_e in an n-$Hg_{0.8}Cd_{0.2}Te$ sample at T_L = 1.5 K when B = 4 T and 6 T

values: $\tau_p = 62.4\,\text{ns}$ for $T_e = 2.5°\text{K}$, $\tau_p = 41.0\,\text{ns}$ for $T_e = 3.0°\text{K}$, $\tau_p = 62.4\,\text{ns}$ for $T_e = 4.5°\text{K}$ at B = 4 T; and $\tau_p = 111.5\,\text{ns}$ for $T_e = 2.5°\text{K}$, $\tau_p = 66.3\,\text{ns}$ for $T_e = 3.0\,\text{K}$, $\tau_p = 55\,\text{ns}$ for $T_e = 4.5\,\text{K}$ at B = 6 T. It shows that τ_p lies in the range of 40–120 ns.

If the momentum relaxation time of the carriers, τ_m, is known, one can also calculate the drift velocity of the hot electrons. The electric field can be calculated from the energy loss rate, P_{ac}, of each electron due to phonon scattering. From (5.251), which takes into account only the hot electron power loss induced by acoustic deformation potential scattering, one can obtain an expression for the electric field:

$$E = \left(\frac{eP_{ac}}{\mu}\right)^{1/2} . \tag{5.285}$$

Therefore an approximation to the drift velocity can be calculated from:

$$v_d = \mu E = \left(e\mu P_{ac}\right)^{1/2} . \tag{5.286}$$

Banerji et al. (1998) calculated the drift velocity of the hot electrons, v_d, at various electron temperatures, T_e, for an n-$Hg_{0.8}Cd_{0.2}Te$ sample with magnetic fields B = 4 T and B = 6 T. When the lattice temperature $T_L = 1.5\text{K}$ and the electron temperature is in the range 2.5–5 K, the calculated results are shown in Fig. 5.48. The two curves in the upper part of the figure are the results without taking into account the contribution from polar optical phonon scattering, while the lower two curves do. Banerji et al. (1998) also calculated the effect of polar optical phonon scattering on the strong field transport, and also studied diffusion in the quantum limit. Wang et al. (1998) added the effect on transport of hot electron induced impurity ionization.

References

Adams EN, Holstein TD (1959) J Phys Chem Solids 10:254
Antcliff GA (1970) Phys Rev B 2:345
Anteliffe GA, Stradling RA (1966) Phys Lett 20:119
Aronzon BA, Tsidilkovskii IM (1990) Phys Status Solidi B 157:17
Averous M, Calas J, et al. (1980) Solid State Commun 34:639
Bajaj J, Shin SH, Bostrup G, et al. (1982) J Vac Sci Technol 21:246
Banerji P, Chattopadyay C, Sarkar CK (1998) Phys Rev B 57:15435

Basu PP, Sarkar CK, Chattopadyay C (1988) J Appl Phys 64:4041
Bate RT, Baxter RD, Reid FJ, et al. (1965) J Phys Chem Solids 26:1205
Bauer G, Kahlert H, Kocevar P (1975) Resonant cooling of hot electrons in high magnetic fields. Phys Rev B 11:968–975
Beck WA, Anderson JR (1987) J Appl Phys 62:541
Blakemore JS (1965) Semiconductor Statistics. Dover Publications, New York
Broom BF (1958) Proc Phys Soc 71:470
Chattopadhyay D, Nag BR (1974) J Appl Phys 45:1463
Chen MC, Colombo L (1993) J Appl Phys 73:2916
Chen MC, Dodge JA (1986) Solid State Commun 59:449
Chen AB, Sher A (1995) Semiconductor Alloys. Plenum, New York
Chen MC, Tregilgas JH (1987) J Appl Phys 61:787
Chu JH, Xu SQ, Tang DY (1982) Chinese Sci Bull 27:403
Chu JH, Xu SC, Tang DY (1983a) Appl Phys Lett 43:1064
Chu JH, Wang RX, Tang DY (1983b) Infrared Res 2:241
Chu JH, Xu SQ, Ji HM, et al. (1985) Infrared Res A4:255
Chu JH, et al. (1993) Appl Phys Lett 43:1064
Dubowski JJ (1978) Phys Status Solidi B 85:663
Dziuba Z, Gorska M (1992) J Appl France III 2:110
Ehrenreich H (1959) J Phys Chem Solids 8:130
Elliott CT (1971) J Phys D: Appl Phys 41:2876
Finkman E, Nemirovsky Y (1979) J Appl Phys 50:4356
Frederikse HPR, Hosler WR (1957) Phys Rev 108:1136
Fu RL, Zheng GZ, Tang DY (1983) Longitudinal oscillatory magnetoresistance in n-type Indium Antimonide. Chinese J Semicond 4:230
Gold MC, Nelson DA (1986) J Vac Sci Technol A 4:2040
Gui YS (1998) PhD Thesis. Shanghai Institute of Technical Physics
Gui YS, Zheng GZ, Chu JH, Guo SL, Zhang XC, Tang DY, Cai Y (1997) Electrical characterization of subbands in the HgCdTe surface layer. J Appl Phys 82:5000–5004
Gui YS, Zheng GZ, Zhang XC, Chu JH (1998a) Chinese J Semicond 19:913
Gui YS, Zheng GZ, Guo SL, Chu JH (1998b) J Infrared Milli Waves 17:327
Hannay NB (1959) Semiconductors. Reinhold Publ Co., New York
Hansen GL, Schmit JL (1983) J Appl Phys 54:1640
Harman TC, Strauss AJ (1961) J Appl Phys 32:2265
Hansen GL, Schmit JL, Casselman TN (1982) J Appl Phys 53:7099
Huang K, Xie XD (1958) Physics of Semiconductor, Chap. 2. Science Press, Beijing
Isaacon RA, Weger M (1964) Bull Am Phys Soc 9:736
Justice RJ, Seiler DG, Zawadzki W, et al. (1988) J Vac Sci Technol A 6:2779
Kahlert H, Bauer G (1973) Magnetophonon Effect in n-Type $Hg_{1-x}Cd_xTe$ (x = 0.212). Phys Rev Lett 30:1211–1214
Kichigin PA, et al. (1981) Solid State Commun 37:345
Kim JS, Seiler DG, Tseng WF (1993) J Appl Phys 73:8324
Kim JS, Seiler DG, Colombo L, et al. (1994) Semicond Sci Technol 9:1696

Kinch MA, Brau MJ, Simmons A (1973) Recombination mechanisms in 8-14-μ HgCdTe. J Appl Phys 44:1649–1663
Kittle C (1963) Quantum Theory of Solids. Wiley, New York
Kittle C (1976) Solid State Physics. Wiley, New York, p. 235
Kowalczyk SP, Cheung JT, Kraut EA, et al. (1986) Phys Rev Lett 56:1605
Krishnamurthy S, Sher A (1993) Proc IRIS Mater 177
Krishnamurthy S, Sher A (1994) Electron mobility in Hg0.78Cd0.22Te alloy. J Appl Phys 75:7904–7909
Krishnamurthy S, Sher A (1995) J Electron Mater 24:641
Long D (1968) Calculation of ionized-impurity scattering mobility of electrons in Hg1-xCdxTe. Phys Rev 176:923–927
Makowski I, Glickman M (1973) J Phys Chem Solids 34:487
Meyer JR, Bartoli FJ (1987) Ionized-impurity scattering in the strong-screening limit. Phys Rev B 36:5989–6000
Meyer JR, Hoffman CA, Bartoli FJ, et al. (1988) J Vac Sci Technol A 6:2775
Meyer JR, Hoffman CA, Bartoli FJ, et al. (1993) Semicond Sci Technol 8:805
Morin FJ, Maita JP (1963) Specific heats of transition metal superconductors. Phys Rev 129:1115–1120
Nemirovsky Y, Finkman E (1979) J Appl Phys 50:8101
Nimtz G, Schlicht B, Dornhaus R (1983) Narrow gap semiconductors, tracts in modern physics Vol. 96. Springer, New York, p. 119
Parat KK, et al. (1990) J Cryst Growth 106:513
Pinchuk II (1977) Semiconductor electrons in a strong electric and a longitudinal quantizing magnetic field. Soviet Physics Solid State 19:383–386
Putley EH (1955) The hall coefficient, electrical conductivity and magneto-resistance effect of lead sulphide, selenide and telluride. Proc Phys Soc B 68:22–34
Putley EH (1960) The Hall Effect. Butterworth, London
Reine MB, Maschhoff KR, et al. (1993) Semicond Sci Technol 8:788
Roth LM, Argyres PN (1966) In: Willardson RK, Beer AC (eds.) Semiconductors and Semimetals Vol. 1, Chap. 6. Academic, New York
Roy CL (1977) Czech J Phys B 27:769
Schmit JL (1970) J Appl Phys 41:2876
Scott W (1972) J Appl Phys 43:1055
Seeger K (1985) Semiconductor Physics, 4th edn. Springer, New York
Seeger K (1989) Temperature dependence of the dielectric constants of semi-insulating III-V compounds. Appl Phys Lett 54:1268–1269
Sher A, Primakoff H (1960) Phys Rev 119:178
Sher A, Primakoff H (1963) Phys Rev 130:1267–1282
Shockley W (1955) Electrons and Holes in Semiconductors. D. Van Nostrand Company, Inc., New York, p. 465
Song SN, Ketterson JB, Choi YH, et al. (1993) Appl Phys Lett 63:964
Stankiewicz J, Giriat W, Bienestock A (1971) Pressure and temperature dependences of CdxHg1-xTe alloy hall mobilities. Phys Rev B 4:4465–4470
Sporken R, Sivananthan S, Faurie JP, et al. (1989) J Vac Sci Technol A 7:427
Stephens AE, Seiler DG, Sybert JR, et al. (1975) Phys Rev B 11:4999

Stephens AE, Miller RE, Sybert JR, et al. (1978) Phys Rev B 18:4394
Subowski JJ, Dietl T, Szymanska W, et al. (1981) J Phys Chem Solids 42:351
Suizu K, Narita S (1972) Solid State Commun 10:627
Szymanska W, Dietl T (1978) J Phys Chem Solids 39:1025
Tang DY (1974) Infrared Phys Technol 3(6):345
Van der Pauw LJ (1958) Philips Tech Rev 20:220
Wamplar WR, Springford M (1972) J Phys C: Solid State Phys 5:2345
Wang XF, Lima IC da C, Lei XL, et al. (1998) Phys Rev B 58:3529
Wiley JD (1975) In: Willardson RK, Beer AC (eds.) Semiconductors and Semimetals, Vol. 10. Academic, New York, p. 91
Yadave RDS, Gupta AK, Warrier AVR (1994) J Electron Mater 22:1359
Ye LX (1987) Semiconductor Physics. Higher Education Press, Beijing, p. 853
Yoo SD, Kwack KD (1997) J Appl Phys 81:719
Zawadzki W (1982) In: Paul W (ed.) Handbook on Semicondutors, Vol. 1. North-Holland, Amsterdam, Chap. 12
Zawadzki W, Szmanska W (1971) J Phys Chem Solids 32:1151
Zheng GZ, Guo SL, Tang DY (1987) Acta Phys Sin 36:114
Атирханов ХИ, Баширов РИ, Закиев ЮЭ (1963) ДАН СССР 148:1279
Атирханов ХИ, Баширов РИ, Габжиалцев ММ (1964) ЖЭТФ 47:2067
Атирханов ХИ, Баширов РИ (1966) ФТТ 8:2189
Бреелер МС, Парфенбев РВ, Шалыт (1966) ФТТ 7:1266
Коренбдит ЛЛ, Штейнберт АА (1956) Ж Т Ф 26:927
Гуревичет ЛЭ, ЭФрос АЛ (1962) ЖЭТФ 43:561
Глуэман НГ, Сабизянова ЛД, Цибилвковский ИМ (1979) ФТП 138:466
Павлов СТ, Парфенбев РВ, Фиреов, et al. (1965) ЖЭТФ 48:1565
Лифшиц ИМ, Косевиц АМ (1955) ЖЭТФ 29:730

6 Lattice Vibrations

6.1 Phonon Spectra

6.1.1 Monatomic Linear Chain

The normal mode vibrations of an oscillating lattice can be viewed as a collection of quasi-particles with energies $\hbar\omega_j(\boldsymbol{q})$ and momenta $\hbar\boldsymbol{q}$. These quasi-particles are called phonons. If the atoms in its surroundings were fixed, the potential well of an atom centered at the lattice sites in which an atom resides closely approximates a quadratic function of its displacement from the minimum. Thus individual atomic vibrations would be nearly harmonic. At higher energies the wells deviate from quadratic functions and individual oscillations would be anharmonic. A phonon's occupation number specifies its a quantum excitation state. In the simple case of phonons propagating along the principal symmetry axes of crystals, two different types of vibration modes can be easily visualized (see Fig. 6.1a,b), namely a longitudinally polarized mode in which the atomic displacements are parallel to the direction of the wave vector, and two orthogonal transverse polarized modes in which the atomic displacements are perpendicular to the direction of the wave vector. In general phonons are Bosons and in three dimensions have an effective spin of one.

We start with the simplest case of a one-dimensional monatomic linear chain with only nearest neighbor interactions (see Fig. 6.2).

N is the number of atoms in the monatomic chain, $l = 1, N$. Then we write the classical Hamilton function of a system of coupled harmonic oscillators as the sum of a kinetic energy T and a potential energy Φ:

$$H = T + \Phi = \sum_l \frac{p_l^2}{2m} + \frac{1}{2}\sum_l \left[\beta(u_{l+1} - u_l)^2 + \beta(u_l - u_{l-1})^2 \right], \tag{6.1}$$

where $p_l = m\dot{u}_l$, and β is the force constant.

If one expands the energy near the equilibrium point for the lth atom in the above quasi-elastic approximation, it produces the Newton equation:

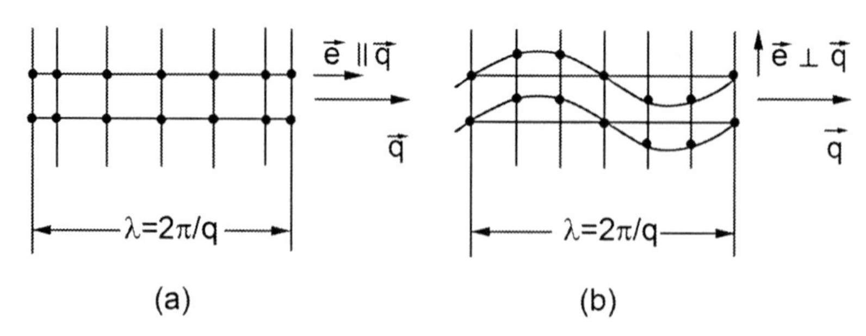

Fig. 6.1. Atomic vibrations for waves propagating along a principal symmetry direction of a crystal with (**a**) longitudinal and (**b**) transverse polarizations in real space

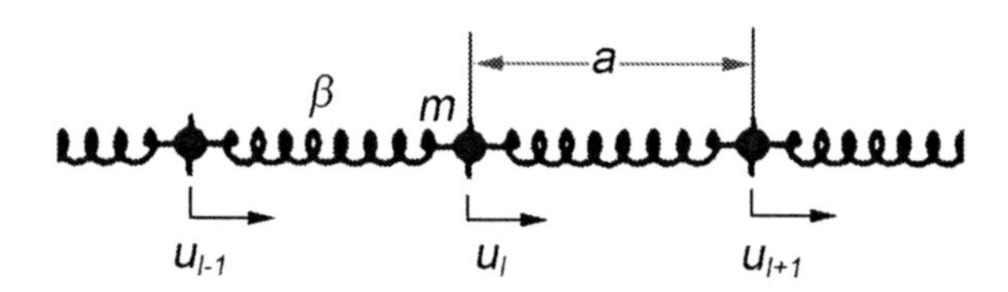

Fig. 6.2. Vibrations of a linear monatomic chain. The displacement of the *l*th atom is u_l, the masses of the atoms are *m*, the lattice spacing is *a*, and the spring constant is β

$$\begin{aligned} m\ddot{u}_l &= -\partial\Phi / \partial u_l \\ &= \beta(u_{l+1} - u_l) - \beta(u_l - u_{l-1}) = \beta(u_{l+1} - 2u_l + u_{l-1}) \cdot \end{aligned} \tag{6.2}$$

This equation is solved by assuming a trial solution: $u_l = A\exp\left[i(\omega t - laq)\right]$, where a is the distance between nearest neighbor atoms, and $q = 2\pi / \lambda$ is a wave number. Immediately we get the phonon dispersion relation, which is the phonon energy dependence of the phonon wave-vector q:

$$\omega = 2\sqrt{\frac{\beta}{m}}\left|\sin\frac{1}{2}aq\right|. \tag{6.3}$$

Periodic boundary conditions require $u_1 = u_N$ or $q_j = \frac{2\pi}{a(N-1)}j$, and $\frac{-\pi}{a(N-1)} < j £ \frac{\pi}{a(N-1)}$. If N is large then q_j is essentially continuous.

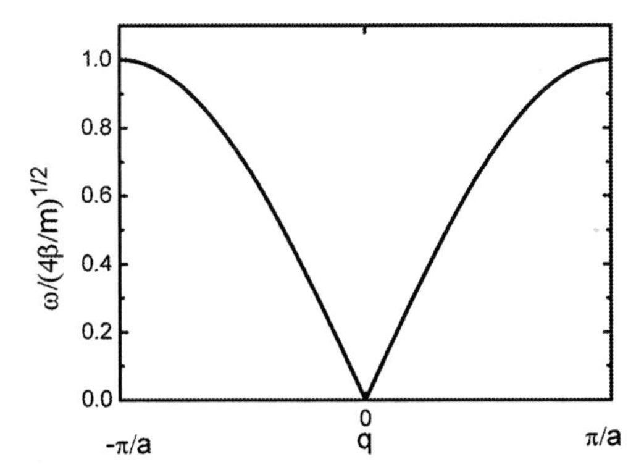

Fig. 6.3. The vibrational spectrum of a linear monatomic chain

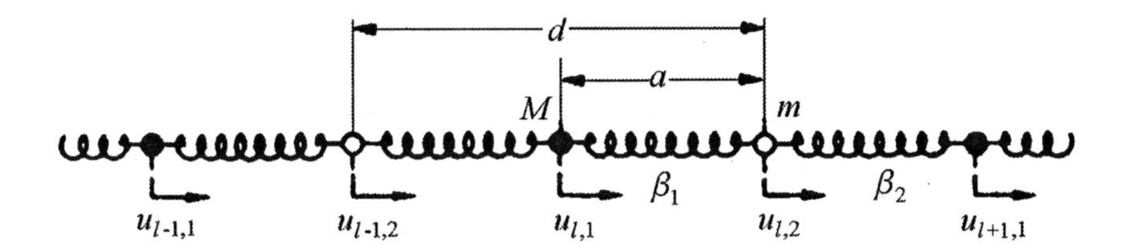

Fig. 6.4. A linear diatomic chain

The relation, (6.3) is shown in Fig. 6.3. In this solution, we tacitly used the assumption of an infinitely large periodic lattice. All useful information is contained in the waves with wave vectors lying between the limits $-\pi/a < q \leq \pi/a$. Because of the periodicity of the lattice q values out of this range repeat the energies in the range. This range of wave vectors is called the first Brillouin zone. In this one-dimensional monatomic case there is only one branch for a each q, which is called an acoustic branch because at small q ($\lambda >> a$), the dispersion relation reduces to $\omega = cq$, where the velocity of sound is defined as $c = \sqrt{\frac{\beta}{m}}a$. This is similar to the case of a homogeneous elastic medium.

Next we discuss the more complex case of vibrations of the one dimensional lattice with two atoms per unit cell that is shown in Fig. 6.4. One can see that each elementary cell contains two atoms. If we assume the force constants to be $\beta_{1,2}$, the masses to be m, M, and $d = 2a$, we have the following equations of motion:

$$\begin{aligned} M\ddot{u}_{l,1} &= \beta_1(u_{l,2} - u_{l,1}) - \beta_2(u_{l,1} - u_{l-1,2}) = \beta_1 u_{l,2} - (\beta_1 + \beta_2)u_{l,1} + \beta_2 u_{l-1,2} \\ m\ddot{u}_{l,2} &= \beta_2(u_{l+1,1} - u_{l,2}) - \beta_1(u_{l,2} - u_{l,1}) = \beta_2 u_{l+1,1} - (\beta_1 + \beta_2)u_{l,2} + \beta_1 u_{l,1} \end{aligned} . \quad (6.4)$$

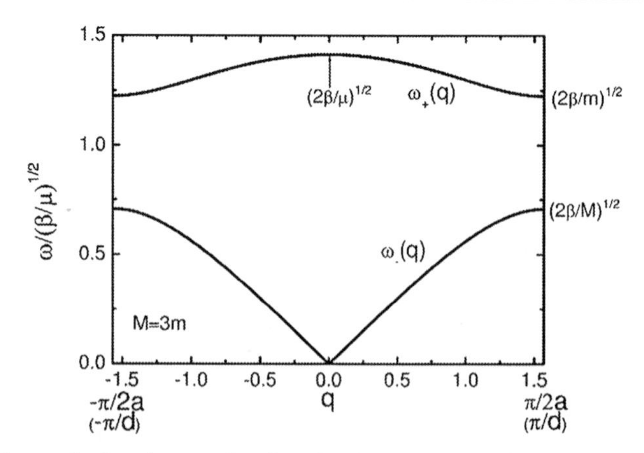

Fig. 6.5. The optical and acoustic vibration branches in a linear diatomic chain

Now chose solutions similar that of the monatomic linear chain: $u_{l,1} = A\exp\left[i(\omega t - 2laq)\right]$ and $u_{l,2} = B\exp\left\{i[wt - (2l+1)aq]\right\}$. Substituting these trial solutions into (6.4), we get:

$$\begin{aligned} &[Mw^2 - (b_1 + b_2)]A + [b_1\exp(-aqi) + b_2\exp(aqi)]B = 0 \\ &[b_2\exp(-aqi) + b_1\exp(aqi)]A + [mw^2 - (b_1 + b_2)]B = 0 \end{aligned} . \tag{6.5}$$

Equations (6.5) have a solution only if the secular determinate vanishes:

$$\begin{vmatrix} M\omega^2 - (\beta_1 + \beta_2) & \beta_1\exp(-aqi) + \beta_2\exp(aqi) \\ \beta_2\exp(-aqi) + \beta_1\exp(aqi) & m\omega^2 - (\beta_1 + \beta_2) \end{vmatrix} = 0 . \tag{6.6}$$

We then find:

$$w_{\pm}^2 = \frac{b_1 + b_2}{2m}\left\{1 \pm \left[1 - \frac{16b_1b_2}{(b_1 + b_2)^2}\frac{m^2}{mM}\sin^2 aq\right]^{1/2}\right\}, \tag{6.7}$$

where $\mu^{-1} = m^{-1} + M^{-1}$. Now for simplicity we assume that $\beta = \beta_1 = \beta_2$. Once a periodic boundary condition is imposed, it is natural to examine the dispersion curve for the wave number q ranging between –π/2a and π/2a. This phonon dispersion relation is shown in Fig. 6.5. The lower branch is called the acoustic branch while the upper one is called the optical branch. For very long waves, in the acoustic mode (ω_-) all the adjacent atoms move synchronously, like in an acoustic wave in homogeneous medium. However, in the optical mode (ω_+), the two sub-lattices move in opposite directions with their center of gravity remaining unperturbed. In an ionic crystal such a vibration produce alternating dipole moments with amplitudes that have the periodicity of the wave. As a result, this mode strongly interacts with the fields of electromagnetic waves, and so is called an optical mode. In the case of 3D lattices, there are three acoustic branches and $3(p-1)$ optical ones if there are p atoms per primitive cell.

6.1.2 Phonon Dispersion Measurement Techniques

Here the main experimental techniques to investigate phonon spectra are briefly described. These include inelastic neutron scattering (INS) and inelastic X-ray scattering (IXS).

Neutrons with energies in the same range as those of lattice vibrations and with wavelengths of the same order as the crystal lattice spacing scatter inelasticly from the bulk phonons. In the INS method, neutrons with energies $E = p^2/2M_n$ (M_n =1.67 × 10^{-24} g) and momentum $\boldsymbol{p}$, are incident upon the crystal. The scattered beam is completely defined by a new momentum p', and energy $E' = p'^2/2M_n$. The phonon dispersion is mapped exploiting the momentum-energy conservation laws:

$$\frac{p'^2}{2M_n} - \frac{p^2}{2M_n} = \pm h\omega(q) \tag{6.8}$$

$$p' - p = \pm hq + hG \tag{6.9}$$

where "+" corresponds to a phonon absorption while "–" correspond to an emission, and $\boldsymbol{G}$ is a reciprocal lattice vector. The incident beam momentum $\boldsymbol{p}$ is fixed and the scattered beam momentum $\boldsymbol{p'}$ is measured in different directions. The phonon dispersion relation is obtained by analyzing the energy and direction of the scattered neutrons. The wavelength resolution of such an instrument follows from the Bragg equation, and reads $\Delta\lambda/\lambda \sim \cot\Theta\Delta\Theta + \Delta d/d$, where Θ is a Bragg angle, and d is a lattice spacing. Near the backscatter direction ($\Theta \to 90°$) $\cot\Theta$ becomes very small, and its contribution to the resolution is determined by the beam divergence $\Delta\Theta$. Using nominally perfect crystals, the uncertainty in the lattice spacing $\Delta d/d$ can also be made very small. Furthermore, $\Delta d/d \sim \Delta G/G$, i.e. this term is proportional to the width of the Bragg reflection of the reciprocal lattice vector G. The wavelength resolution therefore improves for crystal optics operated at higher reflection orders.

Based on this same concept, an X-ray backscattering spectrometer was devised to study the dynamical properties of solids (Burkel 2001; Ruf 2003). With this instrument, the first high-resolution (several tens of meV) inelastic X-ray scattering (IXS) data for phonons were obtained.

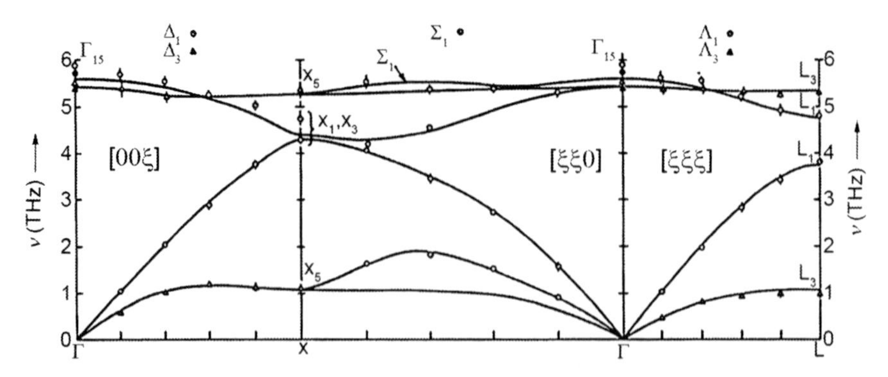

Fig. 6.6. The dispersion relation for InSb at room temperature. The *open symbols* denote neutron measurements, and the *solid curves* result from shell-model calculations (Price et al. 1971)

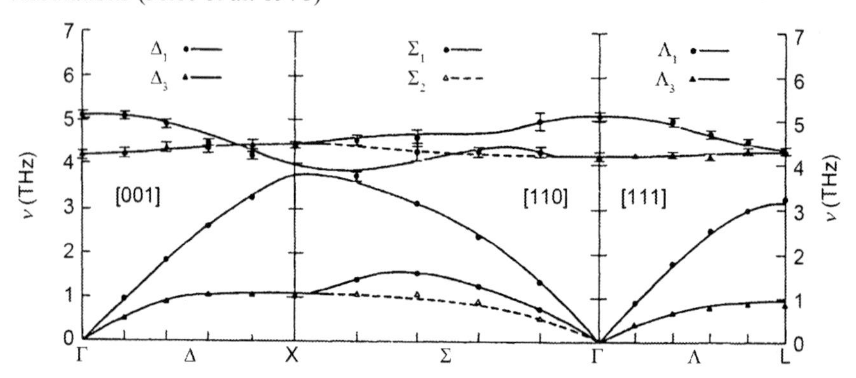

Fig. 6.7. The dispersion relation for CdTe at room temperature. *Symbols* represent neutron measurements, and the *solid curves* result from shell-model calculations (Rowe et al. 1974)

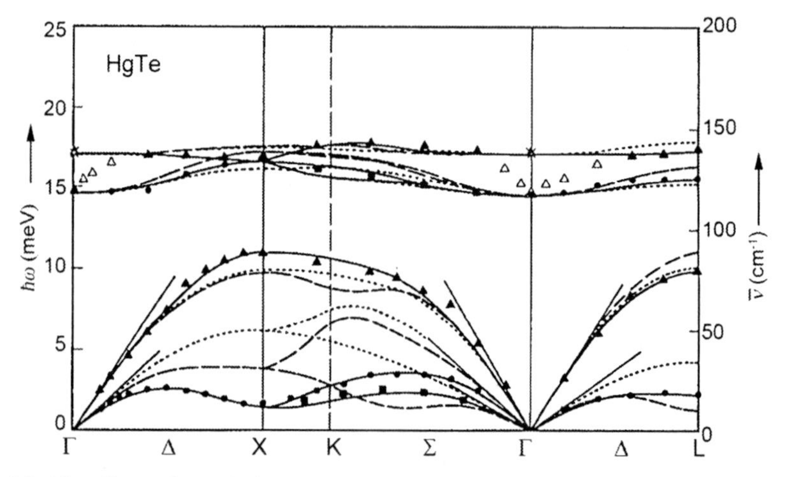

Fig. 6.8. The dispersion relation for HgTe at room temperature. *Symbols* represent neutron measurements, and the *solid curves* result from rigid-ion model calculations (Kepa et al. 1980, 1982)

IXS has a much higher momentum resolution than INS and can be used over a much wider range of energy transfers. The IXS intensity does not depend on the quasi-random (small) INS cross-sections of the individual nuclei and their isotopes, but varies in a well-defined way with the nuclear charge Z, exceeding the coupling strength of neutrons by more than an order of magnitude for $Z > 10$. Due to the high collimation of a synchrotron X-ray beam and the much higher particle (photon) flux, IXS can be applied to very small samples. While the resolution function in INS is affected by the energy and momentum transfer, in IXS it is entirely given by the scattering geometry.

Next we discuss the application of the phonon dispersion relation determination by the INS technique. Price et al. (1971) measured the phonon dispersion relation in symmetry directions for InSb at 300 K by a constant-Q technique. The measured dispersion curves for InSb are indicated by symbols in Fig. 6.6. The continuous curves represent shell-model fits, which will be discussed in next section. There are no experimental results for the $Hg_{1-x}Cd_xTe$ alloy phonon dispersion relations. The experimental data are limited to CdTe and HgTe. Rowe et al. (1974) measured the phonon dispersion relation of CdTe at 300 K using the INS technique. This dispersion relation is shown graphically by symbols in Fig. 6.7. The continuous curves are the fitted shell-model results. Owing to the very high thermal-neutron-absorption cross-section of Cd^{113} found in natural Cd, the experimental measurements were made with a crystal composed predominantly of the isotope Cd^{114}. The difference between the mass of this isotope (113.9) and the average atomic mass of natural Cd (112.4) is expected to have a negligible effect on the frequencies.

Kepa et al. (1980, 1982) measured the phonon dispersion relation of HgTe using the constant-Q technique. The experimental results are presented in Fig. 6.8 as triangles and circles. The main difficulty in experiments with HgTe stems from the high neutron-absorption cross-section of Hg (372×10^{-24} cm^{-3}). However, the very high thermal neutron coherent scattering cross-section of Hg (1.266×10^{-12} cm) is a favorable condition for a thin sample measurement. By using the INS technique, several groups determined the phonon spectra of other II–VI compounds, such as CdSe (Widulle et al. 1999), CdS (Debernardi 1997), HgSe (Łażewski et al. 2003), and β-HgS (Szuszkiewicz et al. 1998). They use materials containing Cd^{116} or Cd^{114} instead of Cd^{113} in the experiments on CdSe and CdS, because of low thermal-neutron-absorption cross-section of Cd^{116} and Cd^{114}.

The IXS technique is insensitive to isotopic variations of the scattering cross-section, and therefore presents an attractive alternative to investigate compounds containing natural Cd rather than materials with very

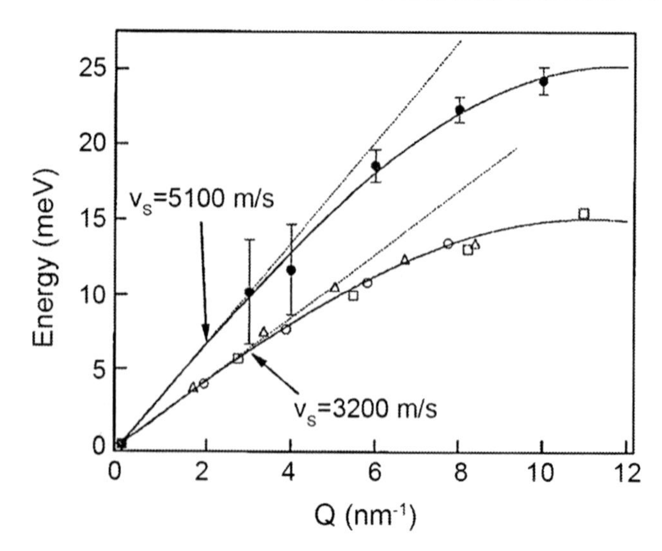

Fig. 6.9. The longitudinal acoustic-phonon dispersion curves of CdTe at 7.5 GPa (*solid symbols*) and at ambient pressure from Rowe et al. (1974) (*open symbols*: *triangles* correspond to the [111], *circles* to the [100], and *squares* to the [110] directions). The *solid curves* through the data points represent the result of a sinus fit, and the *dotted lines* are intended to help visualize the initial slope, yielding the indicated sound velocities (Krisch et al. 1997)

expensive pure isotopes. The dispersion of longitudinal acoustic phonons has been measured by IXS in the sodium-chloride phase of CdTe at 7.5 GPa. The longitudinal acoustic branch could be observed up to 10 nm^{-1}, just beyond the first Brillouin zone. The dispersion curve obtained is shown in Fig. 6.9, together with results from an INS measurement for the zinc-blende phase of CdTe at ambient pressure for the crystallographic [111], [100] and [110] directions. There is a strong increase of the excitation energies in the high-pressure phase compared to the one at ambient pressure.

6.1.3 Theoretical Calculations of the Phonon Spectra

There are several theoretical methods to calculate the phonon spectra, including the "shell model" (Cochran 1959a,b), the "rigid-ion model" (Vetelino and Mitra 1969; Kunc et al. 1974), the "adiabatic bond-charge model" (Weber 1974, 1977; Rajput and Browne 1996) and ab initio calculations (Corso et al. 1993; Łażewski et al. 2003).

We first discuss the shell model.

In the harmonic approximation, the equations of motion are:

$$m_n \ddot{u}_x(l,n) = \sum_{n'} \sum_{l'} \sum_{y} \phi_{xy}(l,n;l',n') u_y(l',n'), \qquad (6.10)$$

where $\phi_{xy}(l,n;l',n')$ is a force constant for the interaction between nuclei (l,n) and (l',n'), where l specifies a lattice site, and n is a type of atom with mass m_n. Using the solution: $u_x(l,n) = U_x \exp[q \times r(l,n) - \omega t]$, the equations of motion lead to:

$$\omega^2 m_n U_x(n) = \sum_{n'} \sum_{y} M_{xy}(nn') U_y(n'), \qquad (6.11)$$

where

$$\mathrm{M}_{xy}(\mathrm{nn'}) = -\sum_{\mathrm{l'}} \varphi_{xy}(\mathrm{l,n;l',n'}) \mathrm{u}_y(\mathrm{l',n'}) \exp\{\mathrm{q} \times [\mathrm{r(l',n')} - \mathrm{r(l,n)}]\}. \qquad (6.12)$$

Then the dispersion relations are given by the solutions of the characteristic equation: $|D - m\omega^2 I| = 0$, where $M_{xy}(nn')$ is an element of the dynamical matrix D.

6.1.3.1 The Physical Meaning of the Shell Model

In this model, an atom consists of a ion core, the nucleus and the inner electrons, and surrounding the core there is a charged shell representing the outer electrons. The shell and core are coupled to one another by an isotropic force constant, and each retains spherical symmetry although a dipole moment may be generated by their relative displacement. The use of a shell model allows the atom to be polarized in an electric field, and experiences "distortion polarizability" under the influence of short-range forces. Some lattice vibration modes, therefore, can induce electric fields in the crystal. We can divide the interaction coefficients into bonding coefficients (depending only on the force constants between nearest neighbor units) and Coulomb coefficients (depending on long ranged electrostatic interaction among units throughout the crystal). A schematic representation of the bonding coefficients in the shell model are depicted in Fig. 6.10. The symbols *F*, *S*, and *D* denote core-shell, shell–shell and core–core bonding coefficients, respectively. The Coulomb coefficients will be discussed later. In the figure, 1 and 2 denote different cores, and 3 and 4 denote electron shells in the unit cell (Hass and Henvis 1962; Bruüesch 1982).

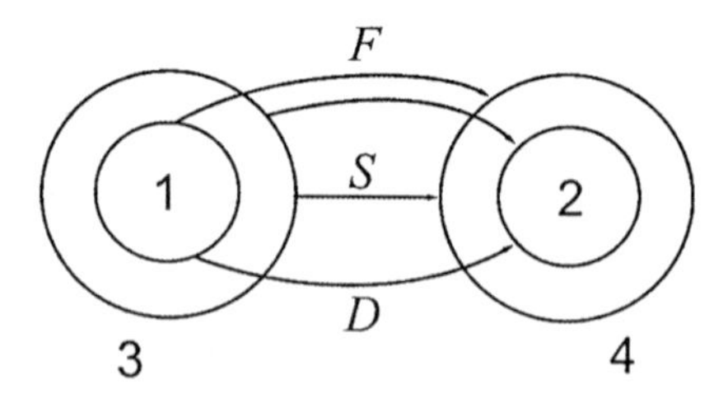

Fig. 6.10. A schematic representation of the bonding coefficients in the shell model. For example F is the coefficient between core 1 and shell 4

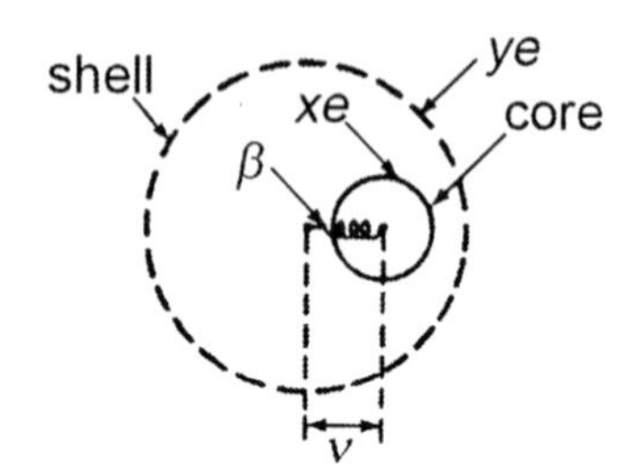

Fig. 6.11. In the shell model, an ion consists of a spherical electronic shell which is isotropically coupled to its rigid ion-core by a spring with force constant β

Next we treat ion polarization within the shell model. In this model the ions are composed of an outer spherical shell of electrons and a core consisting of the nucleus and remaining electrons. In an electric field the shell retains its spherical charge distribution but is allowed to move rigidly with respect to the core. The polarizability is made finite by a harmonic restoring force of spring constant β which acts between the core and shell (Fig. 6.11). v is the displacement of the shell with respect to the core. ye is the charge of shell and xe is the charge of core. (Note: in this discussion x and y are fractional charges not spatial coordinates.) The net charge on the ion is (y+x)e, and x+y is a negative number for anions and positive for cations.

In equilibrium, an external field induced electrostatic force is equal to the elastic restoring force:

$$yeE = \beta v. \tag{6.13}$$

The induced dipole moment is $yev = \chi E$, from which we obtain the free ion polarizability:

$$\chi = \frac{(ye)^2}{\beta}. \tag{6.14}$$

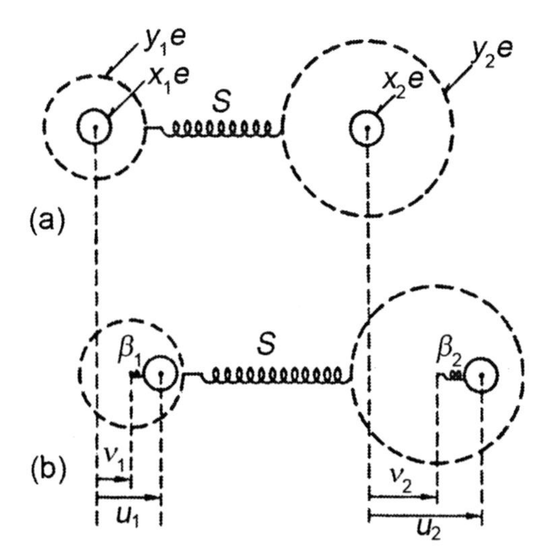

Fig. 6.12. A schematic representation of the coupling of two neighboring ions in the simple shell model. (**a**) Ions in equilibrium positions and (**b**) ions in displaced positions

Now consider a cation and a nearest neighbor anion of a diatomic alkali halide crystal (Fig. 6.12). The labels 1, and 2 represent the positive and negative ions, respectively. β_1 and β_2 are the force constants which couple the shells to the ion cores. y_1e and y_2e are charges of shells, and x_1e and x_2e are charges of cores.

The net charge of the cation is, $Ze = (x_1 + y_1)e$, and the charge of the anion is $-Ze = (x_2 + y_2)e$. The constant S represents the short-range interaction force constant between the shells of neighboring ions.

For a transverse optical branch and at long wave lengths, $q = 0$, the equations of motion are given by:

$$\begin{aligned} m_1\ddot{u}_1 &= \beta_1(v_1 - u_1) + x_1eE^* \\ m_2\ddot{u}_2 &= \beta_2(v_2 - u_2) + x_2eE^* \\ 0 &= S(v_2 - v_1) + \beta_1(u_1 - v_1) + y_1eE^*, \\ 0 &= S(v_1 - v_2) + \beta_2(u_2 - v_2) + y_2eE^* \end{aligned} \tag{6.15}$$

where m_1 and m_2 are the masses of the cores, and E^* is an effective electric field. Using the adiabatic approximation (neglecting the masses of shells), and eliminating the shell coordinates v_1 and v_2, we obtain:

$$\mu\ddot{w} + S^*w = Z^*eE^*, \tag{6.16}$$

where $w = u_1 - u_2$ so:

$$S^* = \frac{S}{1 + (\frac{1}{\beta_1} + \frac{1}{\beta_2})S}, \tag{6.17}$$

$$Z^* = Z + \frac{\frac{y_2}{\beta_2} - \frac{y_1}{\beta_1}}{\frac{1}{\beta_1} + \frac{1}{\beta_2} + \frac{1}{S}}. \tag{6.18}$$

Thus the effective, or often called "dynamic" charge, Z^*e is different from, Ze. Since the anion is usually more polarizable than the cation, the second term in (6.18) is negative, with the result, $Z^*e < Ze$. This result was first obtained by Szigeti, but without using the shell model. The effective charge Z^*e is also called the Szigeti charge (Szigeti 1950). Z^* is often non-zero, even if the binding is pure covalent in crystals with no center of symmetry, or a combination of covalent and ionic. Examples are III–V compounds like the compounds InSb and GaAs (Harrison 1980; Bruüesch 1982). Strong phonon infrared absorption is observed, which in the case of GaAs, is due to an effective charge, $Z^*e = 0.51e$ (Hass and Henvis 1962). From (6.18), in the shell model, $Z^*e = 0$ for neutral and identical atoms ($Z = 0, \frac{y_1}{\beta_1} = \frac{y_2}{\beta_2}$). Z^*e is closely related to the structure and symmetry of the crystal. The shell model takes these features into account and does indeed find $Z^*e = 0$ for crystal with low symmetry such as graphite or Te.

In the absence of an external field, the effective electric field E^* is given by:

$$E^* = \eta P, \tag{6.19}$$

where $\eta = 4\pi/3$ for the transverse optical branch, and $\eta = -8\pi/3$ for the longitudinal optical branch in alkali halides. P is given by:

$$P = \frac{e}{v_a}(x_1 u_1 + x_2 u_2 + y_1 v_1 + y_2 v_2), \tag{6.20}$$

where v_a is the cell volume (Bruüesch 1982). Eliminating v_1 and v_2 again with the help of (6.15), we obtain:

$$P = \frac{1}{v_a}(Z^* ew + \chi^* E^*), \tag{6.21}$$

where

$$\frac{\alpha^*}{e^2} = \frac{y_1^2}{\beta_1} + \frac{y_2^2}{\beta_2} - \frac{\left(\frac{y_1}{\beta_1} + \frac{y_2}{\beta_2}\right)^2}{\frac{1}{S} + \frac{1}{\beta_1} + \frac{1}{\beta_2}}. \tag{6.22}$$

It has been shown that the effective polarizability of the ion couple is not simply the sum of that of the free ions (Bruüesch 1982).

$$\chi = \chi_1 + \chi_2 = \frac{(y_1 e)^2}{\beta_1} + \frac{(y_2 e)^2}{\beta_2} > \chi^*. \tag{6.23}$$

The effective polarizability is less than the corresponding sum of the free ion contributions. From (6.19) and (6.21), we get:

$$E^* = k\eta \frac{Z^* e}{v_a} w, \tag{6.24}$$

where

$$k = \left(1 - \eta \frac{\chi^*}{v_a}\right)^{-1}. \tag{6.25}$$

Substituting (6.24) into (6.16), and putting $w = w_0 \exp(-i\omega t)$, we obtain for the frequencies of the transverse optical branch (TO), and longitudinal optical branch (LO):

$$\begin{aligned} \mu\omega_{TO}^2 &= S^* - \frac{4\pi(Z^* e)^2}{3v_a} k_T \\ \mu\omega_{LO}^2 &= S^* + \frac{8\pi(Z^* e)^2}{3v_a} k_L \end{aligned}, \tag{6.26}$$

where

$$k_T = \left(1 - \frac{4\pi\chi^*}{3v_a}\right)^{-1}$$
$$k_L = \left(1 - \frac{8\pi\chi^*}{3v_a}\right)^{-1} . \tag{6.27}$$

When $k_1, k_2 \to \infty$, we obtain $S^* = S$, $Z^* = Z$, $\alpha^* = 0$, and $k_L = k_T = 1$. Equation (6.26) reduces to the relations:

$$\mu\omega_{TO}^2 = S^* - \frac{4\pi(Z^*e)^2}{3v_a}$$
$$\mu\omega_{LO}^2 = S^* + \frac{8\pi(Z^*e)^2}{3v_a} . \tag{6.28}$$

This is consistent with the results of the rigid-ion model.

6.1.3.2 The Shell Model in the 3D Case

The interaction coefficients in the 3D are given by:

$$M_{xy}(nn') = B_{xy}(nn') + C_{xy}(nn') . \tag{6.29}$$

The $B_{xy}(nn')$ are bonding coefficients between two cores (e.g. 1 and 2), two shells (e.g. 3 and 4), and core-shell (e.g. 1 or 2, to 3 or 4) that for simplicity are written as *B*. $C_{xy}(nn')$ are Coulomb coefficients, that for simplicity are written as C. It is readily shown that $B_{xx}(13) = B_{xx}(24) = -\beta$. When only nearest neighbor units interact, the other bonding coefficients are determined by six independent force constants, $\phi_{xx}^{(B)}(l,n;l,n')$, $\phi_{xy}^{(B)}(l,n;l,n')$, nn'=12, 34, 14, or 24. Then we can define a compact symbol set, $D \equiv B(12)$ (core–core), $S \equiv B(34)$ (shell–shell), and $F \equiv B(14)$ (shell–core). All the other Coulomb coefficients can be expressed in terms of $C(11)$ and $C(12)$. For convenience, we introduce two dimensionless function of $\boldsymbol{q}$, C_1 and C_2, such that

$$(Z^2e^2/v_a)C_1 \equiv C(11) = C(22) = C(33) = C(44) = -C(13) = -C(24)$$
$$(Z^2e^2/v_a)C_2 \equiv C(12) = C(34) = -C(14) = -C(32) . \tag{6.30}$$

Then define, $R \equiv D+S+2F$, and $T \equiv S+F$, where T is the interaction between a shell and an adjacent atom, and R is the net interaction between two adjacent rigid atoms. The shell model may then be written in matrix form as (Cowley 1962):

$$\begin{aligned}\omega^2\mathbf{M}U &= (\mathbf{A}-\mathbf{B}\mathbf{D}^{-1}\mathbf{B}^{+})U\\ \mathbf{A} &= \mathbf{R}+\mathbf{ZCZ}\\ \mathbf{B} &= \mathbf{T}+\mathbf{ZCY}\ ,\\ \mathbf{D} &= \mathbf{S}+\mathbf{YCY}\end{aligned} \tag{6.31}$$

where "+" represents the adjoint matrix, the combination of the complex conjugate and transpose matrix operations. The matrices **M**, **Z**, and **Y** are the mass, core charge (Ze), and shell charge (ye) matrices. The matrices **R**, **T**, and **S** are the dynamical matrices of the short-range ion–ion, ion–shell, and shell–shell interactions, and the **C** matrices are the corresponding Coulomb matrices. The phonon dispersion relations are obtained from (6.31).

The phonon spectrum of germanium is presented as an example. We will restrict the dynamical matrices to 4×4 matrices that refer to a particular mode, with $\boldsymbol{q}$ in a symmetry direction. With this restriction, the equation of motion is given by:

$$m_n\omega^2 U(n) = \sum_{n'=1}^{4} M(nn')U(n')\,, \tag{6.32}$$

where in general $M(nn')$ is a linear combination of $M_{xx}(nn')$, $M_{xy}(nn')$, etc. If for example we are dealing with the longitudinal mode propagated along the [100] direction, in (6.32), $m_1 = m_2 = m$, $m_3 = m_4 = 0$, and $a = Z^2e^2/v_a$, where v_a is the volume of the unit cell. The dispersion relations are given by:

$$m\omega^2 = A_0 \pm A\,, \tag{6.33}$$

where

$$\begin{aligned}A_0 &= R_0 + \frac{-(aC_1+\beta+T_0)(T_0^2+|T|^2)+T_0\left[T(aC_2^*+S^*)+T^*(aC_2+S)\right]}{(aC_1+\beta+T)^2-|aC_2+S|^2}\\ A &= R + \frac{-2TT_0(aC_1+\beta+T_0)+T_0^2(aC_2+S)+T2(aC_2^*+S^*)}{(aC_1+\beta+T)^2-|aC_2+S|^2}\end{aligned} \tag{6.34}$$

Here "*" represents the matrix complex conjugate operation. The subscript 0 refers to the case where q=0. The "+" sign in (6.33) is for the optic modes, and the "–" sign is for the acoustic modes. Using the shell model, Price et al. (1971) fitted the phonon spectra of InSb at 300 K (Fig. 6.6). It is shown that the agreement is good with the exceptions of the LO modes in the Δ_3 direction. Rowe et al. (1974) used the shell model to fit the phonon spectra of CdTe at 300 K (Fig. 6.7). It was found that the agreement is excellent with the exceptions of the TA modes in the Σ_2 and Λ_2 directions. So shell model parameter sets can be found that yield good theoretical fits to phonon spectra.

6.1.3.3 The Adiabatic Bond-Charge Model

Shell models give a reasonable description of the dispersion relations, but the fitting parameters of the models are poorly defined and apparently devoid of physical meaning. So this model cannot be generalized to complicated compound system, such as $Hg_{1-x}Cd_xTe$ and superlattices. Weber (1977) developed an adiabatic bond charge model with six parameters for the lattice dynamics of diamond-type crystals. The simplest empirical lattice-dynamical model correctly describes the phonons and other lattice dynamical quantities such as elastic constants and specific heats for II–VI compounds with the zinc-blende coordination (Rajput and Browne 1996). This model is widely used in the lattice dynamical research of III–V compounds (Tütüncü and Srivastava 2000) and II–VI compounds (Camacho and Cantarero 1999, 2000; Rajput and Browne 1996).

In the bond-charge model the valence electron charge density is represented by massless point particles, the bond charges (BC), that follow the ionic motion adiabatically. The bond-charge model unit cell consists of two ions and four bond charges that are placed along the bonds between the ions, as depicted in Fig. 6.13. If the charge of every ion is $2Ze$, the BC is equal to $-Ze$. In homopolar covalent crystals, the bond charges are placed midway between the neighboring atoms, while in III–V compounds the BC divides the bond length into a typical ratio of 5:3. This is consistent with non-local pseudopotential calculations for the valence electron charge density that indicate that the charge density maximum in III–V compounds shifts toward the group V element. However the 5:3 ratio is only a rough approximation for most III–V compounds and Harrison's bond orbital model (Harrison 1980) more accurately represents the true situation. This shift is even stronger in the case of II–VI compounds, reflecting their more ionic character. The bond-charge model parameter p, which reflects the polarity of the bond, is defined in terms of the ratio into which the BC position divides the bond length. In this model p is an adjustable parameter

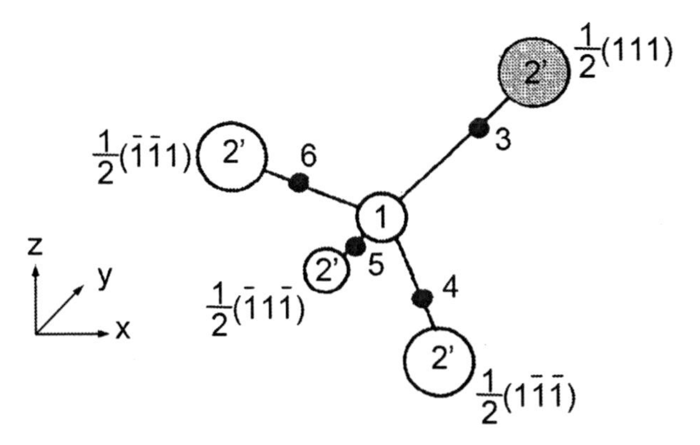

Fig. 6.13. A schematic representation of the adiabatic bond-charge model unit cell with cation 1, anion 2, and BCs 3–6

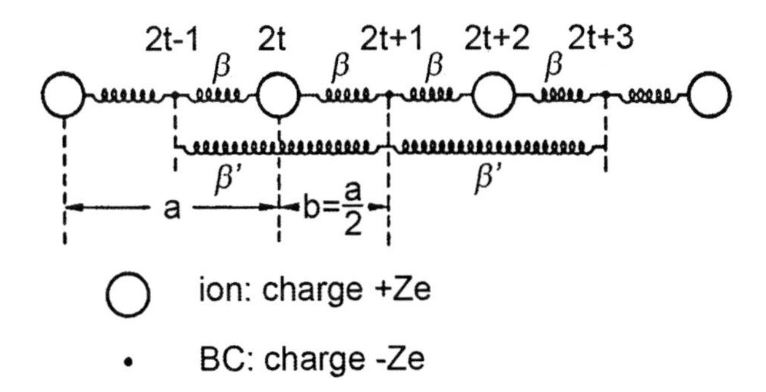

Fig. 6.14. The bond charge model of a linear monatomic chain

that is fit to experiments or other theories and has no fundamental based expression. If l is the bond length and $r_1 = (1 + p)l/2$, $r_2 = (1 - p)l/2$ are the two ion-BC distances, then $p = 0$ ($r_1 = r_2$) for homopolar materials and $p = 0.25$ for III–V compounds. In the extension of the bond-charge model to II–VI materials we have chosen to use $p = 1/3$, corresponding to the ratio $r_1 = 2r_2$, which is based on the average results of 14 microscopic calculations (Chelikowski and Cohen 1976).

We next introduce the short-range ion-BC interaction and long-range Coulomb interaction. The coupling between the nearest neighbor BCs is included. First treat the 1D case, a simple monatomic linear chain illustrated in Fig. 6.14. Here the BCs are assumed to lie in the center of the covalent bonds. β is an ion-BC force constant interaction. β' is the force constant interaction between two nearest neighbor BCs.

Neglecting the short-range and long-range interaction among ions, the equations of motion are given by:

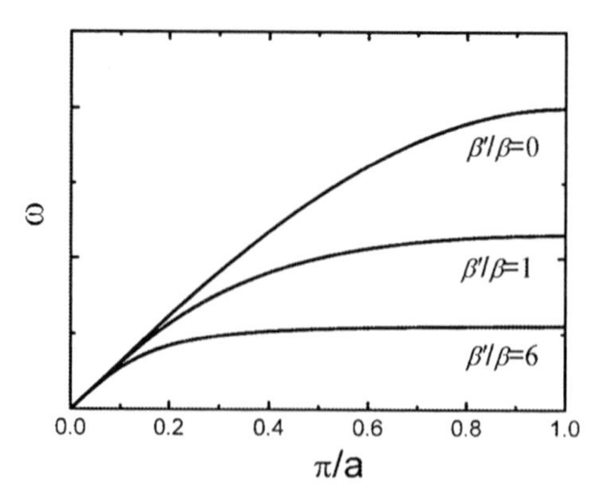

Fig. 6.15. The dispersion curves for a monatomic linear chain with bond charges

$$\begin{aligned} m\ddot{u}_{2t} &= \beta(v_{2t+1} + v_{2t-1} - 2u_{2t}) \\ 0 &= \beta(u_{2t+2} + u_{2t} - 2v_{2t+1}) + \beta'(v_{2t+3} + v_{2t-1} - 2v_{2t+1}) \end{aligned} \tag{6.35}$$

Here we have assumed the mass of the BCs are 0 (the adiabatic approximation). We can solve (6.35) by assuming a trial solution:

$$\begin{aligned} u_{2t} &= \gamma \exp[i(2tbq - \omega t)] \\ v_{2t+1} &= \xi \exp\{i[(2t+1)bq - \omega t]\} \end{aligned} . \tag{6.36}$$

After substitution of this trial solution into (6.35) we get:

$$\begin{aligned} & m\omega^2 \gamma = 2\beta(\gamma - \xi \cos qb) \\ & \beta(\gamma \cos qb - \xi) + \beta' \xi (\cos 2qb - 1) = 0 \end{aligned} . \tag{6.37}$$

Solving the secular equation yields the dispersion equation:

$$m\left[\omega(q)\right]^2 = 2\beta \frac{(\beta + 2\beta')\sin^2\left(\frac{1}{2}aq\right)}{\beta + 2\beta'\sin^2\left(\frac{1}{2}aq\right)} . \tag{6.38}$$

When, $q << \pi/a$, (6.38) becomes:

$$m\left[\omega(q)\right]^2 = \frac{a^2}{2m}(\beta + 2\beta')q^2 = \frac{C}{\rho}q^2 . \tag{6.39}$$

Equation (6.39) shows the elastic constant has the form, $C \sim \beta + 2\beta'$. In Fig. 6.15, various dispersion curves are shown for different β'/β with $\beta + 2\beta'$ held fixed for all curves. If $\beta'/\beta >> 1$, $\omega(q)$ exhibits the typical flattening of the dispersion curve away from $q \approx 0$. In this case, the ions are coupled only weakly to the BCs, which form an almost rigid lattice. The ions vibrate like Einstein oscillators in this lattice, and their frequency is given by the weak ion-BC force constant, β. Only in the long wavelength limit, where the BCs move in phase with the ions, does the strong bond–bond interaction contribute to the dynamics of the ions, to produce a high value of the elastic constant.

6.1.3.4 The Adiabatic Bond-Charge Model in the 3D Case

Four types of interactions in the 3D bond charge model are sketched in Fig. 6.16. The cation and the anion interact with one another and with the bond charges via central potentials $\phi_{ii}(l)$ (ion–ion), $\phi_1(r_1)$ and $\phi_2(r_2)$ (ion-BC). The bond charges centered on a common ion interact via a three-body Keating potential, $V_{bb}^{\sigma} = B_{\sigma}(\vec{X}_i^{\sigma} \cdot \vec{X}_j^{\sigma})/8a_{\sigma}^2$, where $\vec{X}_i^{\sigma}$ is the displacement between ion σ (σ = 1,2) and BC i, β_{σ} is the force constant and a_{σ}^2 is the equilibrium value of $\left|\vec{X}_i^{\sigma} \cdot \vec{X}_j^{\sigma}\right|^2$. $a_1^2 = \frac{1}{3}[\frac{l(1+p)}{2}]^2$, $a_2^2 = \frac{1}{3}[\frac{l(1-p)}{2}]^2$. The bond charges centered on a particular ion also interact directly with one another through a central potential, $\psi_{\sigma}(r_{bb}^{(\sigma)})$, where $r_{bb}^{(\sigma)}$ is the distance between the bond charges centered on the cation (σ = 1) or the anion (σ = 2) (BC–BC). Finally, the ions and the BCs interact via the Coulomb interaction characterized by a single parameter Z^2/ε, where $2Ze$ is the charge of a BC, and ε is the dielectric constant. Each of the ions is presumed to have a charge $+2Ze$ so that the net charge in the unit cell is zero.

To reduce the number of parameters it is assumed that $\partial\psi_1/\partial r_{bb}^1 = \partial\psi_2/\partial r_{bb}^2 = 0$, $\partial^2\psi_1/\partial(r_{bb}^1)^2 = \partial^2\psi_2/\partial(r_{bb}^2)^2 = (\beta_2 - \beta_1)/8$, and $(1+p)\partial\phi_1/\partial r_1 + (1-p)\partial\phi_2/\partial r_2 = 0$. If we use these constraints, the total lattice energy per unit cell it becomes:

$$\begin{aligned}\Phi = 4[\phi_{ii}(t) + \phi_1(r_1) + \phi_2(r_2)] - \alpha_M \frac{(2Z)^2}{\varepsilon}\frac{e^2}{l}, \\ + 6[V_{bb}^1 + V_{bb}^2 + \psi_1(r_{bb}^1) + \psi_2(r_{bb}^2)]\end{aligned} \tag{6.40}$$

where α_M is the Madelung constant. Taken along with the equilibrium conditions, $\partial\Phi/\partial t = 0$, and $\partial\Phi/\partial p = 0$,we find:

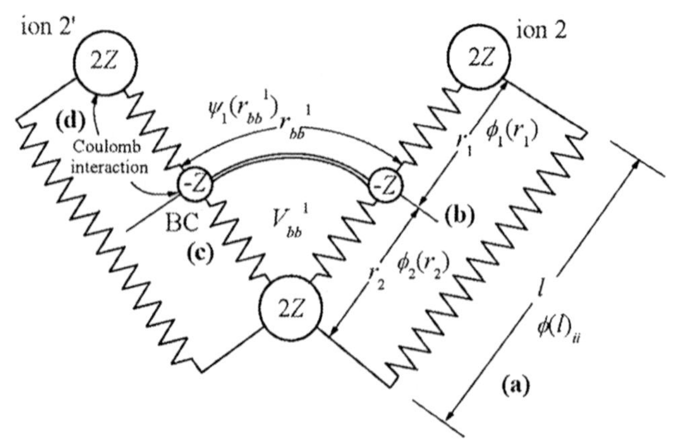

Fig. 6.16. A schematic presentation of the bond charge model (Four types of interaction are taken into account.)

$$\begin{aligned} \frac{d\phi_{ii}}{dl} &= -\alpha_M \frac{Z^2}{\varepsilon}\frac{e^2}{l^2} \\ \frac{\partial\phi_1}{\partial r_1}\frac{1}{r_1} &= 2\frac{d\alpha_M}{dp}\frac{1-p}{1+p}\frac{Z^2}{\varepsilon}\frac{e^2}{l} \\ \frac{\partial\phi_2}{\partial r_2}\frac{1}{r_2} &= -2\frac{d\alpha_M}{dp}\frac{1+p}{1-p}\frac{Z^2}{\varepsilon}\frac{e^2}{l} \end{aligned} \qquad (6.41)$$

The conditions for a stable equilibrium, $\partial^2\Phi/\partial t^2 > 0$, and $\partial^2\Phi/\partial p^2 > 0$, further yield:

$$\begin{aligned} &\frac{4}{3}\frac{d^2\phi_{ii}}{dl^2} + (1+p)^2\left(\frac{1}{3}\frac{\partial^2\phi_1}{\partial r_1^{\,2}} + \frac{\beta_2}{6}\right) \\ &+(1-p)^2\left(\frac{1}{3}\frac{\partial^2\phi_2}{\partial r_2^{\,2}} + \frac{\beta_1}{6}\right) - \frac{128}{9\sqrt{3}}\alpha_m\frac{z^2}{\varepsilon} > 0 \end{aligned} \qquad (6.42)$$

The Coulomb energy per unit cell, $-\alpha_M(2Ze)^2/\varepsilon l$, and β_1, and β_2 are in the units of e^2/v_a. So we have expressions for $d\phi_i/dl$, $\partial\phi_1/\partial r_1$, $\partial\phi_2/\partial r_2$, $\partial\psi_1/\partial r_{bb}^1$, $\partial\psi_2/\partial r_{bb}^2$, $\partial^2\psi_1/\partial(r_{bb}^1)^2$, and $\partial^2\psi_2/\partial(r_{bb}^2)^2$. The six free parameters of the model: $d^2\phi_{ii}/dl^2$, $\partial^2\phi_1/\partial r_1^{\,2}$, $\partial^2\phi_2/\partial r_2^{\,2}$, β_1, β_2 and Z^2/ε, can be fitted to neutron-scattering data and the measured elastic constants.

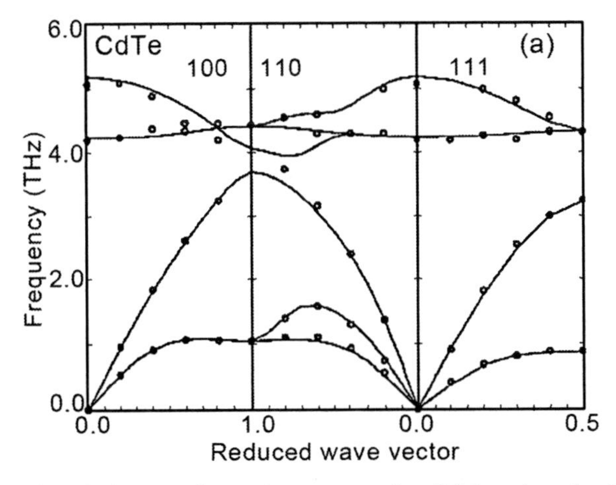

Fig. 6.17. Calculated phonon-dispersion curves for CdTe using the bond charge model (Rajput and Browne 1996)

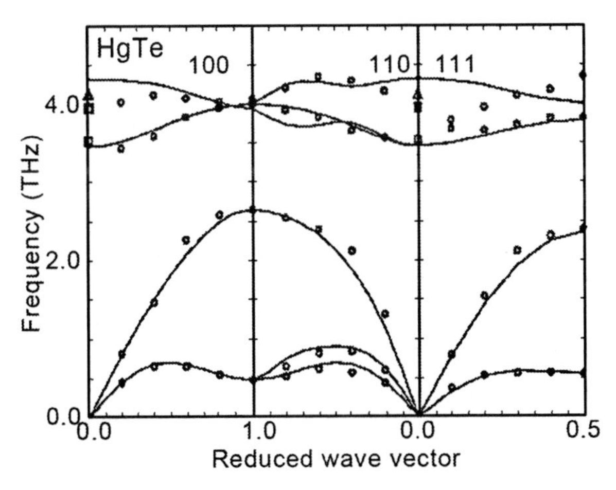

Fig. 6.18. Calculated phonon-dispersion curves for HgTe using the bond charge model (Rajput and Browne 1996)

The phonon eigen-frequencies and eigenvectors are found by diagonalizing the dynamical matrix constructed from the BCM equations of motion:

$$\begin{aligned} \mathbf{M}\omega^2\mathbf{u} &= \mathbf{F}_R\mathbf{u} + \mathbf{F}_T\mathbf{v} \\ 0 &= \mathbf{F}_T^+\mathbf{u} + \mathbf{F}_S\mathbf{v} \end{aligned}, \tag{6.43}$$

with:

$$\mathbf{F}_R = \mathbf{R} + 4\frac{(Ze)^2}{\varepsilon}\mathbf{C}_R, \quad \mathbf{F}_T = \mathbf{T} - 2\frac{(Ze)^2}{\varepsilon}\mathbf{C}_T$$
$$\mathbf{F}_T^+ = \mathbf{T}^+ - 2\frac{(Ze)^2}{\varepsilon}\mathbf{C}_T^+, \quad \mathbf{F}_S = \mathbf{S} + \frac{(Ze)^2}{\varepsilon}\mathbf{C}_S \quad . \tag{6.44}$$

Here **M** is the mass matrix for the ions, and **u** and **v** are the vectors formed by the displacements of the ions and the BCs, respectively. The matrices **R**, **T**, $\mathbf{T}^+$, and **S** are the dynamical matrices for the short-range ion-ion, ion-BC, BC-ion, and BC–BC interactions. The matrices $\mathbf{C}_R$, $\mathbf{C}_T$, $\mathbf{C}_T^+$, and $\mathbf{C}_S$ are the corresponding Coulomb matrices. The matrices $\mathbf{F}_R$, $\mathbf{F}_T$, $\mathbf{F}_T^+$, and $\mathbf{F}_S$ are the corresponding force constant matrices. Rajput and Browne (1996) fitted the dispersion curves for the II–VI materials along with the existing neutron-scattering data. The results for CdTe and HgTe are given in Figs. 6.17 and 6.18. These figures demonstrate that the six-parameter bond-charge model gives a good description of the lattice dynamics of these materials. So it is an effective method to analyze the phonon spectra.

6.2 Reflection Spectra

6.2.1 Two-Mode Model of Lattice Vibrations

Reflection spectra measurements are one of the usual methods to investigate the lattice vibrations of polar crystals. The spectrum is processed by successive applications of dispersion analysis and Kramers–Kronig analysis to deduce the complex dielectric function. The crystal can be considered to be a system consisting of a collection of damped harmonic oscillators. Thus the dielectric function of a mixed crystal can be written as:

$$\varepsilon(\omega) = \varepsilon_\infty + \Delta\varepsilon_{\text{inter}} + \Delta\varepsilon_{\text{phonon}} + \Delta\varepsilon_{\text{intra}} \,. \tag{6.45}$$

where $\Delta\varepsilon_{\text{phonon}}$, $\Delta\varepsilon_{\text{inter}}$ and $\Delta\varepsilon_{\text{intra}}$ represents the contributions of lattice vibrations, inter-band and intra-band transitions, respectively. At low frequencies where the inter-band transitions can be neglected, the infrared response of an alloy can be approximated by equating $\varepsilon(\omega)$ to the sum of two contributions from Lorentzian oscillators corresponding to HgTe-like and CdTe-like TO vibrations plus plasma modes:

$$\varepsilon(\omega) = \varepsilon_\infty^* + \Delta\varepsilon_{\text{phonon}} + \Delta\varepsilon_{\text{intra}}$$
$$= \varepsilon_\infty^* + \sum_j \frac{S_j \omega_{TO,j}^2}{\omega_{TO,j}^2 - \omega^2 - i\omega\Gamma_j} - \omega_P^2 \varepsilon_\infty \cdot \frac{1}{\omega^2 + i\Gamma_P \omega} . \quad (6.46)$$

Then the reflectance can be calculated from:

$$R(\omega) = \left[\frac{\varepsilon^{1/2}(\omega) - 1}{\varepsilon^{1/2}(\omega) + 1}\right]^2 = \frac{[n(\omega) - 1]^2 + k^2(\omega)}{[n(\omega) + 1]^2 + k^2(\omega)}, \quad (6.47)$$

$$\begin{aligned} \varepsilon'(\omega) &= n^2(\omega) - k^2(\omega) \\ \varepsilon''(\omega) &= 2n(\omega)k(\omega) \end{aligned} . \quad (6.48)$$

S_j, $\omega_{\text{TO}\,j}$, and Γ_j represent the strength, frequency and damping constant for the jth oscillator, and ω_P and Γ_p are the frequency and damping constant of the plasma oscillations. ω_P is written as:

$$\omega_P = \sqrt{\frac{4\pi \cdot n_p e^2}{m^* \varepsilon_\infty}}, \quad (6.49)$$

where n_p is the carrier concentration responsible for the plasma. By selecting the oscillator parameters S_j, $\omega_{\text{TO},j}$, Γ_j and Γ_{P}, we can find a model spectrum that approximates the measured reflectivity.

In $Hg_{1-x}Cd_xTe$, Te occupies the anion sub-lattice positions, and Hg and Cd share the cation positions. So we can regard the alloy as a $(CdTe)_x(HgTe)_{1-x}$ mixed crystal. There are CdTe-like and HgTe-like vibrational modes in the crystal. Identification of the two-mode behavior is obvious in the measured reflectivity spectra (Kim and Narita 1971; Baars and Sorger 1972; Mooradian and Harman 1971; Georgitse et al. 1973; Dornhaus et al. 1982). Two strong reflectivity bands are observed at frequencies that vary little with alloy composition x. As x increases, the high-frequency CdTe-like band becomes weaker, departing from the value for pure CdTe; and the low-frequency HgTe-like band strengthens approaching the value for pure HgTe. Baars and Sorger (1972) measured the reflectivity spectra of $Hg_{1-x}Cd_xTe$ single crystals with $0 < x < 0.54$, which are reproduced in Fig. 6.19. Nimtz measured the reflectivity spectra of a $Hg_{0.8}Cd_{0.2}Te$ sample displayed in Fig. 6.20 (Dornhaus 1983). The reflectivity spectra can be fitted by using (6.45). Assuming there are CdTe-like and HgTe-like optical phonon modes, we can deduce phonon frequencies and high-frequency dielectric constant parameters. The high-frequency dielectric constants are presented in Fig. 6.21.

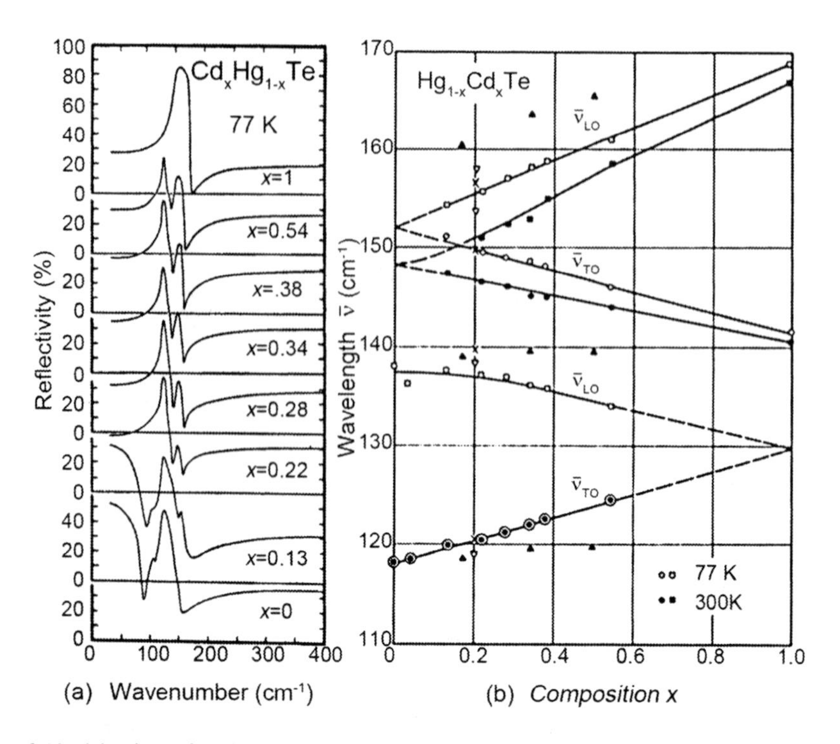

Fig. 6.19. (**a**) The reflectivity of HgCdTe alloys as a function of x at 77 K; and (**b**) The longitudinal and transverse phonon frequencies of HgCdTe as a function of mole fraction x of CdTe at 77 and 300 K

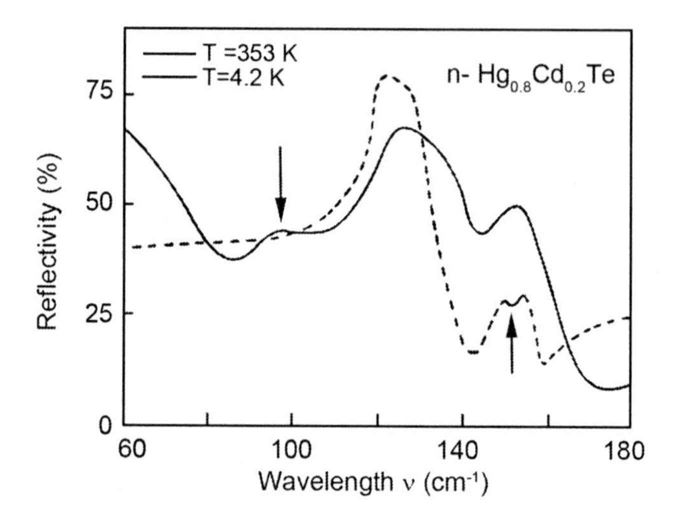

Fig. 6.20. The reflection spectra of an n-$Hg_{0.8}Cd_{0.2}Te$ sample

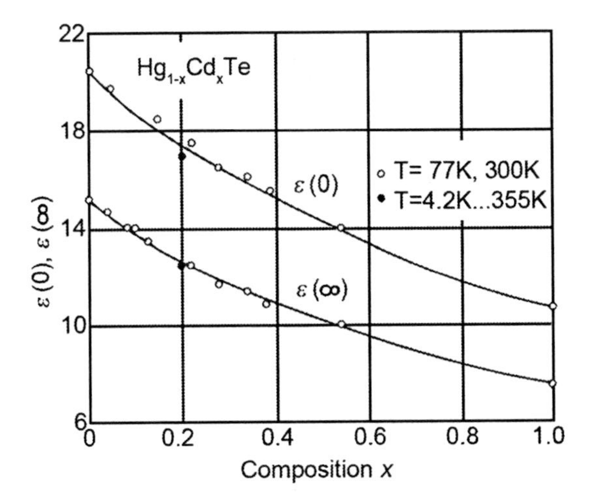

Fig. 6.21. The high-frequency dielectric constant $\varepsilon(\infty)$ of $Hg_{1-x}Cd_x Te$ alloys as a function of x, fits to $\varepsilon(\infty)=16.19-14.52x + 11.06x^2-4.24x^3$, in the ranges $0 < x < 1$, and 4.2 K < T < 300 K

6.2.2 Multi-Mode Model of Lattice Vibration

One can see in Fig. 6.20 that in a $Hg_{0.8}Cd_{0.2}Te$ alloy the CdTe-like reflection spectra mode at low temperature splits. Vodopyanov et al. observed this feature in their experiments (Vodopyanov et al. 1984). One of the most interesting features of the reflection spectra is the existence of this fine structure of the CdTe-like band. This experimentally observed fine structure was interpreted as due to the clustering of like cations (Cd and Hg) around the anions (Te), as shown in Fig. 6.22. There are five types of local clusters surrounding each Te site. If j is the number of Cd atoms on the four adjacent cation sites then $j = 0, 1, 2, 3$, or 4. If an alloy with concentration x is random, the populations of these clusters x_j are distributed in a binominal distribution, $x_j = \frac{4!}{(4-n_j)!n_j!}(1-x)^{4-n_j} x^{n_j}$ (Chen and Sher 1995; Sher et al. 1987, 1991). The effects of correlations on these cluster populations ware first deduced by Sher et al. (1987), and in $Hg_{1-x}Cd_x Te$ alloys is only a small correction. The reflection peaks near 125 cm^{-1} are a HgTe-like band, and the peaks near 155 cm^{-1} are a CdTe-like band. The CdTe-like band slightly shifts to lower frequencies with increasing x, and the HgTe-like band remains stationary. It is seen that the increase in x makes the fine structure of the CdTe-like band more pronounced. The reflection peak intensity of the CdTe-like band is nearly

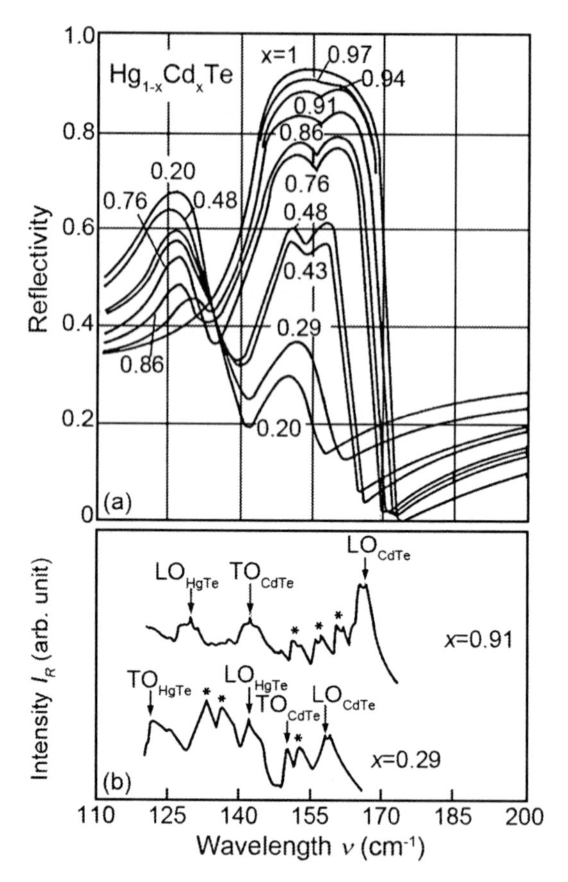

Fig. 6.22. $Hg_{1-x}Cd_xTe$ (**a**) reflectivity, and (**b**) Raman scattering intensity vs. wavenumber for different x at 85 K. The structures in (**b**) marked with an *asterisk* correspond to additional vibrational modes due to clustering

Table 6.1. Mole fraction x, band gap E_g, carrier concentration n of samples

	No. 1	No. 2	No. 3	No. 4	No. 5
X	0.45	0.33	0.27	0.20	0.18
d (μm)	380	290	880	435	190
E_{g300K} (eV)	0.518	0.344	0.260	0.167	0.148
n_{77K} (cm^{-3})	1.14×10^{14}	7.6×10^{15}	1.2×10^{14}	1.6×10^{15}	6.6×10^{15}

the same as that of the HgTe-like band for $x = 0.48$. The HgTe-like band vanished when $x = 1$. Except for the LO phonons of the HgTe-like band, and the LO, and TO phonons of CdTe-like band in the Raman spectra, there are several modes between TO_{CdTe} and LO_{CdTe}, which are due to the slightly different response frequencies of the different cluster types.

Shen and Chu (Shen and Chu 1985; Chu and Shen 1993) investigated the phonon spectra of $Hg_{1-x}Cd_xTe$ alloys with compositions $x = 0.18$ to 0.45 by means of reflection spectroscopy from 4.5 to 300 K. In addition to the fine structures of the CdTe-like optical phonon reflection bands, the complex structure of the HgTe-like band also was observed. The reflection spectra were fitted by using the multi-mode quasi-harmonic oscillator model. The phonon modes, the complex dielectric function spectra, and the complex refractive index spectra have been obtained from a satisfactory fit. They explained the fine structures of the CdTe-like band as a clustering effect, and complex structure of the HgTe-like band as a plasmon-LO phonon coupling effect.

Properties of some of the samples tested are shown in Table 6.1. The reflectance spectra measurements were performed with a Bruker IFS-113 Fourier transform spectrometer in the region 15–400 cm^{-1}, at temperatures 4.5–300 K. The resolution of the spectrometer is 1 cm^{-1}.

Figures 6.23 and 6.24 show the far infra-red reflection spectra for two $Hg_{1-x}Cd_xTe$ samples with compositions $x = 0.45$ and 0.18. In Fig. 6.23 the circles and triangles represent the measured reflection spectra for the sample with composition $x = 0.45$ at 35 and 105 K, and in Fig. 6.24 for $x = 0.18$ at 4.5 and 300 K. The full and broken curves in the figures are the fitted curves based on the multi-mode quasi-harmonic oscillator model. From the figures one can see clearly two characteristic reflection bands for these alloys, i.e. the HgTe-like LO–TO band and the CdTe-like LO–TO band. The fine structure of the CdTe-like reflection band can be observed for the samples with large x values (Fig. 6.23). The lower the measurement temperature is, the clearer the features are. Another significant feature in the reflection band near 130 cm^{-1} for the samples with a small x value reveals some complicated structures at room and higher temperature (Fig. 6.24).

The calculation indicates that the double-oscillator model is a good approximation to record parameters for the two major reflection bands for the alloy, but extra oscillators are needed to fit the fine structures of the CdTe-like and HgTe-like bands. The optimum fitting curve for an $x = 0.45$ sample is achieved in Fig. 6.23 by adding five weaker oscillators to the two main oscillators. The parameters for the seven oscillators are listed in Table 6.2. ω_P at 4.5 and 300 K are given in later discussion. Damping constants are chosen as 10 and 35 from Danielewicz and Coleman (1974) For the sample with $x = 0.18$, after adopting a six-oscillator model listed in Table 6.3, and considering the contribution of the plasma oscillation term, the fitted curve compares well with the experimental curves as shown in Fig. 6.24. The positions and the structures of peaks occurring in the experimental spectra are exactly reproduced by the fitted curves.

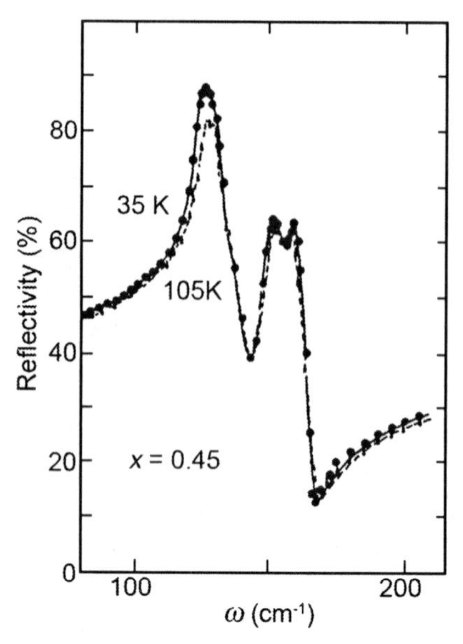

Fig. 6.23. The far-infrared reflection spectra for a $Hg_{1-x}Cd_xTe$ alloy with $x = 0.45$ at 35 K (*solid line* fitted, *filled circle* experimental), and 105 K (*dashed line* fitted, *filled triangle* experimental)

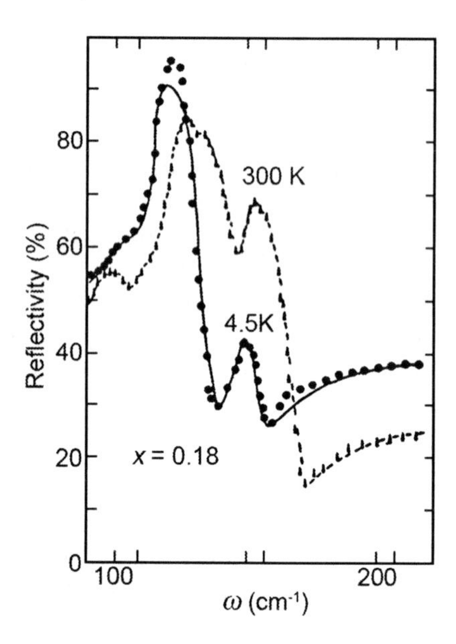

Fig. 6.24. The far-infrared reflection spectra for a $Hg_{1-x}Cd_xTe$ alloy with $x = 0.18$ at 4.5 K (*solid line* fitted, *filled circle* experimental), and 300 K (*dashed line* fitted, *filled triangle* experimental)

Table 6.2. Parameters for the multiple oscillators used to fit the measured reflection spectra of $Hg_{1-x}Cd_xTe$ with $x = 0.45$

	36 K			105 K		
	$\omega_{TO,j}$ (cm^{-1})	S_j	Γ_j (cm^{-1})	$\omega_{TO,j}$ (cm^{-1})	S_j	Γ_j (cm^{-1})
Low-frequency oscillator	110	0.3	10	110	0.3	10
HgTe-like oscillator	123	6.6	2.8	124	6.5	4.5
	134	0.7	9	135	0.6	9
	141	0.15	6	141	0.15	7
CdTe-like oscillator	149	0.9	4	150	0.8	5
	153	0.35	6	153	0.35	7.5
	157	0.15	4	157	0.15	4.8

Table 6.3. Parameters for the multiple oscillators used to fit the measured reflection spectra of $Hg_{1-x}Cd_xTe$ with $x = 0.18$

	4.5 K			300 K		
	$\omega_{TO,j}$ (cm^{-1})	S_j	Γ_j (cm^{-1})	$\omega_{TO,j}$ (cm^{-1})	S_j	Γ_j (cm^{-1})
Low-frequency oscillator	105	2.7	12.5	100	2.7	12.5
	112	1.7	8	113	1.3	10
HgTe-like oscillator	116	5	2.3	122	5	6.6
	135	0.15	8	132	0.18	8
CdTe-like oscillator	142	0.1	5	140	0.1	8
	147	0.45	7.5	149	0.45	9

As discussed above the structure of $Hg_{1-x}Cd_xTe$ alloys is zincblende, with four fold near neighbor coordination. Each Te anion is surrounded by four cation sites that can be occupied by either Hg or Cd ions. The populations of these five atom clusters x_j, with j being the number of Cd atoms in the cluster, $j = 0$, 1, 2, 3, or 4, are in general given by a grand canonical statistical distribution (Sher et al. 1987). When the energy difference between different cluster types is small, then the distribution reduces to the simple binomial distribution given above. While in HgCdTe alloys the deviation from the random alloy result is small the direction of the deviation is toward an ordered crystalline arrangement rather than spinodal decomposition.

There is another way the clustering effect has been described. It introduces a short range order parameter β defined in terms of the probabilities $P_{Cd,Cd}$, and $P_{Hg,Hg}$ of finding adjacent bonds with two Cd ions or two Hg ions respectively, by the expressions:

$$\begin{aligned} P_{\text{Cd,Cd}} &= x + \beta(1-x) \\ P_{\text{Hg,Hg}} &= (1-x) + \beta x \end{aligned} . \tag{6.50}$$

For $\beta = 0$, $P_{\text{Cd,Cd}} = x$, $P_{\text{Hg,Hg}} = 1-x$ and the distribution is random, while for $\beta = 1$, $P_{\text{Cd,Cd}} = P_{\text{Hg,Hg}} = 1$ and a complete clustering is obtained with the material separated into two regions one of CdTe and the other is HgTe (spinodal decomposition). This simpler short range order theory does not allow for the possibility of a long ranged ordered crystalline arrangement.

The equations of motion for the five basic units have been done taking into account only the elastic interactions between nearest neighbors. On this basis the equations were solved and the frequencies and the oscillator strengths of the optical phonon modes obtained for each basic unit (Kozyrev et al. 1983). In HgCdTe lattice matched alloys the elastic energies are small, and the principal interactions distinguishing between the different units are due to charge transfers among the ions (Sher et al. 1987), so care must be taken in interpretations of the data based on this model. If the interactions between second neighbors are included, the combinations of the modes, and the LO phonon-plasmon coupling, produce more than five oscillator frequencies. It is similar to the case of $Ga_xAs_{1-x}P$ (Verleur and Barker 1966; Shen 1984). This is the reason the weakest model oscillators are physically significant, and cannot be omitted from the fitting calculations. The fine structure of the CdTe-like reflection band for the sample with $x = 0.45$ in Fig. 6.23 and the complex structures of HgTe-like reflection band for the sample with $x = 0.18$ in Fig. 6.24 are direct evidence for the need to include the distinctions among the different clusters.

The fitting calculation to the reflection spectra in Figs. 6.23 and 6.24 indicates it is necessary to use the high-frequency dielectric constant as a parameter, and also to take into account the imaginary part of dielectric constant. Nor can we omit the contributions from the inter-band and the intra-band transitions to the dielectric function when investigating the reflection spectra of narrow band gap, high free carrier concentration HgCdTe samples using dielectric function theory. The contribution of intra-band transition to the dielectric function is introduced through the plasmon oscillation term. The contribution from inter-band transitions enters through their contribution to the equivalent high-frequency dielectric constant:

$$\varepsilon_\infty^* = \varepsilon_\infty + \Delta\varepsilon_{\text{inter}} . \tag{6.51}$$

The contribution, $\Delta\varepsilon_{\text{inter}}$, includes both real and imaginary parts. Its imaginary part is written as

$$\Delta\varepsilon''_{\text{inter}} = \frac{4\pi^2 h^2 e^2}{m_0} \cdot \frac{\left|\left(k,c\middle|P\middle|k,V\right)\right|^2}{\left|E_C(k) - E_V(k)\right|^2} \cdot \left\{\delta\left[h\omega - E_C(k) + E_V(k)\right] f_V (1 - f_C) - \delta\left[-h\omega + E_C(k)\right] f_C (1 - fV)\right\}, \quad (6.52)$$

where f_C, f_V are the Fermi functions for the valence band and the conduction band, respectively. The real part of the inter-band transition contribution to the dielectric function can be calculated from the KK relations. The result is almost unrelated to the frequency (Polian et al. 1976) but does vary with the carrier concentration. $\Delta\varepsilon' \approx 6$ at 4.5 K for an $x = 0.18$ sample, which is the same as the estimation made by Polian et al. (1976). $\Delta\varepsilon''$ has been deduced from the adsorption spectra (Shen 1984), and taking into account the density state function in the band tail. It can be approximated by an empirical expression:

$$\Delta\varepsilon'' = \sqrt{B(\omega - \omega_0)}, \quad (6.53)$$

where B and ω_0 are determined by band gap, which are functions of the composition and the measured temperature. For $x = 0.18$ and 4.5 K, we find $B = 1.2$, and $\omega_0 = 148\ \text{cm}^{-1}$. By taking into account the correction to the high-frequency dielectric constant, the fitted curves can be adjusted to be consistent with the experimental results in Figs. 6.23 and 6.24.

6.2.3 Plasmon Oscillation-LO Phonon Coupling Effect

When a solid is irradiated with infrared at a photon frequency near that of a TO phonon band, the photon-TO coupling enters. The coupled TO vibration-photon modes are often called "polaritons". Absorption and emission of a single phonon in the mode is a unit polariton. The photon-TO coupling is described by the dispersion equations:

$$c^2 k^2 = \omega^2 \varepsilon$$
$$\varepsilon = 1 + 4\pi \frac{P}{E} \cdot \quad (6.54)$$

E is the electric field. ω is the frequency of electric field variation for the photon-TO coupling, which is also the frequency of the oscillator. It is taken to be the frequency of the incident light field. P is the polarizability due to the displacements of positive and negative ions in the crystal. $\boldsymbol{k}$ is an electromagnetic wave vector, which is determined by the dielectric function after polarization. $\boldsymbol{k}$ is also the wave vector of polariton. $\boldsymbol{k}$ is different from the incident photon $\boldsymbol{k}_0$ due to the contribution from polarization. Thus (6.54) describes the motion of the oscillator after the photon-TO coupling. Once P is determined, all the physical quantities in (6.54) are set. Let P be:

$$P = Nqu\,, \tag{6.55}$$

where u is the displacement between the positive and negative ion-pair, q is the charge of the ions, and N is the number of ion-pairs per unit volume. u is determined from the equations of motion. We choose the negative ion coordinate as zero and investigate the motion of the positive ions. The motion can be regarded as a forced vibration of an elastic bound charged particle due to the bounding of the ions around the center ion. The equation of motion is given by:

$$M\ddot{u} + M\omega_{TO}^2 u = qE\,, \tag{6.56}$$

where ω_{TO} is the frequency of the transverse optical phonons. By assuming a trial solution: $u = u_0 e^{i\omega t}$, the equation can written:

$$-M\omega^2 u + M\omega_{TO}^2 u = qE\,. \tag{6.57}$$

When the solution for u is substituted into (6.55), it leads to:

$$P = \frac{Nq^2 E}{M(\omega_{TO}^2 - \omega^2)}\,, \tag{6.58}$$

and

$$\varepsilon(\omega) = 1 + \frac{4\pi Nq^2 / M}{\omega_{TO}^2 - \omega^2}\,. \tag{6.59}$$

Equation (6.54) can also rewritten as:

$$c^2 k^2 = \omega^2 \left(1 + \frac{4\pi Nq^2 / M}{\omega_{TO}^2 - \omega^2}\right), \tag{6.60}$$

if $k = 0$ ($\lambda \to \infty$), which characterizes the long wave group vibration modes in the solid. We can solve for the permitted values of ω. It is obvious that $\omega_1^2 = 0$ is one of the solutions, which is the case of no irradiation. Another solution type is:

$$\omega_2^2 = \omega_{TO}^2 + 4\pi Nq^2 / M = \omega_L^2\,. \tag{6.61}$$

This equation is for the group vibration mode with frequency, ω_L. The previous discussion does not take into account the external field polarization effect on the electrons in the ion core. Including this effect, when $\omega \to \infty$, then $\varepsilon(\omega) = \varepsilon(\infty)$, which is the dielectric constant at high-frequency. Then (6.59) is modified to read:

$$\varepsilon(\omega) = \varepsilon(\infty) + \frac{4\pi Nq^2 / M}{\omega_{TO}^2 - \omega^2}. \tag{6.62}$$

When $\omega \to 0$, $\varepsilon(0)$ is the static dielectric constant which is:

$$\varepsilon(0) = \varepsilon(\infty) + \frac{4\pi Nq^2 / M}{\omega_{TO}^2 - \omega^2}, \tag{6.63}$$

and $\varepsilon(0) - \varepsilon(\infty) = \dfrac{4\pi Nq^2 / M}{\omega_{TO}^2 - \omega^2}$.

When $\omega \to \omega_L$, $k = 0$ and $\varepsilon = 0$ then we have:

$$0 = \varepsilon(\infty) + \frac{4\pi Nq^2 / M}{\omega_{TO}^2 - \omega^2}. \tag{6.64}$$

Immediately we get:

$$\frac{\varepsilon(0)}{\varepsilon(\infty)} = \left(\frac{\omega_{LO}}{\omega_{TO}} \right)^2, \tag{6.65}$$

which is called the LST relation. Then we can write the usual dielectric function:

$$\varepsilon(\omega) = \varepsilon(\infty) + \frac{\omega_{TO}^2 [\varepsilon(0) - \varepsilon(\infty)]}{\omega_{TO}^2 - \omega^2}. \tag{6.66}$$

After taking into account the damping term, the equation of motion becomes:

$$M\ddot{u} + M\Gamma \dot{u} + M\omega_{TO}^2 u = qE, \tag{6.67}$$

where Γ is the damping coefficient. Thus the corresponding dielectric constant is:

$$\varepsilon(\omega) = \varepsilon(\infty) + \frac{\omega_{TO}^2 [\varepsilon(0) - \varepsilon(\infty)]}{\omega_{TO}^2 - \omega^2 - i\Gamma\omega}, \tag{6.68}$$

and here:

$$\varepsilon'(\omega) = \varepsilon(\infty) + \frac{[\varepsilon(0) - \varepsilon(\infty)]\omega_{TO}^2(\omega_{TO}^2 - \omega^2)}{\left(\omega_{TO}^2 - \omega^2\right)^2 + \Gamma^2\omega^2}$$
$$\varepsilon''(\omega) = \frac{[\varepsilon(0) - \varepsilon(\infty)]\omega_{TO}^2\omega T}{\left(\omega_{TO}^2 - \omega^2\right)^2 + \Gamma^2\omega^2} \quad (6.69)$$

We can obtain n and κ from $\varepsilon' = n^2 - \kappa^2$ and $\varepsilon'' = 2n\kappa$.

By taking into account the oscillations of the free carriers in a radiation field, we get:

$$\varepsilon(\omega) = \varepsilon(\infty) - \frac{\omega_p^2\varepsilon(\infty)}{\omega^2}, \quad (6.70)$$

where ω_p is the plasmon frequency. If the oscillations of ions and electrons are considered at same time, the dielectric function becomes:

$$\varepsilon(\omega) = \varepsilon(\infty) - \frac{\omega_p^2\varepsilon(\infty)}{\omega^2} + \frac{\omega_{TO}^2[\varepsilon(0) - \varepsilon(\infty)]}{\omega_{TO}^2 - \omega^2}. \quad (6.71)$$

When the frequency of the plasmon oscillations ω_p is near the frequency of the LO-phonon, a coupling becomes important, that affects the collective electron oscillations. It's affect corresponds to the case with $k = 0$, and $\varepsilon(\omega) = 0$. By taking account of (6.65) and with (6.71) set equal to 0, we get the frequency of the LO photon-plasmon coupling.

As can be seen in Fig. 6.24, there exists a low-frequency reflection peak around 100–110 cm^{-1} that is similar to the reflection spectrum of HgTe (Balkanski 1979). All these structures result from the multi-mode behavior of the lattice vibrations and the plasmon-LO phonon coupling effect. The plasma reflection and its temperature variation for the sample with $x = 0.18$ are shown in Fig. 6.25. There exist complicated structures in the HgTe-like reflection band for the sample with $x = 0.18$ at room temperature. The temperature variation of the CdTe-like reflection band of this sample appears to be stronger than that of other samples. In order to explain these phenomena and the measured spectra shown in Fig. 6.24, account must be taken not only of the contribution of the plasmon to the dielectric function, but also the plasmon-LO phonon coupling. When ω_P and ω_{LO} approach closely, the frequencies of coupled modes are (Verga 1965):

$$\Omega_\pm^2 = \frac{1}{2}\left(\omega_P^2 + \omega_{LO}^2\right) \pm \frac{1}{2}\left[\left(\omega_P^2 + \omega_{LO}^2\right)^2 - 4\omega_P^2\omega_{TO}^2\right]^{\frac{1}{2}}, \quad (6.72)$$

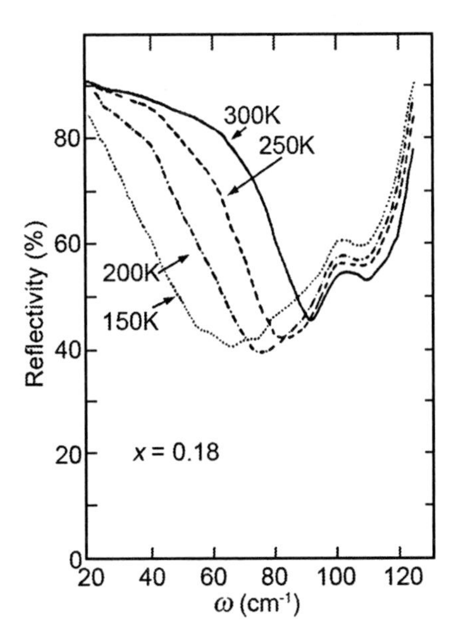

Fig. 6.25. The plasma reflection edge and its temperature variation for the sample with $x = 0.18$ in the far infrared region

where ω_P is the plasmon oscillation frequency, which is a function of the effective mass of electrons and can be calculated from (6.49). It should be noted that the variation of the electron effective mass at the Fermi level with carrier concentration must be taken into account for a degenerate HgCdTe semiconductor with its non-parabolic conduction band. m^* is the effective mass at the Fermi surface. It is related to the carrier concentration, but ω_P^2 is not directly proportional to the carrier concentration as it is in a typical parabolic band semiconductor. In the investigation of the band structure of HgCdTe, Sella et al. deduced the expression (Sella et al. 1967; Dornhaus and Nimtz 1976):

$$\left(\frac{m^*}{m_0 - m^*}\right)^2 = \left(\frac{m_0^*}{m_0}\right)^2 + \frac{3\hbar^2(3\pi^2 n)^{2/3}}{m_0 E_p}, \tag{6.73}$$

where m_0^* is the effective mass at the bottom of conduction band, $m_0^* = 3\hbar^2 E_g / 4P^2$, and $E_p = (2m_0/\hbar^2)P^2$. Under their experimental conditions, $m^* << m_0$, so (6.73) reduces to:

$$\left(\frac{m^*}{m_0}\right)^2 = \left(\frac{m_0^*}{m_0}\right)^2 + \frac{3}{2}\left(\frac{\hbar^2}{m_0 P}\right)^2 \left(3\pi^2 n\right)^{\frac{1}{2}}, \tag{6.74}$$

Table 6.4. Some parameters for a $Hg_{1-x}Cd_xTe$ sample with $x = 0.18$ at different temperatures

T (K)	Eg (ev)	n (cm^{-3})	m^*/m_0	ω_P (cm^{-1})
4.5	0.034	6.55×10^{15}	0.00754	63
200	0.104	2.06×10^{16}	0.0136	85
250	0.122	3.62×10^{16}	0.0162	103
300	0.140	6.54×10^{16}	0.0168	112

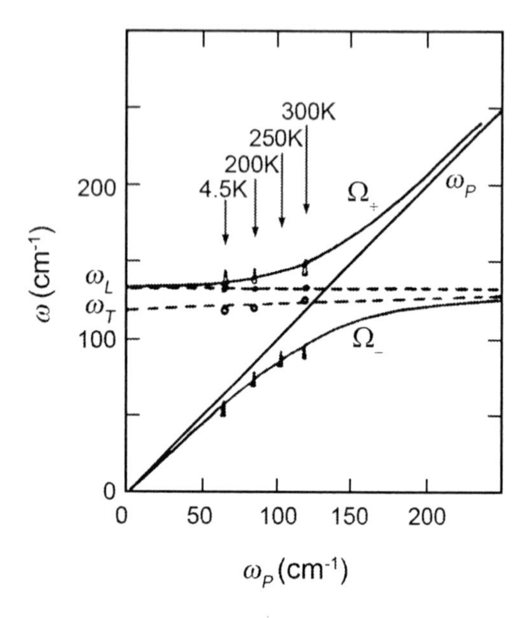

Fig. 6.26. The dispersion relation for the plasmon-LO phonon coupling mode

where $P = 8 \times 10^{-8}$ eV cm (Chu 1983) is the momentum matrix element. The heavy hole effective mass is much lager than that of the conduction band electrons. Thus only the plasmon oscillations of electrons need be considered, that of holes is omitted. The calculated effective mass m^* and the plasmon frequencies ω_P for the sample with $x = 0.18$ at different temperatures are listed in Table 6.4.

In Fig. 6.26 the circles represent the values of ω_{LO} and ω_{TO} deduced from curve-fitting. The curves Ω_- and Ω_+ are the coupled plasmon-LO phonon modes, that are calculated from (6.72). The experimental data for the upper branch Ω_+ of the coupled modes was obtained from the peak positions of the Im[$-1/\varepsilon(\omega)$] spectrum, that is found from a fitting to the measured reflection spectrum. The experimental data on the lower branch $\Omega_-$was read from the reflection spectra of Fig. 6.25 in the far-infrared region.

6.2.4 HgCdTe Far-Infrared Optical Constant

Here we investigate the spectra of the complex dielectric function and the complex refraction coefficient, which are fitted to the reflection spectra using the multi-mode oscillator model. According to the dielectric function theory, the real part $\varepsilon'(\omega)$ and imagery part $\varepsilon''(\omega)$ of the dielectric function are given by:

$$\begin{cases} \varepsilon''(\omega) = \Delta\varepsilon''_{inter} + \sum_j \dfrac{S_j\omega_{TO,j}^2\Gamma_j\omega}{(\omega_{TO,j}^2-\omega^2)+\Gamma_j^2\omega^2} - \dfrac{\omega_P^2\varepsilon_\infty\cdot\omega^2}{\omega^4+\Gamma_P^2\omega^2} \\ \varepsilon'(\omega) = \varepsilon_\infty + \Delta\varepsilon'_{inter} + \sum_j \dfrac{S_j\omega_{TO,j}^2\left(\omega_{TO,j}^2-\omega^2\right)}{\left(\omega_{TO,j}^2-\omega^2\right)+\Gamma_j^2\omega^2} + \dfrac{\omega_P^2\varepsilon_\infty\Gamma_P\omega}{\omega^4+\Gamma_P^2\omega^2} \end{cases} . \quad (6.75)$$

For $\Delta\varepsilon''_{inter}$ we use (6.53), and $\Delta\varepsilon'_{inter}$ is approximated as a constant. S_j, $\omega_{TO,j}$, and Γ_j are the strength, frequency and damping constants for the jth oscillator, that are given in Tables 6.2 and 6.3. These values were obtained from a fit to the reflection spectra. Then we can solve for the spectra of $\varepsilon'(\omega)$ and $\varepsilon''(\omega)$, and the complex index of refraction:

$$\tilde{n} = \sqrt{\tilde{\varepsilon}(\omega)} = n + ik, \text{ or } n = \sqrt{\frac{\sqrt{\varepsilon'^2+\varepsilon''^2}+\varepsilon'}{2}}$$
$$k = \sqrt{\frac{\sqrt{\varepsilon'^2+\varepsilon''^2}-\varepsilon'}{2}} \quad (6.76)$$

In this way we obtain the spectra of the complex dielectric function, and the complex index of refraction of HgCdTe; or the refractive index n and extinction coefficient k. The imagery part of the dielectric function spectra, $\varepsilon''(\omega)$, and $\mathrm{Im}[-1/\varepsilon(\omega)] = \varepsilon''(\omega)/(\varepsilon'^2+\varepsilon''^2)$, for a $Hg_{1-x}Cd_xTe$ sample with $x = 0.18$ are presented in Figs. 6.27 and 6.28, respectively. In Fig. 6.27 the frequencies of the TO phonon modes ($\omega_{TO,j}$), labeled by arrows, are at the peaks of the structures. In Fig. 6.28, the frequencies of the LO phonons ($\omega_{LO,j}$) labeled by arrows, are also at the peaks of these structures. The refractive index n and extinction coefficient k, are presented in Figs. 6.29 and 6.30. The lattice vibration frequencies of HgCdTe, have been use to investigate the lattice vibration modes of the quarternary alloy $Hg_{1-x-y}Cd_xMn_yTe$ (Mazur et al. 1993).

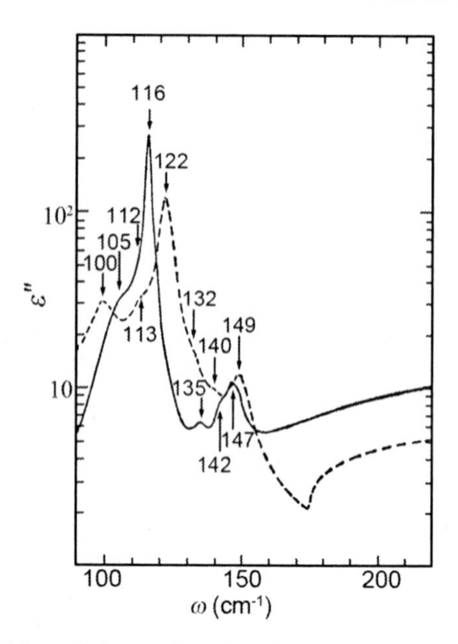

Fig. 6.27. At 4.5 K (*the solid curve*) and at 300 K (*the dash curve*), the imagery part of the dielectric function spectra $\varepsilon''(\omega)$ of a $Hg_{1-x}Cd_xTe$ ($x = 0.18$) sample. The locations of the $\omega_{TO,j}$ are labeled by *arrows*

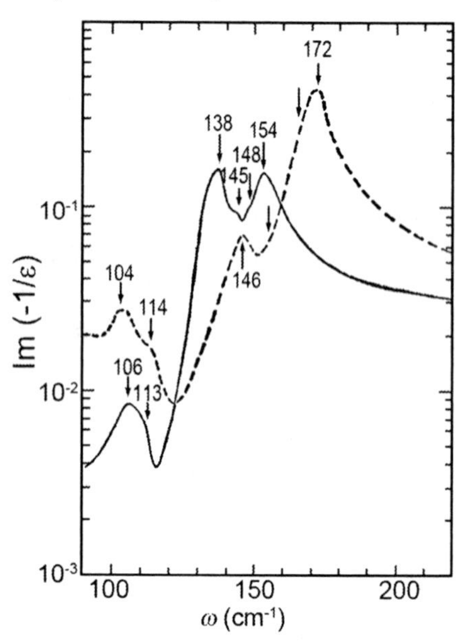

Fig. 6.28. At 4.5 K (*the solid curve*) and at 300 K (*the dash curve*), the imagery part of $-1/\varepsilon(\omega)$ of a $Hg_{1-x}Cd_xTe$ ($x = 0.18$) sample. The locations of the $\omega_{LO,j}$ are labeled by *arrows*

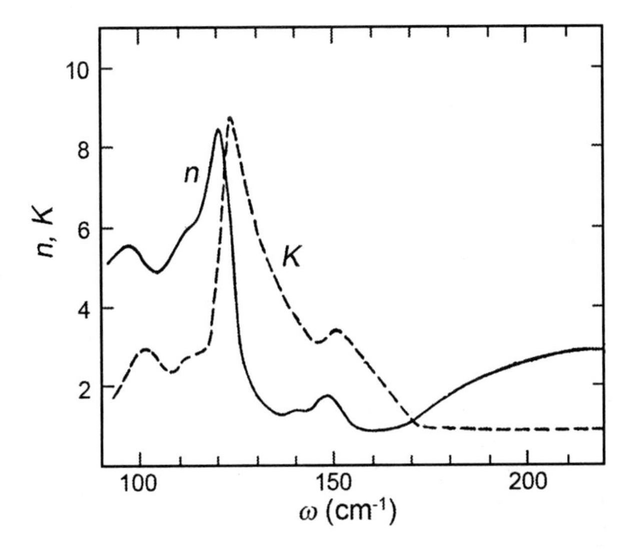

Fig. 6.29. The refractive index n and extinction coefficient k dispersion curves for a $Hg_{1-x}Cd_xTe$ ($x = 0.18$) sample at 300 K

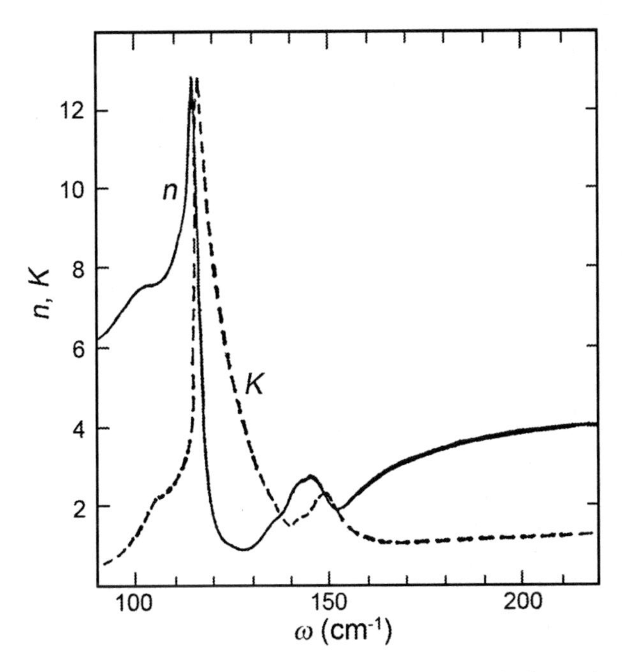

Fig. 6.30. The refractive index n and extinction coefficient k dispersion curves for a $Hg_{1-x}Cd_xTe$ ($x = 0.18$) sample at 4.5 K

6.3 Transmission Spectra

6.3.1 Far-Infrared Transmission Spectra

The main techniques initially used to study the vibrational spectra of $Hg_{1-x}Cd_xTe$ are Raman scattering and far-infrared (FIR) reflectivity. At first little attention was paid to the use of the FIR transmission in the treatment of phonon vibration modes due to the high absorption coefficients of $Hg_{1-x}Cd_xTe$ in the reststrahlen region. However, the FIR transmission gives direct information on phonon modes and impurity states. In addition, some new phenomena, especially at low frequency (~20–70 cm^{-1}), were observed by means of FIR transmission, that were not found in Raman scattering or FIR reflection. The vibrational modes in $Hg_{1-x}Cd_xTe$ epitaxial films grown by LPE and MBE methods and HgTe/CdTe superlattices have been investigated by means of FIR transmittance spectra by Li et al. (1996). The LPE layers, with a typical thickness of 20 μm, were grown on (111) B CdTe substrates from a Te-rich solution by a vertical dipping technique, while the MBE epilayers were grown on (211) B GaAs substrates at 180°C in the FW-III MBE system, with epilayer thicknesses of about 10 μm. To reduce the absorption in the wavelength region of the thick CdTe reststrahlen band for the LPE $Hg_{1-x}Cd_xTe$ samples, some of the samples had their CdTe substrate removed by grinding, and then were glued onto a sapphire substrate. Typical results of measurements for $Hg_{1-x}Cd_xTe$ epilayers and HgTe/CdTe superlattices are displayed in Figs. 6.31–6.33.

Figure 6.31 is the FIR transmission spectra of two *p*-type $Hg_{0.61}Cd_{0.39}Te$ LPE films at various temperatures. The upper solid curves refer to the epilayer on a sapphire substrate with the initial CdTe substrate removed, and the lower dashed curves to the epilayer with a CdTe substrate. The absorption of the sapphire material increases sharply as the temperature rises, resulting in the transmission of epilayer/sapphire hybrid structure decreasing abruptly as the temperature is increased from 4.2 to 50 K. It is obvious that the reststrahlen region is narrower in the absence of the CdTe layer. Labeled absorption features at A_1~63 cm^{-1}, P_1~108 cm^{-1}, TO_2~124 cm^{-1}, TO_1~144 cm^{-1}, LO_{CT}~166 cm^{-1}, TP_1~176 cm^{-1}, IP_2~190 cm^{-1}, and I_1~86 cm^{-1} are present in both samples; while those at A_2~71 cm^{-1}, TP_2~211 cm^{-1}, TP_3~232 cm^{-1}, and TP_4~254 cm^{-1} are apparent only in the sample with the CdTe substrate.

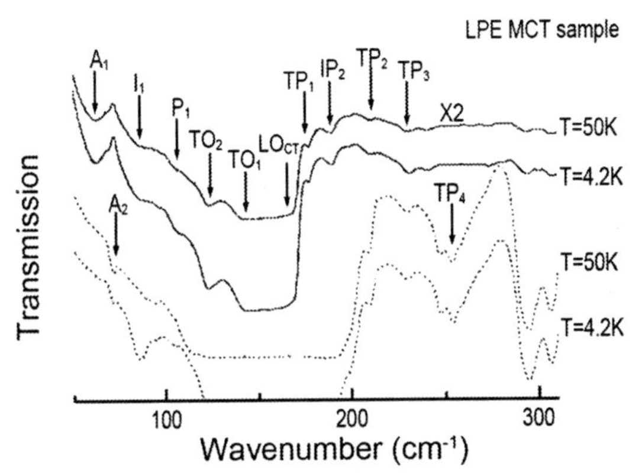

Fig. 6.31. FIR transmission spectra of two $Hg_{0.61}Cd_{0.39}Te$ LPE films at various temperatures (The *solid curves* refer to a sample on a sapphire substrate, while the *dashed curves* are for a sample on a CdTe substrate.)

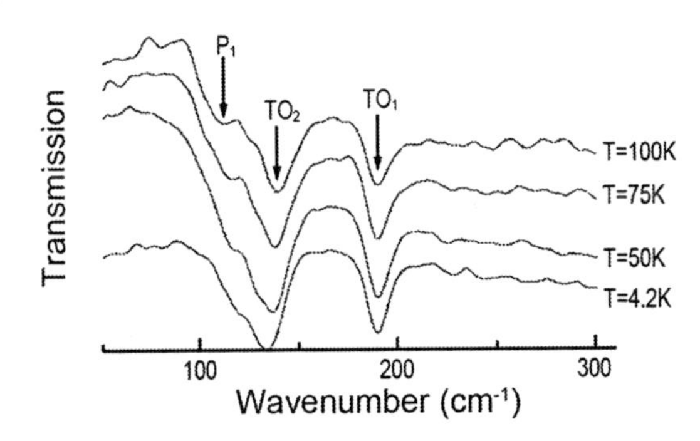

Fig. 6.32. FIR transmission of a HgTe/CdTe superlattice at various temperatures

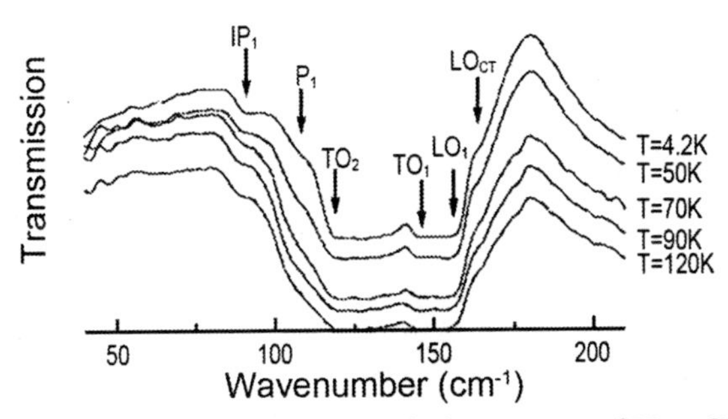

Fig. 6.33. Variable-temperature FIR transmission spectra of $Hg_{1-x}Cd_xTe$ (x = 0.284) MBE film on a GaAs substrate

Figure 6.32 presents the FIR transmission of a HgTe/CdTe superlattice at various temperatures. The phonon modes are distinct in the reststrahlen band region due to the thinness of the superlattice. There are two strong absorption peaks at 120 cm^{-1} (labeled TO_2) and 145 cm^{-1} (labeled TO_1), respectively, and a temperature dependent absorption feature at 107 cm^{-1} (labeled P_1). Figure 6.33 is the FIR transmittance spectra of a $Hg_{0.716}Cd_{0.284}Te$ MBE film from 4.2 to 120 K. The curves display peaks at 92 cm^{-1} (labeled IP_1), 108 cm^{-1} (labeled P_1), 118 cm^{-1} (labeled TO_2), 147 cm^{-1} (labeled TO_1), 156 cm^{-1} (labeled LO_1), and 164 cm^{-1} (labeled LO_{CT}). The temperature-dependent absorption feature IP_1 is notable as seen in Fig. 6.34. Its absorption intensity decreases as the temperature increases from 4.2 to 70 K, but increases as the temperature further increases above 70 K. This alloy displays a mixed-mode behavior, i.e. "HgTe-like" modes (TO_2 and LO_2) and "CdTe-like" modes (TO_1 and LO_1). Peaks TP_1, TP_2, TP_3, TP_4 are two-phonon mode absorptions. The intensities of TP_2 and TP_3 in Fig. 6.31 remain constant while TP_4 disappears after polishing away the CdTe substrate, indicating that TP_4 is due to a CdTe two-phonon process. Moreover, no longitudinal-phonon mode is present in the superlattice sample, Fig. 6.32, because at normal incidence the thin superlattice layer cannot support waves propagating in the normal direction.

The absorption feature P_1 is found in all samples but is not-well-understood. Previously it was assigned to two origins: a native defect and a two-phonon process (Baars and Sorger 1972). Referring to Fig. 6.34, it is clear that the P_1 feature has strong temperature dependence. This favors its originating from the lattice rather than a defect. The P_1 feature is now thought to arise from the transverse acoustic modes of HgTe and CdTe activated by structural disordering of the HgCdTe or HgTe/CdTe lattice (Shen and Chu 1983; Kim et al. 1990; Perkowitz 1986). The lower frequency of the HgTe-type phonon than that of the CdTe type phonon results in the peak position being shifted to the low-frequency side of the spectrum. Further consideration of the A_2 feature suggests that it is due to a Hg-substitution-induced vibrational mode in CdTe. It was seen from earlier work (Talwar and Vandevyver 1984) that no phonon state is present in perfect CdTe in the wave-number range from 50 to 100 cm^{-1}, but a Hg substituted on a Cd site in CdTe would generate a localized vibrational state near 70 cm^{-1}. Additional evidence for this interpretation is that the A_2 feature only occurs in the LPE film. During the homogenization period of LPE growth, Hg atoms from the Hg reservoir source easily penetrate into the CdTe substrate while Cd atoms evaporate. Hence the Cd sites, especially at the substrate surface or near the HgCdTe/CdTe interface, may have a low concentration of isolated Hg atoms which results in the occurrence of the A_2 feature. The I_1 feature has strong temperature dependence and it too only occurs in the LPE sample; it may be related to Hg vacancies. The IP_1 feature in the MBE sample has been attributed to the LA mode of HgTe, activated by structural disordering of the HgCdTe lattice (Mazur et al. 1993). However, the temperature dependence of the absorption intensity (Fig. 6.34) suggests that an impurity may also contribute to the IP_1 feature.

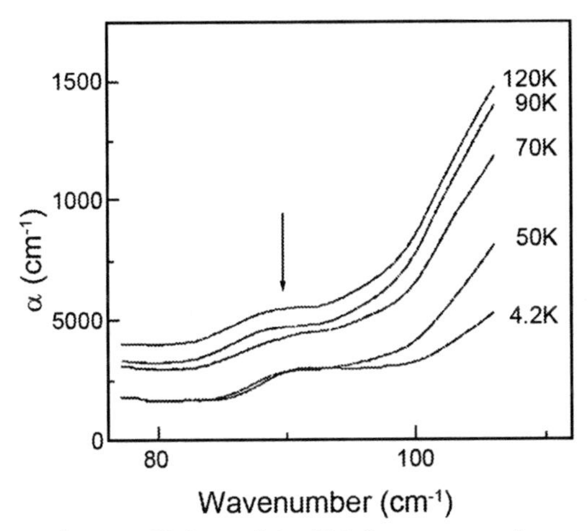

Fig. 6.34. The absorption coefficient of the IP1 feature at various temperatures for the same $Hg_{0.716}Cd_{0.284}Te$ MBE sample reported in Fig. 6.33

6.3.2 The Two-Phonon Process

Several two-phonon absorption processes can be seen along side the reststrahlen band. The samples for this study are $Hg_{1-x}Cd_xTe$ crystals grown by the solid state re-crystallization, semi-molten method. The composition, thickness, band gap at 300 K, and carrier concentration at 77 K are given in Table 6.1.

The transmission spectra measurements were performed at near-normal incidence with a 1 cm^{-1} resolution using a Fourier transform spectrometer (such as a Bruker IFS) in the spectral range, 15–400 cm^{-1}, at temperatures from 4.2 to 300 K. After obtaining the transmission spectra at different temperatures, the absorption coefficient was calculated from:

$$T = \frac{(1-R)^2 \exp(-\alpha d)}{1-R^2 \exp(-2\alpha d)}, \tag{6.77}$$

where d is the thickness of the sample, T is the transmissivity, and R is the reflectivity.

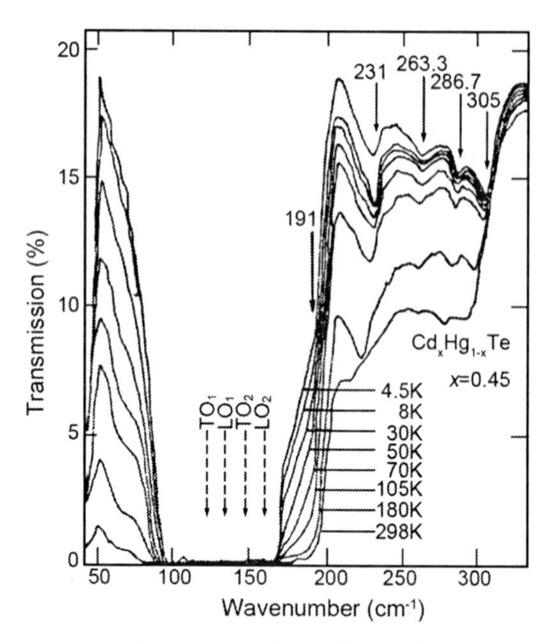

Fig. 6.35. The transmission spectra of $Hg_{1-x}Cd_xTe$ with $x = 0.45$ in the region of 50–350 cm^{-1}

Figure 6.35 shows the transmission spectra adjacent to the reststrahlen band for the sample with composition $x = 0.45$ at different temperatures. The dashed arrows designate the positions of HgTe-like and CdTe-like optical phonons. It is shown there are several weak two-phonon absorption bands outside the reststrahlen band. Two broad absorption bands can be observed near 60 and 100 cm^{-1}. They appear on the shoulders of the reststrahlen band in the absorption and transmission spectra. The absorption band near 60 cm^{-1} becomes weaker as the temperature decreases. The 100 cm^{-1} absorption band also becomes weaker in the same way. However the 100 cm^{-1} band becomes stronger so, it remains recognizable even at 4.2 K. There are several weaker absorption bands in the region below 330 cm^{-1}. They become easier recognized at low temperature.

In order to show the behavior of absorption bands in the high frequency region more clearly, the transmission spectra of a $Hg_{1-x}Cd_xTe$ sample with $x = 0.27$ at 4.5 K is given in Fig. 6.36. It displays weak absorption bands at 308, 287, 278, 263, and 222 cm^{-1}. There is an absorption band superposed on the reststrahlen band near 192 cm^{-1}. The wave number positions of these absorption bands are given in Table 6.5. Figures 6.37 and 6.38 summarize the characteristic wave numbers of these absorption bands as a function of composition and temperature.

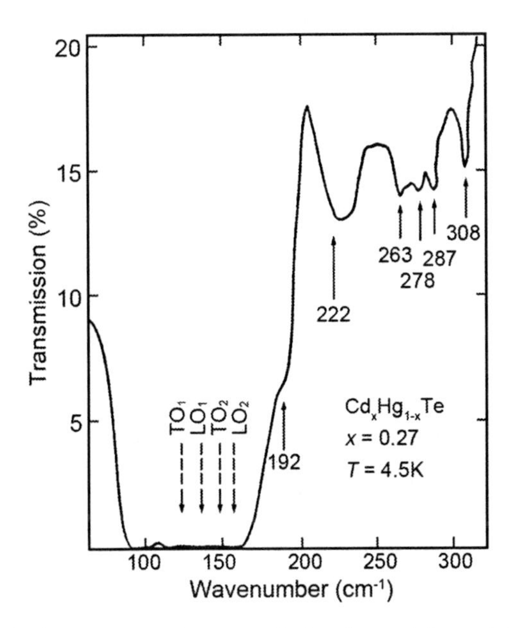

Fig. 6.36. Transmission spectra of a $Hg_{1-x}Cd_xTe$ alloy, with $x = 0.27$ at 4.5 K

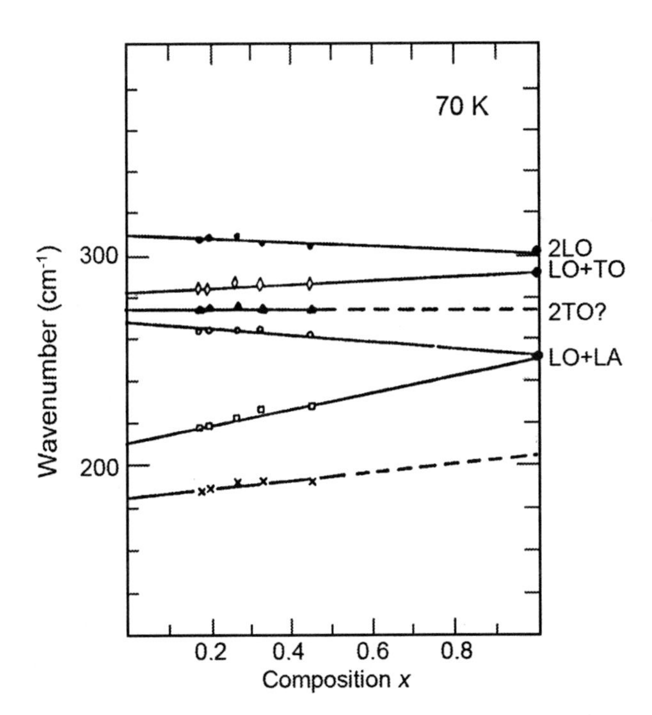

Fig. 6.37. The positions of two-phonon absorption peaks of $Hg_{1-x}Cd_xTe$ alloys as a function of x

Table 6.5. Features of absorption bands near the reststrahlen band of the alloy $Hg_{1-x}Cd_xTe$ with $x = 0.27$ at 4.5 K

Position (cm^{-1})	Probable origin	Comparison with CdTe
60	2TA	72(2TA)
100	TO_1-TA:LO-TA	115(LO-TA)
192	LO+TA:TO+TA	
222	LO_1+LA_1:$2TO_1$?	250(LO+LA)
263	LO_2+LA_2	
278	2TO	
287	LO+TO	290(LO+TO)
308	$2LO_2$:LO_2+TO_2	300(2LO)

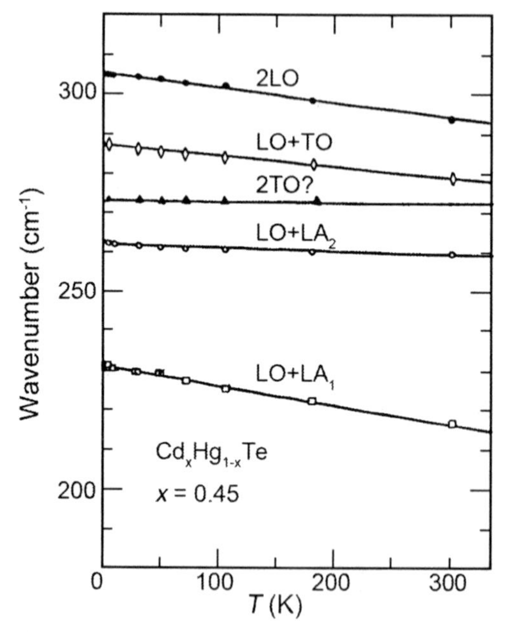

Fig. 6.38. The positions of two-phonon absorption peaks of a $Hg_{0.55}Cd_{0.45}Te$ alloy sample as a function of temperature

The weak absorption band features adjacent to the reststrahlen band in $Hg_{1-x}Cd_xTe$ alloys are displayed in Figs. 6.35 and 6.36. One can see *t* absorption bands at 192 and 278 cm^{-1}. By comparing these with the results from HgTe and CdTe, most of these bands have been interpreted as different combinations of two-phonon processes. This interpretation stems from the fact that the intensities of these bands decreases with a decrease of the temperature, and their positions coincide with the sum frequencies of various two-phonon processes. The probable two-phonon combinations are given in Table 6.5.

An absorption shoulder is found near 100 cm^{-1}, which is recognized as a contributor to the reststrahlen band. The feature, which is characteristic of a transverse mode, also appears in the reflection spectra (Dornhaus et al. 1982; Balkanski 1975) of HgTe, and $Hg_{1-x}Cd_xTe$ and $Hg_{1-x}Mn_xTe$ alloys with small x values. The intensity of these oscillators quickly weakens at lower temperatures and disappears in the reflection spectra. However, the intensity is still visible in the transmission spectra.

A magnetic field can affect the intensity of the absorption bands (Dornhaus et al. 1982). The absorption band of the coupled magnetic plasmon oscillators shifts to the high-frequency region of the reststrahlen band as the magnetic field is increased. The intensity of this band is weaker in the reflection spectra. The probable origins are from a plasmon related TO-TA two-phonon process (Witowski and Grynberg 1979), or a LO-TA two-phonon process (Danielewicz and Coleman 1974; Baars and Sorger 1972).

6.3.3 Low-Frequency Absorption Band of $Hg_{1-x}Cd_xTe$ Alloys

The lattice vibrational spectra of alloys have been investigated by several groups (Barker and Sievers 1975; Shen and 1980a). The short-range disorder of alloys induces one-phonon absorption. The lack of long-range order relaxes the crystal momentum and symmetry selection rules and results in dipole-allowed photon absorption with the creation of a phonon. This process is forbidden in the corresponding crystalline materials. The only selection rule that applies to this absorption process in alloys is energy conservation. Thus one can study these modes using optical measurements. From the far-infrared reflection spectra of $Hg_{1-x}Cd_xTe$ alloys, one observes that there are two main relatively high intensity bands (a CdTe-like band and a HgTe-like band) existing over the entire range of solid solutions. For $x < 0.34$ an additional reflection band was observed around 100 cm^{-1}, which is attributed to a two-phonon subtractive process LO-TA (Kim and Narita 1971; Baars and Sorger 1972; Mooradian and Harman 1971; Georgitse et al. 1973; Dornhaus et al. 1982; Kozyrev et al. 1983; Chu 1993).

There is also an absorption band with a fine structure in the low-frequency range (20–50 cm^{-1}) of the transmission spectra. By using an ab initio interlayer force constant method (Kunc and Martin 1982; Yin and Cohen 1982; Cardona et al. 1982), and the Brout sum rule (Baars and Sorger 1972), the phonon spectra of CdTe was estimated from the observed spectra of CdTe (Bilzand and Kress 1979), HgTe and the characteristic frequencies of $Hg_{1-x}Cd_xTe$. The position and intensity of the low-frequency absorption band are independent of composition x and temperature T. Thus its origin is determined to be a disorder-induced TA mode absorption of the $Hg_{1-x}Cd_xTe$ alloy.

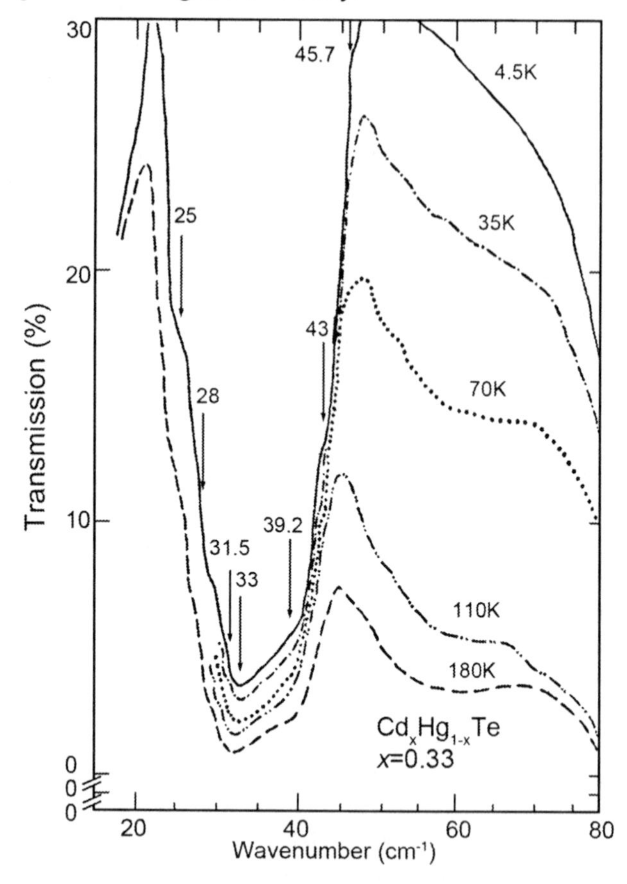

Fig. 6.39. The transmission spectra of a $Hg_{1-x}Cd_xTe$ ($x = 0.33$) alloy in the range of 10–80 μm at different temperatures

The low frequency band energy corresponds to the energy range of the TA phonons in CdTe, HgTe and $Hg_{1-x}Cd_xTe$ crystals. The transmission spectra of a $Hg_{1-x}Cd_xTe$ ($x = 0.33$) alloy in the range of 10–80 cm^{-1} at different temperatures are shown in Fig. 6.39. The coordinates of the transmission curves at $T = 70$, 110, and under 180 K have been rigidly shifted upward by different amounts to fit them onto the same plot as the curves at $T = 4.5$ and 35 K. The main peaks of these absorption bands and other characteristic features are almost invariant to a temperature decrease. By contrast the absorption of the free carrier and plasmon oscillation are suppressed as the temperature decreases. The nearby absorption bands also become weaker, such as the bands near 60 cm^{-1}. As a result, this low-frequency band can be seen clearly. This absorption band has been observed for all samples with compositions from 0.18 to 0.45.

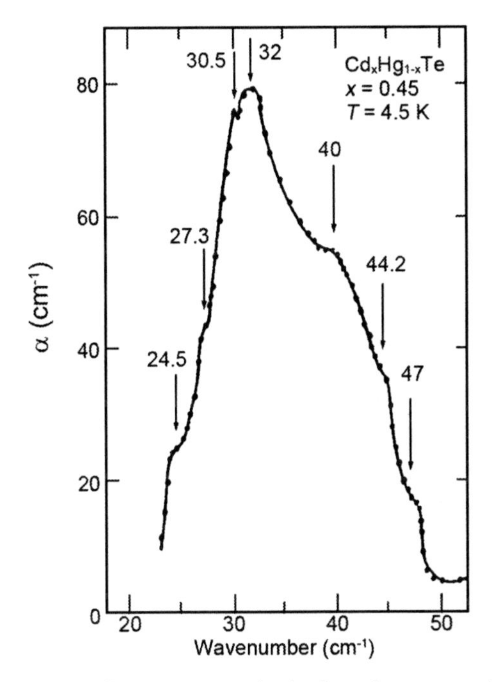

Fig. 6.40. Positions of fine structures in the low-frequency absorption band

In order to gain a more quantitative idea of the intensity of this absorption band, and show its fine structure more clearly, the transmission spectra of a $Hg_{1-x}Cd_xTe$ ($x = 0.45$) alloy at $T = 4.5$ K is presented in Fig. 6.40. In the figure the absorption coefficient α is calculated using (6.77). Several fine structure features in the absorption band are seen clearly in the spectra. In Fig. 6.40 the arrows indicate the positions of these structures.

This absorption band has also been studied by Scott et al. (Scott et al. 1976; Dornhaus 1983; Kimmitt et al. 1985). An absorption peak in the range of 1–7.8 meV, is seen at about 4 meV (32 cm^{-1}) in the far-infrared transmission spectra of a $Hg_{0.8}Cd_{0.2}Te$ sample. They explain it as an electronic transition from the Fermi energy, in this case just above the conduction band minimum, to the Te vacancy at 8 meV above the conduction band edge. They also did a theoretical calculation, the result of which is shown as an insert in Fig. 6.41 along with the experimental data. Further experiments on samples with different compositions and at various temperatures did not support this interpretation. This absorption band feature exists in samples with different compositions. Its line shape is independent of the temperature and its position does not shift with carrier concentration (Fermi energy level). Thus it is not appropriate to attribute the absorption band to an electronic transition process.

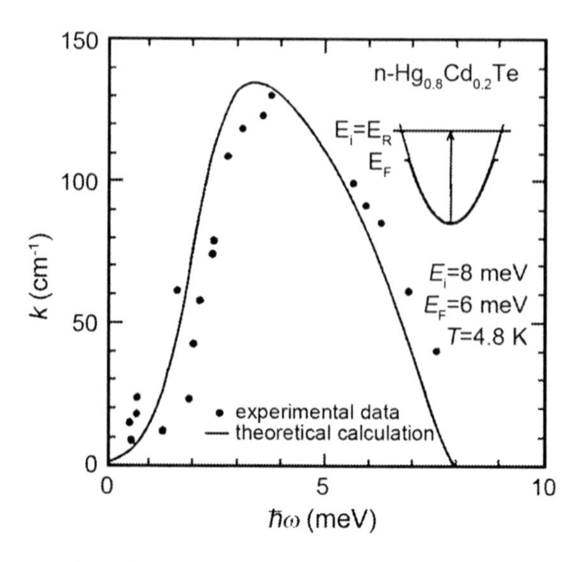

Fig. 6.41. A comparison between a measured absorption coefficient far-infrared spectra and a theoretical result

The existence of this spectral band in $Hg_{1\text{-}x}Cd_xTe$ alloys, where disorder can assist photon absorption to excite the TA phonon band modes, provides an alternative explanation of this spectral feature. The absorption of the disorder-assisted TA phonon band modes was thoroughly investtigated (Brodsky and Lurio 1974; Shen and Cardona 1980b; Shen et al. 1980). In some cases absorption features are observed at energies corresponding to the critical points in the TA bands of pure crystals. The low-frequency absorption bands presented in Figs. 6.39 and 6.40 can be distinguished from these other absorption bands, because their positions are identical to the TA band mode energies of $Hg_{1\text{-}x}Cd_xTe$ alloys. There is also some fine structure in the absorption spectra corresponding approximately to the positions of the critical points in the TA bands. Thus one can identify the low-frequency absorption spectral features arising from the disorder-assisted absorption of the TA phonon band modes. We also can deduce the intensity and equivalent charge of the absorption band. The equivalent charge is:

$$e_T^{*2} = \frac{\mu nc}{2\pi^2 N} \int \alpha \cdot d\omega , \tag{6.78}$$

with e_T^* the equivalent charge, μ the vibrational reduced mass, n the refractive index, N the atom number per unit volume, and c the light velocity. Then we find $e_T^* \approx 0.15e$. This value is close to equivalent charge due to disorder-induced absorption of the TA bands in amorphous semiconductors with the same average coordination as diamond structures, but is slightly larger than that of Ge_xSi_{1-x} alloys. It is due to the iconicity of the chemical bond in $Hg_{1-x}Cd_xTe$. It is an indication that this low-frequency absorption feature in the above three cases have similar physical origins. It originates from the optical activation of the disorder-assisted TA band. However, this type of activation is a partial one so the forbidden transitions are only partially relieved.

6.3.4 Characteristic Estimation of Phonon Spectra

There are no theoretically predicted phonon spectra of $Hg_{1-x}Cd_xTe$. Using the phonon spectra of CdTe (Cardona et al. 1982; Bilzand and Kress 1979), one can estimate the characteristic frequencies of HgTe and speculate about the phonon spectra of $Hg_{1-x}Cd_xTe$ alloys. The phonon density of states (DOS) of crystalline CdTe as a function of frequency is shown in Fig. 6.42. It is calculated using the shell model with 14 parameters, and is identical to the INS experimental results (Bilzand and Kress 1979). In the figure, the L, X, Δ, and Σ labels near and above the TA phonon spectral region represent the characteristic DOS features at these critical points. The TA for CdTe is near 36 cm^{-1}. The frequency positions of the optical phonons were identified using reflection spectra (Balkanski et al. 1975), and an ab initio interlayer force constant method within the local-density-functional formalism (Kunc and Martin 1982; Yin and Cohen 1982; Cardona et al. 1982), and the Brout sum rule (Baars and Sorger 1972). These methods enable calculations of the phonon frequencies and dispersion relations of various symmetry directions in the Brillouin zone of diamond and zinc blende structures.

Consider a wave traveling in the (100) direction, then in terms of the interlayer force constants of the jth layer, the equations of motion layers 1 and 2 are given by:

$$M_1\omega^2 u_1 = \sum_j k_j^{(1)} u_j$$
$$M_2\omega^2 u_2 = \sum_j k_j^{(2)} u_j , \tag{6.79}$$

with $k_0^{(1),(2)} = -\sum_{j\neq 1} k_j^{(1),(2)}$. Due to symmetry, the above equations at the origin of the Brillouin zone (the Γ point) simplifies to:

$$\omega_{TO}^2(\Gamma) = -\frac{1}{\mu}\sum_{j,\text{odd number}} k_j , \tag{6.80}$$

with μ being the reduced mass. For other symmetry points, such as X and L, the eigen frequencies are determined by solutions to the secular equation generated from (6.79). Determining the interlayer force constants is the critical problem of the calculation. After obtaining the charge density of the layers, with use of the Hellman–Feynmann theorem, one can calculate the interlayer force constants. Also a self-consistent method can be used. Here the first-nearest-layer and second-nearest-layer force constants are deduced from the measured characteristic frequencies and elastic constants of HgTe and CdTe. Also the phonon frequencies at the critical points can be

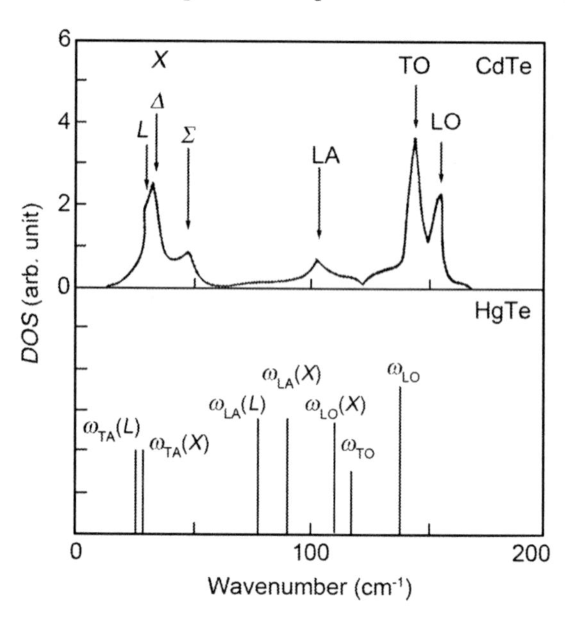

Fig. 6.42. The phonon DOS of CdTe and HgTe

obtained using an ab initio interlayer force constant method. The results can be checked by comparing them to the Brout sum rule:

$$\omega_{LA}^2(K) = \omega_{LO}^2(\Gamma) - \omega_{LO}^2(K) - 2\omega_{TA}^2(K) \tag{6.81}$$

The characteristic frequencies, obtained using the ab initio interlayer force constant method, are indicated in Fig. 6.42 along with some experimental results for CdTe, and HgTe.

CdTe and HgTe can form $Hg_{1-x}Cd_xTe$ random alloys in the zinc blende structure with any composition because the two compounds are almost lattice matched. There is short-range clustering of cations (Cd, Hg) around the anions (Te) in the $Hg_{1-x}Cd_xTe$ alloys. However, the masses of Cd and Hg differ, as do the elastic constants of HgTe and CdTe. The measured reflection spectra show that features like the LO bands are split into CdTe and HgTe bands rather than being a broadened band centered at the average of the two bands. In multiple scattering theory language this is a "strong scattering regime". In strong scattering cases the DOS of the alloy associated with a given band type exhibit split bands that are broadened, but with frequencies that remain resolved. The intensities of the split bands vary proportional to x and $1-x$. By contrast, in the weak scattering limit there is a single broadened band for each band type. The theory that best captures these features is the coherent potential approximation (CPA) (Chen and Sher 1995). While this theory has been applied to phonon spectra in alloys (Grein and Cardona 1992; Gregg and Myles 1985) it has not been used for $Hg_{1-x}Cd_xTe$ alloys. The nearest thing to such an alloy phonon spectra calculation was done by Krishnamurthy et al. (1995) where a virtual crystal approximation calculation was employed.

6.4 Phonon Raman Scattering

6.4.1 Polarizability

Raman scattering studies in semiconductors have proven to be an important means of investigating phonon distributions as well as absorption and reflection spectra. We take ω_i, ω_s to be the frequencies of the incident photon and scattered photon, respectively. The frequency of the scattered photon is unchanged ($\omega_s = \omega_i$) in a reflection spectra measurement, and in an absorption measurement there is no scattered photon. The interaction between light and materials in absorption and reflection is described of the material's polarizability, $\tilde{\chi}(\omega_i)$. However, the frequency of scattered photon ω_s differs from that of the incident photon, ($\omega_s \neq \omega_i$) in Raman scattering. Then the interaction between light and a material must be described by a transition polarizability, $\tilde{\chi}(\omega_i, \omega_s)$. When light of frequency ω_i is incident on a material, an elementary excitation, with the frequency ω_j can exert an influence on the polarizability. The $\tilde{\chi}(\omega_i)$ is modulated with amplitude $\Delta\tilde{\chi}$ and frequency, ω_j. Thus the polarization is time dependent. The change results in scattered electromagnetic radiation with frequencies $\omega_s = \omega_i \pm \omega_j$. The scattered light at frequencies less than the incident frequency (i.e. of the type $\omega_s = \omega_i - \omega_j$) are referred to as Stokes scattering, and those at frequencies greater than the incident frequency (i.e. $\omega_s = \omega_i + \omega_j$) as anti-Stokes scattering. In the range of $\hbar\omega_j$ $<10^{-7}$ eV, the scattering is called Raleigh scattering, which is usually measured by light beating spectroscopy. In the range 10^{-7} eV $< \hbar\omega_j < 10^{-4}$ eV, the scattering is called Brillouin scattering, which is measured by a Fabry–Perot interferometer. In the range of $\hbar\omega_j > 10^{-4}$ eV, the scattering is called Raman scattering, which is measured by a monochromator. It is in the domain of optical phonons. This type of scattering is also named phonon Raman scattering because the typical energy of phonon in semiconductors is around 10^{-2} eV. The frequency of incident light that has been used varies over a broad range from UV, visible to IR radiation. Thus the energies of $\hbar\omega_i$, and $\hbar\omega_s$ are between 1 and 3 eV.

Often the polarizability $\tilde{\chi}$ involves electron interband transitions near critical points. Here $\Delta\tilde{\chi}$ is the change of the polarizability due to a phonon modulation. Viewed this way Raman scattering is like modulation spectroscopy. While in scattering theory language a Stokes shift event is treated in second order perturbation in which the incident photon is first destroyed to create a virtual state in which an electron is excited from a ground state to an intermediate virtual excited state and a phonon is emitted, and then this excited electronic state decays into a photon at a lower energy than that of the incident photon. By measuring the scattering cross-section of the Raman scattering (i.e. getting $\Delta\tilde{\chi}$), the phonon spectral structure can be deduced. When the frequency of the incident light equals the frequency of some transition in the material, the Raman scattering cross-section increases rapidly, and is called resonant Raman scattering.

In the Raman spectra measuring technique one is interested in the frequency difference, $\omega_i - \omega_s$, that determines the energy of elementary excitation $\hbar\omega_j$. The peak of the scattered signal at ω_s changes as the scanning frequency ω_i is varied. However, the frequency difference $\omega_i - \omega_s = \omega_j$ for a given phonon state remains invariant. If we set ω_i to be the zero of the frequency coordinate system, we observe a peak at ω_j as ω_i is varied. When ω_i is the same value as the frequency of an intrinsic electronic transition of the material, the peak reaches its maximum, which corresponds to resonance Raman scattering. The excitation sources in these experiments are often Ar^+ lasers, tunable dye lasers, or monochromators. The detection system consists of an optoelectronic detector, an amplifier, a signal recorder, etc. The phonon frequencies in semiconductors are about 10 meV, which is much less than the important frequencies of the subband interband transitions at Γ, X, and L corresponding to the energies E_0, E_1 and E_2. This also applies to the narrow gap semiconductors. The Raman scattering discussed here is mainly due to phonon deformation potential coupled scattering.

In order to study the polarizability modulation by phonons, we first discuss the concept of polarizability. The polarizability is important in Raman scattering, and is written as:

$$\mathbf{P}(\omega) = \tilde{\chi}(\omega)\boldsymbol{E}(\omega) \tag{6.82}$$

with $\boldsymbol{p}(\omega)$ being the polarization intensity vector, $\boldsymbol{E}(\omega)$ the electric-field intensity vector, and $\tilde{\chi}(\omega)$ the polarizability tensor with nine components.

For an isotropic molecule, the direction of the dipole moment induced by an applied field is identical with that of the field. However, the polarizability χ is different in the x, y, z directions for an anisotropic molecule. As a result, the induced dipole moment is not in general parallel to the field. The dipole moment induced by the electric-field component E_x, will have components in x, y, z directions. Then we get:

$$\begin{aligned} P_x &= \chi_{xx}E_x + \chi_{xy}E_y + \chi_{xz}E_z \\ P_y &= \chi_{yx}E_x + \chi_{yy}E_y + \chi_{yz}E_z \\ P_z &= \chi_{zx}E_x + \chi_{zy}E_y + \chi_{zz}E_z \end{aligned} \tag{6.83}$$

The polarizability is a tensor with these coefficients. If the polarizability tensor is symmetrical, then $\chi_{xy} = \chi_{yx}$, $\chi_{yz} = \chi_{zy}$, and $\chi_{xz} = \chi_{zx}$. For a symmetrical tensor, one can rotate the coordinate system to x', y', z', i.e. make the off-diagonal components zero and leave only the diagonal components $\chi_{x'x'}$, $\chi_{y'y'}$, and $\chi_{z'z'}$. Then the above equation is rewritten as:

$$P_{x'} = \chi_{x'x'} E_{x'},\ P_{y'} = \chi_{y'y'} E_{y'},\ P_{z'} = \chi_{z'z'} E_{z'}, \tag{6.84}$$

and the three primed axes are the principal axes of the polarizability.

We can obtain the polarizability ellipsoid by drawing the $1/\sqrt{\chi}$ three-dimensional curve about the zero of the coordinate system. The lengths of these principal axes, x', y', z', are totally different for an anisotropic molecule. The ellipsoid becomes a rotation ellipsoid when the values of polarizability are the same in two directions. The ellipsoid becomes a sphere for an isotropic molecule. The symmetry of the ellipsoid can be higher than that of the molecule. But in general, the ellipsoid will have the same symmetrical elements as the molecule. When the dimension, shape, and orientation of the ellipsoid are changed due to vibrations, and molecular rotations, the scattered electromagnetic radiation will be affected.

There is a simple relation between the polarizability and dielectric function tensor. Since $\boldsymbol{D} = \tilde{\varepsilon}\boldsymbol{E} = \boldsymbol{E} + 4\pi\mathbf{P} = \boldsymbol{E} + 4\pi\tilde{\chi}\boldsymbol{E} = \tilde{E}(1 + 4\pi\tilde{\chi})$, we have:

$$\begin{gathered} \tilde{\varepsilon} = 1 + 4\pi\tilde{\chi}, \text{or} \\ \tilde{\chi}(\omega) = \frac{\tilde{\varepsilon}(\omega) - 1}{4\pi} \cdot \end{gathered} \tag{6.85}$$

It has been shown that the reflection by the polarizability induced by an electronic transition is the same as that from the dielectric function. From (6.85) it is evident that the real and imaginary parts of the polarizability also follow the KK relations:

$$\begin{aligned} \tilde{\chi}'(\omega) &= \frac{2}{\pi} P \int_0^\infty \frac{\omega_l \tilde{\chi}''(\omega)}{\omega_l^2 - \omega^2} d\omega_l \\ \tilde{\chi}''(\omega) &= -\frac{2\omega}{\pi} P \int_0^\infty \frac{\tilde{\chi}'(\omega)}{\omega_l^2 - \omega^2} d\omega_l \end{aligned} \tag{6.86}$$

We will proceed by calculating the transition polarizability using the one electron model. In the direct transition case, omitting the contribution of intraband transitions, the polarizability is given by (Cardona 1967):

$$\tilde{\chi}(\omega) = \frac{e^2}{m_0^2 \hbar V} \sum_{l,m} \frac{2}{\omega_{l,m}} \frac{\langle m|\hat{P}|l\rangle\langle l|\hat{P}|m\rangle}{\omega_{lm}^2 - (\omega + i\Gamma)^2}, \tag{6.87}$$

with m being an occupied state, and l a vacant state, $\omega_{lm} = \omega_l - \omega_m$, and Γ is the line width convergence parameter. If the life length is infinite,

$\Gamma = 0$; if the life time is finite then Γ is the line width and serves as a convergence constant. $\hat{P}$ is the dipole moment operator. V is the volume of the k space. Assuming the state is undamped, i.e. $\Gamma = 0$, the imagery part of the polarizability is given by:

$$\chi''(\omega) = \frac{\pi e^2}{m_0^2 \hbar V} \sum_{l,m} \frac{\langle m|\hat{P}|l\rangle\langle l|\hat{P}|m\rangle}{\omega_{lm}^2} \delta(\omega - \omega_{lm}). \tag{6.88}$$

The property of the δ function useful in this case is:

$$\int_a^b f(x)\delta(f(x))dx = \sum_{x_0} \delta(x_0) \left|\frac{\partial f}{\partial x}\right|_{x=x_0}^{-1}. \tag{6.89}$$

After rewriting $\iint_{s,k} ds dk$ as $\iint_{\omega,s} \frac{ds_k}{\nabla_k \omega} d\omega$, (6.88) can be written in the form:

$$\chi''(\omega) = \frac{e^2}{4\pi^2 m_0^2} \iint_{\omega_{lm}, S\omega_{lm}=\omega} \frac{\langle m|\hat{P}|l\rangle\langle l|\hat{P}|m\rangle}{\omega_{lm}^2} \frac{ds_k}{\nabla_k \omega_{\mathrm{lm}}} d\omega_{\mathrm{lm}}, \tag{6.90}$$

where the ds_k are surface elements on constant energy difference surfaces near the critical points in k space. For those materials with cubic symmetry, assuming the matrix elements are independent of the frequency ω, one can extract them from the integrals. Introducing the DOS $\rho_d(\omega)$:

$$\rho_d(\omega) = \frac{1}{4\pi^3} \int_{S_{\omega_{lm}=\omega}} \frac{ds_k}{|\nabla_k \omega_{lm}|}, \tag{6.91}$$

allows (6.90) to be written as:

$$\chi''(\omega) = \frac{\pi e^2}{3\omega^2 m_0^2} |\langle l|P|m\rangle|^2 \cdot \rho_d(\omega). \tag{6.92}$$

Here $\chi''(\omega)$ is a scalar, and $|\langle l|P|m\rangle|^2$ is interpreted as an average value. For the point that satisfies the frequency condition, $|\nabla_k \omega_{lm}| = 0$, then from (6.91), $\rho_d(\omega)$ and $\chi''(\omega)$ tend to infinity. Thus the spectral structure is dominated by the critical points in the DOS. $\rho_d(\omega)$ can be obtained from the band structure $E(\boldsymbol{k})$.

We have analyzed the DOS in interband transition theory in Chap. 3. Near the Γ point the non-parabolic nature of the bands should be taken into account. However, here we use a parabolic approximation for simplicity in accounting for contributions from different directions. After expanding the energy bands near the critical points (E_g, k_g) in the k space, we obtain in the parabolic approximation:

$$E_C(\boldsymbol{k}) - E_V(\boldsymbol{k}) = E_g + \frac{\hbar^2 (k_x - k_{gx})^2}{2\mu_x} + \frac{\hbar^2 (k_y - k_{gy})^2}{2\mu_y} + \frac{\hbar^2 (k_z - k_{gz})^2}{2\mu_z} \tag{6.93}$$

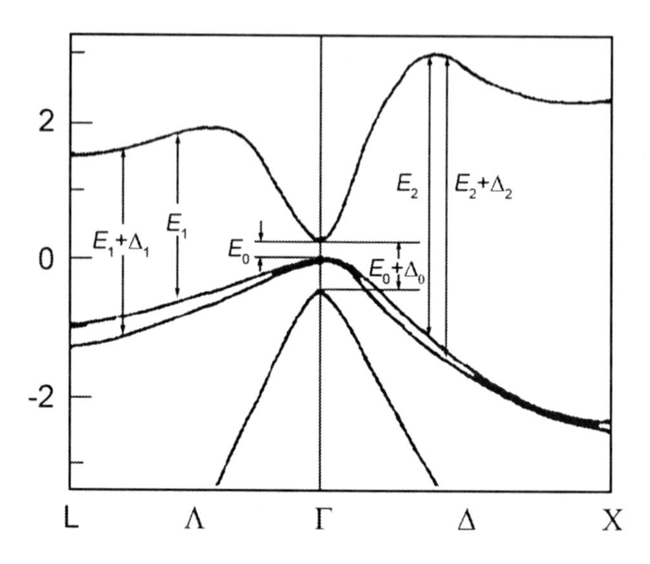

Fig. 6.43. A schematic representation of the band structure of InSb

A term can be omitted if its effective mass is much greater than those of other directions. Then the critical point becomes two-dimensional. If the effective masses of two directions are much greater than that of the remaining direction, both of these can be ignored. Then the critical point becomes one-dimensional. If the effective masses in all the directions are comparable, all the terms must be included and the critical points are three-dimensional. It is obvious that there is a three-dimensional minimum of the energy band near the Γ point. So the corresponding critical point is three-dimensional. Along the <111> direction, the effective masses of the points with the energies E_1 and E_1+Δ_1 are large. In these cases there are two two-dimensional minima in the energy bands. So the corresponding critical points are two-dimensional. Along the <100> direction, the energy has one-dimensional minima at the points with energies E_2 and E_2+Δ_2. So the corresponding critical points are one-dimensional (Fig. 6.43).

Along the <100> direction in InSb, one has $E_2 \approx 5$ eV. Thus for this case it is necessary to use a short wavelength X-ray excitation source. For other critical points it is sufficient to use a laser source. We will focus on discussing the three-dimensional critical point near the Γ point, and the two-dimensional critical points along the <111> direction. Here we assume the conduction band is parabolic.

In a three-dimensional critical point case where $\bar{\mu} = (\mu_x \mu_y \mu_z)^{1/3}$ we have:

(1) when $\hbar\omega > E_g$,

$$\rho_d(\omega) = \frac{(2\bar{\mu})^{3/2}(\hbar\omega - E_g)^{1/2}}{2\pi^2\hbar^3}, \text{ and} \tag{6.94}$$

(2) when $\hbar\omega < E_g$,

$$\rho_d(\omega) = 0. \tag{6.95}$$

In a two-dimensional critical point case where $\bar{\mu} = (\mu_x \mu_y)^{1/2}$ we have:

(1) when $\hbar\omega > E_g$,

$$\rho_d(\omega) = \frac{\bar{\mu}\sqrt{3}}{\pi\hbar^2 a}, \tag{6.96}$$

where a is the lattice constant, and

(2) when $\hbar\omega < E_g$,

$$\rho_d(\omega) = 0. \tag{6.97}$$

For the two-dimensional critical point case, there is a step-like singular point encountered in an integration up to the boundary of the Brillouin zone along the <111> direction.

Thus we can calculate $\tilde{\chi}''$ using (6.92) and $\tilde{\chi}'$ using the KK relation. At the fundamental band gap E_0, a three-dimensional critical point case, we have.

(1) when $\hbar\omega > E_g$,

$$\begin{cases} \chi''(\omega) = \dfrac{(2\bar{\mu})^{3/2} e^2 P^2}{6\pi m_0^2 \omega^2 \hbar^3}(\hbar\omega - E_g)^{1/2} \\ \chi'(\omega) = \dfrac{(2\bar{\mu})^{3/2} e^2 P^2}{6\pi m_0^2 \omega^2 \hbar^3}[2E_g^{1/2} - (E_g + \hbar\omega)^{1/2}] \end{cases}, \text{ and} \tag{6.98}$$

(2) when $\hbar\omega < E_g$,

$$\begin{cases} \chi''(\omega) = 0 \\ \chi'(\omega) = \dfrac{(2\bar{\mu})^{3/2} e^2 P^2}{6\pi m_0^2 \omega^2 \hbar^3} [2E_g^{1/2} - (E_g + \hbar\omega)^{1/2} - (E_g - \hbar\omega)^{1/2}] \end{cases} . \tag{6.99}$$

Since $\chi = \chi' + i\chi''$, (6.98) and (6.99) can be rewritten as:

$$\chi(\omega) = \frac{(2\bar{\mu})^{3/2} e^2 P^2}{6\pi m_0^2 \omega^2 \hbar^3} [2E_g^{1/2} - (E_g + \hbar\omega)^{1/2} - (E_g - \hbar\omega)^{1/2}]. \tag{6.100}$$

When $\hbar\omega > E_g$, the last term in the right side is imagery, corresponding the $\chi''(\omega)$ in (6.98).

By defining $x_0 \equiv \hbar\omega / E_g$ and $f(x_0) \equiv x_0^{-2} \cdot [2 - (1 + x_0)^{1/2} - (1 - x_0)^{1/2}]$, (6.100) can be simplified into:

$$\begin{aligned} \chi &= c_0 \cdot f(x_0) E_g^{-3/2} \\ c_0 &= \frac{(2\bar{\mu})^{3/2} e^2 P^2}{6\pi m_0^2 \hbar} \end{aligned} . \tag{6.101}$$

At the band gap E_1, a two-dimensional critical point case we have:

(1) when $\hbar\omega > E_{1g}$,

$$\begin{cases} \chi''(\omega) = \dfrac{\bar{\mu} e^2 \sqrt{3} P^2}{2\pi m_0^2 \hbar^3 a \omega^2} \\ \chi'(\omega) = -\dfrac{\bar{\mu} e^2 \sqrt{3} P^2}{3\pi m_0^2 \hbar^3 a \omega^2} \ln \left| 1 - \dfrac{\hbar^2 \omega^2}{E_{1g}^2} \right| \end{cases}, \text{ and} \tag{6.102}$$

(2) when $\hbar\omega < E_g$,

$$\begin{cases} \chi''(\omega) = 0 \\ \chi'(\omega) = -\dfrac{\bar{\mu} e^2 \sqrt{3} P^2}{3\pi m_0^2 \hbar^3 a \omega^2} \ln \left| 1 - \dfrac{\hbar^2 \omega^2}{E_{1g}^2} \right| \end{cases} . \tag{6.103}$$

We can combine (6.103) and (6.104) into:

$$\chi' = -c_1 x_1^{-2} \ln(1 - x_1^2)$$
$$x_1 \equiv \frac{\hbar\omega}{E_{1g}^{'}}, c_1 = \frac{\overline{\mu} e^2 \sqrt{3} P^2}{3\pi m_0^2 \hbar^3 a \omega^2}, \tag{6.104}$$
$$\chi'' = \frac{\pi c_1}{x_1^2} \Theta(x_1 - 1)$$

with Θ being a step function.

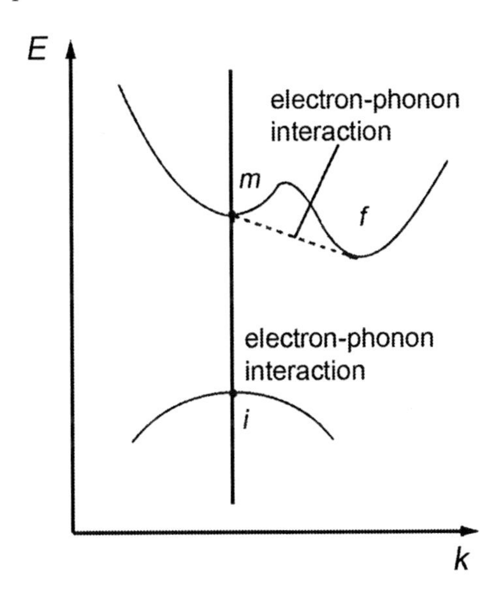

Fig. 6.44. A schematic representation of the indirect photo-excited transition

In the case of an indirect transition, the initial state i is excited by the absorption of a photon to a virtual intermediate state m, $k_i = k_m$, and a phonon with energy $\hbar\omega_j$ and wave vector $k_{i,}$ is emitted (or absorbed) to take the electron from the virtual state m to the final state *f*, as shown in Fig. 6.44. Momentum conservation requires:

$$\boldsymbol{k}_i - \boldsymbol{k}_f = \pm \boldsymbol{k}_j . \tag{6.105}$$

The electron transitions from the maximum of the three-dimensional valence band to the minimum of the three-dimensional conduction band. The approximate expression for χ'' is given by:

(1) when $\hbar\omega > E_g \pm \hbar\omega_j$,

$$\chi''(\omega) \propto |M|^2 \frac{(\hbar\omega - E_g \pm \hbar\omega_j)^2}{\omega^2 (E_f - E_m)}, \text{ and} \tag{6.106}$$

(2) when $\hbar\omega < E_g \pm \hbar\omega_j$,

$$\chi''(\omega) = 0, \tag{6.107}$$

where M is the matrix element of the electron–phonon interaction.

In the case of a direct transition, there is an extra contribution to the polarizability introduced by considering the interaction from excitions.

The above discussion is base on the assumption of the parabolic band. A non-parabolic band correction must be taken into account for narrow gap semiconductors near the band gap E_0. We have described the theory of dielectric function and related optical constant in Chap. 3. It is convenient to introduce this method into the analysis of the polarizability $\tilde{\chi}$.

6.4.2 Scattering Cross-Section

Here we use the classical theory to analyze the scattering cross-section problem. The case treated is one where the frequency of the scattered light ω is identical to the oscillation frequency of the dipole moment, $\boldsymbol{P}$, and the dimension of the dipole is less than the wavelength of the incident light. The emitted power dp into the differential solid angle $d\Omega$ can be written as:

$$dp = \frac{\left\langle \left(\frac{\partial^2 \boldsymbol{P}}{\partial t^2} \right)^2 \right\rangle \sin^2 \varphi}{4\pi c^3} d\Omega, \tag{6.108}$$

with $\langle \ \rangle$ being a time average over a cycle, and φ the included angle between the axis of the dipole moment and the observation direction.

The incident radiation electric-field is:

$$\boldsymbol{E}_i = \boldsymbol{E}_i^0 \exp\left[i(\boldsymbol{k} \cdot \boldsymbol{r} - \omega t)\right]. \tag{6.109}$$

The induced dipole moment is $\boldsymbol{P} = \tilde{\chi} \boldsymbol{E}_i$. Assuming the volume is V, the total dipole moment can be written as:

$$\boldsymbol{P} = \tilde{\chi} \cdot V \cdot \boldsymbol{E}_i^0 \exp\left[i(\boldsymbol{k} \cdot \boldsymbol{r} - \omega t)\right]. \tag{6.110}$$

For a homogenous media, $\tilde{\chi}$ is identical everywhere. The time dependence of $\boldsymbol{P}$ is determined by the time dependence of $\boldsymbol{E}_i$. The oscillation frequency of the dipole moment is ω_i. The incident light experiences the usual reflection and refraction. However, practical media are not homogenous and exhibit specific fluctuations. This results in fluctuations of $\tilde{\chi}$ and the scattering, which is not just the usual reflection and refraction. The fluctuations of $\tilde{\chi}$ are induced by impurities, defects, and different elementary excitations. Phonons are an important class of imperfections, which can be described as a lattice wave:

$$Q_j = Q_{j_0} \exp\left[i(\boldsymbol{k} \cdot \boldsymbol{r} - \omega t)\right]. \tag{6.111}$$

This results in the modulation of the polarizability. The polarizability tensor can be written as:

$$\tilde{\chi} = \tilde{\chi}_0 + \tilde{\chi}(Q_j)\exp\left[i(\boldsymbol{k} \cdot \boldsymbol{r} - \omega t)\right], \tag{6.112}$$

where $\tilde{\chi}_0$ is related to the normal reflection and refraction. $\tilde{\chi}(Q_j)$ is the modulation amplitude introduced by phonons. This term will cause a modified polarization, and is given by:

$$\boldsymbol{P}_s = \tilde{\chi}(Q_j) \cdot V \cdot \boldsymbol{E}_i^0 \exp\left[i[(\boldsymbol{k}_i \pm \boldsymbol{k}_j) \cdot \boldsymbol{r} - (\omega_i \pm \omega_j)t]\right]. \tag{6.113}$$

After substituting this expression into (6.108), the scattered power can be obtained. It is observed to be present not only at the frequency ω_i of the induced dipole moment, but also at the additional frequencies $\omega_i - \omega_j$ and $\omega_i + \omega_j$, which correspond to the Stokes scattering and anti-Stokes scattering, respectively.

Suppose the polarization of the incident light is along the β direction, the wave vector of the scattered light is along γ direction, and the polarization is along α direction. Now we discuss the light scattering introduced by the $P_{s\alpha}$ component of the dipole moment $\boldsymbol{P}_s$. $\boldsymbol{P}_s$ is given by:

$$\boldsymbol{P}_{s\alpha} = \tilde{\chi}_{\alpha\beta}(\omega_i, \omega_s) \cdot V \cdot \boldsymbol{E}_i^0 \exp\left[i[(\boldsymbol{k}_i \pm \boldsymbol{k}_j) \cdot \boldsymbol{r} - (\omega_i \pm \omega_j)t]\right]. \tag{6.114}$$

The incident electric-field with frequency ω_i and polarization β is connected to the polarization field with frequency ω_s and polarization α by $\tilde{\chi}_{\alpha\beta}(\omega_i, \omega_s)$. Following (6.114), we get:

$$\left\langle \ddot{\bar{P}}_{s\alpha} \right\rangle = \left| \tilde{\chi}_{\alpha\beta}(\omega_i, \omega_s) \right|^2 V^2 \omega_s^4 \cdot \frac{4\pi P_i}{A}, \tag{6.115}$$

with $E_{i\beta}^0 = \left(\frac{8\pi P_{i\beta}}{cA} \right)^{1/2}$ and $P_{i\beta}$ is the incident power perpendicular to the area A. Substituting (6.115) into (6.108), one can write $dP_{s\alpha}$ as:

$$dP_{s\alpha} = \frac{\left| \tilde{\chi}_{\alpha\beta}(\omega_i, \omega_s) \right|^2 V^2 \omega_s^4 P_{i\beta}}{c^4 A} d\Omega. \tag{6.116}$$

Define the scattering cross-section (Fig. 6.45) to be:

$$\sigma \equiv \frac{AP_s}{P_i}. \tag{6.117}$$

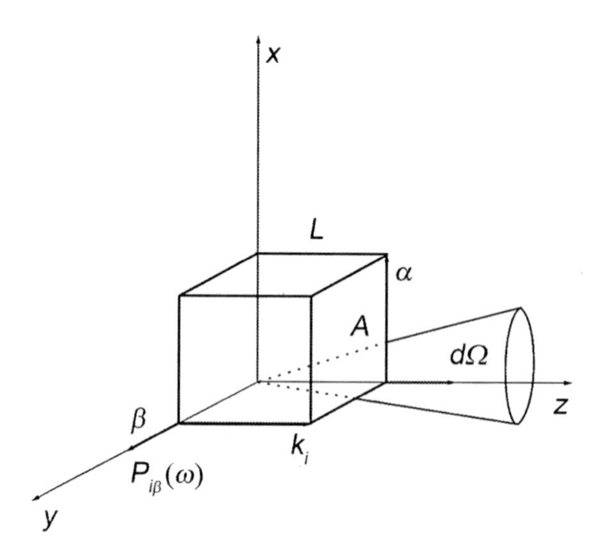

Fig. 6.45. A schematic representation of scattering cross-section

Then the differential scattering cross-section, $d\sigma / d\Omega$, can be written:

$$\frac{d\sigma}{d\Omega} = \frac{A}{P_i}\frac{dP_s}{d\Omega}. \tag{6.118}$$

From (6.116), we can write $d\sigma / d\Omega$ as

$$\frac{d\sigma}{d\Omega} = \frac{\left|\tilde{\chi}_{\alpha\beta}(\omega_i, \omega_s)\right|^2 V^2 \omega_s^4}{c^4}. \tag{6.119}$$

This is the differential scattering cross-section of a scattering center with volume V. Suppose there are N scattering centers in the total volume, then $N = LA/V$, with L being the length of the scattering volume along the k_i direction. Thus the total scattering power is given by

$$dP_{s\alpha} = N \cdot dP_{s\alpha} = \frac{\left|\tilde{\chi}_{\alpha\beta}(\omega_i, \omega_s)\right|^2 V^2 \omega_s^4 P_{i\beta} L}{c^4} d\Omega. \tag{6.120}$$

Take the differential scattering cross-section in the unit volume to be:

$$S_{\alpha\beta} = \frac{1}{V}\frac{d\sigma}{d\Omega} = \frac{dP_{s\alpha}}{LP_{i\beta} d\Omega} = \frac{\left|\tilde{\chi}_{\alpha\beta}(\omega_i, \omega_s)\right|^2 V^2 \omega_s^4}{c^4}, \tag{6.121}$$

where the dimension is cm^{-1}.

In order to obtain the differential scattering cross-section, $S_{\alpha\beta}$, it is necessary to determine $\chi_{\alpha\beta}$. As emphasized earlier, there are three process steps in Raman scattering: (1) The photon with energy $\hbar\omega_i$ is absorbed to generate a virtual electron-hole pair, the state is $|l\rangle$ (2) The electron (or hole) creates(or absorbs) a phonon with energy $\hbar\omega_i$, and the electron-hole pair is scattered from state $|l\rangle$ to state $|m\rangle$; (3) The electron-hole pair recombines to emit a photon with energy $\hbar\omega_s$. In each step of the above three processes momentum conservation is satisfied. The total set of processes from the initial to the final state satisfies energy conservation. The above three processes can happen in various orders. There are six ways to couple the initial to the final state: for finishing the total processes: (1)(2)(3), (1)(3)(2), (2)(1)(3), (2)(3)(1), (3)(1)(2), (3)(2)(1). The polarizability, considering only the lowest order transitions, was derived by Burstein and Pinczuk (1972):

$$\chi_{\alpha\beta}(j)=\frac{e^2}{m_0^2\omega_s^2V}\sum_{l,m}[\frac{\langle 0|P_\beta|m\rangle\langle m|H_{EL}|l\rangle\langle l|P_\alpha|0\rangle}{(E_m-\hbar\omega_s)(E_l-\hbar\omega_i)}$$
$$+\frac{\langle 0|P_\alpha|m\rangle\langle m|H_{EL}|l\rangle\langle l|P_\beta|0\rangle}{(E_m+\hbar\omega_i)(E_l+\hbar\omega_s)}+\frac{\langle 0|P_\alpha|m\rangle\langle m|P_\beta|l\rangle\langle l|H_{EL}|0\rangle}{(E_m-\hbar\omega_s)(E_l-\hbar\omega_j)}$$
$$+\frac{\langle 0|P_\beta|m\rangle\langle m|P_\alpha|l\rangle\langle l|H_{EL}|0\rangle}{(E_m+\hbar\omega_i)(E_l+\hbar\omega_j)}+\frac{\langle 0|H_{EL}|m\rangle\langle m|P_\alpha|l\rangle\langle l|P_\beta|0\rangle}{(E_m-\hbar\omega_j)(E_l-\hbar\omega_i)}$$
$$+\frac{\langle 0|H_{EL}|m\rangle\langle m|P_\beta|l\rangle\langle l|P_\alpha|0\rangle}{(E_m-\hbar\omega_j)(E_l+\hbar\omega_s)}]. \quad (6.122)$$

When $\hbar\omega_i \to E_l$, the denominators of the first and fifth terms in (6.122) tend to zero, causing $\chi'' \to \infty$, which produces an enhancement. This is the type of scattering called resonant Raman scattering. To eliminate the unphysical infinity a refined higher order renormalization theory is required.

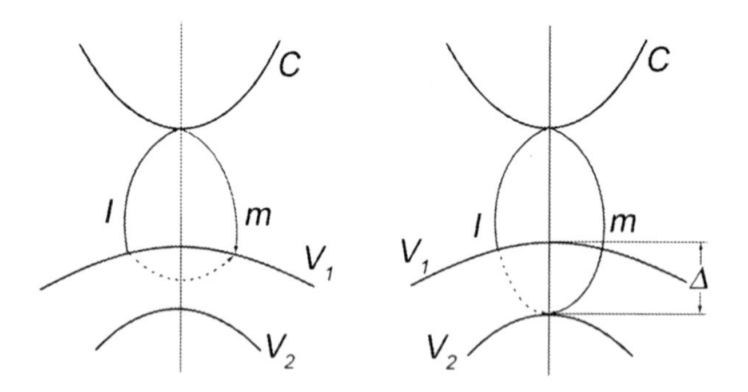

Fig. 6.46. Raman scattering processes in the two bands and the three bands cases

It is seen from (6.122) that $\chi_{\alpha\beta}(j) \propto \omega_s^{-2}$. Then substituting it into (6.121), $S_{\alpha\beta}$ is found to be independent of ω_s. However, when ω_i << the energies of the electronic transitions E_l and E_m, $\chi_{\alpha\beta}$ is constant and $S_{\alpha\beta}$ is proportional to ω_s^4. Many analysis were performed on the relation between the scattering cross-section and the scattering frequency (Zeyher et al. 1976).

Next we examine a simplified discussion of $\chi_{\alpha\beta}$. In the near resonant case, we focus the first term of (6.122), in both the two bands case and three bands case, as shown in Fig. 6.46.

Pincznk and Burstein deduced a simplified expression for the resonant term. By assuming the matrix elements of the electron–phonon interaction, the interband momentum, and the denominators are constants independent of the wave vectors, then the matrix elements can be extracted from the summation to get an approximation to the components of the Raman tensor.

The two band process expression becomes:

$$\chi_{\alpha\beta}(j) = (H_{EL})_{2-b} \frac{\chi(\omega_i) - \chi(\omega_i - \omega_j)}{\hbar\omega_j}. \tag{6.123}$$

The three band process expression becomes:

$$\chi_{\alpha\beta}(j) = (H_{EL})_{3-b} \frac{\chi^+(\omega_i) + \chi^+(\omega_i - \omega_j) - \chi^-(\omega_i) - \chi^-(\omega_i - \omega_j)}{2\Delta}. \tag{6.124}$$

$\chi^+(\omega_i)$ and $\chi^-(\omega_i)$ are the interband transition $V_1 \leftrightarrow C,\ V_2 \leftrightarrow C$ contributions to χ, respectively. Δ is the energy difference between V_1 and V_2.

In the case of two bands, matrix elements of the intra-band transitions result in the electron–phonon interaction contributions to χ:

$$(H_{EL})_{2-b} = \langle C|H_{EL}|C\rangle + \langle V_i|H_{EL}|V_i\rangle;\ (i = 1, 2). \tag{6.125}$$

In the case of three bands, matrix elements of the interband transitions due to H_{EL} contribute terms to χ:

$$(H_{EL})_{3-b} = \langle V_i|H_{EL}|V_i\rangle. \tag{6.126}$$

Since the interband energy is much larger than the phonon energy $\hbar\omega_j$, $\langle V_i|H_{EL}|C\rangle$ can be neglected.

In a quasi-static process, $\omega_j \to 0$, then (6.123) and (6.124) can be written as:

- Two bands:

$$\chi_{\alpha\beta}(j) = (H_{EL})_{2-b} \frac{d\chi}{d(\hbar\omega)}, \tag{6.127}$$

- Three bands:

$$\chi_{\alpha\beta}(j) = (H_{EL})_{3-b} \frac{\chi^{+} - \chi^{-}}{\Delta} \tag{6.128}$$

In (6.128), if $\Delta \to 0$ then we have:

$$\lim_{\Delta \to 0} \frac{\chi^{+} - \chi^{-}}{\Delta} = \lim_{\Delta \to 0} \frac{\chi(\hbar\omega_i, E_g) - \chi(\hbar\omega_i, E_g + \Delta)}{\Delta} \equiv \frac{dx}{dE_g}, \tag{6.129}$$

and

$$\chi_{\alpha\beta}(j) = (H_{EL})_{3-b} \cdot \frac{dx}{dE_g}, \tag{6.130}$$

χ can be found from (6.122), so we can calculate $\frac{dx}{dE_g}$.

Thus we have for the case of three-dimensional critical points:

$$\chi = c_0 f(x_0) E_g^{-3/2}, \tag{6.131}$$

and for the case of two-dimensional critical points:

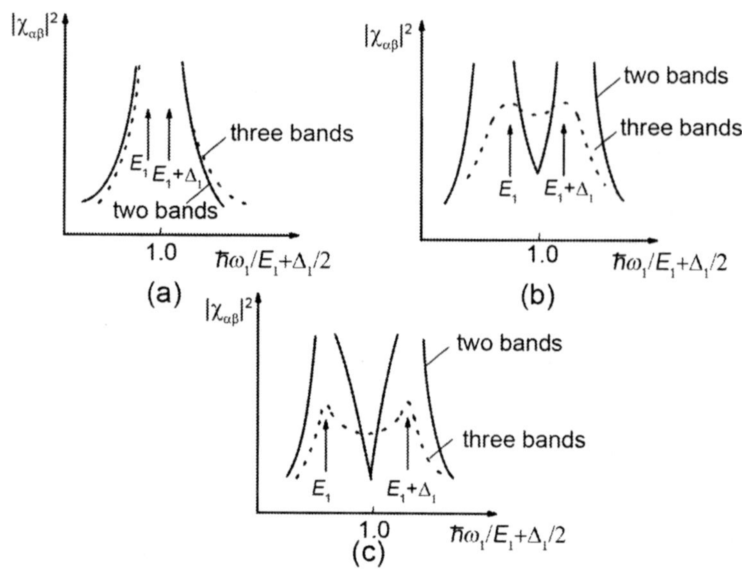

Fig. 6.47. Schematic representations of $|\chi_{\alpha\beta}|^2 \sim \hbar\omega_i / E_1$ at a two-dimensional critical point E_1

$$\chi' = -c_1 x_1^{-2} \ln(1 - x_1^2)$$
$$\chi'' = \frac{\pi c_1}{x_1^2} \Theta(x_1 - 1) \quad . \tag{6.132}$$

The definitions of the terms in (6.131) and (6.132) are the same as those in (6.104). Substituting these expressions into (6.130) and (6.127), we can calculate the dependence of $|\chi_{\alpha\beta}|^2$ on energy in the two bands and three bands approximations. Figure 6.47 shows schematics of the $|\chi_{\alpha\beta}|^2 \sim \hbar\omega_i / E_1$ relationships at a two-dimensional critical point E_1.

It is evident that there is not much difference between the two band and the three band calculations for small Δ_1. However, there is a difference for moderate Δ_1. Since the calculated results of $\frac{dx}{d\hbar\omega}$ and $\frac{dx}{dE_g}$ are different, there is a much larger difference for large Δ_1. The resonance becomes weaker in the three bands model. $|\chi_{\alpha\beta}|^2$ is the sum of the square of the real part and the imagery part of $\chi_{\alpha\beta}$ because the polarizability is a complex number (Fig. 6.48).

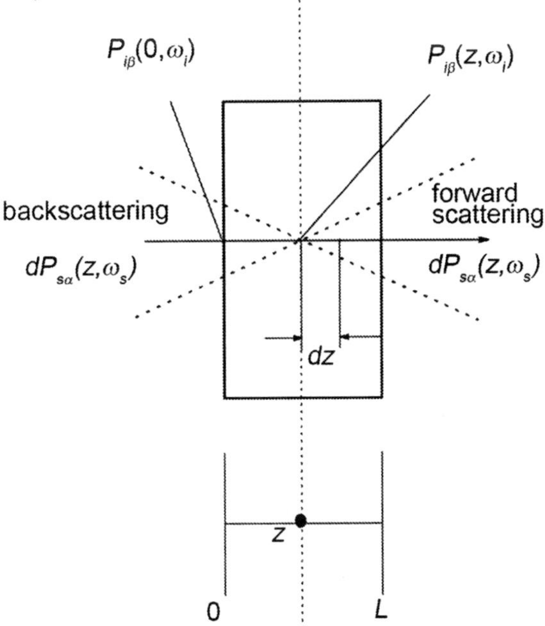

Fig. 6.48. A schematic drawing of the forward and back scattering

It is important to determine the scattering cross-section experimentally. The losses of both the incident light and the reflected light in the system must be considered because it is the net scattering power that is related to the scattering cross-section and optical constants of the tested material. The relation is:

$$d^2P_{s\alpha}(z,\omega_s) = S_{\alpha\beta}d\Omega dzP_{i\beta}(z,\omega_i)\,. \tag{6.133}$$

Taking into account the absorption, and reflection loss of the incident light and the scattered light, described by G_i and G_s, respectively, the above equation can be rewritten as:

$$d^2P_{s\alpha}(0,\omega_s) = S_{\alpha\beta}d\Omega P_{i\beta}(0,\omega_i)G_i(R_i,A_i,z)G_s(R_s,A_s,z)dz\,, \tag{6.134}$$

with R_i and A_i being the reflection coefficient and absorption coefficients of the incident light, respectively. R_s and A_s are the coefficients of the reflected light. Thus the total scattering power of the sample is given by:

$$dP_{s\alpha}(0,\omega_s) = S_{\alpha\beta}P_{i\beta}(0,\omega_i)d\Omega\int_0^L G_i(R_i,A_i,z)G_s(R_s,A_s,z)dz\,. \tag{6.135}$$

Since G_s is independent of z in a rectangular scattering sample, it can be extracted from the integral to obtain:

$$dP_{s\alpha}(0,\omega_s) = S_{\alpha\beta}P_{i\beta}(0,\omega_i)d\Omega G_s(R_s,A_s,z)\int_0^L G_i(R_i,A_i,z)dz\,. \tag{6.136}$$

G_i and G_s are deduced from the optical constants and geometry of the sample. The above equation can be rewritten as:

$$\begin{aligned} dP_{s\alpha}(0,\omega_s) &= S_{\alpha\beta}P_{i\beta}(0,\omega_i)d\Omega T(\omega_i,\omega_s) \\ T(\omega_i,\omega_s) &= G_s(R_s,A_s,z)\int_0^L G_i(R_i,A_i,z)dz \end{aligned}\,. \tag{6.137}$$

If multiple scattering within the sample is neglected, then in the case of forward scattering, $T(\omega_i,\omega_s)$ is given by:

$$T(\omega_i,\omega_s) = \frac{(1-R_i)(1-R_s)}{A_s - A_i}\left[\exp[(A_s - A_i)L]-1\right]. \tag{6.138}$$

In the case of a rectangular sample this becomes:

$$T(\omega_i,\omega_s) = \frac{(1-R_i)(1-R_s)}{A_i}\cdot\exp(-A_s a)\left[1-\exp[-A_i L]\right]. \tag{6.139}$$

In the case of backscattering T is given by:

$$T(\omega_i,\omega_s)=\frac{(1-R_i)(1-R_s)}{A_s+A_i}\left[1-\exp[(A_s+A_i)L]\right]. \tag{6.140}$$

These equations become more complicated once multiple internal reflections are included.

Thus one can deduce the scattering cross-section $S_{\alpha\beta}$ by measuring the scattered power $dP_{s\alpha}$ in a differential solid angle $d\Omega$, the incident power $P_{i\beta}$, and considering the energy loss T from the absorption and the reflection. One can also deduce the scattering cross-section of a sample by using a sample with a known cross-section as a reference sample. The scattering cross-section is usually frequency dependent. To yield accurate results, sample surfaces should be smooth and damage free to reduce diffuse scattering. In addition to changing ω_i to match the electronic energy level difference, one can also change the difference through modifications of the external conditions, such as temperature, magnetic field, uniaxial pressure, and static pressure. Table 6.6 lists the band gaps and phonon energies of some semiconductors with diamond and zinc blende structures.

Table 6.6. Typical band gap and phonon energy of some semiconductor materials with diamond and blende structures

Material	Indirect bandgap (eV)	Direct bandgap (eV)								Phonon wave number (cm^{-1})
		E_0		$E_0+\Delta$		E_1		$E_1+\Delta$		300K
		300 K	77 K	300 K	77 K	300 K	77 K	300 K	77 K	TO LO
InSb		0.180	0.23	1.08	1.1	1.88	1.98	2.38	2.48	180 191
InAs		0.356	0.41	0.77	0.79	2.49	2.61	2.77	2.88	218 243
GaSb		0.70	0.81	1.5	1.56	2.03	2.15	2.48	2.596	231 241
HgTe						2.09	2.21	2.71	2.85	116 131
Ge	0.66	0.796	0.88	1.092	1.164	2.13	2.22	2.33	2.41	318
InP		1.34	1.41	1.55		3.12		3.27		304 345
GaAs		1.429	1.51	1.77	1.85	2.904	3.02	3.314	3.245	269 292
CdTe		1.49	1.59	2.4	2.4	3.28	3.44	3.86	4.01	140 170
AlSb	1.62	2.218		2.97		2.810		3.210		319 340
Si	1.17	3.3				3.4				517
ZnTe		2.25	2.38	3.18		3.58	3.71	4.14	4.28	177 205
GaP	2.26	2.78	2.885	2.90		3.69		3.77		367 403
ZnSe		2.69	2.82	3.12		4.8		6.1		205 250
AlAs	2.13	2.9		319						363 402
ZnS		3.69	3.80	3.76		6.81		6.8		271 352

6.4.3 Application of Selection Rules

Momentum and energy conservation is required in a scattering process. Energy conservation is:

$$\omega_i - \omega_s = \pm\omega_j, \tag{6.141}$$

and momentum conservation is:

$$\boldsymbol{k}_i - \boldsymbol{k}_s = \pm\boldsymbol{k}_j, \tag{6.142}$$

as shown in Fig. 6.49.

The magnitude of (6.142) is:

$$\left|\boldsymbol{k}_j\right| = (\boldsymbol{k}_i^2 + \boldsymbol{k}_s^2 - 2\boldsymbol{k}_i \cdot \boldsymbol{k}_s)^{1/2}. \tag{6.143}$$

For forward scattering we have, θ = 0, and since $k = \frac{\omega}{v} = \frac{\omega}{c/n} = \frac{n\omega}{c}$:

$$\left|\boldsymbol{k}_j\right|_{\min} = \left|\boldsymbol{k}_i\right| - \left|\boldsymbol{k}_s\right| = \frac{n(\omega_i)\omega_i - n(\omega_s)\omega_s}{c}. \tag{6.144}$$

For backscattering we have, θ = 180° and:

$$\left|\boldsymbol{k}_j\right|_{\max} = \left|\boldsymbol{k}_i\right| + \left|\boldsymbol{k}_s\right| = \frac{n(\omega_i)\omega_i + n(\omega_s)\omega_s}{c}. \tag{6.145}$$

For typical narrow band gap semiconductor parameters, $\hbar\omega_i \approx \hbar\omega_s$ = 2.5 eV, $n = 4$, we obtain $k_j \approx 10^6$ cm^{-1}. The scattered phonons are still in the center of the Brillouin zone because the boundary of Brillouin zones occurs in the range of 10^8 cm^{-1}. Backscattering is used in resonant Raman scattering. The ω_j values observed correspond to the standard LO and TO frequencies.

Sometimes the wave vector conservation rule is broken because of the destruction of symmetry around an impurity or a defect. Also, the wave

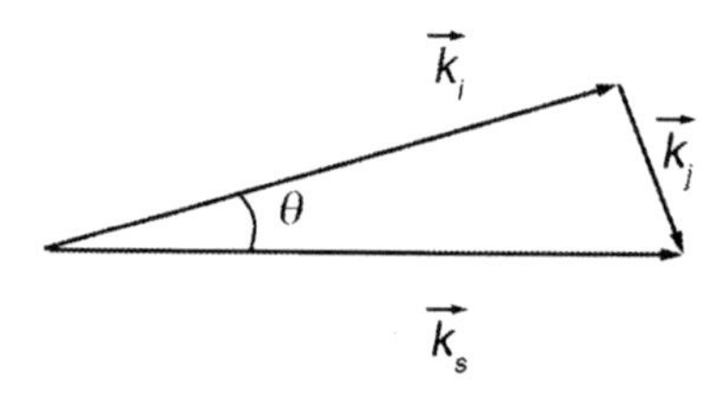

Fig. 6.49. A schematic drawing of moment conservation

vector conservation rule is broken near surfaces or if the volume of the scattering material is small enough. The indeterminacy is given by:

$$\Delta\left|\boldsymbol{k}_j\right| = \frac{1}{d}, \tag{6.146}$$

where d is a characteristic dimension.

From the wave vector conservation, we obtain:

$$\left|\boldsymbol{k}_j\right| = \frac{\left[n^2(\omega_i)\omega_i^2 + n^2(\omega_s)(\omega_i-\omega_j)^2 - 2n(\omega_i)n(\omega_s)\omega_i(\omega_i-\omega_j)\cos\varphi\right]^{1/2}}{c} \tag{6.147}$$

with φ being the scattered angle, and ω_i the frequency of the exciting light. Thus we can deduce $\boldsymbol{k}_j$ by changing φ, and $\hbar\omega_j$ from ω_i and by measuring $\hbar\omega_s$. As a result we can get the dispersion curves, $\hbar\omega_j \sim k_j$.

The selection rules are not only limited by energy and momentum conservations, but also by the symmetry of the crystals. In general in Raman scattering, a monochromatic light source is used to provide the incident light on the samples. The scattered light is collected in the forward, back and side directions. From the information collected different elementary excitations will appear in the spectra if the selection rules are obeyed. In order to specify and help separate the different characteristic elementary excitations, polarization dependent Raman scattering measurements are used. Suppose the incident light wave vector is $\boldsymbol{k}_i$, and the electric-field polarization direction of the incident light is $\boldsymbol{E}_i$. The signs of the elements of the tensor $\boldsymbol{k}_i(\boldsymbol{E}_i,\boldsymbol{E}_s)\boldsymbol{k}_s$ are used to indicate the geometry of the scattered light polarization field $\boldsymbol{E}_s$ its direction $\boldsymbol{k}_s$. Which type of phonon causes the scattering can be determined by analyzing the scattered light geometry.

We next analyze the transition polarizability tensor (Raman tensor) in order to investigate the phonon contribution to the scattering in more detail. If $\boldsymbol{k}_i$ is the wave vector of the incident light with its polarization of the electric-field along the β direction, and $\boldsymbol{k}_s$ is the wave vector of the scattered light with its polarization along the α direction, then the scattering tensor observed for the polarization configuration is $\boldsymbol{k}_i(\beta,\alpha)\boldsymbol{k}_s$, and is due to the $P_{s\alpha}$ component of dipole moment $\boldsymbol{P}_s$ in unit volume written as:

$$P_{s\alpha} = \chi_{\alpha\beta}^{i,s}(\omega_i,\omega_s)E_{i\beta}^0 \exp\left\{i\left[(\boldsymbol{k}_i + \boldsymbol{k}_j)r - (\omega_i \pm \omega_j)t\right]\right\}. \tag{6.148}$$

In the above analysis, the known Raman scattering cross-section is given by:

$$R \propto \left|\chi_{\alpha\beta}(\omega_i,\omega_s)\right|^2 \tag{6.149}$$

Because of features arising from the phonons, $\chi_{\alpha\beta}(\omega_i,\omega_s)$ is dependent on the wave vector $\boldsymbol{k}_j$. As a result $\tilde{\chi}$ can be expanded in terms of $\nabla Q_j = i\boldsymbol{k}_j Q_j$. In addition some external factors, such as pressure, electric-fields, and magnetic fields, effect the phonons and $\chi_{\alpha\beta}$. Thus the transition polarizability can written, $\tilde{\chi} \sim f(Q_j, \nabla Q_j, E_a)$. E_a is the sum of all contributions to the applied electric-field, such as a surface induced electric-field, then

$$\chi_{\alpha\beta}(Q_j,\nabla Q_j,E_a) = \chi_{\alpha\beta}^0(\omega_i,\omega_s) + \chi_{\alpha\beta}^1(\omega_i,\omega_s) + \chi_{\alpha\beta}^2(\omega_i,\omega_s) + \cdots \tag{6.150}$$

with

$$\chi_{\alpha\beta}^0(\omega_i,\omega_s) = \chi_{\alpha\beta}^0(\omega_i,\omega_i) = \chi_{\alpha\beta}^0(\omega_i)$$

$$\chi_{\alpha\beta}^1(\omega_i,\omega_s) = \frac{\partial\chi_{\alpha\beta}}{\partial Q_j}\cdot Q_j + \frac{\partial\chi_{\alpha\beta}}{\partial\nabla Q_j}\cdot iQ_jk_j + \frac{\partial^2\chi_{\alpha\beta}}{\partial Q_j\partial E_a}\cdot E_aQ_j + \cdots \tag{6.151}$$

$$\chi_{\alpha\beta}^2(\omega_i,\omega_s) = \frac{\partial^2\chi_{\alpha\beta}}{\partial Q_j\partial Q_{j'}}\cdot Q_jQ_{j'} + \frac{\partial^2\chi_{\alpha\beta}}{\partial\nabla Q_j\partial Q_j}\cdot ik_jQ_jQ_{j'} + \frac{\partial^3\chi_{\alpha\beta}}{\partial Q_j\partial Q_{j'}\partial E_a}\cdot Q_jQ_{j'}E_a + \cdots$$

$\chi_{\alpha\beta}^0$ is the average polarizability in the absence of phonon transitions. $\chi_{\alpha\beta}^1(\omega_i,\omega_s)$ is the first-order polarizability corresponding to one-phonon assisted transition. It describes the first-order Raman scattering generating the frequencies $\omega_s = \omega_i \pm \omega_j$ in the scattered light. $\chi_{\alpha\beta}^2(\omega_i,\omega_s)$ is the second-order polarizability corresponding to two-phonon assisted transitions. It describes the second-order Raman scattering related to the incident field at frequency ω_i and two-phonons, Q_j, ω_j, and $Q_{j'}$, $\omega_{j'}$ assistance. Equation (6.151) can be written in the simplified forms:

$$\chi^1_{\alpha\beta}(\omega_i,\omega_s)=\chi_{\alpha\beta}(j)+i\chi_{\alpha\beta k}(j)+\chi_{\alpha\beta E}(j)+\cdots, \tag{6.152}$$

$$\chi^2_{\alpha\beta}(\omega_i,\omega_s)=\chi_{\alpha\beta}(jj')+i\chi_{\alpha\beta k}(jj')+\chi_{\alpha\beta E}(jj')+\cdots. \tag{6.153}$$

$\chi^1_{\alpha\beta}(j)$ that only depends on Q_j, describes the normal one-phonon Raman scattering. The rest of the terms describe the higher-order effects found in different conditions. These effects are always weak but become stronger in the case of resonant Raman scattering.

In general (6.152) is used in the analysis of polarization independent resonant Raman scattering in narrow gap semiconductors. In his studies of the Raman scattering of InAs, Rubloff et al. (1973) investigated the Raman scattering selection rules for zinc blende structures. The cross-section of Raman scattering is in direct proportion to the square of the polarizability change $\delta\chi$ induced by a phonon:

$$R\sim\left|\delta\chi\right|^2. \tag{6.154}$$

For a TO phonon, and to lowest order $\delta\chi$ is written as:

$$\delta\chi=\frac{\partial\chi}{\partial u}\cdot u+\frac{\partial^2\chi}{\partial E_s\partial u}\cdot uE_s+\frac{\partial\chi}{\partial(\nabla u)}\cdot\nabla u, \tag{6.155}$$

with u being the magnitude of the atomic displacement at $\boldsymbol{r}$ caused by a phonon, E_s is the surface electric-field, $\nabla u=i\mathbf{q}u$ is the gradient of the atomic displacement, $\boldsymbol{q}$ the scattered wave vector (Burstein and Pinczuk 1972). The first term of (6.155) is the scattering introducing by a $q=0$ phonon, this is a uniform atomic displacement (AD). Phonon scattering occurs even when the wave vector is zero ($q=0$). Even for AD = 0, due to the second and third terms of (6.155) (forbidden terms), TO phonon scattering can occur near the resonant condition. The second term arises from the scattering induced by the surface electric field (SF), caused by the $\frac{\partial\chi}{\partial u}$ dependence on the surface electric field E_s. The third term ($\sim iqu$) is from the linear q induced Raman scattering related to strong absorption (LQ). It is much different from the other two terms.

The above three terms can be modified into a similar analysis for LO phonons by taking into account the effect of the microscopic electric field (Fröhlich interaction) and substituting $\frac{\partial}{\partial u}+\frac{\partial E}{\partial u}\frac{\partial}{\partial E}$ for $\frac{\partial}{\partial u}$.

Thus the Raman scattering cross-section can be written as the quadratic sum of the complex number amplitudes of the above three processes-AD, SF, LQ, i.e.:

$$R \sim \left|AD + SF + iLQ\right|^2 = (AD' + SF' - LQ'')^2 + (AD'' + SF'' + LQ')^2, \quad (6.156)$$

where "and" represent the real part and imagery part, respectively.

For the phonons polarized along x, y, or z axes in a zinc blende structured material, the AD term of the Raman scattering tensor (the $q = 0$ atomic displacement term), is give by:

$$\begin{pmatrix} 0 & 0 & 0 \\ 0 & 0 & a \\ 0 & a & 0 \end{pmatrix} \begin{pmatrix} 0 & 0 & a \\ 0 & 0 & 0 \\ a & 0 & 0 \end{pmatrix} \begin{pmatrix} 0 & a & 0 \\ a & 0 & 0 \\ 0 & 0 & 0 \end{pmatrix}. \quad (6.157)$$

For the backscattering case, the surface electric-field and wave vector $\boldsymbol{q}$ are perpendicular to the surface, so the symmetry of the SF term is the same of that of the LQ term. They have the same scattering selection rules. If $\boldsymbol{q}$ or $\boldsymbol{E}_s$ is along the z axis of the cube, for phonons polarized along x, y, or z axes, the Raman tensor has the form:

$$\begin{pmatrix} 0 & 0 & d \\ 0 & 0 & 0 \\ d & 0 & 0 \end{pmatrix} \begin{pmatrix} 0 & 0 & 0 \\ 0 & 0 & d \\ 0 & d & 0 \end{pmatrix} \begin{pmatrix} c & 0 & 0 \\ 0 & e & 0 \\ 0 & 0 & b \end{pmatrix}, \quad (6.158)$$

and the constants b, c, and d are the three independent elements of the tensor, $\frac{\partial^2}{\partial E_s \partial u}$ in the SF term, or $\frac{\partial x}{\partial(\nabla u)}$ in the LO term in (6.155).

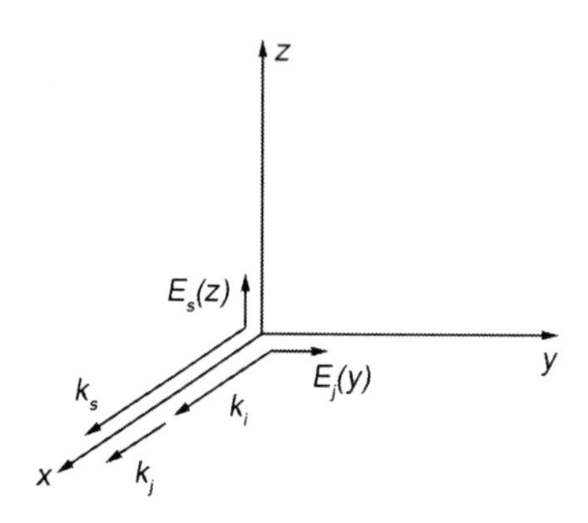

Fig. 6.50. A schematic draw of the backscattering

Since $P_{s\alpha} = \chi_{\alpha\beta} E_{i\beta}$. Due to the introduction of the incident light electric field $E_{i\beta}$, the dipole moment component $P_{s\alpha}$ in the α direction, and scattering cross-section transition polarizability tensor $\tilde{\chi}$ contributions from different phonon polarization directions can be calculated. Thus from the observed scattered spectra, including the polarization, $\boldsymbol{k}_i(\beta,\alpha)\boldsymbol{k}_s$, whether the contribution is from TO or from LO phonons can be determined because of the relation between the polarized direction of involved phonons and the direction of $\boldsymbol{k}_j$.

The polarization configuration of backscattering is illustrated in Fig. 6.50. The incident light $\boldsymbol{k}_i$ is along the x direction, the polarization of the electric-field $E_i(y)$ is along the y direction (0, E_y, 0). Only those polarized phonons in the x direction make a contribution to the tensor according to the transition polarizability tensor (AD term). Because the Raman tensor of the polarized phonons in the x direction is $\begin{pmatrix} 0 & 0 & 0 \\ 0 & 0 & a \\ 0 & a & 0 \end{pmatrix}$, thus from (6.148):

$$\boldsymbol{P}_{s\alpha} = \begin{pmatrix} 0 & 0 & 0 \\ 0 & 0 & a \\ 0 & a & 0 \end{pmatrix} \begin{pmatrix} 0 \\ E_y \\ 0 \end{pmatrix} = aE_y\boldsymbol{z} . \tag{6.159}$$

$\boldsymbol{P}_{s\alpha}$ is not equal to zero and is polarized in the z direction. It can be easily observed in the polarization configuration, $X(YZ)\bar{X}$. Thus the phonons polarized in the x direction, participate in this type of Raman scattering. From wave vector conservation, $\boldsymbol{k}_i \pm \boldsymbol{k}_j = \boldsymbol{k}_s$, $\boldsymbol{k}_j$ lies along the x axis. The phonons involved are LO phonons. However the Raman tensor of the phonons polarized in the y direction is $\begin{pmatrix} 0 & 0 & a \\ 0 & 0 & 0 \\ a & 0 & 0 \end{pmatrix}$ and its product with the electric field $\begin{pmatrix} 0 \\ E_y \\ 0 \end{pmatrix}$ is equal to zero in all directions. Using a similar argument, the Raman tensor of the phonons polarized in the x

direction is $\begin{pmatrix} 0 & a & 0 \\ a & 0 & 0 \\ 0 & 0 & 0 \end{pmatrix}$ and its product with $\begin{pmatrix} 0 \\ E_y \\ 0 \end{pmatrix}$ is also equal to zero in all directions.

Next we analyze the case of the polarization configuration, $X(YY)\bar{X}$. In this case, the incident light is along the x axis, the polarization of the electric-field is along the y direction, the electric-field is (0, E_y, 0). Following the above discussion, the products of the Raman tensors of phonons polarized along the y, or the z directions and the electric-field (0, E_y, 0) are zero. The product of the Raman tensor of polarized phonons along the x direction and (0, E_y, 0) is not equal to zero, but equals $aE_y z$ (6.159). It is the x direction component of the induced polar moment. It cannot be observed in the polarization configuration, $X(YY)\bar{X}$. As indicated by these results, the polarization configuration of the Raman scattering can provide the basis for the experimental results analysis.

For the polarization configuration $Z(XZ)Y$ of the rectangular scattering, the incident light is propagating along the z direction, and the incident light field is polarized along the x axis. Then we can observe the scattered light polarized in the z direction, and propagating in the y direction. In Fig. 6.51, arrows label the directions of the vectors.

For phonons polarized in the x direction, the dipole moment induced by the $\mathbf{E}_{i\beta}$ field is given by:

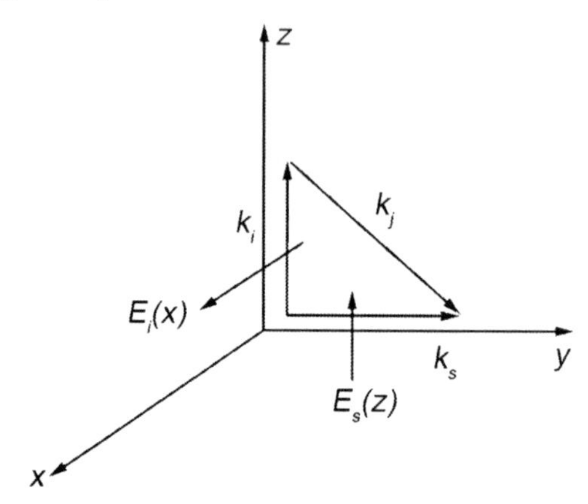

Fig. 6.51. A schematic drawing of the rectangular scattering of an incident light with $\mathbf{E}_{i\beta} = (E_x, 0, 0)$

$$\begin{pmatrix} 0 & 0 & 0 \\ 0 & 0 & a \\ 0 & a & 0 \end{pmatrix} \begin{pmatrix} E_x \\ 0 \\ 0 \end{pmatrix} = 0 . \tag{6.160}$$

For phonons polarized in the y direction, the dipole moment induced by the $\mathbf{E}_{i\beta}$ field is:

$$\begin{pmatrix} 0 & 0 & a \\ 0 & 0 & 0 \\ a & 0 & 0 \end{pmatrix} \begin{pmatrix} E_x \\ 0 \\ 0 \end{pmatrix} = aE_x \boldsymbol{z} . \tag{6.161}$$

For phonons polarized in the z direction, the dipole moment induced by the $\mathbf{E}_{i\beta}$ field is:

$$\begin{pmatrix} 0 & a & 0 \\ a & 0 & 0 \\ 0 & 0 & 0 \end{pmatrix} \begin{pmatrix} E_x \\ 0 \\ 0 \end{pmatrix} = aE_x \boldsymbol{y} . \tag{6.162}$$

So for the polarization configuration $Z(XZ)Y$, we can observe a contribution to the scattering from phonons polarized in the y direction. There is an inclination angle between $\boldsymbol{k}_j$ and the y axis because $\boldsymbol{k}_j$ is in the yz plane. Thus the polarized directions of phonons and $\boldsymbol{k}_j$ have perpendicular and parallel components corresponding to TO phonons and LO phonons. According to (6.162), the dipole moment produced by the phonons polarized along the z direction induced by $E_{i\beta}$, is along the y direction.

One should choose the incident light to be in the x direction or the z direction in order to observe the scattering in the above polarized directions, i.e. the polarization configurations $Z(XY)X$ or $Z(XY)\overline{Z}$. From (6.160) to (6.162), phonons polarized in any direction cannot generate an x directed component of the dipole moment, i.e. we cannot observe scattered light in the polarization configurations $Z(XX)Y$ and $Z(XX)\overline{Z}$. The above analysis considered only the AD term of the Raman tensor. It would be more complicated if account of the SF and the LQ Raman tensor terms were taken into account. However, we can make an initial analysis using above method.

Backscattering is often used as the measurement method for semiconductor materials. The coordinate system is chosen to be a practical orientation. Then the Raman tensor is transformed from the cubic lattice *X*[100], *Y*[010], *Z*[001] components to the selected coordinate system. If we use the *xyz* coordinate system, we must transform the electric-field $E_{i\beta}(x',y',z')$ dependent polarization configuration from the new coordinate system into the $E_{i\beta}(x,y,z)$ in the *xyz* coordinate system. Then we analyze the Raman measurement in the *xyz* coordinate system and the results are transformed back into the polarization configuration in the *x'*, *y'*, *z'* coordinate system. For example, in the $X'[100]$, $Y'[011]$, $Z'[0\bar{1}1]$ system, we consider the polarization configuration $X'(Y'Y')\bar{X}'$:

$$E_{i\beta}(x',y',z') = (0, E_{iy'}, 0), \tag{6.163}$$

When transformed to the *x*, *y*, *z* system:

$$E_{i\beta}(x,y,z) = (0, E_{iy}, E_{iz}), \tag{6.164}$$

or $E_{iy} = \dfrac{E_{iy'}}{\sqrt{2}}$, $E_{iz} = \dfrac{E_{iz'}}{\sqrt{2}}$. Thus for the *x*, *y*, *z* system, the phonons polarized in the *x* direction interact with this $E_{i\beta}$. The dipole moment produced is given by (Fig. 6.52):

$$\boldsymbol{P}_{js\alpha} = \begin{pmatrix} 0 & 0 & 0 \\ 0 & 0 & a \\ 0 & a & 0 \end{pmatrix} \begin{pmatrix} 0 \\ E_y \\ E_z \end{pmatrix} = aE_z\boldsymbol{y} + aE_y\boldsymbol{z} = a\sqrt{E_z^2 + E_y^2} \cdot \boldsymbol{y}'. \tag{6.165}$$

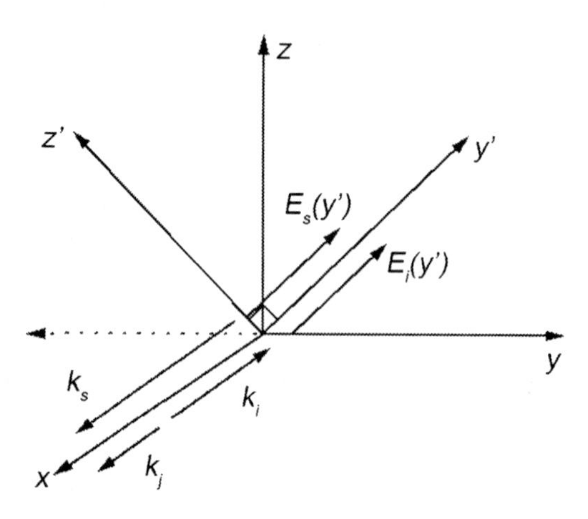

Fig. 6.52. A schematic drawing of the coordinate transformation

The phonons polarized in the y direction also interact with $E_{i\beta}$ to produce a dipole moment written as:

$$\boldsymbol{P}_{s\alpha}=\begin{pmatrix}0&0&a\\0&0&0\\a&0&0\end{pmatrix}\begin{pmatrix}0\\E_y\\E_z\end{pmatrix}=aE_z\boldsymbol{x}=aE_z\boldsymbol{x}'. \tag{6.166}$$

The phonons polarized in the z direction interact with $E_{i\beta}$ to produce a dipole moment given by:

$$\boldsymbol{P}_{s\alpha}=\begin{pmatrix}0&0&a\\a&0&0\\0&0&0\end{pmatrix}\begin{pmatrix}0\\E_y\\E_z\end{pmatrix}=aE_y\boldsymbol{x}=aE_y\boldsymbol{x}'. \tag{6.167}$$

According to momentum conservation, $\boldsymbol{k}_i\pm\boldsymbol{k}_j=\boldsymbol{k}_s$, $\boldsymbol{k}_j$ is along the $\boldsymbol{x}$ or the $\boldsymbol{x}'$ direction. The scattered light observed along the $\boldsymbol{x}'$ direction and polarized in the $\boldsymbol{y}'$ direction, is generated from incident light polarization in the x direction. The relevant phonons propagating in the x direction are LO phonons. However, in (6.166) and (6.167) the dipole moment components in $\boldsymbol{x}'$ direction can not be observed in the x direction, but only in the $\boldsymbol{z}'$ or the $\boldsymbol{y}'$ direction (i.e. the polarization configurations, $X'(Y'Y')Z'$ or $X'(Y'X')Y'$.

Table 6.7. O_h group and T_d group selection rules of Γ_{15} phonons in first-order Raman scattering

No	Coordinate system x y z	Polarization $\alpha\beta$	$\lvert\chi_{\alpha\beta}(\Gamma_{15})\rvert^2$ LO(x)	LO(y)	TO(z)
1	$<100><010><001>$	YY=ZZ	0	0	0
2		YZ	e^2	0	0
3	$<100><01\bar{1}><0\bar{1}1>$	YY=ZZ	e^2	0	0
4		YZ	0	0	0
5	$<110><1\bar{1}0><001>$	YY	0	0	d^2
6		YZ	0	d^2	0
7		ZZ	0	0	0
8	$<111><1\bar{1}0><11\bar{2}>$	YY	$e^2/3$	0	$2d^2/3$
9		YZ	0	$2d^2/3$	0
10		ZZ	$e^2/3$	0	$2d^2/3$
11	$<11\bar{2}><111><1\bar{1}0>$	YY	0	$4d^2/3$	0
12		YZ	0	0	$d^2/3$
13		ZZ	$2e^2/3$	$d^2/3$	0

For the phonons with Γ_{15} symmetries, according to usual crystal orienttation of zinc blende structures and interacting with the atomic displacement term of the Raman tensor, the selection rules and relative intensity of the first-order Raman backscattering in the polarization configuration $X(\alpha\beta)\bar{X}$, is shown in Table 6.7.

Rubloff et al. (1973) investigated the Raman scattering spectra of InAs. He chose $X[100]$ $Y[010]$ $Z[001]$, $X'[110]$ $Y'[\bar{1}10]$ $Z[001]$, $X''[111]$ $Y''[\bar{1}10]$ $Z[\bar{1}\bar{1}2]$ as the coordinate systems and deduced the relative intensities of the AD term, the SF term, and the LQ term in the Raman scattering spectra from TO and LO phonons. One can find a detailed derivation of the selection rules of the Raman scattering tensor in a Zhang reference (Zhang et al. 2001).

6.4.4 Raman Scattering in HgCdTe

Information about many fundamental physical processes can be obtained from Raman scattering studies, such as lattice dynamics, plasmon, coupled-plasmon modes and other excitations. It is an important complement to other spectroscopic techniques for uncovering the behavior of some processes. Raman scattering is mostly used to extract the nature of the vibrational excitations of materials, while other spectroscopic techniques generally investigate the electronic states of the solid. Since Raman scattering is sensitive to the nearest neighbor effects, it probes the crystal structure and its quality. It can also provide detailed information about the properties and quality of both bulk and thin film materials.

Because it is a second order process Raman scattering contains symmetry information that is not obtained from first order interactions. For the zinc blende structure, first order processes are not sensitive to the relative polarization direction of the light and the orientation of the crystal axes. The intensity of the Raman scattering depends on both the incident and scattered light polarizations. Thus it is related to the symmetry of the crystal and it can be used to characterize both fundamental interactions as well as crystal quality.

Only a few studies have been reported on Raman scattering off of HgCdTe alloys in sprite of this alloy's important application as an infrared detector material. Mooradian and Harmon (1971) used Raman scattering to investigate the composition dependence of the TO and LO phonon frequencies. By using the 496.5, 488.0, 501.7, and 514.5 nm lines of an Ar^{+} laser with various polarizations of the incident and the scattered light, the Pollak group (Amirtharaj et al. 1983) investigated the Raman scattering of a $Hg_{0.8}Cd_{0.2}Te$ sample at 77 K near the resonance of the E_1 optical gap. They observed structures due to a symmetry forbidden TO phonon mode, a defect mode, possible clustering effects, and a coupled LO phonon-inter-sub-band excitation in the inversion layer (the S_- mode). Because the S_- mode is extremely sensitive to the surface treatment, Raman scattering is a valuable tool to study surface conditions and preparation.

Their measurements were made in the backscattering geometry at 77 K. The laser power level was less than 250 mW. The samples were p-$Hg_{0.8}Cd_{0.2}Te$ with both the (100) and (111) faces exposed. They had a hole density of 2×10^{15} cm^{-3}. The preparation of the sample surface was crucial to obtain good Raman spectra. The usual method of a mechanical polish (0.05 μm alumina), followed by a ~5% bromine/methanol chemical etch was found to yield unsatisfactory results, i.e. the S_- mode was not observable from such a surface. However, the mode could be observed on a good surface using a polish followed by a 0.1% bromine/methanol chemo-mechanical polish and a final free etch in the same solution.

The Pollak group (Amirtharaj et al. 1983) systematically investigated Raman scattering in HgCdTe. They used the following designations to indicate their scattering geometries:

(a) $<100>$ face, $X_1[100]$, $Y_1[010]$, $Z_1[001]$; or $X_1[100]$, $Y_1^{'}[011]$, $Z_1^{'}[01\bar{1}]$

(b) $<100>$ face, $X_2[1\bar{1}0]$, $Y_2[110]$, $Z_2[001]$; or $X_2[1\bar{1}0]$, $Y_2^{'}[111]$, $Z_2^{'}[\bar{1}\bar{1}2]$

(c) $<100>$ face, $X_3[100]$, $Y_3[1\bar{1}0]$, $Z_3[\bar{1}\bar{1}2]$

For the backscattering geometry, the following selection rules apply for the bulk ($q \approx 0$) LO and TO phonons.

$<100>$ face

(a) The TO phonon is forbidden for all polarization configurations.

(b) The LO phonon is allowed for the $X_1(Y_1Y_1)\bar{X}_1$, the $X_1(Y_1Z_1)\bar{X}_1$, and the $X_1(Y_1^{'}Y_1^{'})X_1$ configurations but is forbidden for the $X_1(Y_1^{'}Y_1^{'})\bar{X}_1$ configuration. In the $X_1(Y_1Y_1)X_1$ polarization, the LO mode is activated by the presence of **q** linear terms: $X_1(100)$, $Y_1(010)$, $Z_1(001)$, $Y_1^{'}(011)$ and $Z_1^{'}(0\bar{1}1)$.

$<111>$ face

(a) The LO and TO phonons are allowed for the $X_3(Y_3Y_3)\bar{X}_3$ configurations.

(b) For the $X_3(Y_3Y_3)\bar{X}_3$ configuration, the TO phonon is allowed but the LO phonon is forbidden (including **q**-linear considerations) X_3 (111), $Y_3(1\bar{1}0)$, and $Z_3(1\bar{1}2)$.

These symmetry selection rules can be used to identify the bulk TO and LO phonons. Pollak et al. have studied polarization-dependent Raman scattering from the three principal faces, i.e. (100), $(1\bar{1}0)$, and (111) of $Hg_{1-x}Cd_xTe$ alloys (Tiong et al. 1984; Amirtharaj et al. 1983, 1985; Ksendzov et al. 1990). The Raman spectra for the (100) and (111) faces recorded for different polarization configurations are displayed in Figs. 6.53 and 6.54. These spectra were measured at 77 K using the 514 nm line of an Ar^+ laser. We can observe the resonant feature in Fig. 6.53 because the 514 nm line is nearly resonant with the E_1 gap (~2.4 eV) of $Hg_{0.8}Cd_{0.2}Te$ at 77 K (Ksendzov et al. 1990).

For $Hg_{0.8}Cd_{0.2}Te$, the features at 122 and 140 cm^{-1} correspond to the HgTe-like TO_2 and LO_2 phonons, respectively. The feature appearing at

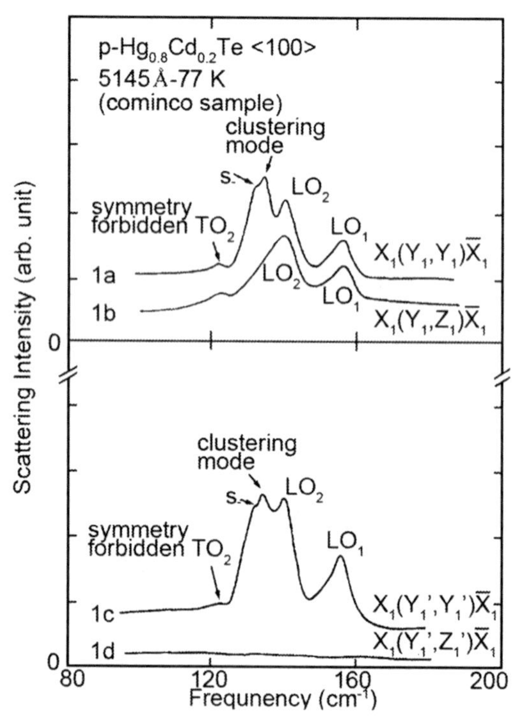

Fig. 6.53. The Raman spectra of $Hg_{0.8}Cd_{0.2}Te$ (100) using the axes designations: $X_1(100)$, $Y_1(010)$, $Z_1(001)$, $Y_1^{'}(011)$, and $Z_1^{'}(0\bar{1}1)$ the configurations are: (**1a**) $X_1(Y_1Y_1)\bar{X}_1$, (**1b**) $X_1(Y_1Z_1)\bar{X}_1$, (**1c**) $X_1(Y_1Y_1)\bar{X}_1$, and (**1d**) $X_1(Y_1^{'}Z_1^{'})\bar{X}_1$

156 cm^{-1} are assigned to the LO_1 (TO_1) phonons (Tiong et al. 1984; Amirtharaj et al. 1983, 1985; Ksendzov et al. 1990). The TO mode of the HgTe-like phonons soften as the temperature is decreased. At $x = 0.2$ the LO (TO) CdTe-like phonon mode behaves like a traveling wave. Thus it does not act like a localized Cd impurity mode in HgTe.

Some visible scattering modes can be observed in addition to the above TO and LO phonons at $q = 0$.

The peak at 135 cm^{-1} seen from both the (100) and the (111) faces may arise from a clustering effect. This mode has a Γ_1 symmetry since it is seen only in spectra where the incident and scattered light are parallel to each other. It has been suggested that this mode is a characteristic of alloy samples and is due to the Te-3Hg-Cd tetrahedral clusters (Tiong et al. 1984; Amirtharaj et al. 1983, 1985; Ksendzov et al. 1990). There is no contribution to this scattering mode from free electrons as indicated from further studies of n-type and p-type HgCdTe samples.

The shoulder at 132 cm^{-1} observed from both the (100) and the (111) faces was found to be extremely sensitive to surface conditions. Thus it

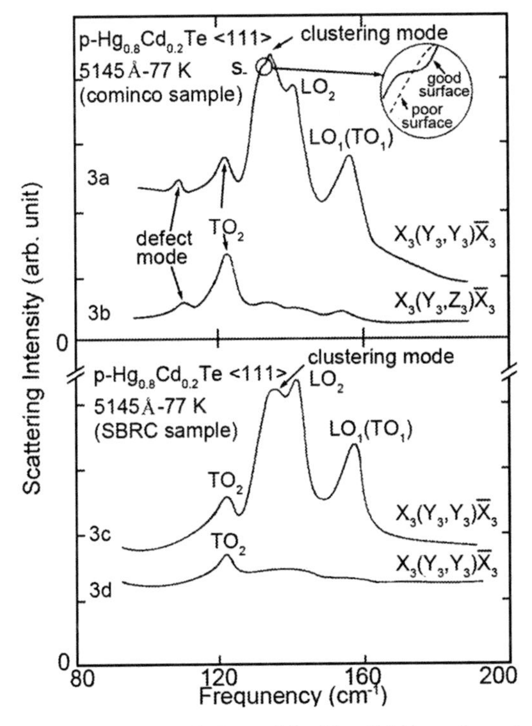

Fig. 6.54. The Raman spectra of $Hg_{0.8}Cd_{0.2}Te$ (111) using the axes labels: $X_3(111)$, $Y_3(110)$, $Z_3(112)$

can be used as a measure of the surface quality of the HgCdTe crystals. This feature arises from the coupled LO phonon-inter-sub-band excitations in the inversion layer (S_- mode) (Tiong et al. 1984; Amirtharaj et al. 1983, 1985). It seemed that there was a different origin of the S_- mode from the S mode, due to the different relative scattering intensities in p-type and n-type samples (Amirtharaj et al. 1990).

The feature at 108 cm^{-1} observed from the (111) face was assigned to a Hg antisite defect (the substitution of a Hg on a Te site). It was also observed in HgTe. This mode is absent in the (100) spectra due to its symmetry character, i.e. TO-like, or due to the difference in the quality between the (100) and (111) samples (Tiong et al. 1984; Amirtharaj et al. 1983, 1985).

For the (100) and the (111) faces of $Hg_{0.8}Cd_{0.2}Te$, a HgTe-like TO_2 mode is symmetry forbidden in the configurations $X'(Z'Z')\bar{X}'$, $X'(1\bar{1}0)$, and $Z'(001)$, including the surface electric-field and the **q**-linear assistance terms. However, the HgTe-like TO_2 mode is observed experimentally in the above normally forbidden configurations. Tiong believed it was due to the existence an internal strain with components (η_{xy}, η_{xz}, and η_{yz}) (Tiong et al. 1984). This kind of internal strain is along the bond directions of the crystal and cannot be produced by an external uniaxial stress. Thus the mode can be used as a measure of the crystalline quality of the HgCdTe (Tiong et al. 1984; Amirtharaj et al. 1983).

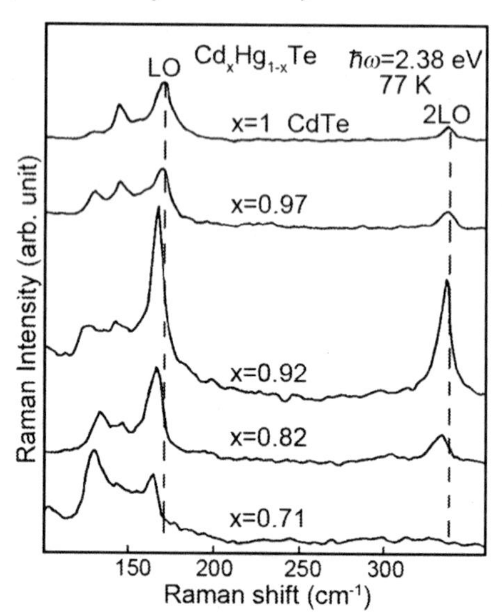

Fig. 6.55. The Raman spectra observed at 77 K of HgCdTe for various compositions x (the spectra were excited at 2.38 eV)

Lusson and Wagner studied LPE grown Cd-rich HgCdTe samples using Raman scattering. For LPE $Hg_{0.29}Cd_{0.71}Te$ thin films grown on $Cd_{0.96}Zn_{0.04}Te$ substrates, the CdTe-like LO-phonon mode is observed at 164.8 cm^{-1} together with the HgTe-like LO-phonon mode at 133 cm^{-1} by using an excitation radiation at 2.38 eV. Then excitation at 2.18 and 1.91 eV results in a drastic enhancement of the CdTe-like LO-phonon line with respect to the CdTe-like TO- and HgTe-like LO-phonon modes. Meanwhile scattering by two CdTe-like LO phonons is seen at 329.5 cm^{-1}. They calculated the one- and two-LO-phonon resonance energies $E_0+\Delta_0$ to be 1.913 and 1.934 eV, respectively, for $x = 0.71$ at 77 K, by using a semi-empirical formula (Lusson and Wagner 1988).

For different compositions $Hg_{1-x}Cd_xTe$ ($0.5 \le x \le 1$) thin films, the CdTe-like LO phonon mode was also studied by Raman spectroscopy (Lusson and Wagner 1988). Figure 6.55 shows the Raman spectra with different compositions but excited at a fixed photon energy of 2.38 eV. A resonance enhancement of the CdTe-like one-and two-LO-phonon modes was found with a maximum at $x = 0.92$. The corresponding resonance energies are 2.342 (1LO) and 2.363 eV (2LO), respectively. The Raman shift of the 1LO and the 2LO phonon lines as a function of composition x are displayed in Fig. 6.56. The experimental curves show the variation of the 1LO and the 2LO phonon frequency changes with x is linear, which is identical to the theoretical prediction (Lusson and Wagner 1988).

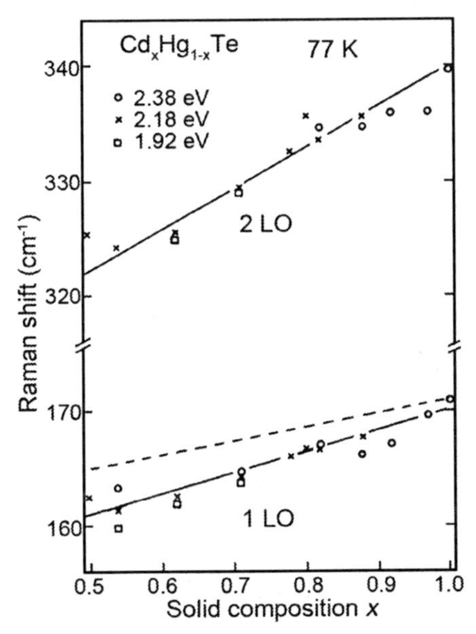

Fig. 6.56. The frequency observed at 77 K of the 1LO and the 2LO phonon Raman lines vs. alloy composition x

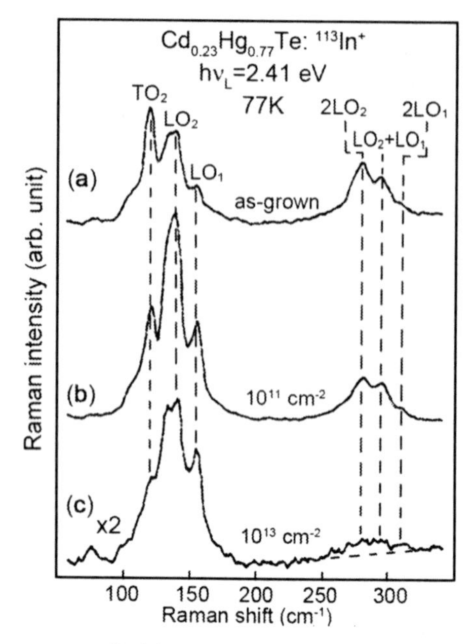

Fig. 6.57. Raman spectra of (**a**) as-grown and (**b**) and (**c**) In implanted $Hg_{0.77}Cd_{0.23}Te$ excited by 2.41 eV radiation (the spectral resolution was 5 cm^{-1})

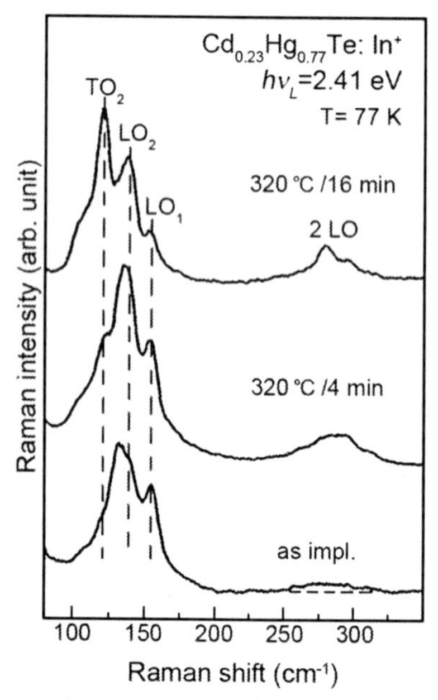

Fig. 6.58. Raman spectra of In^+ implanted $Hg_{0.77}Cd_{0.23}Te$ in the as-implanted state (*bottom curve*) and after annealing at 320°C for 4 and 16 min (*middle and top curve*), respectively

Ion implantation into $Hg_{1-x}Cd_xTe$ is a common technique to prepare n-type conducting layers for the fabrication of p–n junctions in $Hg_{1-x}Cd_xTe$. It has been shown that in polar semiconductors, such as GaAs and CdTe, Raman scattering by LO phonons is a powerful experimental technique to analyze implantation-induced damage and its removal by thermal annealing. The intensity ratio of the 1LO/2LO phonon scattering is a measure of the crystalline quality in polar semiconductors (Lusson et al. 1989; Wagner et al. 1993).

Wagner et al. studied the lattice damage in implanted $Hg_{1-x}Cd_xTe$ using resonant Raman scattering by LO phonons. They also analyzed the lattice damage removal by the type of the implanted ion, the implantation dose, and various annealing techniques (Lusson et al. 1989; Wagner et al. 1993). Figure 6.57 shows the Raman spectra of a $Hg_{0.77}Cd_{0.23}Te$ sample excited at 2.41 eV, that is close to the E_1 gap resonance. It was shown that the 1LO/2LO phonon scattering intensity ratio is a measure of the implantation damage for the implantation doses between 10^{11} and 10^{14} cm^{-2} (Lusson et al. 1989). It was found as expected that the damage produced by the ions with lighter mass is less than that produced by the ions with heavier mass. Lattice damage removal is enabled by using an elevated thermal annealing temperature and increasing the annealing time. Figure 6.58 displays a series Raman spectra from a $Hg_{0.77}Cd_{0.23}Te$ sample implanted with In^+.

Further studies of resonance scattering in HgCdTe can found in the Scepanovic reference (Scepanovic and Jevtic 1998). Also Huang et al. (2001) investigated HgCdTe samples using micro-Raman scattering and micro-photoluminescence.

Reference

Amirtharaj PM, Tiong KK, Pollak FH (1983) J Vac Sci Technol A 1:1744
Amirtharaj PM, Tiong KK, Parayanthal P, et al. (1985) J Vac Sci Technol A 3:226
Amirtharaj PM, Dhar NK, Baars J, et al. (1990) Semicond Sci Technol 5:S68
Baars J, Sorger F (1972) Solid State Commun 10:875
Balkanski M (1979) Narrow Gap Semiconductors, Physics And Applications. Springer, Heidelberg
Balkanski M, Jian KP, Beserman R, et al. (1975) Phys Rev B 22:2913
Barker AS, Sievers AJ (1975) Rev Modern Phys 47(Supp. 2):1
Bilzand H, Kress W (1979) Phonon dispersion relations in insulators Vol 113. Springer-Verlag, Heidelberg
Brodsky MH, Lurio A (1974) Phys Rev B 9:1646
Bruüesch P (1982) Phonon: Theory and Experiments I (Lattice Dynamics and Models of Interatomic Forces). Springer-Verlag, Berlin

Burkel E (2001) J Phys: Condense Matter 13:7627
Burstein E, Pinczuk A (1972) In: Albers WA (ed) The physics of Opto-Electronic Materials. Plenum, New York, p 33
Camacho J, Cantarero A (1999) Phys Stat Sol B 215:181
Camacho J, Cantarero A (2000) Phys Stat Sol B 220:233
Cardona M (1967) In: Semiconductors and Semimetals Vol 3. Academic Press, New York, p 125
Cardona M, Kunc K, Martin RM (1982) Solid State Commun 44:1205
Chelikowski JR, Cohen ML (1976) Phys Rev B 14:556
Chen AB, Sher A (1995) Semiconductor Alloys. Plenum Press, New York
Chu JH (1983) Infrared Research 2:89
Chu JH, Shen XC (1993) Semicond Sci Technol 8:S86–S89
Cochran W (1959a) Proc Roy Soc London A 253:260
Cochran W (1959b) Phys Rev Lett 2:495
Corso AD, Baroni S, Resta R (1993) Phys Rev B 47:3588
Cowley RA (1962) Proc Roy Soc A 268:109
Danielewicz EJ, Coleman PD (1974) Appl Opt 13:1164
Debernardi A, Pyka NM, Göbel A, et al. (1997) Solid State Commun 103:297
Dornhaus R (1983) The properties and applications of the $Hg_{1-x}Cd_xTe$ system. In: Narrow-Gap Semiconductors. Springer-Verlag, Berlin
Dornhaus R, Nimtz G (1976) In: Hohler G (ed) Solid State Phys, Springer Tracts in Modern Phys Vol 78. Springer-Verlag, Heidelberg
Dornhaus R, Nimtz G, Schlabitz W, et al. 1975. Solid State Commun 17:837
Dornhaus R, Faymonville, Bauer G, et al. (1982) Interna Conf on Application of High Magnetic Field in Semiconductors. Grenoble
Georgitse EI, et al. (1973) Sov Phys Semicond 6:1122
Gregg JR, Myles CW (1985) Jour Phys Chem Solids 46:1305
Grein CH, Cardona M (1992) Phys Rev B 45:8328
Harrison WA (1980) Electronic Structure and the Properties of Solids. WH Freeman and Co, San Francisco
Hass M, Henvis BW (1962) J Chem Phys Solids 23:1099
Huang H, Xu JJ, Qiao HJ, et al. (2001) Semicond Sci Technol 16:L85
Kepa H, Gebicki W, Giebultowicz T, et al. (1980) Solid State Commun 34:211
Kepa H, Giebultowicz T, Buras B, et al. (1982) Phys Scr 25:807
Kim RS, Narita S (1971) J Phys Soc Jpn 31:613
Kim LS, Perkowitz S, Wu OK, et al. (1990) Semicond Sci Technol 5:S107
Kimmitt MF, Lopez GL, Röser HP, et al. (1985) Infrared Phys 25:767
Kozyrev SP, Vodopyanov LK, Tribrulet R (1983) Solid State Commun 45:383
Krisch MH, Mermet A, San Miguel A, et al. (1997) Phys Rev B 56:8691
Krishnamurthy S, Chen AB, Sher A, van Schilfgaarde M (1995) J Electronic Mater 24:1121
Ksendzov A, Pollak FH, Amirtharaj PM, et al. (1990) Semicond. Sci. Technol. 5:S78
Kunc K, Martin RM (1982) Phys Rev Lett 48:406
Kunc K, Balkanski M, Nusimovivi (1974) Phys Rev B 12:4346
Lażewski J, Parlinski K, Szuszkiewicz W, et al. (2003) Phys Rev B 67:094305

Li B, Chu JH, Chang Y, et al. (1996) Infrared PhysTechnol 37:525
Lusson A, Wagner J (1988) Phys Rev B 38:10064
Lusson A, Wagner J, Ramsteiner M (1989) Appl Phys Lett 54:1787
Mazur YI, Kriven SI, Tarasov GG, et al. (1993) Semi Sci Technol 8:1187
Mooradian A, Harman TC (1971) J Phys Chem Solid 32 suppl: 297
Perkowitz S, Rajavel D, Sou IK, et al. (1986) Appl Phys Lett 49:806
Polian A, Toullec RL, Balkanski M (1976) Phys Rev B 13:3558
Price DL, Rowe JM, Nicklow RM (1971) Phys Rev B 3:1268
Rajput BD, Browne DA (1996) Phys Rev B 53:9052
Rubloff GW, Anastassakis E, Pollak FH (1973) Solid State Commun 13:1755
Ruf T (2003) Appl Phys A 76:21
Rowe J, Nicklow RM, Price DL, et al. (1974) Phys Rev B 10:671
Scepanovic M, Jevtic M (1998) Appl Phys A 67:317
Scott MW, Stelzer EL, Hager RJ (1976) J Appl Phys 47:1408
Sella C, Cohen-Solal G, Bailly F (1967) Compt Acad Sci B 264:179
Shen SC (1984) Prog Phys 3
Shen SC, Cardona M (1980a) Solid State Commun 36:327
Shen SC, Cardona M (1980b) Phys Rev B 23:5329
Shen SC, Chu JH (1983) Solid State commun 48:1017–1021
Shen SC, Chu JH (1985) Acta Phys Sin 34:56
Shen SC, Fang CJ, Cardona M, et al. (1980) Phys Rev B 22:2913
Sher A, van Schilfgaarde M, Chen AB, Berding MA (1987) Phys Rev B 36:4279
Sher A, van Schilfgaarde M, Berding MA (1991) J Vac Sci Technol B 9:1738
Szigeti B (1950) Proc Roy Soc A 204:51
Szuszkiewicz W, Dybko K, Hennion B, et al. (1998) J Cryst Growth 184–185: 1204
Talwar DN, Vandevyver M (1984) J Appl Phys 56:1601
Tiong KK, Amirtharaj PM, Parayanthal P, et al. (1984) Solid State Commun 50:891
Tütüncü HM, Srivastava GP (2000) Phys Rev B 62:5028
Verga B (1965) Phys Rev 137:1896
Verleur HW, Barker AS (1966) Phys Rev 149:715
Vetelino JF, Mitra SS (1969) Solid State Commun 7:1181
Vodopyanov LK, Kozyrev SP, Aleshchenko, et al. (1984) In: Chadi DJ, Harrison WA (eds) Proc 17th Int Conf Physics of Semiconductors. Springer-Verlag, San Francisco, p 947
Wagner J, Koidl P, Bachem KH, et al. (1993) J Appl Phys 73:2739
Weber W (1974) Phys Rev Lett 33:371
Weber W (1977) Phys Rev B 15:4789
Widulle F, Kramp S, Pyka NM, et al. (1999) Physica B 263–264:448
Witowski AM, Grynberg M (1979) Solid State Commun 30:41
Yin MT, Cohen ML (1982) Phys Rev B 25:4317
Zeyher R, Cardona M, Bilz H (1976) Solid State Commun 19:57
Zhang GY, Lan GX, Wang YF (2001) Lattice Vibration Spectroscopy. Higher Education Press, Beijing

Index

L

M

N

O

P

Q

R

S

W

X

Z

Printed in the United Kingdom
by Lightning Source UK Ltd.
127173UK00007BB/18/A

9 780387 747439